Therapeutic Nutrition in Ayurveda

Nutrition remains the key to the treatment of diseases, in addition to the various evolved medical treatments across the world. The treatment outcome improves to a better extent with a degree of nourishment of the patients.

Therapeutic Nutrition in Ayurveda (TNA) categorizes diseases system-wise and discusses nutrition with references from Ayurveda classics as well as publications from indexed journals in today's world.

This book emerges as a pilot project to discuss the clinical experiences directly and the concept of nutravigilance by experienced authors of respective specialties like hepatology, neurology, dermatology, ophthalmology, oncology, cardiology, gynecology, and so on. It broadly discusses diet and nutrition based on 12 different groups of diet in Ayurveda. Nutrition has been widely discussed for every disease dynamically in Ayurveda, with details of exclusion and inclusion of foods over a stipulated period or entire duration of treatment.

Key Features:

- Presents system-wise and disease-wise therapeutic nutrition
- Includes clinical experience of physicians in therapeutic nutrition
- Contains interdisciplinary discussion on therapeutic nutrition with an integrated approach

The integration of traditional and conventional health systems, along with the multidisciplinary approach, is the emerging trend for inclusive health care in the coming decades. This book serves as a handy guide for health care professionals across the continents, providing interdisciplinary correlations on nutrition.

Therapeutic Nutrition in Ayurveda

Edited by
Pankaj Wanjarkhedkar and Yashwant V. Pathak

CRC Press is an imprint of the
Taylor & Francis Group, an informa business

First edition published 2024
by CRC Press
2385 NW Executive Center Drive, Suite 320, Boca Raton FL 33431

and by CRC Press
4 Park Square, Milton Park, Abingdon, Oxon, OX14 4RN

CRC Press is an imprint of Taylor & Francis Group, LLC

Library of Congress Cataloging-in-Publication Data
Names: Wanjarkhedkar, Pankaj, editor. | Pathak, Yashwant, editor.
Title: Therapeutic nutrition in Ayurveda / edited by Pankaj Wanjarkhedkar and Yashwant Pathak.
Description: First edition. | Boca Raton, FL : CRC Press, 2024. |
Includes bibliographical references and index.
Identifiers: LCCN 2023020526 (print) | LCCN 2023020527 (ebook) |
ISBN 9781032385396 (paperback) | ISBN 9781032385433 (hardback) |
ISBN 9781003345541 (ebook)
Subjects: LCSH: Diet therapy. | Medicine, Ayurvedic. | Nutrition.
Classification: LCC RM217 .T424 2024 (print) |
LCC RM217 (ebook) | DDC 615.8/54–dc23/eng/20230926
LC record available at https://lccn.loc.gov/2023020526
LC ebook record available at https://lccn.loc.gov/2023020527

ISBN: 978-1-032-38543-3 (hbk)
ISBN: 978-1-032-38539-6 (pbk)
ISBN: 978-1-003-34554-1 (ebk)

DOI: 10.1201/9781003345541

Typeset in Times
by Newgen Publishing UK

Dedicated to
the seers and peers

Contents

Foreword

Disease management is not limited to the administration of drugs or procedural therapies. Food management, although neglected, is an inherent part of every disease management. Dean Ornish, an American physician-researcher, in his famous book, *Program for Reversing Heart Disease*, describes food as Doctor Diet, underlining its role as a therapeutic agent in the management of heart disease. Emphasizing its significance profoundly, Ayurvedic classics refer to food as "*Mahabhaishajya*", i.e., Super drug. Organisms including humans are known to survive without ever consuming any medicine, but none is known to survive without consuming food. On this note, the significance of the book *Therapeutic Nutrition in Ayurveda* needs to be appreciated.

Food is life for every living being. It may differ from person to person or animal to animal, but it will always be life for every living being. Therefore, food is referred to as life of every living being in its own nature. Ayurvedic science has elaborated on the significance of food and eating practices in the promotion, protection, and maintenance of health and the alleviation of diseases. These elaborations pertain to the classification of foods according to geographical area, physical and mental constitution of the human body, methods of food processing, effect of food on body and mind, wholesome and unwholesome diets, and digestibility of food.

Conventional science classifies food into proteins, carbohydrates, fats, minerals, and micronutrients. In Ayurveda, the classification of food is dependent on the nature of its use. For example, it is classified into two broad categories on the basis of its effect on the body as a whole: (1) wholesome food and (2) unwholesome food. Wholesome food has a beneficial effect on the body, whereas unwholesome food produces harmful effects. The wholesome or unwholesome effect of any food substance depends on various factors such as quantum of food, time of consumption, processing of food, habitat, body constitution, and physiological functional status of the body.

Food substances are also classified on the basis of their source. In this respect, the source is not only important to know the nature of the food but also important to determine the compatibility and assimilability of food in the human body. In terms of digestibility, the food is classified as *Guru* (heavy) and *Laghu* (light). *Guru* (heavy) is the one which is heavy to digest, whereas *Laghu* (light) is the food which is easy to digest. All these factors need to be considered while determining the programming of food in all disease conditions.

Apart from nutritive value as mentioned in conventional medical science, Ayurvedic science gives more importance to the status of *Agni* (digestive faculty) of the person, referred to as "Metabolic Energy" by the authors of this book. The quantum of the food varies with the status of *Agni* of the individual. Therefore, healthy individuals belonging to similar age groups and equal in physical parameters eat different quantities of food due to variations in the status of their *Agni*. The status of *Agni* is also observed to vary physiologically with environmental upheavals related to change of season, throughout the year. Interestingly, the status of *Agni* is known to adjust itself with the healthy or diseased status of the body. In this context, the *Agni* is observed to be affected adversely due to disease pathology. The status of the *Agni* determines the requirement of quality and quantity of the food. Thus, food and Agni are closely related to each other.

According to Ayurvedic medicine, a thin line of distinction exists between food and drug. Many food articles prolifically used as food material in the conventional world are used as drugs in the Ayurvedic system of medicine due to their therapeutic potential. Use of spices in food and Ayurvedic drugs is a salient example of such use.

Pushing food to the back seat, conventional medicine considers food mainly as a source of energy. Therefore, the patient is always referred to a dietician for managing his diet, whereas Ayurveda accepts food as a therapeutic agent and prescribes food in the first place, followed by the drugs where necessary. Thus, therapeutic nutrition remains the mainstay of all disease management in Ayurveda.

Against the backdrop of the above knowledge, the management of disease in Ayurveda begins with the classification of disease in relation to the gross nutritional status of the patient. In this context, the disease is classified as *Santarpanottha* (produced due to overnourishment) or *Apatarpanottha* (produced due to undernourishment). Based on the above classification, the line of therapeutic intervention with food is determined, which is referred to as "Therapeutic Nutrition" by the present authors.

Sensing the therapeutic significance of food in the management of disease, the editorial team of the book *Therapeutic Nutrition in Ayurveda* has rightly selected the topics for inclusion in this book. The book is divided into four sections. The first section lays down the basic principles of nutrition in therapeutics. Section B is devoted to system-related diseases. Section C describes nutrition in obstetrics and gynecology, whereas the last section, Section D, is related to nutrition in various diseases of importance.

The authors of this book are practicing clinicians, scientists, and academicians. Each one has authored chapters related to their own specialty. Dr. Pankaj Wanjarkhedkar, one of the two editors, has known me for long. His interdisciplinary work on the role of Ayurvedic intervention in oncology in the form of food and drugs has been acclaimed widely in the medical world, crossing the barriers among systems of medicine.

Shri Yashwant Pathak, the second editor of this book, is basically a pharmaceutical scientist working in the field of nanomedicine. His interest in the field of Ayurveda is a welcome step for both systems of medicine.

True and open-minded scientists will come to know that although prima facie concepts appear discordant, the principles laid by all systems of health sciences, traditional or conventional, are not contradictory to each other at baseline. Dr. Pankaj Wanjarkhedkar and his team of expert authors have proved this in their book *Therapeutic Nutrition in Ayurveda*. Therapeutic nutrition is an additional tool in the hands of every physician. I am sure drug therapy accompanied by nutritional therapy will enhance the effectiveness of the medical treatment. In this way, the book will simplify disease management and will definitely prove useful to every medical practitioner across the medical systems. The medical world must welcome such attempts. I wish the editorial team and the authors great success in this venture and hope for more such books in the future.

Prof. Dr. Shriram S. Savrikar M.D. (Ayurveda), Ph.D.
Former Vice-Chancellor
Gujarat Ayurveda University, Jamnagar, Gujarat, India
Former Advisor, Ministry of ISM&H
Union Govt. of India
05/04/2023

Preface

The United Nations General Assembly (UNGA) at its 75th session has declared the year 2023 as the International Year of Millets (IYM 2023). The Indian Prime Minister has the vision to make IYM 2023 a Peoples Movement while positioning India as a Global Hub for millets. Millets are among the first few crops domesticated, with evidence of it being cultivated since 3000 BC by the Indus Valley civilization. Ayurveda, the ancient science of life, has emphasized nutrition for health and discussed the millets across its classics since 1500–2000 BCE.

The IYM 2023 message given by UNGA underlines the much-needed emphasis on nutrition for human health, equal to the need for vaccination for a disease or eradication or treatment programs of a disease. The nutrition at baseline is the deciding factor for the better outcome of treatment. Good nutrition positively contributes to maternal health during antinatal, perinatal, and postnatal outcomes as well as the newborn's immunity. A well-nourished patient withstands surgeries or chemotherapies or radiation treatments in diseases like cancer, where multimodality treatments are involved, in a far better way. The optimum nutrition in geriatric health leads to an increase in lifespan with quality of life.

The trend of evolution has been reflected in the diet and nutrition of the people. The diet can be attributed to ethnicity, which differs with communities, regions, provinces, states, nations, and continents. Multiple patterns of dietary practices have been reported, observed, and studied over the last few years, like the Mediterranean diet and the Flexitarian diet, that emerge from community-based dietary practices over the years.

Nutrition always remained an integral part of treatment in Ayurveda, where it has been discussed alongside herbal treatments for the conservation of health, as well as the restoration of health in patients. Wellness nutrition in Ayurveda has been adopted across the continents over the last four to five decades. It has gained popularity to design a diet on the basis of the *Tridosha Prakriti* for better health.

At times, there are limitations to advise generalized nutrition for a patient undergoing active treatments, as the disease conditions in the patients have multiple aspects to be addressed. In recent years, the need for a wide understanding and practice of advanced disease-specific nutrition mentioned by seers of ancient Ayurveda texts has been discussed for clinical application.

The specialized nutrition for the respective system is the centerpoint of this book, which discusses Therapeutic Nutrition in Ayurveda (TNA). TNA is mainly patient centric, which is focused on multiple factors about the patient's health as well as the disease burden in addition to the calorie value of the food. The schools of Ayurveda choose nutrition based on the physiology of the patient and the biology of the disease at the same time.

The contributing authors are clinicians, academicians, and scientists of Ayurveda who have been working in the field in their respective specialties for over a decade.

The book chapters cover cardiology, hepatology, neurology, and ophthalmology for easy understanding by disciples of all health sciences, from east to west. Nutrition for special physiological conditions like gynecology and obstetrics has been elaborated on under a separate section.

The chapters in this book follow a basic uniform structure for every system, based on the diet classification in *Charaka Samhita* with 12 major diet strata, classified as: *Shuka-dhanya varga* (coarse grains and millets), *Shami-dhanya varga* (pulses and cereals), *Shaka varga* (all vegetables), *Mamsa varga* (meats/animal food), *Phala varga* (fruits), *Harita varga* (spices, herbs, and tubers), *Madya varga* (liquors), *Jala varga* (water), *Gorasa varga* (dairy products), *Ikshu varga* (sugars), *Kritanna varga* (food preparations), and *Aaharupyogi varga* (food processing).

The book is in the format of a diet structure advised in the abovementioned 12 dietary groups, first-hand clinical experiences by authors, and nutravigilance for the respective system. The indications and contraindications of diet have been elaborated on with references from doctrines of Ayurveda as well as citations from present-era publications. The inclusion of first-hand clinical experience has emerged as a unique model in health literature, giving voice to observations in clinics, which may be less likely to get a publication platform formally. "Nutravigilance" is going to be the word of the next decade with the globalization of the diet industry.

This book is a pilot attempt to discuss the first-hand clinical experiences and the concept of nutravigilance, by pushing the limits of formal literature writing. It is better to read out of the box than to be unread.

Nutrition usually remained a point of discussion for the countries, states, or cities having the burden of malnourishment, and maximum efforts from local to international levels are being made to serve this segment. We need to widen the nutrition spectrum, to ensure optimal nutrition for every citizen.

There is a need to generate empirical data through the models of community-based surveys on nutritional behavior for a better understanding of diet spectrums and nutrigenomics. The fundamental research based on genotype and phenotype can play a vital role in nutrition therapy.

The world is well aware of Good Manufacturing Practices, Good Clinical Practices, and many more. The next step is to lay down the policy guidelines for Good Nutrition Practices, involving the governments and the primary health units at the grassroots level, contributing to the ethnicity-based dietary practices as a mission of well-nourished states in every country.

The authors of team TNA have a unanimous understanding that optimal nutrition forms the basis not only for disease treatments or preventive medicine but also for rejuvenation therapies.

Books are sources of knowledge, and authors are the resources. The positive involvement of our team members Dr. Pravin Bhat and Dr. Anagha Ranade consistently supported this task like columns of the book. The eminent authors not only participated but also contributed actively and willingly with added clinical experiences, making this book a benchmark.

We, the editors, extend our gratitude for the kind help and support from Mr. Steve Zollo, Laura Piedrahita, Randy Brehm, and other CRC Press colleagues, publishers, copyeditors, and printer staff, for giving us this opportunity to pen down today's clinical experiences extrapolated from the ancient wisdom of therapeutic nutrition.

About the Editors

Dr. Pankaj Vijaykumar Wanjarkhedkar, M.D. in Ayurveda (the Indian system of medicine), is a Consulting Ayurveda Physician with a wide interest in scientific health research. Since his postgraduation in 2007 from one of the premier Ayurveda institutes run by the Government of Maharashtra, he has been inclined toward building teams for bridging the therapeutic gaps between experience-based medicine and evidence-based medicine. He started the platform of Integrated Remedies, with an integrative approach, catering for a spectrum of metabolic diseases, cancer care, fertility care, and rejuvenation therapies. Dr. Wanjarkhedkar is instrumental in pioneering two Integrative Cancer Care units, where the Ayurveda system of medicine works hand in hand with the conventional oncology arm providing patient-centric care at two multispecialty hospitals in Maharashtra, India; one is Deenanath Mangeshkar Hospital and Research Center, the tertiary care center at Pune, and the other is Vivekanand Hospital and Cancer Center, the rural cancer hospital at Latur. He has published papers and chapters in international journals and delivered keynotes in seminars/workshops at a number of national and international conferences in Health Universities/Hospitals in India, the United States, and Latin America.

Dr. Yashwant Pathak, Ph.D., has over 13 years of versatile administrative experience in an Institution of Higher Education as Dean (and over 30 years as faculty and as a researcher in higher education after his Ph.D.). He now holds the position of Associate Dean for Faculty Affairs and tenured professor of Pharmaceutical Sciences. Dr. Yash Pathak is an internationally recognized scholar, researcher, and educator in the areas of health care education, nanotechnology, drug delivery systems, and Nutraceuticals. His major achievements in the international area include:

Fulbright Senior Scholar Core Fellowship Award 2015–2016 for Indonesia (visiting Ubaya University, Surabaya, Indonesia, from January to July 2017);
Endeavour Executive fellowship by the Australian Government in 2015 in collaboration with Deakin University to work on siRNA delivery;
Prometeo Fellowship award from Ecuador Government, 2015;
CNPQ Brazil Government Fellowship, visiting PUCRS in Porto Alegre every year for one month from 2015 till 2017, working on space pharmaceuticals and microgravity impact on the stability of drug delivery systems;
Outstanding Achievement Award for Global Engagement by University of South Florida, a unique award given to only one faculty/administrator annually;
Fellow of NSF I-Corps USF 2016;
Outstanding Faculty Award from USF March 2017;
Fulbright Specialist Fellowship 2019 for South Africa;
Outstanding faculty award from University of South Florida 2020;
Fellow of AAAS 2021;
Fulbright Specialist University of Cape Coast, Ghana, May to June 2023.

He has published extensively with over 60 edited volumes in the area of nanotechnology, drug delivery systems, artificial neural networks, conflict management, and cultural studies. Elsevier, John Wiley and Sons, Springer, Taylor and Francis, and many other International Publishers publish his books. He has published over 350 research papers, reviews, and chapters in books and presented at many national and international conferences.

He is also actively involved in many nonprofit organizations; to mention a few: Hindu Swayamsevak Sangh, USA; Sewa International, USA; International Accreditation Council for Dharma Schools and Colleges; International Commission for Human Rights and Religious Freedom; and Uberoi Foundation for Religious Studies.

Contributors

Pravin Bhat
Sumatibhai Shah Ayurved Mahavidyalay
Pune, Maharashtra, India

Supriya Bhalerao
Interactive Research School for Health Affairs BVDU
Pune, Maharashtra, India

Swarupa Bhujbal
Ranka Multi-Speciality Hospital
Pune, Maharashtra, India

Mrudul Chitrakar
DYPU School of Ayurveda
Navi Mumbai, Maharashtra, India

Yogesh Deole
G.J. Patel Institute of Ayurvedic Studies and Research
Anand, Gujarat, India

Satwashil Desai
Dr Deepak Patil Ayurved Medical College
Kolhapur, Maharashtra, India

Mangesh Deshpande
Orthoved Hospital
Dombivali, Mumbai, India

Rahul Jadhav
Parul Institute of Ayurved and Research
Vadodara, Gujarat, India

Naresh Kore
Parul Institute of Ayurved and Research
Vadodara, Gujarat, India

Shiva Kumar
SDM College of Ayurveda & Hospital
Hassan, Karnataka, India

Ganesh Malawade
BVK Ayurveda Medical, College & Hospital
Latur, Maharashtra, India

Amit Nakanekar
Govt. Ayurved College & Hospital
Nagpur, Maharashtra, India

P. Nambi Namboodiri
Nagarjuna Ayurveda Centre
Kochi, Kerala, India

Shinsha Puthiyottil
Parul Institute of Ayurved and Research
Vadodara, Gujarat, India

Yashwant Pathak
University of South Florida
Tampa, FL, USA

Deepali Rajput
Striroga & Prasutitantra
Vadodara, Gujarat, India

Anagha Ranade
CCRAS, Govt. of India
New Delhi, Delhi, India

Mukund Sabnis
Jeevanrekha Obesity Solutions
Aurangabad, Maharashtra, India

Pranesh Sanap
Sanap Hospital Sinnar
Nashik, Maharashtra, India

Jui Shahane
Brahma Ayurveda, Multi-Speciality Hospital, Research & Academic Center
Nadiad, Gujarat, India

Narayan Gangadhar Shahane
Brahma Ayurveda, Multi-Speciality Hospital, Research & Academic Center
Nadiad, Gujarat, India

Pankaj Wanjarkhedkar
Deenanath Mangeshkar Hospital & Research Center
Pune, Maharashtra, India

SECTION A

Basic Concepts of Therapeutic Nutrition in Ayurveda

Supriya Bhalerao and Mrudul Chitrakar

INTRODUCTION

Ayurvedic concepts of diet, digestion, metabolism, and nutrition are based on their epistemology. The *Ahara* (diet) is one of the three supportive pillars of life. The nurturing of the body depends not only on its quality and quantity but also on various aspects such as processing techniques utilized, way of consumption, and even the consumer itself. A proper diet not only maintains the physiological homeostasis of the body but also proves helpful in combating diseases. Diet, therefore, remains important for disease prevention and even therapeutics. In this chapter, the basics of diet and good eating practices shall be discussed concerning 'Therapeutic Nutrition' in Ayurveda. For a better understanding of these Concepts, the basic concepts of Ayurveda physiology are discussed initially.

Basic Concepts of *Sharira Kriya* (Ayurveda Physiology)

The concepts of diet and its transformation in the body have their roots in physiology, which makes it necessary to understand the concepts of *Panchamahabhuta, Dosha, Dhatu* & *Ojas, Mala,* and *Agni.*

Panchamahabhuta

These are *Pancha* (five) primordial elements (Sushrut Sutrasthana.41/6-9), which constitute all living and non-living matters. They are namely *Prithvi* (responsible for mass and stability), *Jala* (responsible for fluidity and moistness), *Teja* (responsible for transformations), *Vayu* (responsible for movements and dryness), and *Akasha* (an all-pervading element that occupies hollow spaces). All substances are composed of the *Panchamahabhuta* with a single element predominance. The nomenclature of substances is according to the dominant element, like *Parthiva, Jaliya,* etc. Like all the structural and functional body units, dietary articles are also made up of these same five elements (Sushrut Sutrasthana.46/526). In case of deficiency of any element in the body, the diet primarily composed of that element should be consumed, and vice versa.

DOI: 10.1201/9781003345541-2

TABLE 1 Major Characteristics of *Dosha* (Sushruta Sutrastana.21/8)

DOSHA	MAHABHUTA	*PRIMARY LOCATION*	*MAJOR FUNCTION*	*AGE-WISE PREDOMINANCE*	*DOMINANCE WHILE EATING*
Vata	*Vayu + Akasha*	Lower abdomen and pelvic girdle	Movement	Old age	After
Pitta	*Teja*	Duodenal region	Transformation	Youth	During
Kapha	*Prithvi + Jala*	Upper abdomen, thorax, and head	Cohesion	Juvenile age	Before

TABLE 2 *Dosha* Involved in Digestion and Metabolism

TYPE OF DOSHA	*PRINCIPAL SITE (S)*	*ACTIONS (ASHTANGSANGRAHA.SUTRASTAHANA.1/26-28)*
Prana Vayu	Above sternal region	Facilitates *Annagrahana* (ingestion of food)
Samana Vayu	Duodenum and small intestine	Regulates *Vivechana* (peristaltic movements and appropriate secretion of juices into the digestive system)
Apana Vayu	Pelvic region	Responsible for excretion of *Mala* (elimination of waste products like urine and stool)
Pachaka Pitta	Stomach and duodenum	Responsible for the overall process of *Pachana* (digestion and metabolism)
Bodhak Kapha	Oral cavity	*Rasana* (mixes with food and provides taste perception)
Kledak Kapha	Stomach	*Kledana* (lubricates the food, helps in mixing it with digestive juices, and forms a protective covering over the inner line of gastric mucosa)

Dosha

Tri (three) *Dosha*, viz., *Vata, Pitta,* and *Kapha*, are the functional units originating from the permutations and combinations of the *Panchamahabhuta*. Based on the predominance of constituting elements, the functions of *Dosha* differ. When these *Dosha* are balanced and function in synchrony, body homeostasis is maintained. Even during their balanced state, the *Dosha* tends to fluctuate physiologically. An unwholesome and inappropriate diet vitiates them and diseases are produced in the body. Table 1.1 represents the major characteristics of *Dosha*.

The functions listed above for each of the *Dosha* can be perceived even at the cellular level, e.g., *Vata*—movements such as membrane transport, *Pitta*—cellular metabolism, and *Kapha*—anabolism and energy storage processes.[1] There is a lack of appropriate terms in contemporary physiology that can convey all the attributes of *Dosha*; however, their scientific validity is well established.

Based on their location and specific functions, each *Dosha* is further divided into five subtypes. The *Anna Pachana* (digestion and metabolism) process is carried out by synchronous and articulated actions of these *Dosha* as shown in Table 1.2.

The *Dosha* are modulated by seasonal changes and undergoes states of *Chaya* (accumulation), *Prakopa* (aggravation), and *Prashama* (alleviation) (Table 1.3). These modulations remain within physiological limits only if *Ritucharya* (seasonal dietary and behavioral regimen) is followed properly. These variations, thus, should be monitored and addressed for remaining healthy and disease free (Charaka Sutrasthana.17/118).

The *Dosha* has a close association with the different *Rasa* (tastes) of dietary articles (Figure 1.1). In their imbalanced state, consuming a diet of balancing tastes is advised.

TABLE 3 Seasonal Variations in *Dosha* Status

DOSHA	CHAYA *(ACCUMULATION)*	PRAKOPA *(AGGRAVATION)*	PRASHAMA *(ALLEVIATION)*
Vata	*Grishma* (Summer)	*Varsha* (Rainy season)	*Sharada* (Autumn)
Pitta	*Varsha* (Rainy season)	*Sharada* (Autumn)	*Hemant* (Prewinter)
Kapha	*Shishir* (Winter)	*Vasant* (Spring)	*Grishma* (Summer)

Madhura	Pacifies *Vata, Pitta,* and Increases *Kapha*
Amla	Pacifies *Vata* and Increases *Pitta, Kapha*
Lavana	Pacifies *Vata* and Increases *Pitta, Kapha*
Katu	Pacifies *Kapha* and Increases *Vata, Pitta*
Tikta	Pacifies *Kapha, Pitta* and Increases *Vata*
Kashaya	Pacifies *Vata, Pitta* and Increases *Kapha*

FIGURE 1 Association of *Rasa* (tastes) and *Dosha* (Charaka Sutrastahana.26/11-48).

Sharira and *Manasa Prakriti* (Physical and Psychological Constitution)

Every individual is born with a specific *Sharira* (physical) and *Manasa* (psychological) *Prakriti* (constitution). The *Sharira Prakriti* (physical constitution) is based on the *Dosha* status of gametes at the time of conception, the season of conception, and the diet of a mother throughout pregnancy. The seven types of *Prakriti*, viz., *Vata, Pitta, Kapha, Vata-Kapha, Vata-Pitta, Kapha-Pitta,* and *Sama* (equal balance of *Tridosha*), are based on phenotypic characteristics such as appearance, temperament, and habits.[2,3] This specific constitution does not change throughout one's life. These descriptions point toward the genomic correlation of *Prakriti*. In recent years, through GWAS studies, and epigenetic and gene expression studies, the genomic correlation of *Prakriti* has been established.[4,5] Understanding one's *Prakriti* helps in understanding and selecting appropriate dietary habits and lifestyle to avoid imbalance of the *Dosha* and eventually diseases. The upcoming field of nutrigenomics studies the interaction of nutrition and genes, especially in the prevention or treatment of disease. The concept of *Prakriti* needs to be explored in this context (Ayurnutragenomics).[6]

As the *Dosha* defines *Sharira Prakriti, Sattva, Raja, and Tama*, the *Tri Guna* (three universal components) define *Manas Prakriti* (psychological/spiritual constitution). Based on the predominance of either of the *Guna, Manas Prakriti* can be *Sattvik, Rajasik,* and *Tamasik*. These represent different traits of the mind (Charaka Sharirsthana.3/13). The mind is the faculty responsible for emotions, feelings, and knowledge processing. These functions are carried out by the brain, physiologically. Interestingly, the brain controls the entero-endocrine system and, in turn, the process of digestion.

The brain controls the responses to hunger and satiety, while the endocrine system controls the release of hormones and enzymes required for the digestion of food in the digestive tract. Metabolic signals emanating from the gastrointestinal tract, adipose tissue, and other peripheral organs target the brain to regulate feeding, energy expenditure, and hormones. This relation needs to be considered in the context of diet selection.[7]

Dhatu and *Ojas*

The *Dhatu* (tissues) are structural units that sustain and maintain the body. There are seven *Dhatu*, viz., *Rasa Dhatu* (plasma/lymph), *Rakta Dhatu* (blood or hematocrit), *Mamsa Dhatu* (muscular tissue), *Meda*

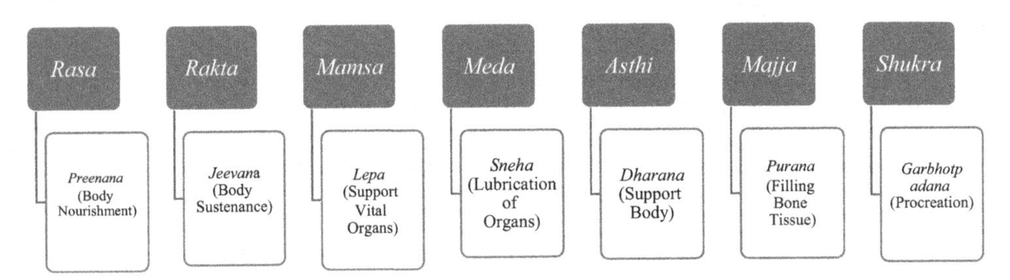

FIGURE 2 Specific functions of seven *Dhatu* (Ashtangsangraha .Sutrastahan.1/31-32).

Dhatu (fat/adipose tissues), which provides lubrication of organs, *Asthi Dhatu* (bones/osseous tissue), *Majja Dhatu* (marrow), and *Shukra Dhatu* (semen) (Ashtang Sangraha Sutrasthana.1/29). Apart from being the building blocks, they also serve certain specific functions as depicted in Figure 1.2.

Ojas is the essence of all seven *Dhatu* that provides physical and psychological strength. It is the force that keeps the body, sense organs, mind, and even the soul together in functional harmony. It plays an important role in protecting the body against adverse conditions (Charaka.Sutrasthana.18/51, Sushrut. Sutrasthana.15/24).

Diet is responsible for the formation of *Dhatu* and, in turn, the *Ojas*. There are specific dietary articles that nourish specific *Dhatu*, like milk, nourishes *Shukra Dhatu* (Charaka.Sharirasthana.30/152).

Mala (Metabolic waste products) (Sushrut.Sutrasthana.21/10)

These are produced during the digestive and metabolic processes of the body. Despite being waste products, in a balanced state, they are needed for the smooth functioning of the body. The three main *Mala* are *Purisha* (feces), *Mutra* (urine), and *Sweda* (sweat). The diet is eventually responsible for the adequate formation of *Mala*. For instance, the ingestion of legumes like chickpeas produces more quantity of fecal matter (Ashtang Hriday Sutrasthana.11/32).

Agni (Energy responsible for transformation) (Charaka.Chikitsasthana.15/38)

Agni is a medium for transformation in the body. The word *Agni* in this context does not mean fire with flame and smoke but the energy that brings out macro- or microlevel transformations in the body. In relation to digestion, metabolism, and nutrition, three different types of *Agni* are described as given in Table 1.4.

Chakrapanidatta, the commentator of Charaka Samhita, has described three independent processes by which the end product after the action of *Jatharagni* is processed further, viz., *Ksheerdadhi Nyaya* (irreversible chemical transformation), *Kedarkulya Nyaya* (permeability), and *Khalekapota Nyaya* (tissue selectivity and specificity) (Charaka.Chikitssthana.15).

Based on the appetite and digestion patterns, the *Jatharagni* is of four types, viz., *Vishama Agni*—unpredictable appetite and digestion, regulated by *Vata Dosha*; *Teekshna Agni*—faster digestion and good appetite, regulated by *Pitta Dosha*; *Manda Agni*—slow digestion and low appetite, regulated by *Kapha Dosha*; and *Sama Agni*—adequate appetite and digestion influenced by the perfect balance of three *Dosha* (Charaka.Chikitsasthana.15/51).

The *Agni* needs to work optimally to ensure proper digestion and nutrition. A healthy diet does not prove beneficial if the *Agni* is unable to digest it properly.

TABLE 4 Types of *Agni* and Their Functions

AGNI	*FUNCTION*
Jatharagni (Charak Chikitsathana.15/ 39-40) (Bio-transformation at GIT Level)	Responsible for the appetite of an individual as well as the digestion of food in the GIT. This is parallel to gut enzymes, neuropeptides, and related transcriptional and translational factors. The gut microbiome can also be considered under this *Agni*.[8] After the action of this *Agni*, complex substances get broken down into smaller particles and converted into bio-compatible forms.
Dhatvagni= *Dhatu + Agni* (Tissue level bio-transformation)	Metabolism takes place at the *Dhatu* (tissue/system) level. They are seven in number and are named after each *Dhatu*. *Dhatvagni* can be considered an intermediate metabolism in a contemporary light. It is the conglomeration of all intracellular chemical processes by which the nutritive material is converted into cellular matter.[9]
Bhutagni (Charak Chikitsasthana15/ 13-14) (End-stage bio-transformation)	Responsible for final bio-transformation. They are five in number and named after each *Mahabhuta*. This is parallel to transformations happening at the cellular or molecular level. From the modern physiological perspective, the action of the *Jatharagni* can be equated with the digestion in the stomach and duodenum, while the action of the *Bhutagni* can be equated with the conversion of digested materials in the liver[10]

TABLE 5 Classification of Diet (Charaka Sutrasthana.27/6-7, SushrutaSutrasthana.46/8, Bhavaprakash-5/ 134-137)

Classification based on source:	**Classification is based on four forms:**	**Classification based on effects:**
- *Shukandhanya* (cereals) - *Shamidhanya* (pulses and legumes) - *Mamsa* (meat) - *Shaka* and *Harita* (vegetables and leafy vegetables) - *Phala* (fruits) - *Harita* (fresh tubers/spices/ herbs) - *Madya* (fermented drinks) - *Ambu* (water from various sources) - *Gorasa* (milk and milk products) - *Iksuvikara* (sugarcane products) - *Kritanna* (food preparations) - *Aharayogi* (spices and condiments)	- *Ashita* (eatable) - *Lidha* (lickables) - *Peeta* (drinkable) - *Khadita* (masticables) **Classification is based on six forms** - *Chushya* (chewable) - *Peya* (drinkables) - *Lehya* (lickables) - *Bhojya* (partially masticable) - *Bhakshya* (snacks/munchies) - *Charvya* (masticable) (Bhavaprakasha)	- *Hitkar* (wholesome) - *Ahitkar* (unwholesome) **Classification based on frequency:** - *Sada Pathyam* (to be consumed regularly) - *Sada Apathyam* (to be avoided regularly) **Classification based on attributes:** - *Rasa* (taste) - *Veerya* (potency) - *Vipak* (post-digestive effect) - *Guna* (qualities) - *Prabhav* (a specific effect that cannot be explained based on other qualities)

Basics of *Ahara* (Diet)

Ahara means any substance ingested via the mouth and passed to the alimentary canal. Among *Tray-Upastambha* (three supportive pillars) of life, *Ahara* (diet) is the foremost. It is *Vrittikaranam Shreshtam* (a major sustainer of life). The pursuits of life depend on good health which in turn depends on *Ahara* (Sushrut.Chikitsasthana 24/68, Charaka.Sutrasthana.27/345-347, Charaka.Sutrasthana.25/40).

Dietary articles are categorized based on their sources, forms, effects on health, frequency of consumption, and attributes as mentioned in Table 1.5.

An emerging branch of 'Nutravigilance' is dedicated to consumer safety by acquiring comprehensive information regarding the adverse effects of dietary articles. On these lines, detailed information on beneficial, as well as deleterious properties, indications, and contra-indications of each dietary article in the above-mentioned classes can be useful.

Various factors influence the effects of diet, such as qualities of dietary articles like *Guru* (heavy to digest) or *Laghu* (easily digestible), environmental factors, a combination of dietary articles, processing/cooking methods, the freshness of articles, time of consumption, and adherence to guidelines for preserving and eating food. The importance of a few of these factors has been highlighted in contemporary research.

These factors are very crucial these days as the newer exotic diets, changed cooking practices such as microwave heating, use of frozen food, and ready-to-use food packets are in practice.[11–13] The availability of fresh and authentic dietary articles is also questionable.

Quantitative Aspects of Diet

The quantity should be individualized and dynamic. It should be suitable to the season and age, physical constitution, digestive capacity, and physical activity profile of an individual. The portion size should be customized for each individual according to the specific needs of the individual (Ashtang Hridaya Sutrasthana.8/2, Charaka Sutrasthana.25/32). The quantity also depends upon the dietary articles, e.g., *Guru* (heavy-to-digest) articles, such as sweets, should be taken up to one-third to one-half of the saturation point of the stomach volume and the remaining part of the stomach should be filled with *Laghu* (light-to-digest) articles, like green gram soup (Ashtang Hridaya Sutrasthana.8/3).

GOOD EATING PRACTICES

Charaka Samhita and *Sushruta Samhita* have provided guidelines regarding health-promoting eating practices under the following headings.

i) *Ashta-Ahara-Vidhi-Vishesh-Ayatanani* (Eight attributes related to diet consumption) (Charaka Vimanasthana.1/21): The factors responsible for deciding wholesomeness or unwholesomeness of the diet.

These factors include (Charaka Vimanasthana.1/21-1-7):

- *Prakriti* (Natural/inherent attributes of the dietary articles), e.g., *Masha* (black gram) is heavy and *Mudga* (green gram) is light for digestion.
- *Karana* (Processing of dietary articles, which results in transformation of the substance and thereby its effects), e.g., curd aggravates conditions such as edema and buttermilk relieves it. Changes induced by heat treatments can produce significant alterations in the chemical composition of the food products, affecting palatability, digestibility, and bioavailability.[14]
- *Samyoga* (Combination of items, which may give rise to new properties which are not seen in individual items), e.g., an equal quantity of honey and ghee is toxic. In a recently reported study, the antioxidant interactions between mostly co-consumed dietary articles such as fruits,

vegetables, grain sources, dairy, and meat products were measured. An antagonism was observed in the combinations of milk with the fruits or green tea extract while a clear synergism was reported in the combination of fruits with breakfast cereal, whole wheat bread, or yogurt.[15]

- *Rashi* (Quantity of food)—The quantity of diet considered in its entirety is *Sarvagraha*, while the quantity of individual dietary articles is *Parigraha*. A healthy eating pattern includes taking all types of dietary articles together (*Sarvagraha*) and not a particular substance (*Parigraha*) at one time. This affects the digestion and metabolism of the diet which finally affects the overall nutrition status. Modern dietetics describes the quantity of food in terms of overall energy requirement (*Sarvagraha*) and quantity of individual components like carbohydrates, fats, and proteins (*Parigraha*).
- *Desha* (Habitat of the dietary articles as well as the consumer)—Certain articles are gathered/grown/produced commonly in a particular locality or region. The local traditional cuisines are generally a reflection of this. For instance, in India, rice-based cuisine is popular in the southern part, whereas wheat-based cuisine is mostly preferred in the northern region.[16] If an individual residing in one region migrates to another region, changes in their diet are most likely. The diet being consumed earlier may not prove useful, while the diet available in the new region may not be easily accepted by the body. Migration for various reasons, such as education or work purposes, has been prevalent since antiquity. In such situations, dietary changes need to be made gradually and sequentially. As mentioned in Charaka Samhita, such a sequential regimen adopts favorable changes in a progressive and health-promoting manner (Charaka Sutrasthana.7/36-38).
- *Kala* (Time—age, diurnal and seasonal variations, stage of disease)—The season as well as the condition of the body (healthy/diseased) influences the choice of diet. Daily, timely consumption of meals is reported to be important to minimize the risk of metabolic disorders.[17] Nutritional needs are different for children, youth, and mature people. The diet pattern is even specific for various stages such as pregnancy and lactation.
- *Upayokta* (Consumer)—As every individual is unique, diet needs to be customized. Various attributes, such as *Prakriti* (constitution), *Agni* (digestive strength), and personal likes and dislikes, need to be addressed along with the aforementioned attributes of diet consumption. *Okasatmya* (habituation toward certain practices) plays an important role in the choice/prescription of diet, as familiarity of an individual with specific dietary articles needs to be considered (Charaka Vimanasthana.1/22). For instance, advising a nonvegetarian diet to a vegan person might be physically beneficial, but psychologically detrimental.
- *Upayoga Samstha* (Guidelines for diet consumption)—These guidelines need to be followed for proper digestion. If they are not observed, consequences like indigestion leading to *Ama* (undigested incompatible food toxins) formation are possibly further resulting in diseases.

Practices like *Ratri Jagarana* (Staying up late at night) and *Diva Swapa* (Daytime sleep) are not recommended due to their *Vata* and *Kapha Dosha* vitiating nature, respectively. Such circadian rhythm disruptions are associated with a variety of metabolic and psychological disorders.[18] Certain conditions, such as occupational needs, require the misalignment of normal sleep patterns. In such unavoidable cases, dietary modifications for predictable *Dosha* vitiation are recommended. For instance, night shift workers, in whom *Vata Dosha* vitiation is imminent, should consume dietary articles such as ghee and other dairy products and wheat-based cuisines frequently.

Although dietary guidelines regarding the choice of diet and its optimum quantity are available in modern dietetics, the Ayurvedic guidelines which consider a range of intrinsic and extrinsic factors are holistic. These eight factors are contextual yet, with a high potential for exploration. They can help to improve conventional dietary guidelines.[19]

ii) *Ahara Vidhi Vidhana* (Healthy eating habits) (Charaka Vimanasthana.1/24)

These are healthy eating guidelines (*Upayog Samstha*) in detail, which should be followed while eating food, to remain healthy and enhance the span of life. These guidelines are described in different texts under different headings like *Bhojana Vidhi* and *Anna Vidhi*. These are applicable to healthy as well as diseased ones.

Certain guidelines pertaining to 'what and how to eat?' are mentioned below (Charaka Vimanasthana.1/25/1-8):

- To eat a freshly cooked diet while it is still warm. To avoid eating overcooked dietary articles.
- To avoid dry roasted, dry and hard dietary articles. It is advisable to utilize an adequate quantity of *Sneha* such as oil, butter, or ghee, while processing or consuming diet.
- To consume a diet in adequate *Pariman* (quantity), as per individual capacity. Even a slight excess of diet can be harmful in the long run, if done habitually.
- To eat only when the previously consumed diet is digested. Munching in between two meals should be avoided. This aligns the meal timings with the biological clock (circadian rhythm). As per the biological clock, the hunger response of the body is set. When the feeding rhythm is altered, e.g., eating when there are no signs of hunger, the body's ability to maintain homeostasis is impaired. This misalignment between physiology and input can give rise to cardio-metabolic diseases.[20]
- To avoid consuming *Veerya Viruddha Ahara* (diet incompatible in terms of its active components or their potencies), such as milk and fish, where milk is of *Sheeta Veerya* (cold potency) and fishes are *Ushna Veerya* (hot potency) by nature.
- To consume a diet at *Ishta Desha* (a suitable and pleasant place). Eating in an open place or direct sun/darkness, lying in bed or under a tree to be avoided.
- To avoid *Atishighra* (too fast) and *Ativilambit* (too slow) eating. The link between eating rate and energy intake has long been a matter of extensive research. A better understanding of the effect of food intake speed on body weight and glycemia in the long term could serve as a means to prevent weight gain and/or dysglycemia.[21]
- To eat in *Mauna* (silently without talking), paying total attention to diet and the eating process. This is similar to 'mindful eating'. Mindful eating (i.e., paying attention to our food, on purpose, moment by moment, without judgment) is an approach to food that focuses on individuals' sensual awareness of the food and their experience of the food. Research has shown that mindful eating can lead to greater psychological well-being, increased pleasure when eating, and body satisfaction.[22]
- To avoid consuming a diet without observing personal cleanliness.
- To avoid consumption of an unaccustomed or unknown diet.
- To observe meal timings by avoiding diet at the wrong time, late in the evening, or very early in the morning.
- To follow the below-mentioned sequence of eating dietary articles even while having a fresh, timely, balanced diet:
- Hard items should be consumed at the beginning, followed by soft and liquids subsequently—Eating and swallowing are complex behaviors involving volitional and reflexive activities of more than 30 nerves and muscles. The movement of the food in the oral cavity and to the oropharynx differs between eating solid food and drinking liquid.[23]
- Sweet-tasting dietary articles should be consumed in the beginning, followed by those with a sour and salty taste. The articles with other tastes like pungent, bitter, and astringent shall be taken at last. Sweets are heavy to digest and eating them, in the beginning, permits the flow of digestive secretion and helps their proper digestion.

Some recommendations for diet consumption are elaborated on in Figure 1.3.

These guidelines further elaborate on meals activities too, such as:

- Water should be drunk in suitable quantities during meals and after meals too.
- After meals, the mouth should be thoroughly cleaned to prevent dental caries and halitosis.

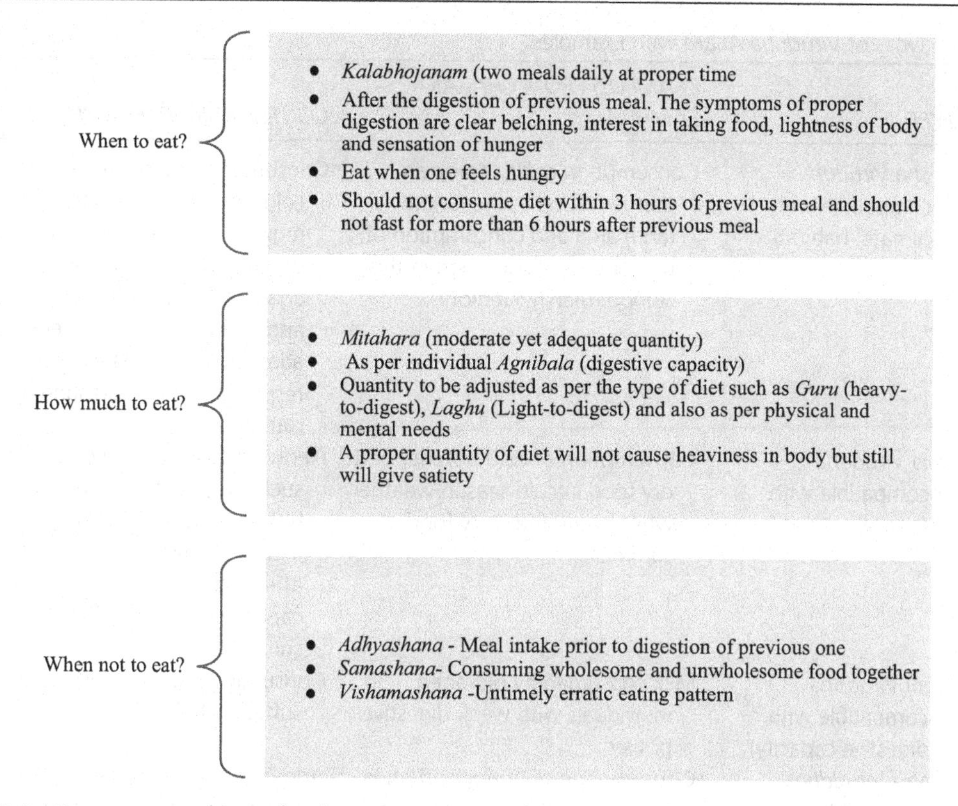

FIGURE 3 Key recommendations for diet consumption (Ashtanghridaya Sutrastahana 8/2, Bhavaprakash. 5/109, Ashtanghridaya. Sutrasthana 8/4-5).

- By nature, *Kapha Dosha* is aggravated after consuming a diet and produces extra salivation. To cleanse the mouth, a mixture of *Puga* (areca nut), *Kankola* (cubeb), *Karpura* (camphor), *Lavanga* (clove), or pungent and astringent fruits like pomegranate should be consumed with betel leaf as a mouth freshener.
- Indulging in strenuous physical activity immediately after meals should be strictly avoided. It is advisable to sit in a comfortable position till postmeal lethargy is over. Slow postprandial walk for a small duration and distance, like only a hundred steps, is recommended. In a study involving type 2 DM patients, the after-meal walk was better in improving postprandial hyperglycemia than walking at any other time.[24] Additionally, prolonged sleeping and sitting, consuming excess liquids, exposure to heat, swimming, and traveling should be avoided immediately after diet intake.

Viruddha-Ahara (Incompatible Diet)

There are 18 types of dietary incompatibility with respect to the consumer's condition, environmental condition, and processing/combination/proportion of the food articles (Charaka Sutrasthana.26/86-87). The incompatible diet disturbs the process of metabolism and nutrition, aggravates *Dosha,* and creates diseases in the body. The different types of incompatibility and their examples are displayed in Table 1.6. Some of these might seem obsolete in current days, but their relevance can be understood principally even today. Novel fusion cuisines are being introduced almost daily. Along with gastronomical attributes, their health-promoting/deteriorating aspects can also be evaluated with the help of this concept.[25] Certain everyday diet combinations, like adding milk to black tea, are also deemed unhealthy.[26] Unfortunately, many people are unaware of these incompatibilities and their grievous effects. A survey of 416 patients

TABLE 6 Types of *Viruddha Ahara* with Examples

NO.	*TYPE*	*EXAMPLES FROM CHARAKA SAMHITA*	*CONTEMPORARY EXAMPLES*
1.	*Desha Viruddha* (Incompatible with climate) habitat	Consumption of dry, spicy, and pungent food in the *Jangal* (arid) area and consumption of unctuous and cold food in the *Anoop* (marshy) region	Consumption of ice cream and cold drinks in coastal and humid regions. People living in a certain geographical region have a natural adaptability toward specific diets and diet practices. This regional adaptability is mostly an analogous response to their surrounding climate.
2.	*Kala Viruddha* (Incompatible with season)	Consumption of cold potency and dry food in cold season/weather while bitter and spicy food in summer	Frequent consumption of dry fruits such as apricots, cashew, and hot beverages in hot weather. Seasonal environment changes affect metabolism. The digestive capacity is naturally low in summer.
3.	*Agni Viruddha* (Incompatible with digestive capacity)	Milk consumed by old-aged individuals with weak digestive power	Consuming junk food even while suffering from diarrhea
4.	*Matra Viruddha* (Incompatible with quantity/proportion of articles)	Consumption of honey and ghee together in the equal quantity	Administration of honey and ghee together in the same quantity has been shown to result in liver toxicity and an increase in oxidative stress, in a rat model[28]
5.	*Satmya Viruddha* (Incompatible with food habits)	Use of sweet and cold substances by a person accustomed to the pungent and hot diet	Sudden switching to the consumption of egg/ nonvegetarian diet by a vegetarian individual
6.	*Dosha Viruddha* (Incompatible with *Dosha*)	Consumption of diet with *Prakriti*—analogous *Dosha* aggravating properties	Consumption of dry food by a *Vata Prakriti* individual causes *Dosha* imbalance by aggravating *Vata Dosha*
7.	*Samskar Viruddha* (Usage of incompatible processing methods)		Deep frying of potatoes can develop toxic substances, such as acrylamide, which can prove to be carcinogenic.[29]
8.	*Veerya Viruddha* (Combination of dietary articles with incompatible potencies)	*Sheeta Veerya* (cold potency) articles mixed with *Ushna Veerya* (hot potency) articles	Consumption of fish and milk together in certain cuisines
9.	*Koshtha Viruddha* (Incompatible with bowel habits)	Consumption of *Mrudu Virechaka Dravya* (mild laxative) by an individual with *Krura Koshtha* (tendency toward constipation)	Consuming sprout-based diet by chronically constipated individuals

TABLE 6 ***(Continued)*** Types of *Viruddha Ahara* with Examples

NO.	*TYPE*	*EXAMPLES FROM CHARAKA SAMHITA*	*CONTEMPORARY EXAMPLES*
10.	*Avastha Viruddha* (Incompatible with state of individual)	Consumption of *Vata* aggravates food by persons who are doing strenuous work daily and consumption of *Kapha* aggravating food by persons who feel sleepy and do not perform any physical activity	Consumption of sprouted pulses by strenuous workers such as porters or construction site workers. Due to their work profile, their *Vata Dosha* is prone to easy aggravation. Similarly, the consumption of heavy-to-digest foods such as boiled rice, and curd by sedentary lifestyle and desk job workers, can cause aggravation of *Kapha Dosha*
11.	*Krama Viruddha* (Consumption of dietary articles in a way contradictory to suggested guidelines)	Consuming meals before emptying the bowel and bladder and without appetite	Snacking, after dinner has been shown to increase the risk of type 2 DM[30]
12.	*Parihaar Viruddha* (Consuming diet already contraindicated in guidelines)	Taking hot articles after consuming pork or cold items after consuming ghee	Consumption of cold water immediately before or after having hot beverages may erode the dental enamel.
13.	*Upchaar Viruddha* (Consuming diet Opposite to recommended treatment)	Cold fomentation in *Vata Dosha* disorders like rheumatoid arthritis	Consumption of cold water immediately after consuming ghee during various therapies
14.	*Paak Viruddha* (Improper cooking)	Improper preparation of dietary articles	Consumption of half-cooked or extra-cooked or burnt food
15.	*Samyoga Viruddha* (Incompatible combinations)	Mixing salty, sour taste articles or fruits with milk	A combination of banana and milk showed significant abnormal changes in the heart, liver, and spleen in an in vivo study.[31] Consuming salt with milk changes the stability of casein, the milk protein.[32] The salt-containing articles, such as toasts, biscuits, and health drink mixes that contain salt, should not be consumed with milk
16.	*Hridaya Viruddha* (Incompatible with individual food choices/likings)	Consumption of diet with disliked tastes	Dislikes can be considered not only by taste but also by other sensory processes of smell, sight, and texture.
17.	*Sampat Viruddha* (Diet deficient in qualities)	Consumption of dietary articles not rich in their best qualities or sub-standard diet	Consumption of immature, over-mature, or damaged fruits. Consuming in-season produce is highly recommended.

(Continued)

TABLE 6 (*Continued*) Types of *Viruddha Ahara* with Examples

NO.	TYPE	EXAMPLES FROM CHARAKA SAMHITA	CONTEMPORARY EXAMPLES
18.	*Vidhi Viruddha* (Contradictory to good eating practices)	Diet not consumed according to good eating practices	Drinking high-temperature diet has been shown to damage the mucosal layer and increase the risk of esophageal cancer[33]

suffering from various disorders revealed that all of them were consuming one or other forms of *Viruddha Ahara*, unaware.[27]

The concept of *Viruddha Ahara* needs extensive research using contemporary methodology. Interactions between certain dietary articles and drugs are well known. The area of food combinations and their possible interactions as trophology of nutrition are under-explored.

Therapeutic Nutrition

'*Ahara*' is the prime essentiality right from birth till the end of life. This fact is intelligently utilized in therapeutics. Importance of diet is understood not only for the nourishment of the body but for prophylactic and interventional purposes too. Therapeutic nutrition, thus, considers aspects of the diet for health maintenance as well as disease management. This section elaborates on concepts of '*Pathya*', '*Satmya*', and '*Langhana*' that collectively define this aspect.

Pathya

Pathya means a diet, which does not cause any harm to body channels due to its wholesome and *Dosha*-balancing nature. The diet requirements differ primarily, as per body status (healthy/diseased). The wholesome diet for healthy individuals is narrated as a part of *Dinacharya* (daily regimen) and *Ritucharya* (seasonal regimen).

The *Dinacharya* defines lifestyle recommendations (diet and activities) to be done on an everyday basis. These recommendations are based on the circadian fluctuations in *Dosha* status. Keeping oneself tuned with this cycle helps in maintaining *Dosha* balance, and thus health. Charaka Samhita has enlisted certain dietary articles, *Sada Pathyam* (beneficial to be consumed frequently), to maintain *Dosha* balance (Charaka Sutrasthana.5/12). There is also a list of certain articles that, on frequent consumption, disrupt *Dosha* balance and destroy health and thus are *Sada Apathyam* (prohibited for consumption) (Charaka Sutrasthana.5/10-11) (Box 1.1). An in vivo study in a rat model established inflammatory changes in the liver and spleen and impairment of cardiac and renal functions with prolonged use of certain prohibited articles.[34]

Along with the above-mentioned list of beneficial and prohibited dietary articles, Charaka Samhita specifically recommends avoiding three dietary articles, viz. *Pippali* (*Piper nigrum*), *Kshara* (alkaline food adjuvants), and *Lavana* (salt). (Charaka Vimanasthana. 1/15). When used in adequate quantities for a short duration, though these articles are beneficial to health, their injudicious and prolonged use can produce detrimental effects on the body. For instance, signs of premature aging such as debility, laxity of body tissues, graying of hair, and baldness are associated with an excessive and habitual high salt diet. In the regions where salt is consumed regularly in excess quantities, a large section of the community is seen as affected by the above-mentioned conditions. Contemporary research has also upheld this concept by associating high salt intake with age-related cardiovascular defects such as myocardial diseases

TABLE 7 Season-wise *'Pathya'* Diet

RITU *(SEASON)*	PATHYA AHARA *(WHOLESOME DIET)*	APATHYA AHARA *(UNWHOLESOME DIET)*
Hemant (Prewinter) *Shishira* (Winter)	Fatty meats, aquatic meat, rodent meat, wine, honey, milk products, sweets, newly harvested rice, warm water	*Vata Dosha* aggravating cold, dry, easily digestible diet
Vasant (Spring)	Barley, wheat, rabbit meat, venison, quail meat, partridge meat, wine	Heavy to digest, sour taste, unctuous and sweet dietary articles
Grishma (Summer)	Light to digest diet, plenty of water and cool beverages, fruit juices, milk with sugar	Salty, sour, spicy diet, alcohol
Varsha (Rainy season)	Salty, sour, unctuous diet, meat soup, gruel	Newly harvested cereals, excessive water, and other beverages
Sharad (Autumn)	Wheat, green gram, sugar candy, honey, the meat of arid habitat animals	Hot, bitter, sweet, oily, and astringent diet, the meat of aquatic animals, curd

BOX 1 *SADA PATHYAM* AND *SADA APATHYAM* DIETARY ARTICLES

Sada Pathyam:

Shashtika rice and *Shali* rice (*Oryza sativa*), *Mudga* (*Vigna radiata*—green gram), rock-salt, *Amalaka* (*Phyllanthus emblica*—Indian gooseberry), *Yava* (*Hordeum vulgare*—barley), rain water, milk, ghee, the flesh of animals from arid habitat, and honey.

Sada Apathyam:

Dried meat, dried vegetables, *Shaluka* (lotus tuber—*Nymphaea alba* Linn.) and *Bisa* (lotus stalk—*Nymphaea alba* Linn.), the meat of the emaciated animal, coagulated milk (curd, cheese, cottage cheese, etc.), cream cheese, pork, beef, fish, black gram, and wild barley.

and hypertension.[35] The upcoming branch of 'Nutritional Epidemiology' studies the relationship between dietary and nutritional factors in a population with disease occurrence. Such a study becomes helpful in planning dietary guidelines for preventing certain diseases.[36] It is interesting to note that the roots of this branch of nutrition can be traced back to Ayurveda.

Seasonal climate variations also bring about changes in the *Dosha* status. The delicate balancing of *Dosha* according to the season can be achieved by following *Ritucharya* (seasonal regimen). Meticulous observance of season-specific activities and diet helps in checking pathological *Dosha* vitiation. The seasonal '*Pathya*' diet according to Charaka Samhita is elaborated in Table 1.7.
The relevance of these seasonal dietary guidelines was compared with changes in human gut flora due to seasonal variations. It concluded that observance of these guidelines was health-promoting by altering gut flora in tune with the season.[37]

The concept of *Pathya* is even more crucial for diseased persons. Consuming *Pathya Ahara* specific for a disease, condition, and *Dosha* status is a unique feature of Ayurveda therapeutics. Diet is an indispensable part of all living beings, including humans. Even before prescribing medicines and other therapies, changes in diet are suggested. For instance, consuming *Yavagu* (medicated gruel) is prescribed in the initial stages of *Jwara* (fever), where even medicines are contraindicated (Charaka Chikitsasthana.3/

TABLE 8 Commonly Used *Pathya Kalpana* (Bhavaprakash.5/134)

TYPE	*METHOD FOR PREPARATION*
Manda	The filtered liquid portion obtained after cooking one part of rice with fourteen parts of water.
Peya	One part of rice and fourteen parts of water, cooked till only one part of water remains. Liquid portion with some cooked rice grains to be consumed.
Vilepi	The thick paste was obtained after cooking one part of the rice and four parts of water.
Yavagu	A thick paste was obtained after cooking one part of the rice and six parts of the water.
Yusha	The thick paste is obtained after cooking one part of pulses like green gram and six parts of water.

TABLE 9 Spices for Pacifying Specific *Dosha*

SPICES FOR PACIFICATION OF VATA DOSHA	*SPICES FOR PACIFICATION OF* PITTA DOSHA	*SPICES FOR PACIFICATION OF* KAPHA DOSHA
Cinnamon, Cardamom, Cumin, Ginger, Cloves, and Garlic	No spices except Coriander, Cinnamon, Cardamom, Turmeric and small amounts of Black Pepper, Green Coriander, Cumin	All spices, especially Cumin, Fenugreek, Sesame, and Ginger

142). Similarly, in the management of *Aamvata* (rheumatoid arthritis), consuming fish, dairy products, and black gram are strongly prohibited (Yog Ratnakar Aamvata Chikitsa.37/3-4). *Pathya* is also an integral part of the *Panchakarma* regimen, which consists of five types of evacuation therapies for eliminating toxins accumulated in the body.[38] *Pathya* in the form of *Samsarjana Krama* (graduated diet protocol) is followed after *Panchakarma* therapies to improve the digestive strength that is weakened during this process of purification (Charaka Sutrasthana.25/45-47). The preparations/recipes mentioned in Table 1.8 are commonly prescribed during this phase.

These recipes are essentially prescribed during the *Samsarjana Krama*. Their easily digestible nature, delicious and satiating taste along with nourishing potential helps to regain body strength after the exhausting *Panchakarma* procedures.

Various spices and condiments are added to these preparations to enhance their aroma and flavor. These spices also impart their digestion-enhancing properties to diet in general. For instance, the addition of *Shunthi* (Ginger—*Zingiber officinale*) helps to convert a heavy-to-digest diet to a lighter state. Antimicrobial and antifungal activities of their active constituents help in eliminating food spoilage organisms and pathogens too.[39] This ability was extensively and effectively utilized during COVID-19 management. An association was observed between higher spice consumption with low mortality and incidence of COVID-19 in different Indian states and union territories.[40]

Like other dietary ingredients, spices, too, pacify/aggravate specific *Dosha*. Their judicious use as per *Dosha* status (pathological/physiological) even in the everyday diet is useful. For instance, in the rainy season, when *Vata Dosha* is naturally aggravated, a diet enriched with spices such as ginger and garlic is pacifying. The variety of spices with specific *Dosha* pacifying properties is enlisted in Table 1.9.

Satmya Ahara (Habitual Diet)

The term *Satmya* denotes the comfortable nature of diet and lifestyle. Owing to the habitat, climate, and familial practices, every individual is adapted or habitual to a certain type of diet. Habituality toward dietary articles of all six *Rasa* is considered best to impart excellent strength and immunity (Charaka Vimanasthana.1/20). Even if any *Satmya Ahara* (dietary habit) is harmful in view of the individual's *Prakriti* (constitution), *Ritu* (climatic conditions), *Desha* (habitat), *Roga* (disease condition), *Oka* (regular habit), and *Jati* (sociocultural factors), the body becomes adjusted to it. Any sudden attempt to change these dietary habits is detrimental to health. For instance, an abrupt change in diet pattern for weight loss does not align with the set point of the body. When such changes are crucial, they must be done in a step-by-step manner as mentioned previously. The concept of '*Satmya*' can be studied in the context of Ayurnutrigenomics. Personalized nutrition based on genetic makeup is conceptualized here. Researchers have associated '*Prakriti*' variation with the effect of diet and nutritional status.[41] Similarly, other aspects can be explored further for disease prevention and management.[42]

Langhana (Therapeutic fasting)

The nutritional needs of the body vary constantly. Sometimes, it becomes necessary to pause or restrict the diet intake to maintain or regain health. The practice of *Langhana* requires this diet intake control. *Langhana* is a basic treatment protocol to bring *Laghuta* (lightness) in the body. This lightness can be a symptomatic feeling or in the form of physical weight loss. Restricted diet is a cardinal aspect of *Langhana*, along with certain other lifestyle modifications, like exercises. The diet restriction can be in the form of complete fasting, intermittent fasting, or therapeutic fasting as per situational demand. *Upavasa* (total/partial diet restriction) is mostly recognized as *Langhana*. Periodic fasting with diet modification has been shown to alter the gut microbiome and immune homeostasis and resulted in a reduction in weight and blood pressure.[43] Evidently, intermittent fasting also increases stress resistance and longevity, by decreasing the risk of diseases such as obesity and even cancer.[44]

Usage of an easily digestible diet is also considered a form of *Langhana*. When *Dosha* are vitiated, the normal functioning of *Agni* is obstructed. This produces *Ama* (metabolic poisons), which is the main source of all illnesses. This *Ama* also blocks *Srotas* (body channels) and prevents proper formation and functioning of body elements. *Langhana* brings about the destruction of these toxins, rekindles the *Jatharagni*, and clears all blockages in body channels. This facilitates the proper elimination of flatus, urine, and feces and results in the feeling of lightness in the body.

Langhana is the prime therapeutic principle for many diseases too. For instance, fasting is recommended in management of diseases characterized with *Agnimandya* (decreased digestive strength) and vitiated *Kapha Dosha*, such as *Jwara* (fever), *Atisara* (diarrhea), and *Amvata* (rheumatoid arthritis).

Basti (medicated enema) is a unique treatment modality, capable of nourishing and strengthening the body tissues (Sushrut. Sutrasthana. 35/4). Drugs administered in lower rectum bypass liver and directly enter systemic circulation.[45] On similar grounds, idea of rectal alimentation was utilized in biomedicine in former times.[46] In a study, ingestion of a milk-based herbal preparation administered both orally and rectally improved physical strength and body weight.[47] Thus, reexploring the '*Basti*' as an alternative to enteral nutrition can be used.

The Nobel Prize–winning concept of 'autophagy' describes self-destructive cellular pathways.[48] The nutrition deprivation of cells can be utilized in managing diseases including cancer, neuropathies, ischemic heart diseases, and so on.[49] Exploring similarities between autophagy and *Langhana* and its applications will be an interesting task.

It should be understood that prolonged fasting by diet restriction just by itself is not suitable for the sustenance of the body. Rigorous *Langhana* observed injudiciously will cause rampant aggravation of *Vata Dosha* and diseases related to it.

SUMMARY

From the stage of hunter-gatherers till today's technologically advanced era, the need and quest of mankind to fulfill hunger have remained the same. Diet has been a dynamic phenomenon with an unchanging quest for satiety, happiness, and also health. Even when medicines are not available, restricting and altering everyday diet helps in disease prevention as well as management. The ancient sages compared it with the '*Brahma*' (ultimate being), owing to this superiority. Although some of the traditional dietary knowledge, conceptions, and practices seem incoherent with today's time, we can still understand them principally. Hopefully, a new renaissance will utilize the total potential of these concepts and provide new applications.

REFERENCES

1 Prasher B, Negi S, Aggarwal S, Mandal AK, Sethi TP, Deshmukh SR, Purohit SG, Sengupta S, Khanna S, Mohammad F, Garg G, Brahmachari SK, Indian Genome Variation Consortium, Mukerji M. Whole genome expression and biochemical correlates of extreme constitutional types defined in Ayurveda. J Transl Med. 2008 Sep 9; 6:48. Doi: 10.1186/1479-5876-6-48. PMID: 18782426; PMCID: PMC2562368.

2 Pt. Shastri K, Chaturvedi G. Agnivesha, Charaka Samhita. Sutrasthana. Vol. 7. Varanasi: Chaukhamba Bharti Academy; 1998. P. 39–40.

3 Shastri A. Sushruta, Sushruta Samhita. Sharirsthana. Vol. 4. Varanasi: Chaukhamba Sanskrit Samsthana; 2001. P. 62.

4 Tiwari P, Kutum R, Sethi T, Shrivastava A, Girase B, Aggarwal S, et al. Recapitulation of Ayurveda constitution types by machine learning of phenotypic traits. PLoS One. 2017; 12(10):e0185380.

5 Govindaraj P, Nizamuddin S, Sharath A, Jyothi V, Rotti H, Raval R, et al. Genome-wide analysis correlates Ayurveda Prakriti. Sci Rep. 2015; 5(1):15786.

6 Banerjee S, Debnath P, Debnath PK. Ayurnutrigenomics: Ayurveda-inspired personalized nutrition from inception to evidence. J Tradit Complement Med. 2015 Mar 24; 5(4):228–233. Doi: 10.1016/j.jtcme.2014.12.009. PMID: 26587393; PMCID: PMC4624353.

7 Ahima RS, Antwi DA. Brain regulation of appetite and satiety. Endocrinol Metab Clin North Am. 2008; 37(4):811–823.

8 Ranade A, Gayakwad S, Chougule S, Shirolkar A, Gaidhani S, Pawar SD. Gut microbiota: metabolic programmers as a lead for deciphering Ayurvedic pharmacokinetics. Current Sci. 2020; 119(3). Doi: 10.18520/cs/v119/i3/451-461

9 Bhojani MK, Sharma R, Joglekar AA. Dhatvagni. In: Deole YS, editor. Charaka Samhita New Edition. ITRA Jamnagar: Charaka Samhita Research, Training and Development Centre; 2022. Doi:10.47468/CSNE.2022.e01.s09.107

10 Agrawal AK, Yadav CR, Meena MS. Physiological aspects of Agni. Ayu. 2010; 31(3):395–398.

11 Wansink B. Environmental factors that increase the food intake and consumption volume of unknowing consumers. Annu Rev Nutr. 2004; 24(1):455–479.

12 Kucukerdonmez O, Rakıcıoglu N. The effect of seasonal variations on food consumption, dietary habits, anthropometric measurements and serum vitamin levels of University Students. Prog Nutr. 2018; 20(2):165–175. Doi: 10.23751/pn.v20i2.5399

13 Lee S, Choi Y, Jeong HS, Lee J, Sung J. Effect of different cooking methods on the content of vitamins and true retention in selected vegetables. Food Sci Biotechnol. 2018; 27(2):333–342. Doi:10.1007/s10068-017-0281-1

14 Zheng J, Xiao H. The effects of food processing on food components and their health functions. Front Nutr. 2022; 9:837956.

15 Cömert ED, Gökmen V. Effect of food combinations and their co-digestion on total antioxidant capacity under simulated gastrointestinal conditions. Curr Res Food Sci. 2022; 5:414–422.

16 Wagh K, Bhalerao S. Traditional foods, Ayurveda, and diet. In: Prakash J, Waisundara V, Prakash V, editors. Nutritional and Health Aspects of Food in South Asian Countries. Cambridge: Academic Press; 2020: pp. 99–110.
17 Prakash N, Mishra V, Bhattacharya B, Kumar A, Raghuram Y, Naik B, et al. Prevention of metabolic risks by Kalabhojanam strategy of Ayurveda. Ann Ayurvedic Med. 2020; 9(2):116.
18 Serin Y, Acar Tek N. Effect of circadian rhythm on metabolic processes and the regulation of energy balance. Ann Nutr Metab. 2019; 74(4):322–330. Doi: 10.1159/000500071. Epub 2019 Apr 23. PMID: 31013492.
19 Laska MN, Hearst MO, Lust K, Lytle LA, Story M. How we eat what we eat: identifying meal routines and practices most strongly associated with healthy and unhealthy dietary factors among young adults. Public Health Nutr. 2015; 18(12):2135–2145.
20 Boege HL, Bhatti MZ, St-Onge M-P. Circadian rhythms and meal timing: impact on energy balance and body weight. Curr Opin Biotechnol. 2021; 70:1–6. Doi: 10.1016/j.copbio.2020.08.009
21 Argyrakopoulou G, Simati S, Dimitriadis G, Kokkinos A. How important is eating rate in the physiological response to food intake, control of body weight, and glycemia? Nutrients. 2020; 12(6). Doi:10.3390/nu12061734
22 Nelson JB. Mindful eating: the art of presence while you eat. Diabetes Spectr. 2017; 30(3):171–174. Doi:10.2337/ds17-0015
23 Matsuo K, Palmer JB. Anatomy and physiology of feeding and swallowing: normal and abnormal. Phys Med Rehabil Clin N Am. 2008;19(4):691–707, vii.
24 Reynolds AN, Mann JI, Williams S, Venn BJ. Advice to walk after meals is more effective for lowering postprandial glycaemia in type 2 diabetes mellitus than advice that does not specify timing: a randomised crossover study. Diabetologia. 2016; 59(12):2572–2578. Doi: 10.1007/s00125-016-4085-2
25 Sabnis M. Viruddha Ahara: a critical view. Ayu. 2012; 33(3):332–336.
26 Brown PJ, Wright WB. An investigation of the interactions between milk proteins and tea polyphenols. J Chromatogr. 1963; 11:504–514. Doi: 10.1016/s0021-9673(01)80953-5
27 Talekar M. et al. Prevalence of Viruddha Ahara in patients attending Arogyashala of N.I.A and its effects on health. IJAAYUSH. 2015; 4(1):297–303
28 Aditi P, Srivastava S, Pandey H, Tripathi YB. Toxicity profile of honey and ghee, when taken together in equal ratio. Toxicol Rep. 2020; 7: 624–636. Doi: 10.1016/j.toxrep.2020.04.002
29 Carrieri G, Anese M, Quarta B, De Bonis MV, Ruocco G. Evaluation of acrylamide formation in potatoes during deep-frying: the effect of operation and configuration. J Food Eng. 2010; 98(2):141–149.
30 Mekary RA, Giovannucci E, Willett WC, van Dam RM, Hu FB. Eating patterns and type 2 diabetes risk in men: breakfast omission, eating frequency, and snacking. Am J Clin Nutr. 2012; 95(5):1182–1189. Doi: 10.3945/ajcn.111.028209
31 K S, Sudhakar, Bhat K S. Toxicological evaluation of banana and milk combination as incompatible diet—An experimental exploration of Samyoga viruddha concept. J Ayurveda Integr Med. 2021 Jul–Sep;12(3):427–434.
32 Huppertz T, Fox PF. Effect of NaCl on some physico-chemical properties of concentrated bovine milk. Int Dairy J. 2006; 16(10):1142–1148.
33 Islami F, Boffetta P, Ren J-S, Pedoeim L, Khatib D, Kamangar F. High-temperature beverages and foods and esophageal cancer risk—A systematic review. Int J Cancer. 2009; 125(3):491–524 Doi: 10.1002/ijc.24445
34 Deshmukh S, Vyas M, Nariya MKB. An experimental study to evaluate the effect of Nitya Sevaniya (daily consumable) and Nitya Asevaniya (daily non-consumable) food items on albino rats. Ayu. 2019; 40(4):247–55. Doi: 10.4103/ayu.AYU_288_18
35 Wen DT, Wang WQ, Hou WQ, Cai SX, Zhai SS. Endurance exercise protects aging Drosophila from high-salt diet (HSD)-induced climbing capacity decline and lifespan decrease by enhancing antioxidant capacity. Biol Open. 2020 May 29;9(5). doi: 10.1242/bio.045260. PMID: 32414766; PMCID: PMC7272356.
36 McCullough M, Giovannucci, E. Nutritional epidemiology. In: Heber D, editor, Nutritional Oncology, 2nd ed. Amsterdam: Academic Press Books Elsevier; 2006: pp. 85–96. www.sciencedirect.com/science/article/pii/B9780120883936500622
37 R D, M VR, Robin DT, Dileep A. Adopting seasonal regimen (Ritucharya) to modulate the seasonal variation in gut microbiome. J Ethn Foods. 2021;8(1):1–9.
38 Sharma H, Chandola HM, Singh G, Basisht G. Utilization of Ayurveda in health care: an approach for prevention, health promotion, and treatment of disease. Part 1 – Ayurveda, the science of life. J Altern Complement Med. 2007; 13(9):1011–1019. Doi: 10.1089/acm.2007.7017-A

39 Gottardi D, Bukvicki D, Prasad S, Tyagi AK. Beneficial effects of spices in food preservation and safety. Front Microbiol. 2016; 7:1394.
40 Bhapkar V, Bhalerao S. Relation of spice consumption with covid-19 first wave statistics (infection, recovery and mortality) across India. Int J Ayurvedic Med. 2022; 13(3):699–705.
41 Banerjee S, Debnath P, Debnath PK. Ayurnutrigenomics: Ayurveda-inspired personalized nutrition from inception to evidence. J Tradit Complement Med. 2015 Mar 24; 5(4):228–233. Doi: 10.1016/j.jtcme.2014.12.009
42 Dey S, Pahwa P. Prakriti and its associations with metabolism, chronic diseases, and genotypes: possibilities of new born screening and a lifetime of personalized prevention. J Ayurveda Integr Med 2014; 5:15–24.
43 Maifeld A, Bartolomaeus H, Löber U, Avery EG, Steckhan N, Markó L, et al. Fasting alters the gut microbiome reducing blood pressure and body weight in metabolic syndrome patients. Nat Commun. 2021; 12(1).
44 de Cabo R, Mattson MP. Effects of intermittent fasting on health, aging, and disease. N Engl J Med. 2019; 381(26):2541–2551.
45 Paul A. Drug absorption and bioavailability. In: Raj GM, Raveendran R, editors. Introduction to Basics of Pharmacology and Toxicology: Volume 1: General and Molecular Pharmacology: Principles of Drug Action. 1st ed. Singapore: Springer; 2019.
46 Barr J, Gulrajani NB, Hurst A, Pappas TN. Bottoms up: A history of rectal nutrition from 1870 to 1920. Ann Surg Open. 2021; 2(1):1–8.
47 Rajoria K, Singh SK, Sharma R. Clinical evaluation of Ksira Basti and KsiraPaka of Balya drugs on Karshya. J Ayurveda. 2015; 9(1):38–50.
48 Mizushima N, Levine B, Cuervo AM, Klionsky DJ. Autophagy fights disease through cellular self-digestion. Nature. 2008; 451(7182):1069–1075. Doi: 10.1038/nature06639
49 Glick D, Barth S, Macleod KF. Autophagy: cellular and molecular mechanisms. J Pathol. 2010; 221(1):3–12. Doi: 10.1002/path.2697

Essentials of Therapeutic Nutrition in Ayurveda: Part – I

Pankaj Wanjarkhedkar and Yashwant V. Pathak

Therapeutic nutrition in Ayurveda has been centered around the *Agni,* which is the metabolic energy of the patient at the center with respect to the other factors, as discussed in Figure 2.1.

Agni is the cardinal factor, which has been labeled as metabolic energy for translational understanding throughout the present book.

Metabolism is the complete series of reactions occurring in the body from the system to each cell, providing energy to the body tissues. The rate of energy generation through metabolism is considered basal metabolic rate (BMR),[1] which is the basic unit of human physiology.

Similarly, *Agni* is termed as the basic fraction of Ayurveda biology. The assessment of *Agni* is based on *Jarana-shakti* (the complete digestion and assimilation process).[2]

Agni is person-specific phenomenon, which is the reason all emphasis has been given to the conservation of *Agni* in healthy individuals and the earliest possible reinstating normalcy of *Agni* in the diseased individual.

The *Agni* controls the microorganism's play, energy extraction, and energy regulation. *Agni* may have a specific role that individual gut microbes play in energy harvest. The individualized aspect of gut microbes is attracting the attention of scientists and researchers in view of the traditional systems of medicine with an individualized approach.

Novel nutritional treatment strategies may emerge with a better understanding of host-microbe and microbe-microbe interactions.[3]

The demonstration of the physiological steps of the metabolism in Ayurveda can be correlated with the physiology of energy metabolism, as discussed below for carbohydrate metabolism.[4]

The *Avasthapaaka* and the *Vipaka* are two important steps of digestion. *Avasthapaaka* is the first stage of metabolism, which begins with the *Agni* (metabolic energy) as food processing starts in *Amashaya* (part of stomach and duodenum). The *Vipaka* is the later stage of digestion where complete conversion of food occurs through the *Agni* (metabolic energy) to a highly assimilable form.[6]

The digestion is based on the following types of *Agni*.[7]

Samagni

The *Agni* that leads to the timely and complete assimilation of food and that is a uniformly regulated form of metabolic energy is called *Samagni*. It keeps a person healthy.

DOI: 10.1201/9781003345541-3

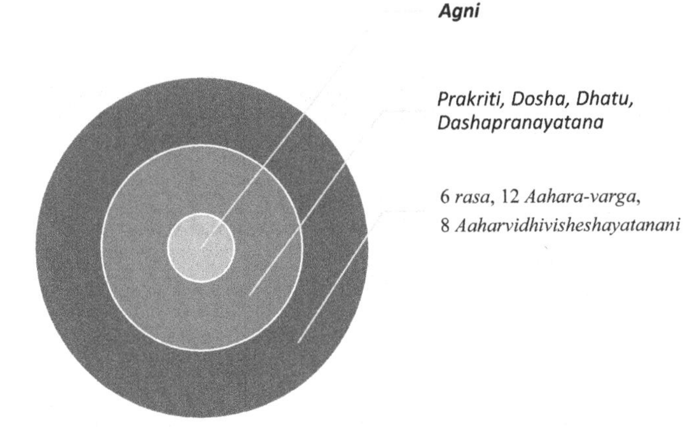

FIGURE 2.1 Nutrition principle of Ayurveda.

TABLE 2.1 Extrapolated interpretation of Metabolism[5]

STEPS OF METABOLISM IN AYURVEDA	STEPS OF CARBOHYDRATE METABOLISM
Madhur Avasthapaaka (First phase)	Combustible Energy
Amla Avasthapaaka (Second phase)	Digestible Energy
Katu Avasthapaaka (Third phase)	Metabolizable Energy
Vipaka or *Nishthapaka*	Net Metabolizable Energy

Vishamagni

The metabolic energy, when misregulated by the influence of *Vata Dosha*, creates a digestion process of uncertain potential and sequence, which can be termed *Vishamagni*.

Tikshnagni

Tikshnagni is high-intensity upregulated metabolic energy, which is a deviation from normalcy, causing repetitive hunger and frequent desire for food.

Mandagni

"*Manda*" is slow. Slowed down or down-regulated metabolic energy causing a delay in digestion can be termed *Mandagni*, which is the general cause of all diseases.

Therapeutic nutrition broadly has been understood as a treatment based on nutrition. The treatment based on nutrition is practiced for enteral and parenteral nutrition as well. Nutritional treatments contribute not only to preventing malnutrition but also to reducing morbidity and mortality in patients undergoing medical management in chronic illness.[8]

The conventional approach for nutrition treatments has been summarized in Table 2.2.
Nutritive intake values (NIVs) are observed mainly for assessing the adequacy of nutrient intakes and for planning diets.[9] The nutritional management of various diseases is largely based on NIVs for respective systems like the cardiovascular system, obesity, and diabetes.

Conventionally, the nutrition approach is quantitative, while from the Ayurveda perspective, it is predominantly qualitative.

Therapeutic nutrition is an integral part of Ayurveda treatment for every disease, irrespective of the system involved. The basis of Ayurveda nutritional treatments is discussed in brief in Table 2.3.

TABLE 2.2 Nutritional basis as per conventional science

CLASSIFICATIONS	*TYPES*
3 Types of diet	Therapeutic Diet, Maintenance Diet, Experimental diet
3 Primary macronutrients	Carbohydrates, fat, protein
3 Ways of feeding	Oral, Tube Feeding, Parenteral
4 Vital nutrients	Calcium, Potassium, Dietary fiber, Vitamin D
4 Main therapeutic diets	Gluten-free diet, clear liquid diet, full liquid diet, diabetic (calorie-controlled diet), low-fat diet, high fiber diet, renal diet
6 Major nutrients	Water, Carbohydrates, proteins, fats, vitamins, minerals
6 Types of therapeutic diet	Clear liquid diet, full liquid diet, low-fat diet, gluten-free diet, low FODMAP diet, diabetic diet, renal diet, heart-healthy diet
17 Essential nutrients	Nitrogen, Phosphorus, Potassium, Calcium, Magnesium, Sulfur, Boron, chlorine, iron, manganese, zinc, copper, nickel, molybdenum

TABLE 2.3 Nutrition basis as per Ayurveda

CLASSIFICATION	*TYPES*
1 Cardinal *Agni*	*Jatharagni* (Metabolic Energy)[7]
3 Principal *Dosha*	*Vata, Pitta, Kapha*[10]
5 *Mahabhutas* (Basic elements)	*Prithvi, Aap, Teja, Vayu, Akash*
6 Types of *Rasa* (Tastes)	*Madhura, Amla, Lavana, Katu, Tikta, Kashaya*[11]
7 Principal *Dhatu*	*Rasa, Rakta, Mamsa, Meda, Asthi, Majja, Shukra*
7 *Prakriti*	*Vata, Pitta, Kapha, Vata + Pitta, Pitta + Kapha, Vata + Kapha*
8 *Aaharvidhivisheshaayatanani*	*Prakriti* (nature of food), *Karan* (food processing), *Samyoga* (food combination), *Rashi* (food quantity), *Desha* (food habitat), *Kala* (season), *Upyog-Sanstha* (rules of diet), *Upayokta* (the person/ patient)[12]
12 *Aahara Varga (Charaka samhita)*[13]	*Shuka-dhanya, Shimbi dhanya, Mamsa, Shaka, Phala, Harita, Madya, Ambu (Jala), Gorasa, Ikshu, Krittanna, Aaharopayogi*
2 *Aahara Varga (Shushruta Samhita* and *Ashtanga Hridayam)*[14]	*Drava varga, Anna varga*
22 *Aahara Varga (Shushruta Samhita* and *Ashtanga Hridayam)*[15]	*Drava varga* (10), *Anna varga* (12) *Drava – Jala-varga* (Water), *Ksheer-varga* (Milk), *Ikshu-varga* (Sugarcane), *Thaila-varga* (Oil), *Madya-varga* (liquor), *Mutra-varga* (Urine), *Dadhi-varga* (Curds), *Takra-varga* (Buttermilk), *Ghrita-varga* (Ghee), *Madhu-varga* (Honey) *Anna – Shukadhanya-varga* (Grains and Millets), *Shimbidhanya-varga* (Pulses), *Kritann-avarga* (Processed Food), *Mamsa-varga* (Meat), *Shaka-varga* (Vegetables), *Phala-varga* (Fruits), *Aushadha-varga* (Herbs), *Pushpa-varga* (Flowers), *Kandavarga* (Tubers and Rhizomes), *Lavana-varga* (Salt), *Bhakshya-varga* (Solid/Hard Food), *Anupana-varga* (Vehicle for Food/Herb)

REFERENCES

1. Sánchez López de Nava A, Raja A. Physiology, metabolism. In: StatPearls [Internet]. Treasure Island (FL): StatPearls Publishing; 2023; PMID: 31536296.
2. Anonymous. Charak Samhita with Ayurved Dipika commentary, Vimanasthana, Chapter 5, verse 8, e-Samhita Designed and Developed by National Institute of Indian Medical Heritage, Hyderabad, Central Council for Research in Ayurveda and Siddha (CCRAS), New Delhi. 2010. https://niimh.nic.in/ebooks/ecaraka/:
3. Krajmalnik-Brown R, Ilhan ZE, Kang DW, DiBaise JK. Effects of gut microbes on nutrient absorption and energy regulation. Nutr Clin Pract. 2012 Apr;27(2):201–14. doi: 10.1177/0884533 611436116. Epub 2012 Feb 24. PMID: 22367888; PMCID: PMC3601187.
4. Elia M, Cummings JH. Physiological aspects of energy metabolism and gastrointestinal effects of carbohydrates. Eur J Clin Nutr. 2007;61 Suppl 1:S40–S74. doi: 10.1038/sj.ejcn.1602938. PMID: 17992186.
5. Dwivedi, L. (ed.), Charak Samhita, commentary of Chakrapani. Chikitsasthana, Chap 15, verse 9–11, Chaukhambha Prakashana, Varanasi, 2013, 1st edn, p. 511.
6. J.L.N. Shastri, Dravyaguna Vijnana, Volume I, Chaukhamba Orientalia, 2002, 1st edn; Chapter 6: Concept of Vipaka, pp. 131–133.
7. Anonymous. Charak Samhita with Ayurved Dipika commentary, Chikitsasthan, Chapter 15, verse 51, e-Samhita Designed and Developed by National Institute of Indian Medical Heritage, Hyderabad, Central Council for Research in Ayurveda and Siddha (CCRAS), New Delhi. 2010. https://niimh.nic.in/ebooks/ecaraka/
8. Reber E, Gomes F, Bally L, Schuetz P, Stanga Z. Nutritional management of medical inpatients. J Clin Med. 2019 Jul 30;8(8):1130. doi: 10.3390/jcm8081130. PMID: 31366042; PMCID: PMC6 722626.
9. King JC, Garza C. Harmonization of nutrient intake values. Food Nutr Bull. 2007;28(1 Suppl International):S3–12. doi: 10.1177/15648265070281S101. PMID: 17521115.
10. Payyappallimana U, Venkatasubramanian P. Exploring Ayurvedic knowledge on food and health for providing innovative solutions to contemporary healthcare. Front Public Health. 2016 Mar 31;4–57. doi: 10.3389/fpubh.2016.00057. PMID: 27066472; PMCID: PMC4815005.
11. Srikanta Murthy, K.R. Ashtang Hrdaya of Vagbhata, Sutrasthan; Chaukambha Krishnadas Academy: Varanasi, India, 2004; Volume 1, p. 10.
12. Susruta Samhitas Shastri Ambika Dutta editors, Chaukhambha Sanskrit Sansthan, Varanasi, reprint edition 2007, Sutra sthana, Chapter 46.
13. Acharya Charaka, Charaka Samhita, 27 chapter Ayurveda-dipika commentary of Chakrapanidatta, edited by Yadvaji Trikamji, Sutrasthana Varanasi, Choukamba Vishwabharati, reprint 1984.
14. Acharya Sushruta, Sushruta Samhita commentary of Dalhanacharya, Sutrasthana Chapter 45/46, edited by Acharya Yadavji and Trikamji, 7th edition Varanasi, Chaukhambha Orientalia, 2002.
15. Acharya Vagbhata, Astanga Hridaya with Sarvangasundari of Arunadatta and Ayurveda Rasayana of Hemadri commentary, edited by Pt. Bhishagacharya Harishastri Paradkar Vaidya, Varanasi, Krishnadas Academy, reprint 2000.

Essentials of Therapeutic Nutrition in Ayurveda: Part – II

Shiva Kumar

Nutrition is applicable to healthy people as well as those suffering from any disease. It helps in maintaining health and as treatment or complementary to primary treatment in patients.

There are mainly two types of diseases when it comes to nutrition. One is *Santarpana* (overnutrition) and *Apatarpana* (undernutrition).

Therapeutic Nutrition in Ayurveda is mainly based on *Jatharagni* and *Dhatwagni* (cellular level). The preparation method and proportion mentioned in *Ayurveda* play an important role in treating a health or disease condition.[1]

The 12 *Ahara Varga*,[2] *Nitya Sevaneeya Ahara*,[3] *Pathya-apathya*,[4] and preparations mentioned in *Krithanna Varga*,[5] all these fulfill the recommended dietary allowance of proximal principles of food, i.e., carbohydrates, proteins, fats, vitamins, and minerals.

There are mainly two sources of foods consumed daily, which are of either plant or animal origin. In Ayurveda, 12 *Ahara Varga* (classification of food sources) are mentioned. They are the sources of pulses, different varieties of cereals, millets, nuts and oils, roots, tubers, and a variety of meat sources. The different foods available in each *Ahara Varga* are explained in detail in terms of Ayurveda nutritive principles. These 12 *Varga* fulfill the contemporary recommended dietary allowances of carbohydrates, proteins, fats, vitamins, and minerals when consumed as per requirements. The food available in 12 *Varga*s influences the three *Doshas* present in the body, which are responsible for both health and disease. Table 2.4 lists the food based on *Tridosha.*

In a therapeutic diet, the form of preparation is very important because a healthy person can consume any form of diet, but it's not the same for the diseased. They may require food that they can swallow easily. Ayurveda mentions different forms of food preparations like *Ashita* (eatables), *Khadita* (chewable), *Peeta* (drinkables), and *Leedha* (lickable).[6]

Kalabhojana means consuming food on time which is very essential in planning meals for the diseased. It has been suggested to ensure the complete digestion of a previously taken meal, before taking a subsequent meal; that is *Jeerna Ahara Lakshana.*[7] *Kalabhojana* is based on individual physiology, which is based on *Agni,* the digestive fire. *Ayurveda* recommends a one-time meal (*Eka Kaala*)[8] or a two-time meal (*Dwikala*),[9] which can be modified according to individual needs based on occupation and requirements. While employing the same, we can term it as time-restricted feeding in diseased again considering the *Agni* and *Matra*.

One should not take food within three hours (1 *Yama*) of food consumption as it leads to *Rasodvega* and should not fast for more than six hours (i.e., 2 *Yama*) as it leads to loss of strength.

DOI: 10.1201/9781003345541-4

TABLE 2.4 Food advice as per Dosha

FOODS	VATA	PITTA	KAPHA
CEREALS			
Wheat	YES	YES	X
Rice (Raw)	X	X	YES
Rice (Boiled)	YES	X	YES
Rice (Old)	X	YES	X
Millets	YES	YES	YES
Maize	X	YES	X
Corn	X	X	YES
Barley	X	X	YES
Grains			
Black Gram	YES	X	X
Green Gram	YES	YES	YES
Red Gram	X	X	YES
Horse Gram	X	X	YES
Chickpeas	X	X	YES
Peanuts	X	X	YES
Bengal Gram	X	YES	YES
Pigeon Peas	X	YES	YES
Sesame Seeds	YES	YES	YES
Vegetables			
Drumstick	X	YES	YES
Radish	YES	X	X
Brinjal	YES	X	X
Tapioca	YES	X	X
Bitter Gourd	X	YES	YES
Pumpkin	YES	X	X
Yam	X	YES	X
Cabbage	X	YES	X
Potato	X	YES	X
Tomato	X	X	YES
Bottle Gourd	YES	YES	YES
Onion	YES	X	X
Grey Gourd	YES	X	YES
Ribbed Gourd	X	YES	YES
Carrot	YES	X	YES
Okra	X	YES	X
Beetroot	YES	X	X
Ash Gourd	YES	YES	X
Cucumber	X	YES	YES
Snake Gourd	YES	YES	X
Spinach	X	YES	X
Mushroom	X	YES	X
Cauliflower	X	YES	YES
Garlic	YES	X	YES

TABLE 2.4 (***Continued***) Food advice as per Dosha

FOODS	VATA	PITTA	KAPHA
Spices			
Ginger	YES	X	YES
Pepper	YES	X	YES
Long Pepper	YES	YES	YES
Mustard	YES	X	YES
Clove	X	YES	YES
Cinnamon	YES	X	YES
Cardamom	X	X	YES
Red Chilies	X	X	YES
Green Chilies	Little	X	YES
Coriander	YES	YES	YES
Cumin	YES	X	YES
Asafetida	YES	X	YES
Fruits			
Mango	YES	YES	X
Mango (Raw)	X	X	YES
Papaya	YES	X	X
Orange	YES	X	X
Pineapple	YES	X	YES
Apple	YES	YES	X
Banana	YES	YES	X
Grapes	YES	YES	YES
Coconut	YES	X	X
Strawberry	YES	X	X
Custard Apple	YES	YES	X
Gooseberry	YES	YES	YES
Cheery Plum	YES	X	X
Lemon	YES	X	X
Pomegranate	YES	YES	YES
Guava	YES	X	YES
Jackfruit (Ripened)	X	YES	X
Dates	YES	YES	X
Cashew	YES	X	YES
Oils			
Sesame Oil	YES	YES	YES
Coconut Oil	YES	X	X
Mustard Oil	YES	X	YES
Sunflower Oil	X	X	YES
Milk And Milk Products			
Cow Milk	YES	YES	YES
Buffalo Milk	YES	YES	X
Goat Milk	X	YES	X
Butter	YES	YES	X
Ghee	YES	YES	X

(*Continued*)

TABLE 2.4 (*Continued*) Food advice as per Dosha

FOODS	VATA	PITTA	KAPHA
Cream	YES	YES	X
Butter/Cheese	YES	X	X
Curd	YES	X	X
Non-Vegetarian Foods			
Seafood	YES	X	X
Mutton	YES	YES	YES
Beef	YES	X	X
Chicken	YES	X	YES
Eggs	YES	X	YES
Dry Meat/Fishes	X	X	X
Pork	YES	YES	X
Turkey	YES	YES	YES
Fowls	YES	X	X

Paryushita Ahara – *Ayurveda* doesn't recommend eating stale food or the food prepared on the previous day because it has lost qualities and will have an adverse effect on *Doshas* and aggravate the condition. The diet should be freshly prepared and served. Nowadays, bakery foods are very much recommended for the diseased, which is not a good choice.

IMPORTANCE OF WATER

Water is very important in food. The source and quality of water are essential factors while planning the diet. Though *Ayurveda* mentions many sources of water for drinking from bore well, river water is largely used nowadays as potable water should be free from pathogenic organisms and chemical impurities. Water has an influence on *Tridosha* in the body. *Sheeta Jala* is good for *Pitta*-dominant people, and hot water is suitable for *Vata* and *Kapha* people. But boiled and cooled water is good in *Pitta* conditions.[10]

Water consumed during meals keeps the body healthy; if taken at the beginning, it hampers digestion and makes the body lean, and if consumed at the end, it causes obesity.[11] Drink water only when thirsty till satiety.

Water boiled and reduced to ¼th part is *Kaphahara*, that reduced to 1/3rd part is *Vatahara,* and that reduced to ½nd part is *Pittahara*.[12]

Taste and Diet – *Ayurveda* recommends a diet that comprises all six *Rasas* (tastes). One should not consume a diet that is predominant in only one *Rasa* because all six tastes are required for the maintenance of the body. The order of consumption of *Rasa* is important as it influences *Tridoshas*. Food with *Madhura Rasa* is consumed first; *Amla Rasa Ahara* in the middle; and the remaining *Tikta, Katu,* and *Kasaya* at the end of the meal.[13]

- **Sweet -** pacifies aggravated *Vata* due to excess hunger.
- **Pungent, Sour, and Salt -** boost *Agni* so that digestion of food is easy.
- **Bitter and Astringent -** in the end, reduces *Pitta* production in excess.

The six flavors can be divided into two main categories according to their effect on the body, determined by the prevalence of *Rasas*.

- **Increasing the body mass –** Sweet, sour, and salty (*Madhur, Amla, Lavana*)
- **Decreasing the body mass –** Spicy, bitter, and astringent (*Katu, Tikta, Kashaya*)

AAHAR PARINAAMKAR BHAVA (FACTORS RESPONSIBLE FOR THE TRANSFORMATION OF FOOD)[14]

The factors leading to the transformation of food are *Ushma, Vayu, Kleda, Sneha, Kala,* and *Samyoga* (appropriate administration), described as follows:

- *Ushma* (Heat) digests.
- *Vata* (*Vayu*) transports food nearer to *Pitta* (*Jatharagni*) for digestion.
- *Kleda* (Moisture) loosens the food particles.
- *Sneha* (Unctuousness) refers to the softness of the ingredients.
- *Kala* (Time) brings about the maturity of the process of digestion.
- *Samyoga* refers to the appropriate administration of food, which brings about the equilibrium of *Dhatus*.

The diet should be wholesome. Mixing a wholesome and unwholesome diet is not recommended (***Samashana***), and eating food before the previously taken meal is digested (***Adhyashana***) is also not recommended, because overeating and unwholesome food consumption will aggravate the disease and deteriorate its condition.[15]

Anupana* and *Sahapana – *Anupana* or *Sahapana* refers to consuming liquids along with food or medicine.[16]

Actually, buttermilk is the best *Anupana* or post-prandial drink that can be used after having food. *Anupana* aids easy digestion and assimilation and enhances food quality and bioavailability of nutrients from the nutrients. Normal or warm water is the most preferred *Anupana* after digestion of meal, whereas buttermilk is an exception and can be taken as after-meal drink.

Meat soup, honey with water, milk, and wine are other substances that can be used as *Anupana* as per the disease condition.

The processing of food plays a key role in food preparation. The processing mentioned in *Ayurveda* is very different from the methods followed in the present day. Nutrients are lost while hulling, milling, blanching, overheating, washing too many times, peeling, lipid oxidation, and using enzymes in browning. Usually, food processing should retain or fortify the nutrition in food, but there is a loss of the same.

In Ayurveda, many preparations that can be used as a therapeutic diet are explained. These preparations can be used in diet plans for many diseases. They fulfill the *Ayurvedic* principles of a therapeutic diet.

The 12 *Ahara Varga, Nitya Sevaneeya Ahara, Pathya-apathya,* and preparations mentioned in *Kritanna Varga* (*Ayurvedic* therapeutic preparations) are very essential in therapeutic nutrition. All these fulfill the recommended dietary allowance of proximal principles of food i.e., Carbohydrates, Proteins, Fats, Vitamins, and Minerals.

Ashta Ahara Vidhi Visheshaayathana are eight important things to be followed in therapeutic nutrition.[17]

Nature of the diet and *Ahara* is very important as the food we use has its own nature like being light, heavy, cold, etc.

Karana or *Samskara* plays a key role in therapeutic nutrition. The processing of food according to the needs of the patient and disease condition is the basic concept described as *Samskara*. Processing of food brings about the quality of the diet. By heating *Guru* or heavy food ingredients can be transformed into light ingredients, or those with *Laghu* quality. Time makes food substances naturally fit for a particular use. Old rice naturally becomes light after some time.

Rational combinations of food considering the nutritive values and its complementary role to the primary treatment are essential factors in therapeutic nutrition.

Samyoga or a combination of food ingredients brings synergistic action to the preparations.

One more important aspect of therapeutic nutrition is the quantity of food to be given to the diseased. This is explained as *Rashi*. As we intend to correct the *Agni,* the amount of food is to be decided by an expert. In patients, the quantity of food depends on the appetite mainly, while in healthy individuals, it depends upon the overall digestive capacity.

The quality of the preparation depends on the food grown in specific regions. Usually, the diet comprises food grown in the locality. But sometimes, we may need ingredients grown in a particular region for a particular disease. Dates grown in the desert are considered to be a more nutritious and healthy food.

As per Ayurveda nutritional guidelines, freshly sourced food and freshly cooked recipes shall be consumed over frozen or stale food. Liquid, semisolid, or solid diets are advised, considering the general condition of the patient and the stage of the disease. These preparations are mentioned in *Ayurveda* under *Kritanna Varga*. These preparations are advised again based on the *Avastha* of a patient. These preparations help in providing required nutrition in terms of Ayurvedic principles like improving the *Agni* or pacifying increased *Dosha* or *Dhatus*. Precisely, these preparations help in treating all the diseases that derange the dosha and *Dhatus* because of overnourishment or undernourishment.

The important thing about a therapeutic diet is that it should be freshly prepared and served to harvest the best results. *Paryushita,* or stale or foods prepared on the previous day, are not recommended. But Western people mainly depend on baked foods like bread, buns, and biscuits, which are extensively used. Nowadays, these foods are fortified with particular nutrients to overcome the deficiency. So these preparations can also be used judiciously in therapeutic nutrition accordingly.

Finally, therapeutic nutrition aims to pacify the *Vata*, *Pitta*, and *Kapha* involved in causing the disease and correction of *Agni,* which is essential in the treatment of a disease. Ayurveda texts explain that it is easy to prepare *Kritanna Kalpana*, which can be prepared from simple food sources available, and is beneficial as a nutritional supplement.

Therapeutic nutrition is all about pacifying *Doshas*, repairing *Dhatus,* and supplementation of nutrients in the required form (protein, carbohydrates, fats, vitamins, minerals, and water).

Manda, Peya, Yavagu, Anna, Vilepi: The first five formulations are prepared by using rice and water in varied proportions. Rice is a cereal that is a source of carbohydrates made up of glucose. The thin consistency of the preparation supplies enough water required for the body and rice as a simple form of glucose, which is required during periods of increased basal metabolic rate.

Yusha is a preparation that consists of pulses and water as main ingredients as well as spices (*Trikatu*) and ghee whenever necessary. Usually, *Mudga* (green gram), *Masha* (black gram), *Masura* (lentils),. pulses are used, which are good sources of protein. It's a thin-consistency preparation that supplies protein. *Yusha* is an easy-to-digest replenishing recipe for infectious diseases and malnourishment.

Among all pulses, *Mudga* is given prime importance in Ayurveda as it nourishes the body well. Green gram soup is an energy drink that relieves thirst, is good for the eyes, relieves fever, and strengthens the body.[18]

Many conditions demand energy, body-building elements, and water, and *Krishara* is one such diet form that supplies all these. *Krishara* is a semisolid preparation made of pulses, rice, spices, and water and is a rich source of protein, carbohydrates, and water. When it is seasoned with spices and ghee, it supplies fat, vitamins, and minerals. The *Mamsarasa* (clear soup) and *Veshavara* (thick soup) prepared from Goat meat are nourishing.

It is prepared with meat (with bone or without bone), water, and spices (*Trikatu*). It is a source of amino acids, the energy required to pacify inflammation. These are highly nutritious and aphrodisiac too.

Takra is preparation mentioned in Ayurveda. It can be used as an after-meal drink (*Anupana*) and adjuvant along with medicine. It is very useful in gastrointestinal disorders and digestive disorders as it

TABLE 2.5 *Kritanna Kalpana* and its preparation method

Kritanna	Ingredients	Preparation Method
Laja Manda	*Laja, Pippali, Shunti*	1:14 1 part of *Laja* (parched rice) + 14 parts of water. Till the *Laja* is completely cooked
Peya	Gruel, Water	1:14 Solid rice and liquid portion are taken in equal proportion
Vilepi	Gruel, Water	1:4 Maximum solid portion with little liquid is taken
Krutha Vilepi	Gruel, Water, Vegetables (*Shaka*), Meat (*Mamsa*), Fruits (*Phala*)	1:4
Payasa	Rice (*Tandula*), Milk (*Khseera*)	A sort of porridge cooked by boiling rice with milk and sugar
Krushara	Sesame, Rice, Black gram	1:6 *Tandula* 1 part, *Dal* ¼ or ½ part Gruel is cooked by boiling rice with sesame and black gram
Soupa	Pulse, Water	Any of the *Shimbi Dhanya* are soaked, dehusked, and boiled with enough quantity of water
Ullupta	Meat, Ghee	*Parisushka Mamsa* (*Mamsa* is fried in more quantity of ghee, sprinkled with hot water often) minced and made into cake form
Saurava	Meat, Water	Chopped *Mamsa* is boiled with 2/4/6/8 times of water. It is boiled till one-fourth remains. *Saurava* – The Supernatant clear portion of meat soup
Veshavara	Boneless meat, Water, *Pippali, Shunthi, Maricha*, Jaggery, Ghee	Boneless meat steamed first, then pounded on a stony slab and cooked after mixing with *Pippali, Maricha, Shunti*, jaggery, and ghee
Mudga Yoosha	Green gram (*Mudga*), Water	1:18 Cooking 1 part of grain in 18 parts of water
Raga Shadava	Green gram (*Mudga*), Water, Pomegranate (*Dadima*), *Mrudwika* (Grapes)	All ingredients combined together
Mulaka Yoosha	Radish (*Mulaka*), Water	1:18
Kulatta Yoosha	Horse gram (*Kulatta*), Water	1:18
Dadimamalaka Yoosha	Pomegranate (*Dadima*), *Amalaka*, Green gram, Water	1:18
Rasala	Curd, Jaggery	Curd is added with the required quantity of jaggery and churned properly
Mantha	*Saktu* (Parched grain flour), Ghee, and Coldwater	1:4 *Saktu* well mixed with ghee added with cold water and in not too thin or too thick form

contains lactobacillus, which is essential for gut health. It is also helpful in treating obesity, indigestion, and inflammatory conditions.[19]

CLASSIFICATION OF DIET IN 12 GROUPS

- ***Shooka Dhanya*** **(Cereals)** – Rice is the most commonly used cereal as a major source of carbohydrates in major Asian countries, Africa, and Latin America, whereas wheat is used in various forms in Europe, Australia, and North America. Unpolished red rice is the best rice for therapeutic preparations. *Yava* (*Hordeum vulgare* L) is preferred in diseases due to overnutrition like obesity.
- ***Shami Dhanya*** **(Pulses)** – Green gram, black gram, kidney beans, chickpeas, green peas, black-eyed peas, and horse gram are used as major sources of pulse proteins in different parts of the world, but Ayurveda recommends green gram as the best protein source in both routine use and therapeutic diet.
- ***Shaka*** **and** ***Harita*** **(Vegetables)** – Green leafy vegetables are sources of fiber, vitamins, and minerals. *Palak* (*Spinacia oleracea*) leaves and *Tanduliayaka* (*Amaranthus*) are best sources of iron and fibers, and other green leaves that can be used are *Manuka Parni* (*Bacopa monnieri*) and dill leaves. Fenugreek leaves (*Trigonella foenum-graecum*) are best used in diabetes mellitus. Vegetables like bitter gourd, ridge gourd, ash gourd, bean pods, onion, and garlic are extensively used vegetables in nutritional treatments.
- ***Mamsa*** **(Meat)** - Though many varieties of meat are being used, Ayurveda recommends only goat meat for therapeutic preparations.
- ***Phala*** **Fruits** – *Amla* (Indian gooseberry) is a fruit that can be used for therapeutic purposes in diseases caused by an imbalance in *Tridosha*. Similarly, to balance *Tridoshas*, pomegranate (*Punica granatum*) in gastrointestinal disorders, vrikshamla (*Garcinia morella*) in obesity, and matulunga (*Citrus medica*) in anorexia, constipation, nausea, and vomiting are advised.
- ***Ambu*** **Water** – Water in moderate quantity is good for therapeutic purposes and regular consumption.
- ***Gorasa*** **(Milk and Milk Products)** – Cow's milk, ghee, and buttermilk more preferred.
- ***Ikshuvikara*** **(Products of Sugarcane Juice)** – Sugarcane juice is an aphrodisiac, a simple source of sugar, and jaggery
- ***Madya*** **(Alcoholic Preparations)** – *Asava* and *Arista* are therapeutic (alcoholic) preparations.
- ***Taila Varga*** **(Fats and Oils)** – *Tila Taila* is the best among oils and fats for therapeutic uses. Coconut oil, groundnut oil, and olive oil can also be considered vegetable oils that are rich sources of essential fatty acids.

Functional foods are also known as nutraceuticals. These foods offer health benefits that extend beyond their nutritional value. They are a combination of ingredients designed to improve health. They can also be fortified foods made of fruits, vegetables, nuts, seeds, and grains that contain vitamins, minerals, probiotics, or fiber. They are also rich in antioxidants. In Ayurveda, different *Rasayana* preparations can be considered fortified preparations as they contain many ingredients of the abovementioned quality. Usually, whole food ingredients are natural functional foods, but in the market, modified functional foods are also available in the form of fortified juices, dairy products, cereal preparations, etc. Fortified foods offer protection against specific nutrient deficiencies. Ready-to-eat therapeutic foods have become a recent practice to overcome specific deficiency disorders. Ready-to-use therapeutic foods (RUTFs) are again combinations of foods fortified with nutrients. RUTFs are preferred while treating nutrition deficiency disorders for larger population. Ayurveda recommends freshly prepared foods for treating diseases, so they are of less importance.

An adulteration is an intentional act of reducing the quality of food either by admixture or substitution of an inferior substance or by removal of some valuable ingredient. Such adulterated foods derange the *Doshas* and aggravate the condition. Milk is a food that is commonly adulterated with urea, chalk, etc. Pulses, cereals, and spice powder, which are consumed daily, are adulterated with clay, pebbles, sand, artificial colors, and food-mimicking products. Growing demand and the economy is the main cause of food adulteration.

REFERENCES

1. Bhisagacarya Harisastri Paradakara Vaidya edited, Vagbhata, Ashtanga Hrudayam, Sarvanga Sundara commentary of Arunadatta and Ayurveda Rasayana Commentary of Hemadri. Edition: Reprint, 2012. Varanasi, Choukambha Sanskrit Sansthan: 2012. 394 p.
2. Vaidya Jadavaji trikanji edited, Agnivesa, Charaka Samhitha revised by Charaka and Dridhabala, with the Ayurveda-Dipika Commentary. Edition reprint 2013. Varanasi: Choukambha Prakashan: 2013. 153 p.
3. Vaidya Jadavaji trikanji edited, Agnivesa, Charaka Samhitha revised by Charaka and Dridhabala, with the Ayurveda-Dipika Commentary. Edition reprint 2013. Varanasi: Choukambha Prakashan: 2013. 38 p.
4. Vaidya Jadavaji trikanji edited, Agnivesa, Charaka Samhitha revised by Charaka and Dridhabala, with the Ayurveda-Dipika Commentary. Edition reprint 2013. Varanasi: Choukambha Prakashan: 2013. 133 p.
5. Vaidya Jadavaji trikanji edited, Agnivesa, Charaka Samhitha revised by Charaka and Dridhabala, with the Ayurveda-Dipika Commentary. Edition reprint 2013. Varanasi: Choukambha Prakashan: 2013. 167 p.
6. Vaidya Jadavaji trikanji edited, Agnivesa, Charaka Samhitha revised by Charaka and Dridhabala, with the Ayurveda-Dipika Commentary. Edition reprint 2013. Varanasi: Choukambha Prakashan: 2013. 130 p.
7. Misra B, Vaisya R edited, Sri Bhavamisra, Bhavaprakasa edited with Vidyotini Hindi Commentary. 11th Edition. Varanasi: Choukambha Sanskrit Sansthan; 2004. 121 p.
8. Bhisagacarya Harisastri Paradakara Vaidya edited, Vagbhata, Ashtanga Hrudayam, Sarvanga Sundara commentary of Arunadatta and Ayurveda Rasayana Commentary of Hemadri. Edition: Reprint, 2012. Varanasi, Choukambha Sanskrit Sansthan: 2012. 7 p.
9. Vaidya Kashinath Samagandi. A Text book of Swasthavritta and Yoga Swasthavrittamrtam. First edition 2019. Jaipur: Ayurveda Sanskrit Hindi Pustak Bhandar; 2019. 5 p.
10. Bhisagacarya Harisastri Paradakara Vaidya edited, Vagbhata, Ashtanga Hrudayam, Sarvanga Sundara commentary of Arunadatta and Ayurveda Rasayana Commentary of Hemadri. Edition: Reprint, 2012. Varanasi, Choukambha Sanskrit Sansthan: 2012. 65 p.
11. Bhisagacarya Harisastri Paradakara Vaidya edited, Vagbhata, Ashtanga Hrudayam, Sarvanga Sundara commentary of Arunadatta and Ayurveda Rasayana Commentary of Hemadri. Edition: Reprint, 2012. Varanasi, Choukambha Sanskrit Sansthan: 2012. 65 p.
12. Bhisagacarya Harisastri Paradakara Vaidya edited, Vagbhata, Ashtanga Hrudayam, Sarvanga Sundara commentary of Arunadatta and Ayurveda Rasayana Commentary of Hemadri. Edition: Reprint, 2012. Varanasi, Choukambha Sanskrit Sansthan: 2012. 66 p.
13. Misra B, Vaisya R edited, Sri Bhavamisra, Bhavaprakasa edited with Vidyotini Hindi Commentary. 11th Edition. Varanasi: Choukambha Sanskrit Sansthan; 2004. 124 p.
14. Vaidya Srivastava S edited, Vriddavagbhata, Ashtanga Samgraha Jeevan edited with Hindi Commentary of Dr. Srivastava S. 1st edition, Varanasi, Choukambha Orientalia: 2006. 210 p.
15. Vaidya Jadavaji trikanji edited, Agnivesa, Charaka Samhitha revised by Charaka and Dridhabala, with the Ayurveda-Dipika Commentary. Edition reprint 2013. Varanasi: Choukambha Prakashan: 2013. 525 p.
16. Bhisagacarya Harisastri Paradakara Vaidya edited, Vagbhata, Ashtanga Hrudayam, Sarvanga Sundara commentary of Arunadatta and Ayurveda Rasayana Commentary of Hemadri. Edition: Reprint, 2012. Varanasi, Choukambha Sanskrit Sansthan: 2012. 158p.

17. Vaidya Jadavaji trikanji edited, Agnivesa, Charaka Samhitha revised by Charaka and Dridhabala, with the Ayurveda-Dipika Commentary. Edition reprint 2013. Varanasi: Choukambha Prakashan: 2013. 235 p.
18. Misra B, Vaisya R edited, Sri Bhavamisra, Bhavaprakasa edited with Vidyotini Hindi Commentary. 11th Edition. Varanasi: Choukambha Sanskrit Sansthan; 2004. 635 p.
19. Vaidya Jadavaji trikanji edited, Agnivesa, Charaka Samhitha revised by Charaka and Dridhabala, with the Ayurveda-Dipika Commentary. Edition reprint 2013. Varanasi: Choukambha Prakashan: 2013. 166 p.

SECTION B

Therapeutic Nutrition in Ayurveda for Metabolic Disorders

Mukund Sabnis

1. INTRODUCTION

Metabolic disorders are one of the biggest challenges for medical systems all over the world.[1] The advancement in technology, automation, food availability and food habits, and lack of physical activity are some of the causative factors/etiological factors for these disorders.[2] Obesity is one such[3] disorder that presents with comorbidities related to the lifestyle of an individual. Considering the increasing prevalence of obesity and metabolic syndrome it's most important to put light on the nutritional trends regarding food all over the world. Ayurved has significantly focused on food style and promotes appropriate food consumption to maintain the health of an individual. Nutrition is one of the most important tools for controlling metabolic disorders. Knowledge about the caloric value of food is insufficient to control metabolic disorders. Factors like inflammation, advanced glycation end products, the glycemic index of food, digestibility of food, portion size, and oxidation of food are also some of the important aspects of food that should be made aware to society.

Ayurved science has elaborated disease with different approaches like

Nija – Endogenous and *Agantuja* - exogenous depending on the cause of the disease. One of the other classifications is:

Santarpanjanya (Anabolic disorder)
Aptarpanjanya (catabolic disorder).

These two types of classifications are important when considering nutrition in metabolic diseases:

Ayurveda defines the health of the person where the body tries to maintain homeostasis between the *Dosha. Dhatus* (body tissues), *Mala* (Excreta), and *Agni* (bio metabolic energy), along with harmony between mind, body, soul, and sensory organs, keeps an individual healthy. Here. *Dosha*s can be defined as the physiological regulators and are factors responsible for maintaining homeostasis and if vitiated create diseases.

- To maintain this homeostasis body has to constantly undergo anabolic and catabolic activity at different cell levels of different tissues. Metabolism comprises anabolism and catabolism, and in

DOI: 10.1201/9781003345541-6

metabolic disorders the imbalance between these two leads to more anabolism, in a particular tissue, and at the same time catabolic activity in other tissues.

- In the concern of *Sthaulya* (obesity) or *Prameha* (diabetic syndrome) *Charak* has clearly mentioned that *Sthaulya* (obesity) is a quantitative increase in adipose tissues of the body and other body tissues are malnourished. The constant imbalance between tissue nourishment and growth leads to metabolic syndrome.

2. Therapeutic nutrition in metabolic disorders

Nutrition as a whole completely depends on food type and micro- and macronutrients available in food its bioavailability in each and every individual. The nutrient uptake depends on the status of the gastrointestinal system, starting from the mouth to the large intestine. The normalcy of the gastrointestinal tract helps in the uptake of macro and micronutrients in food. Factors like gut inflammation, altered gut flora, and disease of the gut affect nutrient uptake leading to many anabolic diseases. *Charak* elaborates on nutrient uptake in a very peculiar manner.[4]

The essence part or *Aahara Rasa* (principal digestive juice) nourishes *Dhatu* (body issues), the basic components of five sense organs, joints, ligaments, tendons, and mucilage parts in the body.[5] Nutrition and its uptake are elaborated by three different rules[6]:

1) *Khalekapot Nyaya* - It is a rule of doves alighting from a threshing floor[7] which explains the direct uptake of nutrients from the gut and nourishes the tissue.
2) *Kedarikulyaya Nyaya* - *Kedari* means small fields and *Kulya* means small creeks. As the small creeks in the fields, one by one, supply water to the whole field and correspondingly the food after digestion gradually nourishes the tissues starting from *Rasa Dhatu* to *Shukra Dhatu* and further essential components of the body.
3) *Kshira Dadhi Nyaya* - This rule is an analogy of food getting digested and further metabolized into different entities and ultimately nourishes particular tissues as milk gets converted into curds, curds into butter, and butter into ghee which ultimately helps for tissue growth.

Food is an integral part of life and it is also contemplated as one of the important pillars on which life is dependent.

Charak explains food in the context of wholesome and unwholesome food.

Food that brings homeostasis in the body tissue maintains normalcy, and corrects the imbalance in body tissue is called whole food. Food that disturbs the homeostasis of body tissue is unwholesome food.[8] Irrespective of the calorific value and nutrients the way homeostasis is achieved and maintained in the body is of utmost importance.

Consumption of wholesome food keeps an individual in a normal state of health. Nutrition is considered in the context of wholesome food as food high in nutrient value will not necessarily be helpful in maintaining homeostasis in the body. There are certain guidelines that are important for the wholesomeness of food in metabolic disorders.

Food should be consumed half or one-third the capacity of the stomach.[9]

Food should be freshly cooked.

It should not be too dry, too cold, or too hot moreover should not be consumed too slowly or too fast.

Too fast or too slow food consumption can have an impact on insulin surge. Metabolic disorders are more overrelated to the status of insulin in the body. Insulin surge is observed twice after consumption of every meal. The first surge of insulin starts from the 5th minute and lasts up to the 10th minute. The next surge depends on the time duration of eating and may last up to 100 minutes. In a slow eater individual, the postprandial insulin surge can be prolonged chronically, leading to insulin resistance.[10]

An individual landing into metabolic disorder or anabolic disorder should consider mainly the digestibility of food. As the status of *Agni* is low in people prone to metabolic disorders food that is very light to digest should be consumed.[11] Ideally, for a healthy individual, food that is heavy to digest should be consumed at half the status of satiety.

The dose or portion size of food should be decided in light of the status of *Agni*. The quality of the food is an important concern regarding digestion and it should be monitored very closely. The quantity of food consumption depends on the quality of food, that is, its digestibility index. High-calorie or dense food if consumed in a regular fashion in metabolic syndrome can lead to major pathological cascades. It's very important to decide the portion size and digestive factor of food.

Each and every individual has a different digestive power as per his phenotype makeup and status of *Agni*. Every person should identify the status of *Agni* and decide the portion size of the food.

The status of *Agni* can be identified by the time taken by the body for food digestion, the time required for the feeling of lightness in the body after food consumption, and the portion size of food a person can digest. Individuals having downregulated status of *Agni* do not feel hungry or their intensity of hunger is low and they cannot digest big portion sizes. Excess food consumption leads to various pathological conditions due to the vitiation of *Dosha*. This leads to difficulty in digestion and vitiation of all the *Dosha* at a single given time. The undigested food and vitiated *Dosha* obstruct the channels of the body, causing food stagnation in the abdomen, further leading to the formation of *Ama*. A normal individual or a person landing in metabolic syndrome should never have an excess of food consumption which leads to pathological conditions at gastrointestinal and cell levels.

Recurrent and excess food consumption upsets the digestive capacity of an individual leading not only to complaints of the gastrointestinal (GI) tract but also to systemic disturbances like *Jwara* (fever), *Daha* (burning), *Murchha* (fainting) *Anga Gaurav* (heaviness in the body), and *Vakasanga* (inability to express words properly).[12]

All such type of unwholesome food leads to pathogenesis, creating metabolic diseases. Following are the factors that govern digestion and metabolic disorders.

- Portion size of food
- Food Properties
- Time of eating
- *Agni* of an individual
- *Dosha* predominance in the body
- Digestive strength or capacity
- Time intervals between two meals
- *Viruddhanna* (noncompatible food) *Abhishayandi* food
- Place of eating
- Mindful eating or eating with concentration.
- Proper mental status while eating
- Age and physical activity of an individual

3. Nutrition in metabolism from classical text *Charak*

Classifies diseases as *Nija* (endogenous) and *Agantuja* (exogenous) diseases. *Nija* (endogenous) and *Agantuja* (exogenous) classification is made on the basis of the cause of disease. These metabolic disorders are mostly endogenous, where the cause of disease and the pathology of the disorder initiates itself in the body because of certain reasons. To identify the exact cause of metabolic disease there are that is to be understood that can help elaborating proper nutrition in such diseases.

These factors are:

- *Agni* (digestive and metabolic factor)
- *Viruddhanna* (incompatible food)
- *Ama* (disease-creating factor)

- *Strotorodha* (Obstructive factors in microchannels of the body)
- *Kleda* (dampness or moisture which helps in digestion but if in excess causes tissue function decline)
- *Abhishandya* (inflammation or swelling)

Agni, misunderstood as digestive fire, is an important metabolic tool that is responsible for the transportation process of energy all over the body right from digestion to cells.

Malfunction or downregulation of *Agni* can disturb tissue homeostasis, thereby manifesting *Ama*.[13] The effect of downregulation of *Agni* on food digestion leads to many metabolic unwanted substances that can injure the cell and cell organ, mainly mitochondria, leading to mitochondrial dysfunction and endoplasmic reticulum (ER) stress.

Food that, when consumed, creates heaviness in the abdomen and the body creates bloating in the abdomen and that is difficult to get digested easily should be considered as *Guru* (heavy). The substances that are heavy for digestion are bakery products, processed milk, excess sweets, and oily substances. Food substances that are too heavy to get digested can create disturbances at a cellular level, but also substances that are very dry, hot, and pungent, and incompatible food combinations are responsible for creating vitiation in the digestive process of food. Examples are frozen food, which is dry and very cold, highly processed and dried food, and deep-fried food or eaten at night hours. These are all causes of *Nijavyadhi* (endogenous diseases), creating disturbances at the cellular level leading to low-grade inflammation.

Kleda - Moisture, referred to as *Kleda* in Ayurveda, plays a crucial role in the process of food digestion. Its primary function is to regulate the proper moisture levels during digestion, which is essential for the breakdown of food and the efficient absorption of nutrients. However, excessive *Kleda* levels can lead to disrupted digestion and negatively impact tissue function. At the tissue level, *Kleda* is characterized as an excess of dampness, and it is believed to be excreted through urine. An excessive buildup of *Kleda* can result in various anabolic disorders, including Prameha, a condition similar to diabetes, and other metabolic syndromes.[14]

Abhishyanda - This is a characteristic of food that, if in excess, blocks the channels of *Rasa Dhatu* (principal nutrient juice). This *Abhishyandi* property of substance creates swelling and obstruction which hampers active nourishment of the tissues. *Abhishyandi* can be compared with low-grade inflammation at cell and tissue levels. Food substances that create an excess of *Kleda* in the body channels come under the category of *Abhishyandi*[15] and are advised not to be consumed in a regular fashion. Curd is one of the important examples of *Abhishyandi* food.[16]

Strotorodha – *Strotas* means to channel and *Avarodha* is obstruction. *Charak* explores in very unique ways of tissue formation and tissue nourishment. All the tissues are nourished through *Ahar ras* (principal nutrient juice) from *Ras Dhatu* to *Shukra Dhatu* in a consecutive manner. This internal channel of nourishment which transforms the *Dhatu* into the next *Dhatu* is called a *Strotas* (channels of nourishment).[17] In the disease process, *Charak* explains that obstruction of small channels in the body can vitiate *Dosha* and further effects *Dhatu* (tissue) nourishment. *Ama* is said to be one of the important causes behind the obstruction of channels. Channels can be thought to be micro- and macrochannels that get affected by factors like *Ama*. A complete transformation from one *Dhatu* to another decides the metabolic status of the body. In this context, any disturbance in the transformation of *Dhatu* can be due to *Strotorodha*. In the milieu of metabolic disorders tribulations in insulin release, insulin acceptance by the tissue at different levels or any disturbance in the release or acceptance of hormone or an enzyme must be considered *Strotorodha*. [18]

List of anabolic disorders elaborated in *Charak samhita*:

"प्रमेहपिडकाकोठकण्डूपाण्ड्वामयज्वराः॥५॥
कुष्ठान्यामप्रदोषाश्चमूत्रकृच्छ्रमरोचकः।
तन्द्राक्लैब्यमतिस्थौल्यमालस्यंगुरुगात्रता॥६॥
इन्द्रियस्रोतसांलेपोबुद्धेर्मोहःप्रमीलकः।
शोफाश्चैवंविधाश्चान्ये शीघ्रमप्रतिकुर्वतः॥७॥[19]

TABLE 1 Table shows the classification of anabolic and catabolic disorders mentioned in Ayurved

ANABOLIC DISORDERS AS PER AYURVED	*CATABOLIC DISORDERS AS PER AYURVED*
Prameha (diabetes)	Emaciated body
carbuncles, urticaria, itching	Reduced digestive capacity
Pandu	Reduced strength
Kushtha Metabolic skin disorders	Reduced *Ojas*
Jwara (Fever)	Reduced semen and muscle strength
Mutrakruchha (Dysuria)	Fever associated with cough
Anorexia of metabolic origin	*Parshvashool* (Pain in cardiac region)
Tandra (Drowsiness)	Hearing loss
Erectile dysfunction (metabolic origin)	*Unmad* (insanity)
Obesity	Pain in calf, thigh, and lumbar region
laziness, heaviness	Retention of urine and stool
Delusion	Cracking sound in fingers, bones, and joints
Edema over the body [20]	Vitiation of *Vata dosha* [21]

4. Obesity – an alarming metabolic disorder

Charak explains obesity as one of the eight alarming disorders and elaborates on obesity in a very systemic manner. *Charak* has classified obesity in different ways like *Sthula* (over weight) and *Ati sthula* (Obese) depending on weight of person. Another type of classification is *Bija Dosha* (Genetic) and *Santarpanotha* (Lifestyle induced[22]) depending on underlying cause of obesity.

Obesity is also classified as per the area of deposition of fats in the body.[23]

Android - Fat distribution in the visceral and upper thoracic area

Gynecoid type of obesity - Fats distribution in the lower part of the body, mainly in the hips and thighs.

Hyperplastic fats - Number of fat cells per square centimeter. These types are seen in arms and thighs. Hyperplastic fats are prevalent in the prepubertal age.

Hypertrophic fats - The number of fat cells is less but the size of fat cells is large. These types of fat are observed in the belly, hips, and breasts. Hypertrophic obesity leads more to inflammation and metabolic complications as compared to hyperplastic obesity. [24]

5. Causes of anabolic disorders

Giving the dietary causes of anabolic disorders, *Charak* explains about excess consumption of fatty food and sweet food, both of which are heavy to digest, newly harvested food,[25] new alcoholic beverages, meat from animals living in marshy areas, milk and its products, and sweets prepared out of milk. People indulging in sedentary activities and inactive lifestyles can develop anabolic or metabolic disorders.

It is very important to observe the relationship between diet and anabolic disease in light of molecular biology as well, which can very well explain the effect of food on metabolic disorders.

The recurrent use of food like carbohydrate- or fat-rich food which is explained to be a cause of the metabolic disorder can cause damage to cell metabolism and energy transport, which leads to infinite pathologies in a number of body systems at the different cell levels.

It is essential to know that not only high-fat and high-carbohydrate food leads to metabolic syndrome through varied pathologies, but along with that, the following also do:

1) Untimely eating
2) Recurrent eating
3) Overeating or big portion size
4) Excess of dry food
5) Incompatible food
6) Dry meat
7) Dry vegetables
8) Regular consumption of curds
9) Regular consumption of milk
10) Newly harvested grains
11) New alcoholic beverages
12) Meat of animals from marshy land

High-fat and high-carbohydrate diets are more emphasized, leading to pathologies creating metabolic syndrome. Food substances listed above are regular consumption by general population. Unfortunately, not much research in the context of metabolic disorders with above food habits is observed.

High-fat, high-carbohydrate, and wrongly processed foods like excess use of temperature or microwave, mixing of wines and ice creams, and heating or improper combinations of food like meat or fish consumed along with milk products like ice creams and cheese can lead to metabolic downregulation, leading to multiple disorders.[26] Digestibility of the food is directly correlated to the faculty of *Agni* which is responsible for the complete transformation of food into energy.[27]

6. Classification of *Agni*

Agni has been classified into four types on the basis of its digestibility. *Mandagni*, *Teekshnagni, Sama Agni,* and *Vishamagni*. It is very important to note that *Mandagni* and *Teekshnagni* are many times physiologically inherited in individuals. *Mandagni* people require less portion size of food. *Teeksna Agni* is physiologically inherited in some people and these people can digest a good amount of food and do not land into indigestion very easily. *Vishamagni* is a pathological condition that creates disturbed digestion leading to abnormal nourishment of tissue. Those individuals having inherited *Mandagni* (downregulated digestive faculty) are more prone to anabolic disorders. Obesity is prone to those having *Mandagni.* In the pathological condition of obesity, *Mandagni* which is inherited gets converted into *Teekshnagni* (upregulated digestive faculty) which should be considered pathological. In the management of such disorders of *Agni,* we have to focus on pathological change at the tissue level and maintaining *Agni* at an original status.

Mandagni - Downregulated digestive faculty

TABLE 2 Types of Agni prone to metabolic disorders

MANDAGNI	TEEKSHNAGNI	SAMAGNI	VISHAMAGNI
Downregulated digestive faculty	Upregulated digestive faculty	Normal digestive faculty	Dysregulated digestive faculty
Can be physiological status	Can be physiological status	Physiological status	Pathological status
Prone for anabolic disorders	Highly prone to metabolic disorder	Maintains homeostasis	Highly prone to metabolic or any other diseases.

Individuals of this type have low appetite and cannot digest food that is heavy to digest. Most people having metabolic syndrome land into this category. These types of people have low digestive capacity. In the context of downregulated *Agni,* digestion is very weak and tissues are not nourished in the right manner because of low digestibility of food. This stage leads to maldigestion, and malnourishment of tissue occurs. Improper nourishment is an initiation of a disturbed Krebs cycle where food is not completely digested and proper energy is not derived from food. This stage is the initiation of endogenous injuries to cells and tissue, leading to inflammatory changes at the cell level. If this stage continues for a longer period energy distribution and cell turnover at various tissue levels are hampered, landing to cell function decline.

Mandagni is supposed to be one of the vital and decisive factors which lead to metabolic disorders like obesity, type 2 DM, atherosclerosis, dyslipidemia, and fatty liver.

It is evident that every tissue or *Dhatu* also has its own *Agni*. These are called *Dhatwagnis.* These are tissue-specific *Agni*. Conceptually if *Jatharagni* is weak then corresponding *Bhutagnis* and *Dhatwagnis* are also weak. *Bhutagnis* are responsible for nutrient transport and *Dhatwagnis* are responsible for energy reception and cell turnover.

In the milieu of downregulated digestive faculty other digestive faculties like *Bhutagnis* and *Dhatwagnis* are affected up to the cell level, imparting many metabolic cascades.

The overall principle of *Agni* explains the concept of digestion of food from a gastrointestinal (GI) till the energy turnover up to the cell level.

Mandagni is responsible for *Ama* formation and food that is heavy for digestion, soaked in water for a longer period while processing (*Abhishyandi*), and sweet in taste can aggravate the condition *of Mand Agni*. In anabolic disorder majority of diseases are due to the downregulation of *Agni* and the majority of pathologies initiate the formation of *Ama.*

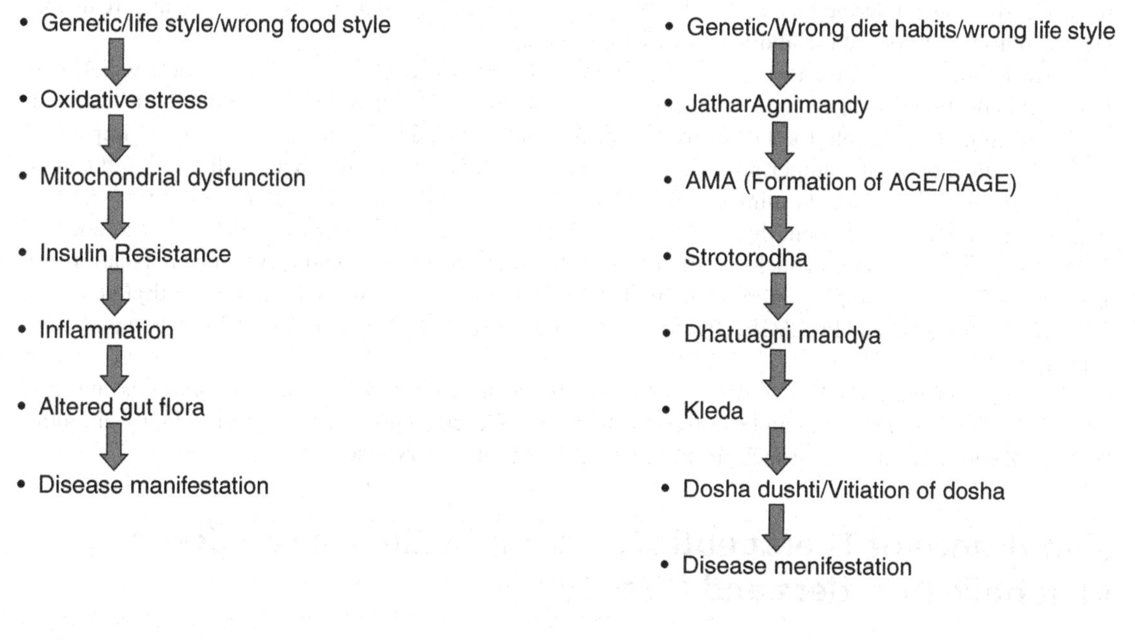

FIGURE 1 Comparative chart showing the difference in the pathogenesis of metabolic disorders in modern medicine and Ayurved.

Major causative factor of metabolic disorder is incompatibility of food which is called *Virruddhanna*. Food with opposing properties if consumed or improper food processing hampers the nutritional quality of food. The concept of *Viruddhanna* is very broad and should be discussed in the context of metabolic disorders. *Viruddhanna,* or incompatible food combination, or improper food processing can lead to formation of toxic substances in vitro or can create toxicity in vivo if consumed regularly. Identification of *Viruddhanna* in modern recipes is an important task. Heating of honey,[28] reheating of food, reheating of oils or repeatedly heated oil for frying,[29] and reheating of milk[30] are all considered to be *Viruddha.* Knowledge of incompatibility of food is important in view of preventing the pandemic of metabolic syndrome. Majority of the population only consider the caloric context of food and important factors like *Viruddhanna* (incompatibility of food) are ignored.

Important incompatible foods as observed in clinics are:

1) Tea with milk.
2) Coffee with milk.
3) Excessive and recurrent heating of milk.
4) Consumption of fruit juices with milk or milkshakes.
5) Consumption of meat with fruits; for example, during breakfast people consume fruit juices and chicken sausages.
6) Excess consumption of bakery products along with meat.
7) Marination of meat with curds.
8) Deep-fried potato chips.
9) Smoothies made out of fruits and vegetables and mixed together.
10) Food processing like grilling, barbequing, roasting, baking, frying, broiling, searing, and roasting.

The majority of this food or food processing changes the properties of food ingredients.

Viruddhanna (incompatibility of food) is a chief concept that is a prime cause of metabolic diseases. *Viruddhanna* is defined as food that vitiates the *Dosha* but does not expel out of the body. Incompatibility of food leads to such intermediate metabolic end products which are uptaken by the tissues and are unable to come out via the normal metabolic pathway for elimination.

This definition[31] is close to pathogenesis with AGE (advanced glycation end product) which gets deposited in the tissue and gets lodged but cannot come out of the cell, disturbing the cell milieu. The toxin coming from noncompatible food disturbs the normal metabolic pathway and can have a bad impact on the *Agni* of an individual, hampering the digestive process, thus creating the factor called *Ama.* It can be said that *Viruddhanna* is an alarming combination of food and is the most important causative factor for metabolic disorders. It is evident that such types of food combinations or incorrect processes on food can create many toxic substances which are all known to deposit in body tissue called Advanced glycation end products (AGEs, for example, carboxy methyl lysine, carboxy ethyl lysine, acrylamide, methyl glycosyl, effective free fructose, and Fru-AGE, (fructose-associated advanced glycation end product), to name some of them.[32]

This type of food not only leads to metabolic disorders but also lands a person in autoimmune and systemic disorders. *Viruddhanna* can be compared with the AGE concept which can lead to many alarming diseases like asthma, chronic bronchitis, arthritis, and coronary artery disease.

Significance of Therapeutic Nutrition in Clinical Practice in Metabolic Disorders and Obesity

Agni is considered a prime factor in metabolism. The low status of an *Agni* is unable to digest even light food. This can lead to different unwanted digestive activities like fermentation and putrefaction which can generate more toxic substances.

Charak says that due to excess vitiation of all the *Dosha* and their interlinkage, the reciprocal outcome can be highly toxic which is called *Ama*.

The *Tridosha*, when it comes in proximity to *Ama*, jointly vitiates the body tissues or *Dhatu*. These are called *Dushyas* and are in association with *Ama*. When the tissues or *Dhatus* get associated with *Ama* the tissue function or cell function gets vitiated and the disease process starts.[33]

Once tissue gets affiliated with *Ama*, toxins are generated. These toxins remain undigested in the tissues or cells, making them all aplastic. Further sclerotic changes take place at the cellular level. This *Ama* spreads all over the body and remains attached to the cell or tissue and does get displaced out of the cell. This creates obstruction in the microchannels of the cell, hindering its function.

While considering nutrition in metabolic disorders the food which is advised should not provoke the pathogenesis of *Ama* formation. Food substances that disturb the digestive process as well as unindicated food processing like heating of honey or reheating of food are known to initiate *Ama*-like pathology, creating aggravation in the disease process.

The complete process of *Ama* formation and its lodgment in the cell and not getting out of the cell in the process of normal metabolic pathways can be compared with the concept of AGEs.

To understand the concept of *Ama*, the concept of advanced glycation end products in metabolic disorders must be understood. Formation of *Ama* can be endogenous (due to the vitiation of *Dosha* ascribed to aging or other factors) or can be exogenous due to food or exposure to the wrong environment and inappropriate mental conditions.[34] Majority of the time exogenous *Ama* is through food consumption.[35] High processing of food, high temperature while cooking, and consumption of fried food lead to a high amount of AGE.[36] It is evident that up to 30% of ingested AGE is absorbed in circulation.[37] This absorbed AGE initiates the pathology of inflammation. It is also evident that the absorbed AGE is known to disturb gut microbiota. An alteration in the proportion of sulfate-reducing bacteria (SRB) and a decrease in *Bifidobacteria* is observed.[38] High AGE is known to induce variations in microbial composition. Changes in microbial composition lead to consequent changes in the production of microbial metabolites. Short-chain fatty acid production gets altered, leading to inflammation.[39] Some of the research has shown that dietary AGE correlate with AGE and inflammatory markers.

Endogenous AGE-induced pathology can be seen in populations who are genetically predisposed to metabolic disorders like diabetes, obesity, and atherosclerosis. If food with overprocessing is consumed in a regular manner then the disease process can hasten to create more sclerotic changes in the tissues.

The theory of *Ama* elaborated by *Charak* clearly indicates the formation of *Ama* at two stages. *Ama* is generated by improper food style which affects *Jatharagni*. Satisfactory metabolism of food by *Jatharagni* consequently leads to appropriate nourishment of tissue. Constant improper processing of food accordingly leads to dAGE. Energy generated by the food through the Krebs cycle should be accepted by peripheral tissue, which is a function of *Dhatvagnis*.[40] As discussed earlier *Ama* can lodge inside the cell structure and does not get dislodged easily; similarly, AGE can also take entry into the tissue and create inflammation. AGEs are known to accumulate intracellularly and modify the intracellular protein signals.[41] AGE alters enzymatic activity and interferes with receptor recognition. These changes in receptor recognition can be compared with a loss of nutrient recognition at the cell level, disturbing proper cell nourishment.[42] *Ama* at a *Dhatwagni* level obstructs cell nutrition, further leading to cell function decline. The extent of pathological events created at the cellular level by *Ama* or AGE depends on multiple factors and healthiness of tissue. Some of the important factors are antioxidants at particular cell levels, fructosamine kinases, and receptors for advanced glycation end products (RAGEs).

Ama formation, in the overall process of food digestion and its undesirable association with tissue, is defined not only by the rate of formation of *Ama* but also by the speed and the rate of removal of *Ama* from the tissues. There are many methods and nutritional principles by which breakdown of *Ama* and its capacity to come out of cell is possible. Application of certain nutritional concepts like fasting and consumption of lukewarm water can properly digest *Ama* and bring it to the normal cellular pathway.

Uptake of AGE depends on certain receptors at the tissue level. These receptors are called receptors for advanced glycation end products (RAGEs). In fact, the binding of AGE with RAGE stimulates many

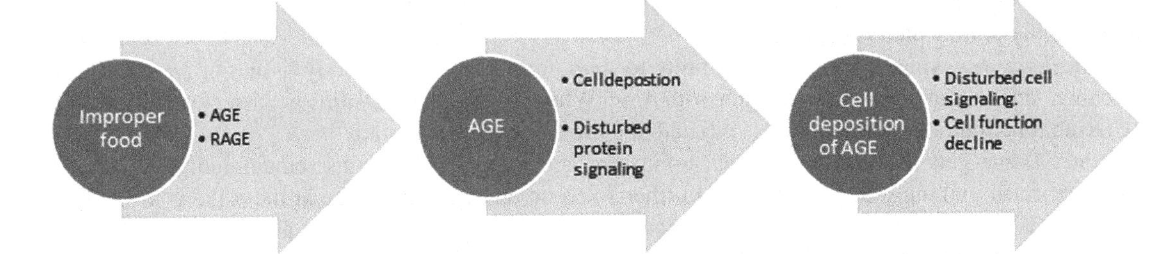

FIGURE 2 AGE leading to cell function decline.

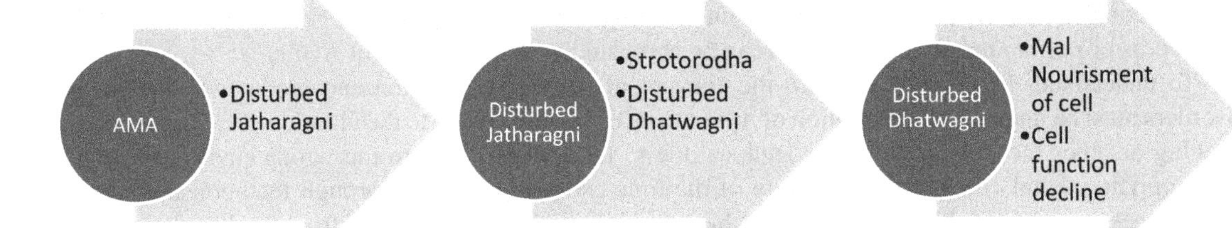

FIGURE 3 *Ama l*eading cell decline.

enzymatic reactions which activates inflammatory markers, mainly nuclear factor kappa B (NFkB).[43] NFkB formation further can increase oxidative stress and increase expression of oxidized low-density lipoproteins (OxLDL) that further leads to complications of metabolic syndrome.[44]
Metabolic disorders are attributed to vitiation of *Kapha dosha.*

The list of diseases quoted in Ayurveda in the category of *Santarpan Vyadhi* (anabolic disorders) is so unique that it seems like metabolic disorders are often anabolic in nature and this anabolism is associated with adipose tissue.

Diseases are the main cause of undigested digestive capacity that is, downregulation of *Jatharagni* at a gross level. The effect of downregulation of *Agni* on the digestion of food leads to many unwanted metabolic functions, which can injure the cell organs, mainly mitochondria, leading to mitochondrial dysfunction.

Ama, AGE formation, NFkB, and oxidized LDL are causes of endogenous disorders creating disturbance at the cell level. The above factors can cause cell injury leading to low-grade inflammation (LGI).

In the view of principles of nutrition in metabolic disorders, it is very important to establish the concept of diet and nutrition in the vision of cellular sciences.

Unindicated food style can lead to improper food digestion. This improperly digested food gets absorbed from intestine and is circulated as intermediary metabolite product circulating along with blood, leading to improper turnover of cell metabolism at the respective tissue. This leads to tissue-specific mitochondrial dysfunction which is more commonly seen in metabolic disorders or commonly seen in noncommunicable diseases.[45] Though modern medicine has identified insulin resistance as basic pathology of noncommunicable diseases[46] (NCDs) or metabolic disorders (MDs) its comorbidities like type 2 DM, cancer and polycystic ovarian syndrome (PCOS) could not be overlooked.

Majority of these metabolic disorders are due to insulin resistance (IR) or inflammation at a cellular level. Considering the lifestyle, food is the prime cause of both IR and inflammation. It is very important to note that undisciplined food consumption superimposes sedentary life style in creating IR or inflammation.

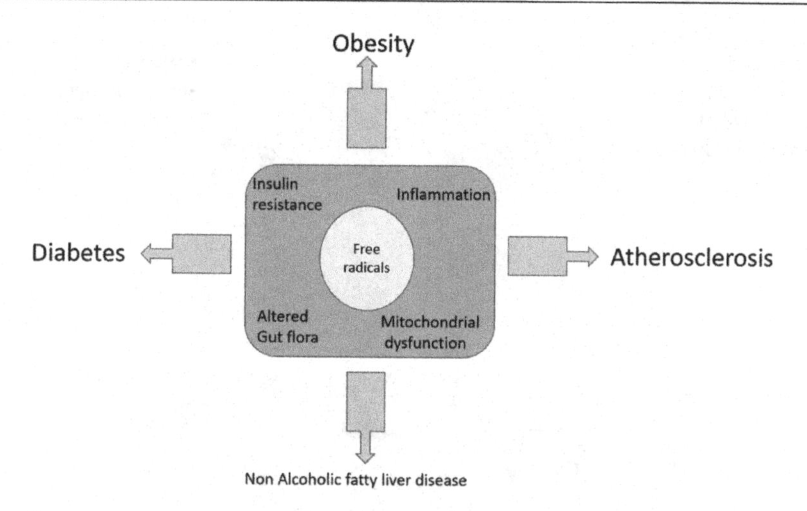

FIGURE 4 The effect of ROS on the basic pathology of metabolic disorders leading to the disease process.

Inflammation is basically induced through cellular injuries. It is well elaborated that a big-portion size of food and other anabolic factors create cell stress, which damages mitochondrial structure. The damaged mitochondrial structure gets discharged into the cytoplasm and extracellular fluids, where they attract immune signals which are recognized by the innate immune system. All such pathological factors initiate free radicals and attenuate endogenous antioxidants like superoxide dismutase, glutathione peroxidase, and catalase. A change in extracellular milieus leads to inflammation, and that is considered endogenous pathology or *Nija Hetu* and can be compared with *Kleda* formation as described in ancient texts.[47]

The damaged cytoplasmic component which comprises modified proteins and amyloid bodies disturbs mitochondrial functions and does not have the capacity to come out of the cell for further metabolic process of elimination. Factors contributing to disturbed cellular function can be called *Ama* or disease-creating factors at the cellular level.

One of the important pathologies in noncommunicable diseases (NCDs) or metabolic disorders, which is again related to food as well as exercise, is insulin resistance (IR). The causes elaborated by *Charak* regarding fatty food and recipes coming from sugar and jaggery are told to provoke metabolic syndrome. It has been observed that obesity and metabolic disorders aggravate due to cellular glucose concentration and fatty acid concentration in skeletal muscle and plasma.[48]

Another important pathology that plays a basic role in the majority of metabolic disorders is free radical generation due to high fat or high carbohydrate diet that leads to lipid peroxidation.

Raised reactive oxygen species (ROS) has an impact on cellular energy metabolism, nutrient transport, cell signaling, and many more biological activities.[49]

7. Investigations and approaches in obesity management

Management of obesity as a disease is a crucial task, as the disease is more patient driven rather than physician driven. *Charak* has elaborated on obesity as a genetic as well as lifestyle disorder. Classification of obesity and metabolic syndrome is identified for the purpose of dietary management. Lipid profile is the investigation that is frequently carried out while exploring the possible cause of obesity. Lipid profile is not supposed to be a specific marker to investigate the cause of obesity or metabolic syndrome. Obesity management and treatment aim to treat underlying pathology along with weight reduction. In obesity along with disturbed metabolism, hormones are also disturbed which badly makes the person prone for cardiac and diabetic disorders. Obesity also makes an individual susceptible for cancer.

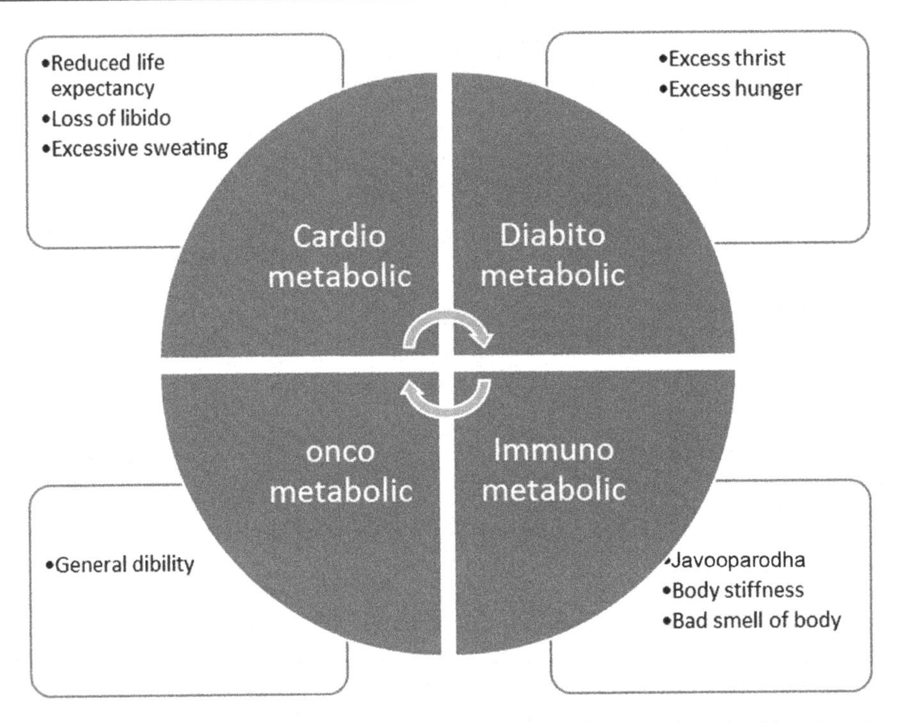

FIGURE 5 Obesity and metabolic complications and the main complications described by *Charak*.

Body is made of different types of tissues called *Dhatus*. In the disease progression *Dhatus* gets vitiated by *Dosha*. It is observed that in the disease process not all the tissues are affected. This is due to tissue excellence which is called *Dhatusarata*. In obesity if adipose tissues show *Meda Sarata* (adipose tissue excellence), then overweight of the patient does not create any comorbidity. It is a key factor to identify tissue excellence in adipose tissue and obesity. Charak elaborates that if there is tissue excellence of *Meda Dhatu,* then adipose tissue is considered healthy tissue and not much management is required in such condition.

> वर्णस्वरनेत्रकेशलोमनखदन्तौष्ठमूत्रपुरीषेषुविशेषतःस्नेहोमेदःसाराणाम्। सासारतावित्तैश्वर्यसुखोपभोगप्रदा
> नान्यार्जवंसुकुमारोपचारतांचाचष्टे॥१०६॥

Obese people usually present themselves with some or other co-morbidity. In such conditions, we propose investigations like serum Adiponectin level for obese patients which is inversely proportional to the amount of weight. Adiponectin is secreted by adipose tissue. Like lipid and carbohydrate metabolism, appetite control, and insulin resistance are attributed to Adiponectin. The more the weight, the less Adiponectin levels, or it can show Adiponectin resistance.[50] Low serum Adiponectin is observed in overweight individuals. Normal serum Adiponectin along with normal fasting insulin in obese individuals suggests tissue excellence at an adipocyte level.[51] The condition suggests *Meda Sarata* (wellbeing of adipose tissue) which restricts the person landing into comorbidity. Such persons ideally do not need weight reduction unless it is tolling the body mechanically.

Obesity is classified for clinical purposes in a different manner. The thin or obese status of the body as per their appearance or BMI does not reveal the disease status. Obesity is classified clinically as:

1. Thin fat obese
2. Obese

TABLE 3 The investigation and its interpretation in the view of obesity management

S. NO	*INVESTIGATION*	*INTERPRETATION IN THE ANGLE OF OBESITY*	*DIETARY MANAGEMENT AS PER THE INVESTIGATION*
1	Serum Insulin	Suggests Insulin resistance at the Musculoskeletal level.	Reduces *Madhur Rasa* (sweet tasting food), increases with pulses and lentils
2	Homa IR	Shows Insulin resistance and fatty liver.	Helps to decide the consumption of fibers and proteins in daily diet
3	HsCRP	Suggests inflammation at cellular and adipocyte level also non-specific Cardiometabolic risk factors.	Avoids inflammatory food such as highly processed, deep fried, or frozen nonvegetarian food
4	Apo lipoproteins	Specific marker for cardiometabolic comorbidity in obesity.	This can suggest the need for aggressive management, nonvegetarian food, and AGE-induced food leading to atherosclerosis
5	Thyroid function test	Helps to rule out associated hormonal issues leading to low basal metabolic rate.	Helps to decide the proportion of carbohydrates and proteins in the diet. A Low T3 level suggests the carbohydrate portion to be advised.
6	FSH, LH	Helps to rule out allied hormonal issues further leading to gynecological complications.	1. Suggests to advise diet with maximum proteins leading to develop insulin sensitivity. 2. Add some dietary substances affecting IR at ovarian level like cinnamon.
7	Serum Adiponectin	Most important marker of obesity, suggesting insulin resistance, inflammation, and blood coagulation.	1. Helps to decide the need for weight reduction. 2. Very specific and sensitive marker to decide healthiness of adipose tissue.
8	Vitamin B12	1. Suggests the reuptake of vitamins. 2. Moreover can suggest gut inflammation if recurrently low.	1. Suggests food supplements which will improve gut uptake example addition of ginger, cumin in food.
9	Hemogram	Suggests anemia of inflammation	Suggests renal or hepatic pathology associated with obesity
10	HbA1c	Suggests about glycation status	Helps to decide about the carbohydrate load and proportion of simple and complex carbohydrates in food
11	Estradiol	Suggest about menopausal status of women	Helps to decide regarding aromatization in fat cells which can change obesity management

Normal weight obesity or the thin fat phenotype is defined as the presence of an increased body fat percentage in an individual with normal body mass index.[52] In thin obese individuals there is more fatty infiltration in organs and skeletal muscles. Ayurved also specifies that obesity is not only the disease of fats but also the disease of muscles. In Asian countries and specifically in India there is more prevalence of thin obesity.[53] In obese category BMI is above 27 as people look obese with more fat in the belly and thighs. Thin obese people are more prone to metabolic insults as there is more vitiation *of Vata dosha* as compared to obese individuals. Depending on serum high sensitive C reactive protein (HsCRP) levels obese individuals are classified as inflammatory obese. Such individuals can reduce their weight faster when specific non-inflammatory food is advised to them. Following advised diet controls these people reduce their risk of inflammatory obesity. In inflammatory obesity refined carbohydrates, bakery products, and deep-fried and stored food should be avoided.

Charak classifies diabetes as thin diabetes and obese diabetes.[54] Thin diabetes exhibit muscle wasting, and in this category Charak explains management which helps to treat muscle wasting and improve muscle quality in thin diabetics. Thin diabetics are managed and promoted with food which helps to nourish muscles. Meat of animals, *Moong bean* recipes along with old rice processed with vegetables, and flax seed oil are advised in thin diabetics.[55] Flax seed oil is rich in omega-3 fatty acids which exerts anti-inflammatory activity at the cell level.[56]

8. Appetite in Obesity

Charak explains the increased appetite for obesity. Explaining the pathology of obesity the upregulated *Agni* makes the person eat more food which gets converted into only fats and other tissues are malnourished.[57] These changes are due to deranged appetite control system which is governed by the hormone leptin and ghrelin. The hormone leptin is a hormone secreted by peripheral adipose tissues and ghrelin is secreted by parietal cells of abdomen. A balance between these hormones maintains the satiety of an individual. In obesity a high level of insulin and leptin resistance is found in the limbic area of the brain which disturbs the satiety of an individual. In such conditions mangosteen juice is advised successfully, which reduces insulin resistance and concurrently leptin resistance and controls appetite of an individual.[58]

Time of Eating

Time of eating has an impact on food metabolism. Circadian rhythm plays an important role in time of eating. It is observed that daily time of eating affects weight control in individuals. Circadian rhythm controls gene expression and also controls lipid and carbohydrate metabolism, thus controlling weight. Suprachiasmatic nucleus controls the master clock in brain and this master clock controls the oscillating clocks which control lipid and carbohydrate metabolism.[59] Same food ingested at different times shows a change in digestion pattern and even lipid oxidation. This clearly suggests the discipline to be followed in the eating time of an individual. Eating time also affects insulin sensitivity and time interval between two meals can create changes in insulin response towards food leading to metabolic syndrome.[60] *Yogratnakar* offers twice-a-day eating for normal individuals. The time interval between two meals should not be below six hours and a very small portion of food can be consumed after 3 hours.[61] In metabolic disorders it is very important to check digestive capacity before having food. Food should not be consumed unless the previous food is completely digested and body becomes light, and there is passing of flatus, urine, and no bloating or fullness in stomach. Food consumed in such condition land to severe indigestion leading to *Ama* formation. *Chakrapani* explains one day a meal which can bring longevity.[62] In metabolic disease fasting is advised to reduce toxicity of *Ama*. Fasting can be advised in obesity and in all such metabolic disorders where insulin resistance is observed. Fasting should never be aimed at reducing weight but to bring down the underlying pathological changes.

Fasting = Fasting is one of the nonpharmacological tools in management of obesity. Fasting refers to having no calorie intake to limited calorie intake. Fasting is advised in obesity as well as all *Santarpanjanya Vyadhis* (anabolic disorders). Fasting is nonpharmacological and most effective management in treating obesity. The main effect of fasting is on serum insulin levels and reduced inflammation. It is expected to have an insulin to attain low level while fasting and this can be achieved by intermittent fasting. Ayurved for the healthy person advises to have single-day meals which is known to improve longevity.[63]

In obesity at least overnight fasting of 16 hours is expected. Fasting or calorie-restricted diet is mainly useful in obese diabetics which improves glycation, C peptide level, and inflammation.[64] Intermittent fasting is a continuous process which also has a positive effect on longevity as explained by SIR2 gene theory.[65]

9. Food properties

As observed, metabolic disorders are anabolic disorders and the properties of food which promote *Santarpan* (anabolism) should be controlled. Food which is *Ruksha* (rough in nature), *Ushna* (hot in nature), *Laghu* (light for digestion) should be advised in obesity. Food having excess of *Madhur Rasa* (sweet taste) *Lavan Rasa* (salty), and excess *Amla* taste (sour food) should be avoided.

Excess sweet, salty, and sour food can promote anabolic activity in metabolic disorders. Foodstuffs like ghee, milk, wheat, and some oils are classified as sweet taste. All carbohydrate-rich foods and fruits are *guru* (heavy to digest) and are also *Madhur* (sweet) in taste. Food having heavy digestibility and sweet taste fruits are not advised in metabolic disorders. The taste of food is one of the cardinal properties of food and many activities of food depend on its taste. It is known that taste receptors are spread all over the body in the form of a diffuse chemosensory system. These extra gustatory taste receptors immediately respond to the taste of food and further affect metabolism.

Milk - Milk has been advised in neonates, pregnant women, aged people, and those who exercise regularly. Milk in general is sweet in taste and anabolic in nature, aggravating *Kapha Dosha.* Milk and milk products are not recommended for metabolic disorders. Milk and milk product consumption is increasing in Asian and Western countries.[66] Increase in milk consumption has shown an increase in metabolic syndrome in some of the researches.[67] Change in diet pattern and the addition of milk has brought an escalation in the occurrence of metabolic syndrome and cancer. People all over the world consume milk and milk products as a source of calcium and other nutrients.[68] Practically milk is consumed after heating or in the form of yogurt or curd. Milk products are also prepared after completely evaporating water out of them, which is called *Khava.* This heating of milk generates receptor for advanced glycation end products which are the cause of obesity, metabolic syndrome, and low-grade inflammation.[69] In clinical practice of metabolic disorders milk is strictly avoided in obesity, hypothyroid, diabetes, PCOD, dyslipidemia, and fatty liver.

Grains – Grains are the main ingredients of food platter. Charak explains about spiked grains which are called *Shuka Dhanya*. In *Shuka Dhanya* or spiked grains rice and wheat are main grains. In metabolic disorder a special category of rice that is *Shashitika Shali* rice is advised. This is rice which is harvested in 60 days.

Shashtika Shali is a variety of rice ripening within 60 days of its cultivation. *Shashtika shali* rice is sweet in taste, *Snigdha* (oleaginous) in nature, and not heavy to digest. It pacifies all the three *Dosha* and stabilizes them in a state of equilibrium. This unique characteristic of this rice allows its use in metabolic disorders. Among the rice red rice, that is, *Rakta Shali* is the superior one, which can be used in metabolic disorders. White rice is not advised for regular use In metabolic disorder it is advised to roast rice before cooking, rice should not be cooked under pressure or in pressure cookers. It is suggested to prepare rice in open vessel.[70] While preparing in open vessel water should be strained out . This makes it very light for digestion.[71]

Godhuma (wheat) is sweet in taste and cold in temperament. It keeps the tissues bound together and promotes binding of tissues. *Godhuma* (wheat) is known to deplete *Vata Dosha*. It is *Snigdha* (oleaginous) and *Guru* (heavy) in nature. It is *Sthairyakara* (stabilizes the body). It is *Jeevaniya* (life sustaining), *Bruhana*(growth promoting), and *Vajikar* (virility enhancing) in action. *Godhuma* is sweet in taste and cold in temperament and *Snigdha* (oleaginous) in nature.

Gavedhuka is known as *Ghulunca* (Job's tears: *Coixbarbata* Roxb.). Two varieties of *Gavedhuka* (Job's tears) are found *Gramya* (domestic) and *Aranyak* (wild spikes are red in color). It is advised in obesity and metabolic disorders.

Millets – The term millet is applied to any grass which bears small round seeds. These come under the class *Kudhanya*. *Kudhanya* is a class of cereal grains that are lesser in quality with respect to their nutrient value. *Kodrava* (Kodo millet: *Paspalum scrobiculatum* Linn.) and *Shyamaka* (finger millet: *Eleusine coracana*), commonly known as *Ragi* or *Nachani* in vernacular, all these grains belong to the group of millets. Pearl millet (*Pennisetum glaucum* (L.) R.Br.), commonly known as *Bajra* in India, is the most important millet in India. The remaining millets often are termed minor millets. Among them proso millet (*P. miliaceum* L.) is the most widespread. Proso millet and barnyard millet are both referred to as *Varai* in India. Millets have high antinutritional characteristics and hence are advised to be used in metabolic disorders.[72]

Yava (barley) is sweet and astringent in taste. Astringent taste is secondary to sweet taste. *Yava* (barley) is dry in nature and cold in temperament and *Aguru* (not heavy) to digest. According to Sushrut it is *Atishushka* (extremely dry) and *Laghu* (light). *Yava* (barely) increases *Vata Dosha* and amount of feces. It is strengthening and stabilizing in action. Barley is known to alleviate diseases of *Kapha Dosha*.[73] Barley is supposed to be one of the effective grains for diabetes and obesity. Barley has very low glycemic index. Fresh grains are strictly contraindicated in obesity and metabolic disorders as they are heavy for digestion and are said to be *Abhishyandi*. Grains 1 year postharvest become light for digestion and are indicated in metabolic disorders.[74]

Pulses - Pulses are mostly sweet and astringent in nature. *Moong* beans are mostly advised for regular use because of their very soft digestibility. *Mudga* (green gram: *Phaseolus radiatus*) are astringent and sweet in taste, cold in temperament, and dry in nature. Taste of their digestive end product is *Katu* (pungent) and *Laghu* (light). *Sushrut* describes *Mudga* (green gram) to be nutritious for the visual sensory organ. It has a cleansing effect on the body. Mostly the pulses are astringent and sweet in taste and so can be advised in obesity and metabolic disorders. It is observed to deplete *Kapha Dosha* and *Pitta Dosha*. Apart from *Moong* beans horse gram, red grams are also advised in metabolic disorders. *Rajma*, black gram, and kidney beans are supposed to be heavy for digestion and hence are not advised for regular use in metabolic disorders.

The concept of prohibiting fruits in anabolic disorder is contradictory to the contemporary concept of fruits in nutritional science. Fruits are rich in secondary metabolites like polyphenols, carotenoids, phytosterols, and antioxidants.[75] Fructose present in fruits undergoes a different metabolic pathway in the liver. In metabolic syndrome IR and inflammation are the main pathological factors. Fructose gets metabolized through fructokinase which yields ADP, which is lower-quality energy than ATP with uric acid as a byproduct.[76] Fruits are classified as *Guru* (heavy for digestion). This makes sense to avoid fruits in metabolic disorders. Many a time glycemic index of fruits is high, which can induce a spike of insulin, creating IR at the hepatocyte level. The advice given by *Charak* as fruits are *Guru* (heavy to digest) in nature seems to be correct and should be avoided in regular fashion in metabolic disorders.

Consumption of water is also very important, considering metabolic disorders and obesity. Excess consumption of water can worsen the pathology of *Santarpan Vyadhi* (anabolic disorder) and vitiates *Dosha*. Properties of water in bore wells, ponds, and lakes depend on the corresponding soil and geographical conditions in which they are located. In downregulation of *Agni*, drinking excess water is contraindicated. In metabolic disorder the status of *Agni* is low at all levels, that is, *Jatharagni, Bhutagni,* and *Dhatwagni,* so water which is heavy for digestion, that is, water from ponds, lakes, bore and wells; cold water, refrigerated water; and frequent water consumption must be avoided.[77] It is advised to consume lukewarm water when one feels thirsty. Lukewarm water has *Pachana*, that is, digestive property, and it helps for proper micturition. Warm water is more advisable in metabolic syndrome, as it is known to pacify *Ama*.[78]

Curd is promoted as a probiotic and is used exclusively in Indian *Thali* (plate). Curd is used as a probiotic and a source of calcium. Curd is not promoted on a regular basis, that too in metabolic disorders like obesity.[79] Curd is strictly contraindicated in most of the anabolic disorders. Curds are said to be *Mahabhishyandi*; that is, it has the property to create swelling on the body and block the channel of nutrient juices, that is, *Rasa Dhatu*. Curd is mainly sour in taste, which is known to increase *Apa* (water) *Mahabhuta*. A sour taste in excess use creates edema over the body and creates a laxity in muscles. It is heavy for digestion and increases fats and *Kapha Dosha* and vitiates *Agni*.

Curd due to its *Abhishyandi* (inflammatory) property should not be consumed at night and should never be heated. Curd is known to promote *Ama* and should not be consumed regularly by a person having metabolic disorders.

Though the inflammatory properties of curds are not proven it has been a clinical experience that obese people or people with metabolic disorders, digestive complaints, and rhinitis show an increase in their symptoms after consumption of curds.

Buttermilk – *Takra* (buttermilk) is prepared by churning the curd. It is known as *Chasa* in India. Indian butter milk is different from the buttermilk used in the western world. Indian buttermilk is prepared from curd and used as a regular beverage in India. Buttermilk is *Ruksha* (rough) and *Laghu* (light for digestion) and is known to stimulate *Agni* (digestive faculty).[80] Bhavaprakash mentions five different types of buttermilk depending on the proportion of water added to it. One of the types is *Chhacchika* which is prepared by churning curd with abundant water and removing fat[81]. Buttermilk is useful in *Srotorodha* (obstruction in body channels) and in individuals having downregulated status of *Agni.* It is widely used in obesity and other anabolic disorders as it is known to pacify *Kapha dosha* and low status of *Agni.* It is also useful in reducing edema and in urinary disorders.[82]

Coffee and tea - Coffee is a widely used beverage which has a positive role in metabolic syndrome, especially in fatty liver and obesity. Regular coffee consumption without milk and sugar is known to be very useful in reducing insulin resistance and improving homa index. Nonalcoholic fatty liver disease is most common in obesity and metabolic syndrome. Coffee is known to inhibit the incidence-e and severity of NAFLD.[83] These effects of coffee are due to its aromatic extracts and its antioxidant activity specific to the liver. However, how much coffee consumption can help to inhibit liver changes in metabolic disorders is not known. Obesity and diabetes are diseases having liver involvement. Coffee consumption is found to be beneficial in both of these conditions and prevents liver pathology.[84] Tea is made from leaves of the plant *Camellia sinensis*, family of Theaceae. Tea is used in the form of green and black tea which vary in their phytochemical constituent. Black tea seems to be more beneficial as compared to green tea in obesity and allied disorders.[85] Regular consumption of black tea is known to reduce adipocyte differentiation and thus control weight gain. Tea is known to increase thermogenesis and imparts positive effect on gut microbiota. Tea is known to control appetite in obese individuals.[86] Several observational studies have lately highlighted possible beneficial effects of coffee and green tea on liver and metabolic health

10. Oils and Fats

Oils are a class of unctuous property. Excess of oils is anyways not advisable in metabolic disorders.

Vegetable oils are unsaturated fatty acids which are prone to more oxidation.

The properties of oil depend on the properties of the source of seeds from which it is extracted., Ayurved explains properties of vegetable oil as *Tikshna* (sharp) in action and *Vyavayi* (disseminating) means, it spreads fast into the organs.

Oils coming from sesame do not vitiate *Kapha Dosha* though sesame is unctuous in nature. Sesame oil is advised for obese people and metabolic disorders. Sesame oil has a characteristic that it nourishes the lean persons and emaciates the obese. Sesame oil is rich in sesamol and sesamin, alpha tocoferol, and some lignans which have a very strong antioxidant and anti-inflammatory activity. Processed sesame oil exhibits weight-reducing activity when used in a proper way. In clinical experience, the processing of sesame oil with herbs exhibits very strong antioxidant and weight reduction activity.[87] Processed sesame oil is known to increase its antioxidant activity which proves beneficial in obese individuals for weight reduction.[88]

Mustard oil is said to have *Tikshna* (sharp) properties and is pungent in taste. It is known to pacify *Vata* and *Kapha* dosha and can be recommended for metabolic disorders. It is *Laghu* (light) for digestion and so is useful in *Kustha,* that is, skin disorders, ulcers, and parasites.

Safflower oil is promoted by modern nutritionists for cardiac and coronary artery disease.[89] Many Ayurved scholars like *Charak* and *Bhavaprakash* strictly contraindicates safflower oil and state that it can create some diseases of the eyes and is contraindicated in metabolic disorder. It is known to vitiate *Kapha* and *Pitta dosha*.[90] Safflower oil is a rich source of unsaturated fatty acids. The amount of unsaturated fatty acid in safflower oil is 90.33%.[91] Due to its highest concentration of unsaturated fatty acids it is prone to early oxidation. Flaxseed oil is known to be rich in omega fatty acids and is recommended in metabolic diseases, mainly cardiometabolic as it is known to reduce inflammation. Safflower oil is highest in unsaturated fats and creates more oxidation in cooking process. and so not advised for regular use in metabolic diseases. Free radicals play a major role in pathogenesis of metabolic disorders and food substances which induce lipid peroxidation are to be avoided.

Ghee or clarified butter is again a category of animal fats which is widely used in the Indian subcontinent. Due to some of the important properties of ghee like having a good effect on memory, *Agni* and *Shukra Dhatus*[92] it is widely used in the treatment of many ailments.

Ghee or clarified butter is widely used all over the world. *Ghee* is known to have an anabolic activity on fats and is known to increase the dominance of *Kapha Dosha*. *Ghee* is promoted in *Pitta* predominance disease ghee is not recommended in metabolic disorders.

Chakrapani elaborates contraindication of ghee. As per *Chakrapani* the commentator of *Charak Samhita* states that people having *Kapha Dosha* predominance, sedentary life style, leaving in marshy land near sea and rivers and people leaving in urban areas should not consume ghee.[93] Ghee is complete saturated fat and undergoes very less oxidation.

11. Animal Meat and Flesh in Metabolic Disorder

Charak and *Sushrut* both have elaborated properties of animal meat. Depending on the habitat, their nature, and their living patterns, animals which live in *Jangal Desha* (dry habitat) are light to get digested. Meat of animals that move in jungles and in dry habitats is supposed to be light for digestion.

The flesh of goat balances the body metabolism and keeps the *Dosha* in harmony with body tissue. It nourishes the body tissues without creating untoward effects.[94]

It is very clearly indicated that animals or birds which have restricted movements or are brought up in the domestic atmosphere are heavy to digest. When movements of animals are limited they lose their habitat. The meat of such animals becomes heavy for digestion and is not recommended for metabolic disorder. Meat which is slaughtered and stored in cold storage is not recommended for metabolic disorders. *Charak* does not recommend use of fish in regular diet.[95] Fish is unctuous and vitiates all the three *Dosha*.[96] Considering its unctuous nature we do not recommend fish in high portion in obesity and metabolic disorders.

Beef and Pork

Meat coming from buffalo and pig is said to be excessively heavy for digestion. These provide good nourishment to tissue but due to their heavy digestive property should not be consumed in metabolic syndrome.[97] Meat of animals from dry habitats, which are domesticated, meat that is freshly slaughtered, clean, and meat of young animals can be consumed. Meat from diseased animals and emancipated animals, fat-rich meat, and that from animals that are poisoned are not indicated for consumption.

12. Vegetables

Vegetables are routinely supposed as low calories and micronutrient-rich food by people all over the world.[98] Salads are supposed to be the healthiest food and are used as filler to achieve satiety. *Charak* has promoted vegetables as per their taste and properties and has explained their effects in a very systematic manner. It explains certain vegetables like leafy vegetables which are heavy to digest, are sweet in taste, and have a cooling effect on the body. These vegetables should be boiled and consumed.[99] Vegetables should be processed with unctuous substances before eating.[100] It is very important to note that raw vegetables should not be consumed. Vegetable salads are heavy to digest and should undergo the same processing. Vegetables when sautéed with little oil become light for digestion and should be consumed.

Charak does not recommend regular consumption of vegetables.[101] Indian food is rich in fibers. Vegetables are richer in fibers and it is evident that the source of fibers is cellulose which is carbohydrate. In regular clinical practice of obesity consumption of vegetables is completely restricted till the patient achieves desired target of weight loss. Vegetables are strongly promoted in metabolic disorders. It is observed that obese patients have altered gut flora. In obesity it is observed that the gut flora of obese people are completely deranged when BMI goes above 23.[102] Altered gut flora shows an increase in firmicutes bacteria which have an energy harvesting tendency. The altered bacteria in the gut, especially in obesity, develop an energy-harvesting tendency that disintegrates the fibers from vegetables and uptake carbohydrates out of them. Inflammation in the gut and large intestine wall leads to generate more short-chain fatty acids (SCFAs) which are transferred to the liver. Excess generation of SCFA worsens the pathology of obesity.[103] To control this breakdown of fibers in the gut in the initial phase of obesity management vegetables are restricted. In the latter stage of management vegetables having soluble fibers are advised in limited portions. Certain vegetables having soluble fibers are advised while managing weight reduction. Boiled gaur gum, boiled drumsticks, and sautéed spring onions with pepper powder and salt are some of the vegetables which usually do not interfere with weight reduction.

Condiments like garlic and ginger are more useful in metabolic disorders. Garlic is extremely *Teekshna* (sharp), hot, and pungent in nature. It is useful for heart disease. It is also useful in diabetes and diseases having *Kapha Dosha* predominance. Ginger is an appetizer, enhances the taste, and is known to pacify *Kapha* and *Vata Dosha*.

13. Alcoholic Beverages

Alcoholic beverages are explored by *Charak* as well as *Sushrut*. As alcohol provides a high number of calories, that is, 7 kilocalories per gram, it creates a positive energy balance, ultimately leading to obesity. Alcohol metabolizes differently between age groups, sex, and pattern of drinking. Heavy and binge drinking of alcohol is directly associated with weight gain.[104] Consuming alcohol more than seven times a week is associated with weight gain and obesity. It has been observed that moderate drinking or less drinking is less risky in metabolic disorders as compared with heavy drinking.[105] Charak has elaborated on certain properties of alcohol beverages, where beverages made out of grape juice have been advised. It is called as *Mardvika madya*. It has *Lekhana* (scaping) and *Hridya* activities. This type of alcohol beverage is known to augment *Pitta* and *Vata* Dosha and is used in *Prameha* (diabetes).[106]

14. Artificial Sweeteners or Intense Sweeteners

Artificial sweeteners are excessively sweet in taste and are known to stimulate extra gustatory taste receptors. Intense sweeteners (ISs) are strictly contraindicated in obesity and metabolic disorders. These sweeteners are known to affect glucose metabolism, gut flora, and appetite control system. Intense sweeteners are said to be thirty times sweet than sucralose or normal white sugar.[107] Intense sweeteners (IS) have a very high affinity toward sweet taste receptors. Stimulation of sweet taste receptors imparts a negative effect on obesity and metabolic disorders and aggravates obesity.[108] Physiologically IS affects weight gain by stimulating glucagon-like peptide-1. Glucagon-like peptide-1 (GLP-1) hormone is secreted from intestinal epithelial cells and plays a very important role in insulin secretion.[109] The physiological impact of the GLP-1 study helps in exploring the concept of Ayurved regarding sweet taste and metabolic disorders. Charak clearly says that excess sweet-tasting substances lead to obesity and metabolic disorders irrespective of their calorie value.[110]

15. Food Processing

Heating, drying, and roasting food or grains make them very light for digestion. In metabolic disorders at many places *Laja* is advised. Grains when roasted in a specific manner turn into puffed grains which are called *Laja*. Puffed or parched grains are advised for metabolic disorders. Roasting and heating of grains changes photochemical constituents and digestibility. Similarly soaking and sprouting of grains also brings many changes in their properties. Sprouts have been defined by *Sushrut* and *Dalhan* in different ways.

1. Seeds which have lost germinating capacity.[111]
2. Seeds after roasting which lose germinating capacity.[112]
3. Seeds which retain germinating capacity in pulses like moong beans and horse gram.[113]

The above types of sprouts show unalike characteristics and have diverse indications.

Sprouts are also called *Virudha*. Beans which have the capacity of germinating and are used processed just after germination create *Abhishyanda* (inflammation) and are *Vidhahi* (that causes burning). In this context, beans when subjected to sprouting become heavy for digestion, when they are roasted and cooked can be advised in metabolic disorders. Pulses when subjected to sprouting are not recommended for regular use.[114] Sprouting of grains brings physical and biochemical changes. Sprouting reduces protein content and increases the carbohydrate content of grains.[115] Controlled heating over food can bring good changes in digestibility, but excess heating on food, that is, the late stages of the Milliard reaction, can lead to advanced glycation end products. Excessive heating of food is an exogenous source of AGE and can land a person in obesity, inflammation, and metabolic syndrome.[116]

Sattu is one of the recipes which can be used for metabolic disorders. *Sattu* is prepared by making flour out of roasted grains. For example, wheat and horse gram can be roasted, mixed, and subjected to flour and consumed with hot water, adding some salts and condiments. *Sattu* is highly nutritious as it's made out of chickpeas and wheat flour. Roasting of wheat prior to subjecting it to flour makes it light for

TABLE 4 Food Recommendations in obesity and metabolic disorders

1	Cereals and Grains	Puffed grains and cereals, Roasted grains are advised
2	Pulses	All pulses except black gram, Rajmaha
3	Flesh, meat, and fish	Domestic animals, meat, pork, beef are heavy and not recommended in obesity and metabolic disorders Fish not recommended in metabolic disorders
4	Vegetables	Vegetables with (*Madhurras*) sweet taste, non-soluble vegetables and salads not recommended in metabolic disorders
5	Fruits	Not recommended in metabolic disorder and obesity
6	Oils	Sesame oil strongly recommended in obesity and metabolic disorders Safflower oil not recommended
7	Ghee	Completely restricted
8	Milk and milk products	Completely restricted
9	Alcoholic beverages	Aged alcoholic beverages made from grapes moderately recommended. No other alcoholic drinks recommended
10	Sprouted food	Not advised in metabolic disorders and obesity as they are heavy for digestion
11	Water	Moderate consumption. Chilled water not advised
12	Buttermilk	Curds churned with many times water and made thin is advised

digestion. It is highly recommended by *Charak* for obesity and allied metabolic disorders.[117] Roasting of grains changes the total phytochemistry making it easy for digestion.

Soups made out of moong beans and other pulses are recommended for obesity. Soups can be added with cinnamon, cumin, and black salt and consumed as a meal which helps for correcting metabolism and reducing weight.

Low digestibility food is advised in obesity. Clinically vegetarian sources of protein are advocated in the initial management of weight reduction along with complex carbohydrates coming from millet or whole wheat flour. Moong bean curry along with whole wheat roti (bread) is generally the best meal for reducing weight. Homemade buttermilk can be advised which can be prepared by adding six times water to curds and churning it properly. It has been observed that tetra-pack buttermilk contains a lot of preservatives and is not diluted as expected which creates much inflammation if consumed regularly.[118]

In obesity and metabolic disorders appetite is very high due to deranged appetite control system. Mangosteen juice added with black salt is used to suppress hunger and induce satiety.[119]

Obesity management completely depends on the underlying pathology of obesity. A difference is observed while managing exogenous and endogenous obesity. Obesity associated with hypothyroidism or polycystic ovarian disease needs respective management.

In obesity as a general guideline, high protein, and low carbohydrate diets are advised. In obesity associated with hypothyroid high protein diet along with carbohydrates is essential. In thyroid metabolism, carbohydrates play an important role in the conversion of hormone T4 to T3. If carbohydrates are advised in very low portions in obesity-associated hypothyroidism the conversion of T4 to T3 can be affected.[120] In PCOD insulin resistance at the ovarian level, hepatocyte level, and musculoskeletal level is observed. Insulin resistance is the basic pathology of PCOD. In such hormonal disorder the main cause can be a high portion of carbohydrate intake. In such disorders, reduced carbohydrate portion to a greater extent is advised which increases insulin sensitivity. This difference in the diet pattern of obesity associated with different hormonal changes must be focused.

In patients having high triglyceride levels more use of fenugreek seed in diet is observed to be beneficial. Fenugreek seeds can be used in the form of powder or as a whole in curries. Fenugreek powders can be added to breads and consumed. Similarly, cinnamon, ginger, Amla, and garlic can be used in the form of pickles or used in tea preparation. These herbs can be commonly used in all metabolic disorders. Garlic is known to have aphrodisiac activity.[121] This activity of garlic is due to nitric oxide synthesis stimulating activity.[122] Men having obesity and metabolic disorders suffer from erectile dysfunction. Garlic can be useful in such conditions. Garlic should not be used concurrently with aspirin or anticoagulant.[123]

In prevention of metabolic disorder patient education is very important. People should be encouraged to hold packed food, refrigerated food, and food containing high preservatives. Family members should be trained while grocery shopping and should be advised to keep themselves apart from sugary and fatty substances. People should be taught to avoid excess use of oil, butter, cheese, sugar, and alcohol. The importance of timely eating, intervals between two meals, and portion size of meal should be explained. It is of prime importance to educate school-going children in regard to food habits.

Summary of advisable and nonadvisable food in anabolic disorders.

Non-advisable food:

1. Excess of sweet, sour, and salty-tasting food aggravates metabolic disorders.
2. Food which is stale,
   Refrigerated, frozen is heavy for digestion.
3. Food having high AGE like cookies, carbonated beverages, and deep-fried food.
4. Incompatible food like milk and fruits, tea with milk.
5. Inflammatory food like deep-fried food, reheated food, and fermented food.
6. Overconsumption of unctuous food like ghee, cheese, butter, and oils.
7. Excess consumption of sugar, jaggery, and fruits having high glycemic index and intense sweeteners.
8. Regular consumption of packed food and rancid food.

9. Frequent eating and extra portion size and untimely eating.
10. Excess water consumption and frequent chilled water consumption.

Advisable food style in metabolic disorders:

1. Low carbohydrate, medium protein, and very low-fat diet.
2. Vegetarian sources of proteins like pulses, lentils, and legumes like moong beans, drumsticks, and gaur gum.
3. Vegetables with soluble fibers.
4. Freshly cooked food with oils like sesame and mustard and moderate to very low consumption of coconut oil and ghee.
5. Preferably millets like sorghum, *Ragi, Bajara,* or *Barley*. Brown rice or red rice is preferred to white rice.
6. Meat of goat and chicken preferred after proper marinating with turmeric and lemon.
7. Fruits with a very low glycemic index like pears, amla, and pomegranate.
9. Roasted flours like *Sattu*, parched and puffed grains, and chickpeas as finger food.
10. Mangosteen juice, buttermilk, black coffee, black tea, and green tea can be consumed intermittently.
11. Use of food condiments like cinnamon, pepper, asafetida, turmeric, garlic, mint, black salt, clove, fenugreek seeds, and funnel seeds is preferred in food recipes.
12. Lukewarm water as regular water consumption only when thirsty.

16. Nutravigilance

Food supplements are commonly consumed for obesity and metabolic disorders. It is a routine practice of consuming honey and warm water for weight reduction early morning which is not recommended by Sushrut.[124] Honey is used in Ayurved therapeutic along with certain food or medicines due to its property as a vehicle. Honey is never to be heated or added to hot water.

Many food supplements are fortified with herbs like *Ashwagndha* (*Withania somnifera*), *Shatavari* (*Asparagus racemosus*), or licorice. These types of herbs are strictly contraindicated in obesity and metabolic diseases in general. Majority of the time the herbs which are used as supplements are sweet in taste, heavy for digestion, and strengthen the tissues. Herbs having these characteristics are contraindicated in obesity and other anabolic disorders. Whey proteins are consumed in a regular manner by exercising people. Certain proteins are insulinogenic and should be avoided in obesity and metabolic disorders. Proteins coming from whey are insulinogenic proteins and are contraindicated in obesity. People having morning exercise or walking are also seen to have medicinal herbs juices of *Neem (Azadirachta indica), Karel (Momordica charantia), Amla (Emblica officinalis), Tulsi (Ocimum sanctum)*, and *Giloya (Tinospora cordifolia)* which is not indicated. These types of medicinal herbs with self-prescription can show untoward effects in long-term consumption.

REFERENCES

1 Saklayen MG. The global epidemic of the metabolic syndrome. Curr Hypertens Rep. 2018 Feb 26;20(2):12. doi: 10.1007/s11906-018-0812-z. PMID: 29480368; PMCID: PMC5866840.

2 Sun, J.; Buys, N.J.; Hills, A.P. Dietary pattern and its association with the prevalence of obesity, hypertension and other cardiovascular risk factors among Chinese older adults. Int. J. Environ. Res. Publ. Health 2014, 11, 3956–3971.

3 Vaidya Jadavaji trikamaji edited, Agnivesa, Charaka Samhitha revised by Charaka and Dridhabala, with the Ayurveda-Dipika Commentary. Edition reprint 2013. Varanasi: Choukambha Prakashan: Vimansthan 6/3.
4 Vaidya Jadavaji trikamaji edited, Agnivesa, Charaka Samhitha revised by Charaka and Dridhabala, with the Ayurveda-Dipika Commentary. Edition reprint 2013. Varanasi: Choukambha Prakashan: Chikitsasthan 15/12.
5 Vaidya Jadavaji trikamaji edited, Agnivesa, Charaka Samhitha revised by Charaka and Dridhabala, with the Ayurveda-Dipika Commentary. Edition reprint 2013. Varanasi: Choukambha Prakashan: Sutrasthan 28/4.
6 Vaidya Jadavaji trikamaji edited, Agnivesa, Charaka Samhitha revised by Charaka and Dridhabala, with the Ayurveda-Dipika Commentary. Edition reprint 2013. Varanasi: Choukambha Prakashan: Chikitsasthan 15/16.
7 www.learnsanskrit.cc/
8 Vaidya Jadavaji trikamaji edited, Agnivesa, Charaka Samhitha revised by Charaka and Dridhabala, with the Ayurveda-Dipika Commentary. Edition reprint 2013. Varanasi: Choukambha Prakashan: Sutrasthan 25/34.
9 Vaidya Jadavaji trikamaji edited, Agnivesa, Charaka Samhitha revised by Charaka and Dridhabala, with the Ayurveda-Dipika Commentary. Edition reprint 2013. Varanasi: Choukambha Prakashan: Sutrasthan 5/7.
10 Mann, Elizabeth. Bellin, Melena D. (2016). Secretion of Insulin in Response to Diet and Hormones. Pancreapedia: Exocrine Pancreas Knowledge Base, DOI: 10.3998/panc.2016.3.
11 Bhisagacarya Harisastri Paradakara Vaidya edited, Vagbhata, Ashtanga Hrudayam, Sarvanga Sundara commentary of Arunadatta and Ayurveda Rasayana Commentary of Hemadri. Edition: Reprint, 2012. Varanasi, Choukambha Sanskrit Sansthan Sutra sthan 8/12.
12 Bhisagacarya Harisastri Paradakara Vaidya edited, Vagbhata, Ashtanga Hrudayam, Sarvanga Sundara commentary of Arunadatta and Ayurveda Rasayana Commentary of Hemadri. Edition: Reprint, 2012. Varanasi, Choukambha Sanskrit Sansthan Sutrasthan 9/9.
13 Vaidya Jadavaji trikamaji edited, Agnivesa, Charaka Samhitha revised by Charaka and Dridhabala, with the Ayurveda-Dipika Commentary. Edition reprint 2013. Varanasi: Choukambha Prakashan: Chiistsasthan 15/44.
14 Vaidya Jadavaji trikamaji edited, Agnivesa, Charaka Samhitha revised by Charaka and Dridhabala, with the Ayurveda-Dipika Commentary. Edition reprint 2013. Varanasi: Choukambha Prakashan: Sutrasthan Chapter 27 verse 351 352.
15 Yadavji Trikamji Acharya, Sushruta Samhita Dalhana Commentary Choukhamba Surabharati Publication Sushrut sutrasthan 21/23.
16 Bhavaprakash Nighantu Purvakhanda/Mishra Varga 243.
17 Vaidya Jadavaji trikamaji edited, Agnivesa, Charaka Samhitha revised by Charaka and Dridhabala, with the Ayurveda-Dipika Commentary. Edition reprint 2013. Varanasi: Choukambha Prakashan: Vimansthan 5/4.
18 Vaidya Jadavaji trikamaji edited, Agnivesa, Charaka Samhitha revised by Charaka and Dridhabala, with the Ayurveda-Dipika Commentary. Edition reprint 2013. Varanasi: Choukambha Prakashan: Vimansthan 5/24.
19 Vaidya Jadavaji trikamaji edited, Agnivesa, Charaka Samhitha revised by Charaka and Dridhabala, with the Ayurveda-Dipika Commentary. Edition reprint 2013. Varanasi: Choukambha Prakashan: Sutrasthan 23/5-6.
20 Vaidya Jadavaji trikamaji edited, Agnivesa, Charaka Samhitha revised by Charaka and Dridhabala, with the Ayurveda-Dipika Commentary. Edition reprint 2013. Varanasi: Choukambha Prakashan: Sutrasthan 23/7.
21 Vaidya Jadavaji trikamaji edited, Agnivesa, Charaka Samhitha revised by Charaka and Dridhabala, with the Ayurveda-Dipika Commentary. Edition reprint 2013. Varanasi: Choukambha Prakashan: Sutrasthan 23/29.
22 Vaidya Jadavaji trikamaji edited, Agnivesa, Charaka Samhitha revised by Charaka and Dridhabala, with the Ayurveda-Dipika Commentary. Edition reprint 2013. Varanasi: Choukambha Prakashan: Sutrasthan 21/4.
23 Okosun, I., Seale, J. & Lyn, R. Commingling effect of gynoid and android fat patterns on cardiometabolic dysregulation in normal weight American adults. Nutr & Diabetes 5, e155 (2015).
24 Identification and characterization of metabolically benign obesity in humans N Stefan, K Kantartzis, J Machann, F Schick, C Thamer, K Rittig, ... Archives of internal medicine 168 (15), 1609–1616
25 Yadavji Trikamji Acharya, Sushruta Samhita Dalhana Commentry Choukhamba Surabharati Publication Sutrasthan 43/51.
26 Sabnis M. Viruddha *Aaharaa*: A critical view. Ayu. 2012 Jul;33(3):332–6. doi: 10.4103/0974-8520.108817.
27 Vaidya Jadavaji trikamaji edited, Agnivesa, Charaka Samhitha revised by Charaka and Dridhabala, with the Ayurveda-Dipika Commentary. Edition reprint 2013. Varanasi: Choukambha Prakashan: Sutra sthan 5/3.
28 Chenzhipeng Nie, Yan Li, Haifeng Qian, Hao Ying & Li Wang (2022) Advanced glycation end products in food and their effects on intestinal tract, Critical Reviews in Food Science and Nutrition, 62:11, 3103–3115.
29 Srivastava S. Singh M. George J. Bhui K. Murari Saxena. A. Shukla Y. 2010 Genotoxic and carcinogenic risks associated with the dietary consumption of repeatedly heated coconut oil. Br. J. Nutr.. 2010 Nov;104(9):1343-52.

30 Ahmed N, Mirshekar-Syahkal B, Kennish L, Karachalias N, Babaei-Jadidi R, Thornalley PJ. Assay of advanced glycation endproducts in selected beverages and food by liquid chromatography with tandem mass spectrometric detection. Mol Nutr Food Res. 2005;49:691–699.
31 Vaidya Jadavaji trikamaji edited, Agnivesa, Charaka Samhitha revised by Charaka and Dridhabala, with the Ayurveda-Dipika Commentary. Edition reprint 2013. Varanasi: Choukambha Prakashan: 2013 Sutra sthan 26/85.
32 Sabnis M. Viruddha *Aaharaa*: A critical view. Ayu. 2012 Jul;33(3):332–6. doi: 10.4103/0974-8520.108817. PMID: 23723637; PMCID: PMC3665091.
33 Bhisagacarya Harisastri Paradakara Vaidya edited, Vagbhata, Ashtanga Hrudayam, Sarvanga Sundara commentary of Arunadatta and Ayurveda Rasayana Commentary of Hemadri. Edition: Reprint, 2012. Varanasi, Choukambha Sanskrit Sansthan Sutrasthan 13/26.
34 Vaidya Jadavaji trikamaji edited, Agnivesa, Charaka Samhitha revised by Charaka and Dridhabala, with the Ayurveda-Dipika Commentary. Edition reprint 2013. Varanasi: Choukambha Prakashan: Vimansthan 2/8.
35 Vaidya Jadavaji trikamaji edited, Agnivesa, Charaka Samhitha revised by Charaka and Dridhabala, with the Ayurveda-Dipika Commentary. Edition reprint 2013. Varanasi: Choukambha Prakashan: Vimansthan 2/9.
36 Goldberg T, Cai W, Peppa M, et al. Advanced glycoxidation end products in commonly consumed foods. J Am Diet Assoc. 2004;104(8):1287–1291.
37 Koschinsky T., He C.-J., Mitsuhashi T., Bucala R., Liu C., Buenting C., Heitmann K., Vlassara H. Orally absorbed reactive glycation products (glycotoxins): An environmental risk factor in diabetic nephropathy. Proc. Natl. Acad. Sci. USA. 1997;94:6474–6479. doi: 10.1073/pnas.94.12.6474.
38 Mills D.J.S., Tuohy K.M., Booth J., Buck M., Crabbe M.J.C., Gibson G.R., Ames J.M. Dietary glycated protein modulates the colonic microbiota towards a more detrimental composition in ulcerative colitis patients and non-ulcerative colitis subjects. J. Appl. Microbiol. 2008;105:706–714. doi: 10.1111/j.1365-2672.2008.03783.x.
39 Qu W., Yuan X., Zhao J., Zhang Y., Hu J., Wang J., Li J. Dietary advanced glycation end products modify gut microbial composition and partially increase colon permeability in rats. Mol. Nutr. Food Res. 2017;61:1700118. doi: 10.1002/mnfr.201700118.
40 Appel LJ, Moore TJ, Obarzanek E, et al. A clinical trial of the effects of dietary patterns on blood pressure. N Engl J Med. 1997;336(16):1117–1124.
41 Giardino I, Edelstein D, Brownlee M. Nonenzymatic glycosylation in vitro and in bovine endothelial cells alters basic fibroblast growth factor activity. A model for intracellular glycosylation in diabetes. J Clin Invest. 1994;94:110–117.
42 Hsieh CL, Yang MH, Chyau CC, Chiu CH, Wang HE, Lin YC, Chiu WT, Peng RY. Kinetic analysis on the sensitivity of glucose- or glyoxal-induced LDL glycation to the inhibitory effect of Psidium guajava extract in a physiomimic system. Biosystems. 2007;88:92–100.
43 Li J, Schmidt AM. Characterization and functional analysis of the promoter of RAGE, the receptor for advanced glycation end products. J Biol Chem.. Jun 27;272(26):16498-506.
44 Iwashima Y, Eto M, Hata A, Kaku K, Horiuchi S, Ushikubi F, Sano H. Advanced glycation end products-induced gene expression of scavenger receptors in cultured human monocyte-derived macrophages. Biochem Biophys Res Commun. 2000; 277: 368–380.
45 avid Sebastián, Rebeca Acín-Pérez, Katsutaro Morino, "Mitochondrial Health in Aging and Age-Related Metabolic Disease", Oxidative Medicine and Cellular Longevity, vol. 2016, Article ID 5831538, 2 pages, 2016. https://doi.org/10.1155/2016/5831538
46 David B. Savage, Kitt F. Petersen and Gerald I. Shulman Mechanisms of Insulin Resistance in Humans and Possible Links With Inflammation Originally published11 Apr 2005 https://doi.org/10.1161/01.HYP.0000163475.04421.e4Hypertension. 2005;45:828–833.
47 Yadavji Trikamji Acharya, Sushruta Samhita Dalhana Commentry Choukhamba Surabharati Publication Sutrasthan 21/23.
48 Wells, G., Noseworthy, M., Hamilton, J., Tarnopolski, M., & Tein, I. (2008). Skeletal Muscle Metabolic Dysfunction in Obesity and Metabolic Syndrome. Canadian Journal of Neurological Sciences/Journal Canadien Des Sciences Neurologiques, 35(1), 31–40. doi:10.1017/S0317167100007538
49 Mark P. Mattson, Roles of the lipid peroxidation product 4-hydroxynonenal in obesity, the metabolic syndrome, and associated vascular and neurodegenerative disorders, Experimental Gerontology, Volume 44, Issue 10, 2009, Pages 625–633, ISSN 0531-5565,
50 Matsuzawa, Y. (2006) The metabolic syndrome and adipocytokines. FEBS Lett. 580: 2917–2921.

51 Yamauchi, T., Kamon, J., Minokoshi, Y., et al (2002) Adiponectin stimulates glucose utilization and fatty-acid oxidation by activating AMP-activated protein kinase. Nat Med. 8: 1288– 1295.

52 Kapoor N. Thin Fat Obesity: The Tropical Phenotype of Obesity. [Updated 2021 Mar 14]. In: Feingold KR, Anawalt B, Boyce A, et al., editors. Endotext [Internet]. South Dartmouth (MA): MDText.com, Inc.; 2000. Available from: www.ncbi.nlm.nih.gov/books/NBK568563/

53 Kapoor N. Thin Fat Obesity: The Tropical Phenotype of Obesity. [Updated 2021 Mar 14]. In: Feingold KR, Anawalt B, Boyce A, et al., editors. Endotext [Internet]. South Dartmouth (MA): MDText.com, Inc.; 2000. Available from: www.ncbi.nlm.nih.gov/books/NBK568563/

54 Vaidya Jadavaji trikamaji edited, Agnivesa, Charaka Samhitha revised by Charaka and Dridhabala, with the Ayurveda-Dipika Commentary. Edition reprint 2013. Varanasi: Choukambha Prakashan: Chikitsasthan 6/15.

55 Vaidya Jadavaji trikamaji edited, Agnivesa, Charaka Samhitha revised by Charaka and Dridhabala, with the Ayurveda-Dipika Commentary. Edition reprint 2013. Varanasi: Choukambha Prakashan: Chikitsasthan 6/21.

56 Rodriguez-Leyva D, Dupasquier CM, McCullough R, Pierce GN. The cardiovascular effects of flaxseed and its omega-3 fatty acid, alpha-linolenic acid. Can J Cardiol. 2010 Nov;26(9):489–96. doi: 10.1016/s0828-282x(10)70455-4. PMID: 21076723; PMCID: PMC2989356.

57 Vaidya Jadavaji trikamaji edited, Agnivesa, Charaka Samhitha revised by Charaka and Dridhabala, with the Ayurveda-Dipika Commentary. Edition reprint 2013. Varanasi: Choukambha Prakashan: Sutrasthan 21/5.

58 Watanabe M, Gangitano E, Francomano D, Addessi E, Toscano R, Costantini D, Tuccinardi D, Mariani S, Basciani S, Spera G, Gnessi L, Lubrano C. Mangosteen Extract Shows a Potent Insulin Sensitizing Effect in Obese Female Patients: A Prospective Randomized Controlled Pilot Study. Nutrients. 2018 May 9;10(5):586.

59 Reppert SM, Weaver DR. Coordination of circadian timing in mammals. Nature. 2002;418(6901): 935–941. pmid:12198538

60 Wefers J, van Moorsel D, Hansen J, Connell NJ, Havekes B, Hoeks J, et al. (2018). Circadian misalignment induces fatty acid metabolism gene profiles and compromises insulin sensitivity in human skeletal muscle. Proceedings of the National Academy of Sciences. 2018;115(30): 7789–7794. pmid:29987027

61 Yogratnakar Purvardha Nityapravritti prakara 108/109.

62 Vaidya Jadavaji trikamaji edited, Agnivesa, Charaka Samhitha revised by Charaka and Dridhabala, with the Ayurveda-Dipika Commentary. Edition reprint 2013. Varanasi: Choukambha Prakashan: Sutrasthan 24/40.

63 Vaidya Jadavaji trikamaji edited, Agnivesa, Charaka Samhitha revised by Charaka and Dridhabala, with the Ayurveda-Dipika Commentary. Edition reprint 2013. Varanasi: Choukambha Prakashan: Sutrasthan 25/40.

64 Kahleova H, Belinova L, Malinska H, Oliyarnyk O, Trnovska J, Skop V, et al. Eating two larger meals a day (breakfast and lunch) is more effective than six smaller meals in a reduced-energy regimen for patients with type 2 diabetes: a randomised crossover study. Diabetologia. 2014;57(8):1552–60. Epub 2014 May 18. Erratum in: *Diabetologia* 2015;58(1):205.

65 Leonard Guarente, Frédéric Picard, Calorie Restriction— the SIR2 Connection, Cell, Volume 120, Issue 4, 2005, Pages 473–482, ISSN 0092-8674.

66 OECD-FAO Agricultural Outlook 2018-2027.

67 A.A. Çerman et al. Dietary glycemic factors, insulin resistance, and adiponectin levels in acne vulgaris J Am Acad Dermatol (Volume 75, Issue 1, 2016, p155-162,

68 Górska-Warsewicz H, Rejman K, Laskowski W, Czeczotko M. Milk and Dairy Products and Their Nutritional Contribution to the Average Polish Diet. Nutrients. 2019 Aug 1;11(8):1771. doi: 10.3390/nu11081771. PMID: 31374893; PMCID: PMC6723869.

69 L. J. Sparvero, D. Asafu-Adjei, R. Kang, D. Tang, N. Amin, J. Im, R. Rutledge, B. Lin, A. A. Amoscato and H. J. Zeh, RAGE (Receptor for Advanced Glycation Endproducts), RAGE ligands, and their role in cancer and inflammation. *J Transl Med.* 2009 Mar 17;7:17. doi: 10.1186/1479-5876-7-17.

70 Gunathilaka, Thilina Ekanayake, Sagarika2015/06/13 – Effect of different cooking methods on glycaemic index of Indian and Pakistani basmati rice varieties VL – 60 DO – 10.4038/cmj.v60i2.7545 JO – Ceylon Medical Journal -. 2015 Jun;60(2):57-61. doi: 10.4038/cmj.v60i2.7545. PMID: 26132185.

71 Vaidya Jadavaji trikamaji edited, Agnivesa, Charaka Samhitha revised by Charaka and Dridhabala, with the Ayurveda-Dipika Commentary. Edition reprint 2013. Varanasi: Choukambha Prakashan: Vimansthan 1/22.

72 Vaidya Jadavaji trikamaji edited, Agnivesa, Charaka Samhitha revised by Charaka and Dridhabala, with the Ayurveda-Dipika Commentary. Edition reprint 2013. Varanasi: Choukambha Prakashan: Sutrasthan 27/18.

73 Yadavji Trikamji Acharya, Sushruta Samhita Dalhana Commentry Choukhamba Surabharati Publication Sutrasthan 46.

74 Bhisagacarya Harisastri Paradakara Vaidya edited, Vagbhata, Ashtanga Hrudayam, Sarvanga Sundara commentary of Arunadatta and Ayurveda Rasayana Commentary of Hemadri. Edition: Reprint, 2012. Varanasi, Choukambha Sanskrit Sansthan Sutrsthan 6/24.
75 Santos DCD, Oliveira Filho JG, Sousa TL, Ribeiro CB, Egea MB. Ameliorating effects of metabolic syndrome with the consumption of rich-bioactive compounds fruits from Brazilian Cerrado: a narrative review. Crit Rev Food Sci Nutr. 2022;62(27):7632–7649. doi: 10.1080/10408398.2021.1916430. Epub 2021 May 12. PMID: 33977838.
76 Sebastian Stricker, Dr. med.,1,* et al Dtsch Arztebl Fructose Consumption—Free Sugars and Their Health Effects Int. 2021 Feb; 118(5): 71–80. Published online 2021 Feb 5. doi: 10.3238/arztebl.m2021.0010 PMCID: PMC8188419 PMID: 33785129
77 Yadavji Trikamji Acharya, Sushruta Samhita Dalhana Commentry Choukhamba Surabharati Publication Sutrasthan 20/17.
78 Yogratnakar Purvrdha Ushnavarigunaha 1.
79 Bhisagacarya Harisastri Paradakara Vaidya edited, Vagbhata, Ashtanga Hrudayam, Sarvanga Sundara commentary of Arunadatta and Ayurveda Rasayana Commentary of Hemadri. Edition: Reprint, 2012. Varanasi, Choukambha Sanskrit Sansthan Sutra sthan 5/37.
80 Yadavji Trikamji Acharya, Sushruta Samhita Dalhana Commentry Choukhamba Surabharati Publication Sutrasthan 45/87.
81 Bhavaprakash nighantu Mishra prakaran 16/1.
82 Vaidya Jadavaji trikamaji edited, Agnivesa, Charaka Samhitha revised by Charaka and Dridhabala, with the Ayurveda-Dipika Commentary. Edition reprint 2013. Varanasi: Choukambha Prakashan: Sutrasthan 27/229.
83 Catalano, D., Martines, G.F., Tonzuso, A., et al. Protective Role of Coffee in Non-alcoholic Fatty Liver Disease (NAFLD). *Dig Dis Sci* **55**, 3200–3206 (2010).
84 Tillmann, Hans & Suzuki, A. & Pang, H. & Dellinger, Andrew & Guy, Cynthia & Moylan, Cynthia & Piercy, Dawn & Smith, M. & Hauser, Michael & Diehl, A. & Abdelmalek, Manal. (2011). Coffee consumption increases hepatic expression of cytochrome P450s and significantly reduces liver fibrosis in patients with nonalcoholic fatty liver disease (NAFLD). Journal of Hepatology - J HEPATOL. 54. 10.1016/S0168-8278(11)60871-4.
85 - Mohamed Hédi, Hamdaoui,- Snoussi, Chahira – Dhaouadi, Karima – Sami, Fattouch – Ducroc, Robert – Le Gall, Maude – Bado, André PY – 2016/07/01 T1 – Tea decoctions prevent body weight gain in rats fed high-fat diet; black tea being more efficient than green tea VL – 6DO – 10.1016/j.jnim.2016.07.002JO – Journal of Nutrition & Intermediary Metabolism.
86 Thien Chu Dinh, et al The effects of green tea on lipid metabolism and its potential applications for obesity and related metabolic disorders – An existing update, Diabetes & Metabolic Syndrome: Clinical Research & Reviews, Volume 13, Issue 2, 2 (2019) 1667-1673
87 Bhisagacarya Harisastri Paradakara Vaidya edited, Vagbhata, Ashtanga Hrudayam, Sarvanga Sundara commentary of Arunadatta and Ayurveda Rasayana Commentary of Hemadri. Edition: Reprint, 2012. Varanasi, Choukambha Sanskrit Sansthan Sutrasthan 5/60-61.
88 Amulya Murty, Ashok Patil, Mukund Sabnis International Journal of research in Ayurved and Pharmacy.2022(13)4,95-98 Staulya, Ama and meda dhatu dysfunctional nutritional metabolism with special reference to oxidative stress.
89 Maede Ruyvaran, Ali Zamani, Alireza Mohamadian, et al Safflower (*Carthamus tinctorius* L.) oil could improve abdominal obesity, blood pressure, and insulin resistance in patients with metabolic syndrome: A randomized, double-blind, placebo-controlled clinical trial, Journal of Ethnopharmacology, Volume 282,2022, 2022 Jan 10;282:114590. doi: 10.1016/j.jep.2021.114590. Epub 2021 Sep 4. PMID: 34487844. ,https://doi.org/10.1016/j.jep.2021.114590.(www.sciencedirect.com/science/article/pii/S0378874121008199)
90 Bhisagacarya Harisastri Paradakara Vaidya edited, Vagbhata, Ashtanga Hrudayam, Sarvanga Sundara commentary of Arunadatta and Ayurveda Rasayana Commentary of Hemadri. Edition: Reprint, 2012. Varanasi, Choukambha Sanskrit Sansthan Sutrasthan 5/67.
91 Fatty acid profile and quality assessment of safflower (*Carthamus tinctorius*) oil MB Katkade, HM Syed, RR Andhale and MD Sontakke Journal of Pharmacognosy and Phytochemistry 2018; 7(2): 3581–3585.
92 Vaidya Jadavaji trikamaji edited, Agnivesa, Charaka Samhitha revised by Charaka and Dridhabala, with the Ayurveda-Dipika Commentary. Edition reprint 2013. Varanasi: Choukambha Prakashan: Sutrasthan 21/23.
93 Vaidya Jadavaji trikamaji edited, Agnivesa, Charaka Samhitha revised by Charaka and Dridhabala, with the Ayurveda-Dipika Commentary. Edition reprint 2013. Varanasi: Choukambha Prakashan: Sutrasthan 25/40.

94 Vaidya Jadavaji trikamaji edited, Agnivesa, Charaka Samhitha revised by Charaka and Dridhabala, with the Ayurveda-Dipika Commentary. Edition reprint 2013. Varanasi: Choukambha Prakashan: Sutrasthan 26/61.
95 Vaidya Jadavaji trikamaji edited, Agnivesa, Charaka Samhitha revised by Charaka and Dridhabala, with the Ayurveda-Dipika Commentary. Edition reprint 2013. Varanasi: Choukambha Prakashan: Sutrasthan 5/11.
96 Vaidya Jadavaji trikamaji edited, Agnivesa, Charaka Samhitha revised by Charaka and Dridhabala, with the Ayurveda-Dipika Commentary. Edition reprint 2013. Varanasi: Choukambha Prakashan: Sutrasthan 27/82.
97 Ashtangasangraha sutrasthan 6/67.
98 Naik S, Mahalle N, Greibe E, Ostenfeld MS, Heegaard CW, Nexo E, et al.. Hydroxo-B12 for supplementation in B12 deficient lactovegetarians. (2019) 12:1–14. 10.3390/nu11102382.
99 Vaidya Jadavaji trikamaji edited, Agnivesa, Charaka Samhitha revised by Charaka and Dridhabala, with the Ayurveda-Dipika Commentary. Edition reprint 2013. Varanasi: Choukambha Prakashan: Sutrasthan 27/166.
100 Asthanga sangraha sutrasthan 6/96.
101 Vaidya Jadavaji trikamaji edited, Agnivesa, Charaka Samhitha revised by Charaka and Dridhabala, with the Ayurveda-Dipika Commentary. Edition reprint 2013. Varanasi: Choukambha Prakashan: Sutrasthan 5/12.
102 Bradlow HL. Obesity and the gut microbiome: pathophysiological aspects. Horm Mol Biol Clin Investig. 2014 Jan;17(1):53-61. doi: 10.1515/hmbci-2013-0063. PMID: 25372730
103 An obesity-associated gut microbiome with increased capacity for energy harvest Peter J. Turnbaugh, Ruth E. Ley, Michael A. Ahowald, Vincent Magrini, Elaine R. Mardis & Jeffrey I. Gordon *Nature* volume 444, pages1027–1031 (2006).
104 Sayon-Orea C, Martinez-Gonzalez MA, Bes-Rastrollo M. Alcohol consumption and body weight: a systematic review. *Nutr Rev.* 2011;69:419–31. doi: 10.1111/j.1753-4887.2011.00403.x.
105 MacInnis RJ, Hodge AM, Dixon HG, et al. Predictors of increased body weight and waist circumference for middle-aged adults. Public Health Nutr. 2014;17:1087–97.
106 Ashtangasangraha sutrasthan 5/72.
107 Food Standards Australia and New Zealand Intense Sweeteners. [(Accessed on 30 May 2020)]; Available online: www.foodstandards.gov.au/consumer/additives/Pages/Sweeteners.aspx [Ref list]
108 Turner A., Veysey M., Keely S., Scarlett C.J., Lucock M., Beckett E.L. Interactions between Bitter Taste, Diet and Dysbiosis: Consequences for Appetite and Obesity. *Nutrients.* 2018;**10**:1336. doi: 10.3390/nu10101336.
109 Holst J.J. The Physiology of Glucagon-like Peptide 1. *Physiol. Rev.* 2007;**87**:1409–1439. doi: 10.1152/physrev.00034.2006.
110 Vaidya Jadavaji trikamaji edited, Agnivesa, Charaka Samhitha revised by Charaka and Dridhabala, with the Ayurveda-Dipika Commentary. Edition reprint 2013. Varanasi: Choukambha Prakashan: Sutrasthan 26/43/1.
111 Yadavji Trikamji Acharya, Sushruta Samhita Dalhana Commentry Choukhamba Surabharati Publication Sutrastahan 46/51.
112 Yadavji Trikamji Acharya, Sushruta Samhita Dalhana Commentry Choukhamba Surabharati Publication Sutrastahan 46/51.
113 Yadavji Trikamji Acharya, Sushruta Samhita Dalhana Commentry Choukhamba Surabharati Publication Sutrastahan 46/404.
114 Yadavji Trikamji Acharya, Sushruta Samhita Dalhana Commentry Choukhamba Surabharati Publication Sutrasthan 46/51.
115 Agu R.C., Chiba Y., Goodfellow V., MacKinlay J., Brosnan J.M., Bringhurst T.A., Jack F.R., Harrison B., Pearson S.Y., Bryce J.H., et al. Effect of germination temperatures on proteolysis of the gluten-free grains rice and buckwheat during malting and mashing. J. Agric. Food Chem. 2012;60:10147–10154. doi: 10.1021/jf3028039.
116 M. B. Sukkar, M. A. Ullah, W. J. Gan, P. A. Wark, K. F. Chung, J. M. Hughes, C. L. Armour and S. Phipps, *Br. J. Pharmacol.*, 2012, **167**, 1161–1176.
117 Vaidya Jadavaji trikamaji edited, Agnivesa, Charaka Samhitha revised by Charaka and Dridhabala, with the Ayurveda-Dipika Commentary. Edition reprint 2013. Varanasi: Choukambha Prakashan: Chikitssthan 6/48.
118 Preservation and bottling of buttermilk.: Bhanumurthi, J. L.; Trehan, K. S. Author Affiliation: Nat. Dairy Res. Inst., Karnal, India. Indian Dairyman 1970 Vol.22 No.11 pp.275–278.
119 Muhamad Adyab, N.S., Rahmat, A., Abdul Kadir, N.A.A. et al. Mangosteen (*Garcinia mangostana*) flesh supplementation attenuates biochemical and morphological changes in the liver and kidney of high fat diet-induced obese rats. BMC Complement Altern Med 19, 344 (2019). https://doi.org/10.1186/s12906-019-2764-5

120 Effects of low-carbohydrate diet therapy in overweight subjects with autoimmune thyroiditis: Possible synergism with ChREBP September 2016 Drug Design, Development and Therapy Volume 10:2939–2946.
121 Vaidya Jadavaji trikamaji edited, Agnivesa, Charaka Samhitha revised by Charaka and Dridhabala, with the Ayurveda-Dipika Commentary. Edition reprint 2013. Varanasi: Choukambha Prakashan: Sutrasthan 27/176.
122 Modulation of Cytokine Secretion by Garlic Oil Derivatives Is Associated with Suppressed Nitric Oxide Production in Stimulated Macrophages Hsiao-Pei Chang, Shih-Yi Huang, and Yue-Hwa Chen *Journal of Agricultural and Food Chemistry* 2005 53(7), 2530–2534.
123 Mohammed Abdul MI, Jiang X, Williams KM, Day RO, Roufogalis BD, Liauw WS, Xu H, McLachlan AJ. Pharmacodynamic interaction of warfarin with cranberry but not with garlic in healthy subjects. Br J Pharmacol. 2008 Aug;154(8):1691–700. doi: 10.1038/bjp.2008.210. Epub 2008 Jun 2. PMID: 18516070; PMCID: PMC2518459.
124 Yadavji Trikamji Acharya, Sushruta Samhita Dalhana Commentry Choukhamba Surabharati Publication Sutrastahan Sutrasthan 20/15.

Therapeutic Nutrition in Ayurveda for Inflammatory Disorders

Amit Nakanekar and Pravin Bhat

1. INTRODUCTION

Improvement in nutrition as therapy delivers metabolic health. It is rightly quoted that "let the food be your medicine and the medicine be your food."[1] A better diet can have a positive impact on metabolic health, and food can be used to heal and maintain the body, just as medicine can be. This underscores the significance of paying close attention to nutrition and making appropriate adjustments to diet to achieve better health outcomes.

Inflammation is a host immune response to any antigen. Inflammation is caused by pathogens, external injury, chemicals, or radiations.[2] Inflammation can be of various organs such as cystitis (inflammation of bladder), bronchitis (inflammation of bronchus), otitis media (inflammation of middle ear), dermatitis (inflammation of skin), etc. In inflammatory conditions, immune systems release different inflammatory mediators such as cysteinyl-LT, eicosanoids, interleukins (ILs) IL-10, IL-12, IL-13, IL-beta, IL-4, IL-5, IL-6, IL-8, while low adiponectin, etc., are observed. [3]

It has the ability to dilate the small vessels and increase the blood flow toward the inflamed area so that immune system cells move toward the infected area. Long-term inflammation results in chronic inflammatory diseases such as rheumatoid arthritis, psoriasis, inflammatory bowel disease, and metabolic diseases.[4] Postprandial inflammation is a normal response of body cells to stress during food ingestion and metabolism.[5] Thus nutrition can modulate the inflammatory response, e.g. fermented dairy products, mainly bioactive peptides and glycans, can produce inflammation through gut microbiota. Inflammation is an emerging research topic.[6]

Diet plays a significant role in the regulation of inflammation in the body. Pro inflammatory diets, which are high in saturated fats, refined carbohydrates, and processed foods, have been linked to the development and exacerbation of chronic inflammatory conditions such as obesity, type 2 diabetes, and cardiovascular disease.[7]

On the other hand, an anti-inflammatory diet, which is rich in fruits, vegetables, whole grains, lean proteins, and omega-3 fatty acids, has been associated with a reduction in inflammation and a lower risk of chronic diseases. The consumption of omega-3 fatty acids, specifically eicosapentaenoic acid (EPA) and docosahexaenoic acid (DHA), has been shown to have anti-inflammatory properties; and the high antioxidant content of fruits and vegetables can also help to reduce inflammation.[8]

The gut microbiome is composed of communities of bacteria like lactobacillus, bacillus, clostridium, enterococcus, ruminococcus, etc., which carry specific functions such as nutrient metabolism and immune

DOI: 10.1201/9781003345541-7

function.[9] Changes in gut microbiome influence inflammation. Dietary modulation alters the absorption of nutrients from the gut. Individuals consuming meat possess a different gut microbiota in comparison with individuals eating a plant-based diet. Diet and gut microbiome and host interaction play an important role in human health.[10]

There is evidence that the gut microbiome plays a role in the link between diet and inflammation. Diets high in processed foods and low in fibers have been associated with an imbalance in the gut microbiome, known as dysbiosis, which can lead to increased inflammation.[11] [12]

Chrononutrition is a branch of science that includes meal intake frequency, calorie count, duration between eating and fasting, and the role of these factors in metabolic health and risk for chronic inflammations. In Ayurveda, the concept of timely food and physiological processes has been aligned to the natural cycles of the day, thereby giving specific guidelines for food intake.[13] Timing of diet intake as per circadian rhythm also affects inflammation. Time-restricted eating reduces inflammatory markers such as tumor necrosis factor-alpha and interleukin-1-β.[14]

The main cascade of various metabolic and chronic diseases is inflammation.[15] Drug therapy has various limitations such as adverse effects, drug failure, and worsened metabolic health.[16]

Nutritional therapy also includes the way in which food is prepared and administered to the host (individuals). Nutritional therapy involves various terms such as nutritional behavior counseling, external and parenteral nutritional supports, and various cultural barriers during nutritional counseling. The classical texts of Ayurveda encompass a variety of topics related to food, including the different natural sources available, their properties in relation to the seasons and geographical locations, and their specific functions in both healthy and diseased states of the body.[17] Nutritional therapy is aimed at the avoidance of antinutrients (food that is edible but has no nutritional value such as processed food, high fructose, artificial sweeteners), eating a whole food diet, and giving importance to the therapeutic value of whole food diet.[18]

It is important to note that diet is just one of the many factors that contribute to the development and progression of inflammatory conditions, and it should be considered alongside other lifestyle and genetic factors.[19]

Improvement in nutrition as therapy, i.e. therapeutic nutrition, gives a new therapeutic approach in terms of metabolic health. Therapeutic nutrition will overcome the limitation of drug therapy.[20]

2. NUTRITION IN INFLAMMATORY DISORDERS FROM CLASSICAL TEXTS

Ayurveda considers nutrition in solid diet (*Anna*) or *Drava* (liquids). Various dietary preparations such as *Peya, Ushna Jala, Khichdi,* etc. are described that are very effective in reducing inflammation. Excessive nutrition and undernutrition both cause inflammatory disorders, and hence appropriate quantity of nutrition for each individual can be determined on the basis of *Agni, Dosha,* season, and disease.[21]

Eight factors that have an impact on the food metabolism are *Ashtavidh "Ahara Vidhivisheshayatan."*[22] These eight factors have an impact on modulating inflammatory responses.

A few eating conditions that can have anti-inflammatory effects are given as follows: food should be taken freshly prepared and lukewarm; food should be taken in appropriate quantity after the digestion of previously taken food. Food should be taken in a pleasant state of mind with appropriate speed.[23]

There are three different faulty eating methods. The three types of faulty eating can create diseases that are difficult to treat. The faulty types of eating are *Adhyashan* (eating before the digestion of previous food), and *Vishamshan* (irregularity in quantity and timing of meal). *Samashan* means it is a mixture of healthy and unhealthy diets. All of these create inflammatory responses by modulating gut-mediated immunity. When healthy and unhealthy diets are both mixed together and eaten, it is called as *Samashna. Pathya* and *Apthya* diet mixed together causes diseases.[24]

TABLE 4.1 Impact of Eight Factors on Inflammation

SR. NO	AASHTAVIDHA AHAR VIDHIVISHESHAYATAN	*MEANING*	*IMPACT ON INFLAMMATION*
1	*Prakriti*	Inherited attribute of dietary material	Some dietary elements have natural tendency to produce inflammatory response in the body, e.g. oral intake of chili produces inflammation in gastrointestinal tract.
2	*Karan*	A process of making food/ diet that has inflammatory or anti-inflammatory effect	For example, process of making bakery products makes these products inflammatory.
3	*Sanyog*	*Sanyog* is mixing of two or more different dietary products	Chronic consumption of equal quantity of honey and cow ghee or fruits and milk, fish, and milk (*Viruddha Ahaar*) can produce chronic inflammation.
4	*Rashi*	Quantity of food	Excessive consumption of *Guru* (difficult to digest), *Tikshna* food can produce inflammation.
5	*Desha*	Geographical condition in which food is consumed	Consumption of coconut water is anti-inflammatory in desserts.
6	*Kala*	Time frame in which food is consumed	Having hot water and drinking water processed with dry ginger is anti-inflammatory in spring and winter.
7	*Upoyogsanstha*	These are dietary rules that are to be followed. E.g. Dietary rule – food should not be consumed before digestion of previously consumed food	Frequent food consumption in indigestion can produce chronic inflammation.
8	*Upayogta*	It is an individual response	For example, *Okasatmya* of inflammatory food for a particular individual can nullify the inflammatory effects of that particular food.

Vishamashan means untimely erratic eating patterns: taking food at inappropriate timings, that is, before initiation of hunger or not taking food at excessive hunger. *Adhyashan* means eating before the digestion of a meal and eating again in a very short time after the meals. All the above-mentioned faulty eating methods can alter metabolism and induce chronic inflammation.[25]

Ayurnutrigenomics is an Ayurveda-based individualized nutrition that is suitable for specific genetic compositions considering lifestyle, diet, seasonal regimen, and individual *Prakriti*. It is an emerging field in research that may provide numerous possibilities for evidence-based *Ayurvedic* therapeutic nutrition for specified disease conditions in an individualized approach to *Ayurveda*.[26]

Acharya *Charaka* has categorized dietary items into various *Anna Varga* and *Drava Varga*. *Anna Varga* includes all solid diets, while *Drava Varga* includes various liquids. *Anna Varga* is divided into 12 types, while *Drava Varga* is divided into 10 different types, including *Jala Varga* (different types of water), *Dugdha Varga* (Milk and milk products from different animals), and *Ikshu Varga* (different types of sugarcane and its products). Ayurveda has also described various food preparatory methods like *Peya, Manda, and Mamsarasa*.

Concept of *Aam*

"*Aam*" is a concept in *Ayurveda*, which refers to the accumulation of toxins or metabolic waste products in the body. This accumulation can lead to various health issues, including indigestion, constipation, and general malaise.

According to *Ayurveda*, *Aam* is the primary cause of many diseases and the imbalance of *Tridosha* (*Vata, Pitta,* and *Kapha*) in the body. It can be caused by poor digestion, improper diet, and a sedentary lifestyle. The accumulation of *Aam* can lead to the aggravation of the *Vata Dosha*, which governs movement and circulation in the body, leading to symptoms such as bloating, gas, and constipation.

The concept of *Aam* is referenced in several *Ayurvedic* texts, including the *Charaka Samhita*, one of the foundational texts of *Ayurveda*. *Aam* is described as a "disease caused by the accumulation of undigested food." The treatment of *Aam* involves "removing the accumulated *Aam* and restoring the balance of the *Doshas*."

Sushruta Samhita and the *Ashtanga Hridayam* also mention the concept of *Aam* and its role in the development of various diseases. The treatment of *Aam* involves a combination of dietary and lifestyle changes, as well as the use of herbal remedies to support digestion and eliminate toxins from the body.

Correlation of *Aam* with Inflammation

Inflammation and the concept of "*Srotorodh*" are closely related and refer to the body's response to imbalance or injury. Inflammation is a natural response of the immune system to injury or infection, characterized by redness, swelling, heat, and pain. The response is known as "*Srotorodh*," or "blockage," as it refers to the body's attempt to protect itself from further harm by sealing off the affected area.

According to *Ayurvedic* principles, inflammation occurs when there is an imbalance in the body's *Dosha*, or elemental energies. The three *Doshas* – *Vata, Pitta*, and *Kapha* – are responsible for maintaining the body's balance and health. When one or more of the three *Dosha* become imbalanced, it can lead to inflammation and other health problems.

An excess of *Pitta Dosha*, which governs digestion and metabolism, can lead to inflammation in the digestive system. It can manifest as conditions such as ulcers, acid reflux, and irritable bowel syndrome. Similarly, an excess of *Vata Dosha*, which governs movement and circulation, can lead to inflammation in the joints and muscles, resulting in conditions such as arthritis and fibromyalgia.

Inflammation and *Srotorodh*, both can be addressed by the use of a combination of herbal remedies, dietary changes, and lifestyle modifications. Herbs such as turmeric and dry ginger, have anti-inflammatory properties and are commonly used in *Ayurvedic* medicine to reduce inflammation. Additionally, a diet balanced with all six tastes – sweet, sour, salty, pungent, bitter, and astringent – can help to balance the *Doshas* and reduce inflammation.

Lifestyle changes, such as regular exercise, yoga, and meditation, can also help to balance the *Dosha* and reduce inflammation. Specific yoga postures and breathing exercises help ease inflammation in specific areas of the body.[27–30]

3. NUTRITION IN INFLAMMATORY DISORDER FROM CONVENTIONAL SCIENCE

Studies indicate that, there is a strong association between the nutritional aspects of individual and various acute and chronic inflammatory disorders. Modulation in nutritional status is also fundamental to modulating inflammatory and oxidative stress processes, which are all interrelated with the immune system.[31]

In terms of nutrition, health biomarkers help in understanding the relationship between the functional effect of nutrition and diseases.[32] Different studies show results of different types of diet on biomarkers. Ketogenic therapy act as anti-inflammatory in Parkinson's disease[33] Whole grain diet reduces IL-6, IL-10, and TNF-alpha.[34] A rheumatoid arthritis diet, which reduces the C-reactive protein, includes fish oil, eicosapentaenoic acid (EPA), docosahexaenoic acid (DHA), evening primrose oil (EPO), fasting, sub-total fasting followed by gluten diet, vegan and gluten-free diet, and Mediterranean diet.[35] Mediterranean diet improves the lipid profile, endothelial function and blood pressure, protection from oxidative stress, inflammation, and platelet aggregation.[36] High antioxidant-rich diet attenuates the oxidative stress signs.[37]

Dietary and nutritional constituents known to exert anti-inflammatory and antioxidant properties include omega-3 fatty acids, vitamin A, and vitamin C, as well as a variety of phytochemicals, such as polyphenols and carotenoids; these are widely present in plant-based foods. The dietary fiber present in plant-based food items is associated with various health benefits, including anti-inflammatory properties.[38]

The meat-based dietary pattern is associated with low-grade inflammation. Healthy dietary patterns include various components such as high intake of fibers, vitamins, antioxidants, minerals, polyphenols, and monounsaturated and polyunsaturated fatty acids and low intake of salt, refined sugar, saturated and trans fat, and high intake of carbohydrates of low glycemic load. Fruits and vegetables have anti-inflammatory activity on TNF-alpha, TNFR-60, IL-1-beta, IL-4, IL-6, fibrinogen, and sE-selectin and antioxidant activity on F2-isoprostanes and 2,3-dinor-5,6-dihydro-15F2t-Isop.

Nuts have anti-inflammatory activity on CRP, IL-6, TNF-alpha, TNF-beta, TNF-R2, sICAM-1, fibrinogen, resistin, and antioxidant activity on oxLDL.[39]

Concept of Proinflammatory and Anti-inflammatory Diet

Proinflammatory diets are characterized by a high intake of saturated fats, refined carbohydrates, and processed foods, as well as a low intake of fruits, vegetables, and omega-3 fatty acids. This type of diet has been linked to the development of chronic inflammatory conditions such as obesity, type 2 diabetes, and cardiovascular disease.[40]

On the other hand, an anti-inflammatory diet is characterized by a high intake of fruits, vegetables, whole grains, lean proteins, and omega-3 fatty acids, as well as a low intake of saturated fats, refined carbohydrates, and processed foods. This type of diet has been associated with a reduction in inflammation and a lower risk of chronic diseases.[41]

Proinflammatory foods are those that are believed to contribute to inflammation in the body, while anti-inflammatory foods are those that are believed to have anti-inflammatory properties and can help to reduce inflammation.

Proinflammatory foods include:

Refined carbohydrates: White bread, pasta, and pastries are high in refined carbohydrates, which can cause a spike in blood sugar levels and contribute to inflammation.

Fried foods: Fried foods are high in omega-6 fatty acids, which can contribute to inflammation when consumed in excess.

Processed meats: Processed meats, such as bacon, sausages, and deli meats, are high in saturated fats and sodium, which can contribute to inflammation.

Trans fats: Trans fats, which are found in some processed foods and baked goods, can contribute to inflammation.

High-fat dairy product: High-fat dairy products, such as cheese, butter, and cream, can contribute to inflammation when consumed in excess.

Anti-inflammatory food includes:

Fruits and vegetables: Fruits and vegetables are rich in antioxidants and anti-inflammatory compounds, such as vitamin C, vitamin E, and beta-carotene.

Whole grains: Whole grains, such as quinoa, brown rice, and oats, are rich in antioxidants and anti-inflammatory compounds.

Fish: Fish, particularly salmon, tuna, and sardines, are rich in omega-3 fatty acids, which have anti-inflammatory properties.

Nuts and seeds: Nuts and seeds, such as almonds, walnuts, and flaxseeds, are rich in antioxidants and anti-inflammatory compounds.

Olive oil: Olive oil is rich in monounsaturated fats, which have anti-inflammatory properties.

4. CLASSIFICATION OF INFLAMMATORY DISORDERS ACCORDING TO *AVASTHA* (DISEASE CONDITIONS)

The inflammation in Ayurveda can be studied with fundamental concepts like *Jwara, Vranashoth, Shotha, Aamvata, Grahani, Vatarakta,* and *Visarpa.*

Inflammation is considered the involvement of *Tridoshaj Vyadhi* along with *Rakta Dushti.*[42]

Vata is responsible for pain generation, pitta causing rubor, dolor, and calor, and transition occurs with *Rakta*. Vitiated *Vata* along with vitiated *Pitta, Kapha,* and *Rakta* is spread all over the body and occupies the place where *Khavaigunya* is present. *Khavaigunya* is the susceptible body tissue/system for the occurrence of particular pathophysiological processes for the generation of symptoms.[43]

When vitiated *Doshas* with *Rakta* occupies joints of the body it results in inflammation of joints that is Arthritis; in skin it causes skin inflammation; in stomach it causes gastritis; in colon it results in colitis; in bronchus it causes bronchitis; and many more. Similarly, when it occupies different *Dhatus* of the body like *Rasa, Rakta, Mansa, Meda, Asthi, Majja,* and *Shukra*, inflammation of that particular *Dhatu* occurs. The *Updhatus* can also get involved and can cause diseases like breast abscess and arteritis.

The involvement of various *Dhatus* in inflammation is also possible. Various *Dhatu* involvement shows different symptoms.[44] When *Dosha* occupies different *Sthana* (body systems)[45] it shows different symptoms.

Ayurveda-based understanding of inflammations can be summarized as in Figure 4.1. Individualized therapeutic nutrition strategies are planned after considering all these aspects while planning nutrition for the particular patient.

On the basis of *Dosha* predominance, symptoms of inflammation are as follows:

1. ***Vata*-predominant inflammation** – It includes numbness in the limb with swelling, severe pain along with nonpitting edema, and diminished during night.[44] Therapeutic nutrition strategies

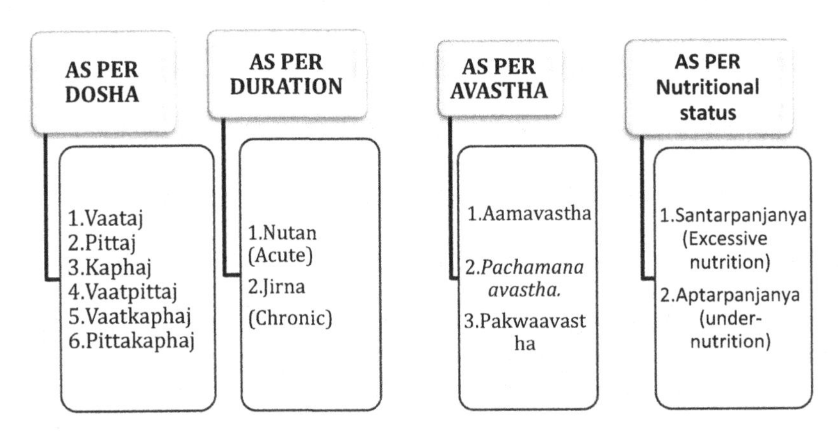

FIGURE 4.1 Classification of inflammation on basis of Ayurveda.

include drinking of camel milk and normal-temperature water after boiling as well as eating of moringa leaves, *soup made from moong bean, and fruits of pomegranate and musk melon.*[46]

2. ***Pitta*-predominant inflammation** – When patients have fever, pain, burning, sweating, thirst, tenderness, and foul smell along with swelling with or without yellowish discoloration of skin and/or eyes.[24] Therapeutic nutrition strategies include the consumption of fruits of pomegranate, dates, and dry grapes; drinking of cow milk; consumption of lukewarm water processed in dry ginger (*Sunthi*); consumption of cow ghee and butter made through traditional Indian method; and avoidance of hot spices, chilis, and daytime sleeping just after consuming the meal.[46]
3. ***Kapha*-predominant inflammation** –In this type Swelling is fixed, itching i, pitting edema, and sticky discharge is present.[44] Therapeutic nutrition strategies includes the consumption of lukewarm food and lukewarm water processed in *Trikatu*, consumption of honey, and consumption of fat-removed buttermilk (*Takra*) made with traditional Indian methods, as well as avoidance of daytime sleeping just after consuming meals and sweet milk products.[46]
4. ***Vata*- and *Pitta*-predominant inflammation** – It has symptoms of both *Vataj* and *Pittaj* inflammation.[47] Therapeutic nutrition strategies include strategies described in *Vata*-predominant as well as *Pitta*-predominant inflammation.
5. ***Vata*- and *Kapha*-predominant inflammation** – It has symptoms of both *Vataj* and *Kaphaj* inflammation.[45] Therapeutic nutrition strategies include strategies described in *Vata*-predominant as well as *Kapha*-predominant inflammation. It includes the consumption of lukewarm food and water and the consumption of oils made with traditional Indian methods.[47]
6. ***Pitta*- and *Kapha*-predominant inflammation** – It has symptoms of both *Pittaj* and *Kaphaj* inflammation.[45] Therapeutic nutrition strategies include strategies described in *Pitta*-predominant as well as *Kapha*-predominant inflammation. It includes the consumption of dry types of food like *Chanaka* and the consumption of dry ginger.[46]
7. ***Sannipata*** – When symptoms of all three *Doshas* are present, that is, of *Vataj*, *Pittaj,* and *Kaphaj*.[47] Therapeutic nutrition strategies require intermittent use of therapeutic nutrition strategies described in *Vata-*, *Pitta-*, and *Kapha*-predominant inflammation.[46]

Inflammation on the Basis of Duration

- ***Nutan* (acute) inflammation** – Inflammation that occurs suddenly and that is present over the body for a short duration of time, that is, less than 3 weeks, is called *Nutan* inflammation. Therapeutic nutrition strategies include the consumption of hot water and consumption of *Peya* (0–7 days), herbal tea as per *Dosha* predominance (8–14 days), and processed milk as per *Dosha* predominance (15–21 days). [44]
- ***Jirna* (chronic) inflammation** – Inflammation that is present over the body for a long duration of time, more than 21 days, is called *Jirna inflammation*. Therapeutic nutrition strategies include the use of processed milk, butter, and buttermilk and the consumption of medicated cow ghee.[44]

Inflammation on the Basis of *Awastha*

Samvastha of inflammation can be identified through the following symptoms: stiffness in the body, *Aruchi* (altered taste), heaviness of the body and abdomen, indigestion, constipation, laziness, and fatigue without any work. *Pachyaman awastha* can be identified by increased body temperature (fever), increased thirst, vertigo, loose motion, nausea, and breathlessness. ***Niramvastha*** can be identified by weakness in the body, weight loss, and the generation of hunger.[43]

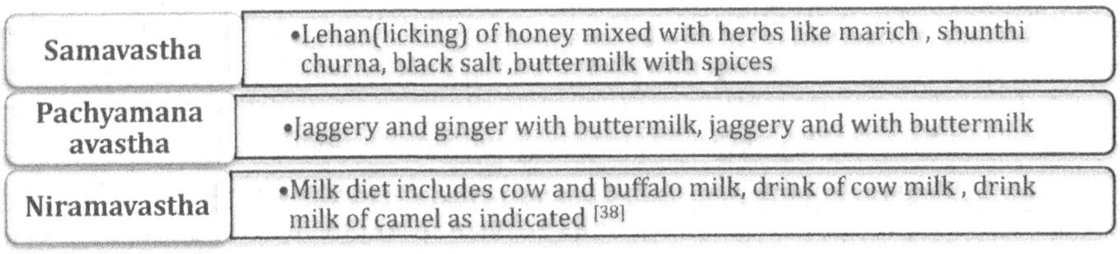

Samavastha	•Lehan(licking) of honey mixed with herbs like marich , shunthi churna, black salt ,buttermilk with spices
Pachyamana avastha	•Jaggery and ginger with buttermilk, jaggery and with buttermilk
Niramavastha	•Milk diet includes cow and buffalo milk, drink of cow milk , drink milk of camel as indicated [38]

FIGURE 4.2 Therapeutic nutrition in various states of inflammation.

Inflammation on the Basis of *Santarpana*

- *Santarpana*[28] – In this condition inflammation occurs due to nutrition. Therapeutic nutrition strategies include intermittent fasting, the use of easy-to-digest food like *Rajgira*, drinking of hot water after consumption of honey, and consumption of *Peya, Manda, Vilepi, Yawa* (barley, *Hordeum vulgare* L.), *Jawas* (flaxseed,), *Aadhaki* (pigeon pea, *Cajanus cajan*), *Kodrawa* (kodo millet, *Paspalum scrobiculatum* L.), *Mudga* (green gram, *Vigna radiata*), *Kulatha* (Horse gram, *Dolichos biflorus* Linn.), and *Brinjal* (Eggplant, *Solanum melongena*).
- *Apatarpan*[46] – This is a condition in which inflammation is due to undernutrition. Therapeutic nutrition strategies include the consumption of cow ghee, cow milk, seafood, jaggery, and sugarcane products and sleeping after the meals.

TABLE 4.2 Types of Inflammation and Causes

1	**Inflammation on Basis of Causes**	**1. *Aagantuja* (caused due to trauma)**[48] occurs due to cut, laceration, sharp or blunt trauma, insect bite, contact with poisonous leaves, nail injury, poisonous wind, due to abortion or miscarriages from improper postnatal gynecological care, and defects. **2. *Nija* (caused due to *Doshas*)**[46] occurs due to improper technique of *Snehana, Swedan, Vamna, Virechan, Niruh Basti, Anuvasana Basti, Alsaka* (diarrhea), excessive weakness, fever, due to heavy work, fasting, due heavy meal intake, eating more salt, salty food, fruits, vegetables, salad, pickles, curd, green vegetable, due to abortion.
2	**On basis of duration of inflammation**	**1. *Nutan Shoth* –** Any inflammation which occurs suddenly. Present over body for short duration of time, that is, less than 3 weeks. **2. *Jirna Shoth* –** Any inflammation which presents over body for long duration of time.
3	**On basis of *Santarpan* and *Apatarpan***	**1. *Santarpana***[49]is condition of *Bruhan* (overweight). *Prameha* (type II diabetes), *Aamdosha* (rheumatoid arthritis), *Jwar* (fever), *Kushtharog* (skin disorders), *Visarpa* (herpes zoster), *Vidhradhi* (abscess), *Pilharogi* (splenomegaly), *Shirorogi* (headache), etc., are considered to be under *Santarpan Vyadhi*. **2. *Apatarpana***[50] is condition of *Langhan* (underweight). *Vatavyadhi* and *Uraskhat* (hemothorax) are considered under it.
4	**On Basis of *Kala* of inflammation**[31]	***1. Diwabali* (strong symptoms during daytime) –** When inflammation occurs during daytime. It is due to *Vata Dosha* predominance. ***2. Raatribali* (strong symptoms during night time)** – When inflammation occurs at night time. It is due to *Kapha Dosha* predominance.

Nija and *Agantuj* Inflammation[44]

Nija and *Agantuj* inflammation are two different types of inflammation described in Ayurvedic medicine.

Nija inflammation, also known as "internal inflammation," refers to inflammation that originates from within the body. It' is believed to be related to an internal accumulation of toxins and imbalances in the body's natural metabolic processes.

Agantuj inflammation, also known as "inflammation due to external factors," refers to inflammation that is caused by external factors such as injury, infection, or exposure to toxins. It is thought to be caused by the body's response to external irritants/trauma and can manifest in various forms such as wounds, infections, and allergic reactions.

Any inflammation is a complex process that occurs in response to injury or infection, which is mediated by a variety of cellular and molecular pathways. The body's immune system responds to an injury by releasing various chemical mediators, such as histamine, prostaglandins, and leukotrienes, that cause blood vessels to dilate, increase blood flow, and attract immune cells to the site of injury.

This results in the characteristic signs of inflammation such as redness, warmth, swelling, and pain. Inflammation is a necessary and beneficial process that helps the body to fight off infection and repair tissue damage. However, chronic or excessive inflammation can lead to tissue damage and contribute to the development of various diseases such as arthritis, cancer, and heart disease.

The distinction between *Nija* and *Agantuj* inflammation in *Ayurvedic* medicine needs more fundamental research to understand the underlying mechanisms of these types of inflammation.

5. GENERAL THERAPEUTIC NUTRITION IN INFLAMMATORY DISORDERS

Ayurveda terminologies that involve inflammatory pathologies at different sites are described as follows.

Amavata

It is a chronic immune-inflammatory disease. It is due to *Aam* with vitiated *Vata* at *Kaphasthana* (involving all joints of the body). Generally, indigestion of food along with heavy exercises results in the formation and mitigation of *Aam* all over the body. *Aam* moves all over the body along with *Vyana Vata* and gets deposited in various *Kaphasthana* all over the body (thorax, throat, head, joints, stomach, body fluids, fat, nose, tongue).[49]

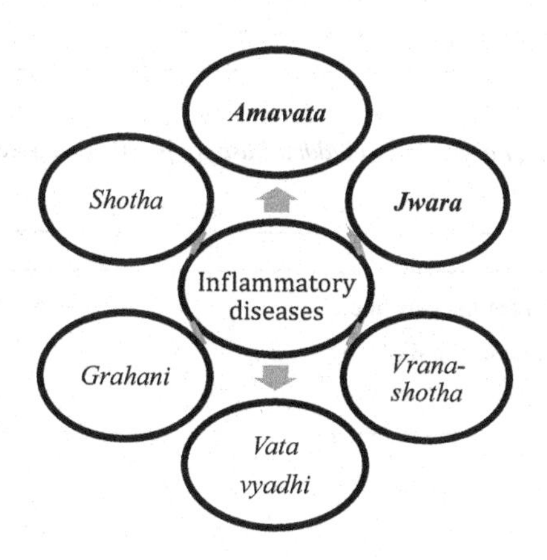

FIGURE 4.3 Inflammatory diseases in Ayurveda.

Therapeutic nutrition strategies include the consumption of *Yawa, Kulatha, Raktsahali, Moringa,* bitter gourd, dry ginger, *Piper nigrum, Cuminum cyminum,* garlic, hot water, buttermilk, and the meat of wild animals processed in various spices and the avoidance of cold water, yogurt, curd, fish, milk, *Vishamshan,* heavy-to-digest food, jaggery, and related products.[46]

Shotha

Shotha is generally considered edema at any part of the body. Vitiated *Vata* enters external blood vessels and contaminates *Kapha, Rakta,* and *Pitta*. It results in an increase in vitiated *Kapha, Rakta,* and *Pitta* and blocks the path of *Vata,* causing edema (inflammation) in that area.[48,50] Therapeutic nutrition strategies include eating of legumes, moringa, chicken, aged ghee, and different rice preparations like ginger flavored *Peya* and avoiding seafood, salty food, jaggery products, curd alcohol, fermented food, fish, and curd.[51]

Vranshotha[52]

Vrana is injury. Inflammatory condition at the site of *Vrana* is termed *Vranashotha*. It can be better understood with its three *Avastha* stages) *(Aamvastha, Pachyamana Avastha, Pakvavstha)* Different nutritional strategies are required for different *Avastha (stages),* in the *Aam* stage easy-to-digestible food is essential. In the *Pachyamana* state easy-to-digest, bitter-predominant food is essential. Avoiding milk products is essential.[43] Details are explained in the chapter related to the musculoskeletal system. Figure 4.3 summarizes nutrition in a state of inflammation related to *Vranshotha.*

Grahani[47]

Grahani (part of the gastrointestinal tract dealing with digestion) is considered one of the eight major diseases (*Ashtamahagada*). It is closely related to *Agni*. *Grahani* diseases are caused due to derangement of *Agni* situated in *Grahani,* causing inflammation in the stomach and intestine.[6] Alteration in the consistency of stool is the symptom. Buttermilk consumption is peculiarity in this type of inflammation of GIT. The use of ghee preparations and spices is also indicated as a dietary regimen.[44]

Jwara[53]

Jwara (fever) is the cause of inflammatory conditions and vice versa. It is observed in the respiratory, cardiovascular, and musculoskeletal systems inflammations. Therapeutic nutrition mentioned in *Jwara* is important in inflammatory conditions.

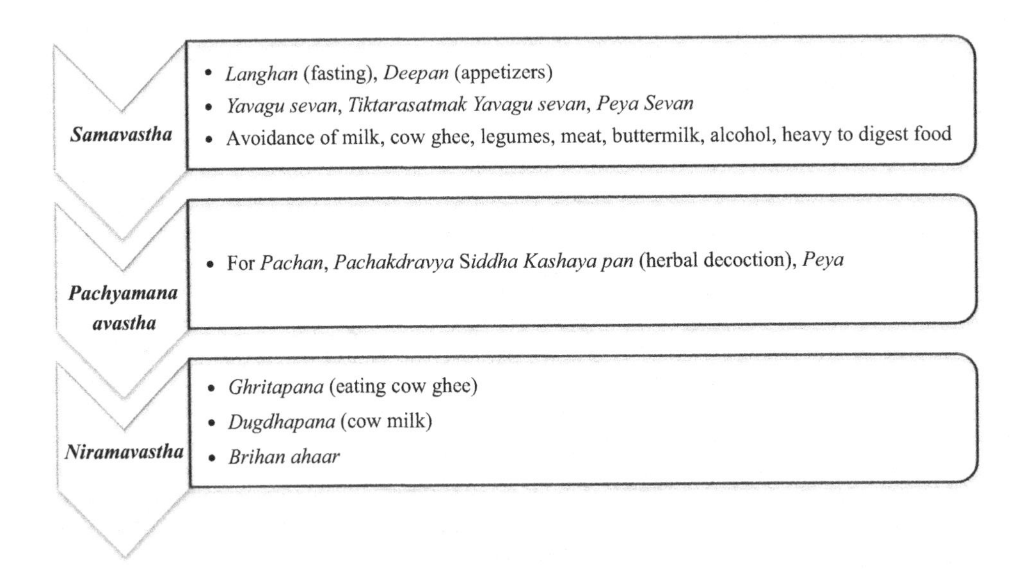

FIGURE 4.4 Therapeutic nutrition in various stages of *Jwara*.

AGANTUJ JWARA – Therapeutic nutrition indicated in this type is Consumption of cow ghee, wine preparation, meat soup, *Shastishali*, bitter gourd, *Mudga, Tandulja, Harandodi.*[48]

6. NUTRITION IN INFLAMMATORY DISORDERS AS PER *DWADASHA AAHAR VARGA*

TABLE 4.3 Nutrition Categories for Inflammatory Disorders

S. NO.	*DIET CATEGORY*	*INGREDIENT NAME (DRAVYA NAME)*	*MEDICINAL QUALITY*	*METHOD OF PREPARATIONS*
A.	*Shukadhanya Varga* (Millets)	Rice[26] i) Red rice ii) Brown rice iii) *Shashthishali* (rice harvested in 60 days)	*Laghu* (easy to digest), *Mutral* (removes accumulated water in body through urine), *Grahi* (absorbent)	It must be prepared in open pot. Froth produced on top of it should be removed while cooking. Gruel, soup, soupy rice, and rice plate can be made from it.
		Sawa millet[47] (*Echinochloa frumentacea*)	*Laghu* (Easy to digest), *Grahi* (absorbent)	It must be prepared in open pot.
		Barley/*Yava*/*Java* (*Hordeum vulgare*)	*Ruksha* (dry in nature), light to digest, Pungent, vitiates *Vata* (so should not be used more where dolor symptom is dominant or where fluid accumulation is less)	Can be consumed whole or used after coarsely grinding. Gruel soup can be made.
		Wheat[47] (*Triticum*)	*Guru* (heavy to digest), *Snigdha* (contains moisture), cold in potency **Contraindications**: Newly developed inflammation, acute conditions, if fluid accumulation is more (*Kaphaj Shotha*)	Gruel, soup, *chapati*, roti, Indian flatbread can be made from wheat flour. It can be consumed by coarsely grinding it like semolina.
		Foxtail millet (*Setaria italica*)	Decreases *Kapha*, used in the healing of fractured bones	Prepared like rice by boiling in water or decoction.
		Great Millet Sorghum/*Jowar* (*Sorghum bicolor*)	Astringent, light to digest, cold potency	Gruel, soup, *chapati* from its flour should be used; 1- or 2-year-old grains should be used.

(*Continued*)

TABLE 4.3 *(Continued)* Nutrition Categories for Inflammatory Disorders

S. NO.	*DIET CATEGORY*	*INGREDIENT NAME* (DRAVYA *NAME*)	*MEDICINAL QUALITY*	*METHOD OF PREPARATIONS*
B.	*Shamidhanya Varga* (Legumes and Pulses)	Green gram[47] *Mung* bean (*Vigna radiata*)	Easy to digest, reduces *Kapha*, *Pitta*, reduces fever	Gruel, soup, *Khichadi*, and risotto (without cheese) can be made from it with or without spices and condiments. Used whole or split.
		Mat bean[47] Moth bean (*Vigna aconitifolia*)	Reduces *Kapha, Pitta, Grahi* (absorbent). **Contraindication:** inflammation due to worm infestation	Gruel, soup with or without spices and condiments. Used whole or split.
		Lentils[47] Masoor (*Lens culinaris*)	Vitiates *Vata*, reduces *Kapha*, *Pitta,* absorbent, easily digestible; if dolor sign is more use it less	Gruel, soup with or without spices and condiments. Used whole or split.
		Pigeon pea[47] Toor arhar (*Cajanus cajan*)	Vitiates *Vata*, easily digestible, reduces *Kapha*	Gruel, Soup with or without spices and condiments. Used whole or split.
		Bengal gram [47] Chickpea (*Cicer arietinum*)	Astringent, dry, easily digestible, cause constipation, vitiates Vata, reduces *Kapha*	Gruel, Soup with or without spices and condiments. Used whole or split.
		Horse gram[47] *Kulitha* (*Macrotyloma uniflorum*)	Hot potency, Astringent, sour after digestion, reduces *Kapha*, *Vata* absorbent (*Grahi*); when calor and rubor are more, don't use it; when burning is more don't use it	Gruel, Soup, with or without spices and condiments. Used whole or split.
		Linseed[54] Flaxseed Javas Tisi (*Linum usitatissimum*)	Provides moisture, Softness, and reduces *Vata* and Pain	For local application of poultices made by cooking it lightly and making rice balls putting it in cloth and tying it and use it warm in poultices form.
C.	Vegetables/ *Shaak Varga*	Wild spinach, Bathua *Chakavat*[54] (Lambs Quarters)	Pungent after digestion, increases digestive fire, easy to digest (*Laghu*)	Consumed by cooking the leaves (Must be consumed by cooking, not in raw form)
		White jute/ *Kaalshaak*[54] (*Corchorus capsularis*)	Reduces *Kapha*, Increases urine output when given in fever	A hot infusion of leaves should be consumed.
		Garden Purslane/ Ghola/[32] Little hogweed (*Portulaca oleracea*)	Dry in nature, reduces *Kapha*, improves digestive fire, reduces inflammation	Cooked leaves and seeds are used in salads. Leaves are crushed and used as poultices.

TABLE 4.3 *(Continued)* Nutrition Categories for Inflammatory Disorders

S. NO.	*DIET CATEGORY*	*INGREDIENT NAME* (DRAVYA *NAME*)	*MEDICINAL QUALITY*	*METHOD OF PREPARATIONS*
		Bladder dock Chuka[54] (*Rumex vesicarius*)	Reduces pain, reduces inflammation, reduces *Vata*, easy to digest	Cooked leaves are used in salad form in colon inflammation, seeds are used in gastritis, external application of its crushed leaves on inflammation.
		Harkuch shaak/ Helencha[54] (*Enhydra fluctuans*)	Oily in nature, reduces *Vata* and pitta	Cooked leaves are used in salad or as is
		Head Leucas *Guma/Kumbha/* Shetwad/ *Dronapushapi*[54]/ *Leucas cephalotes*	Reduces *shotha*, reduces *Aam*, Sweet after digestion, hot in potency	Cooked vegetables/leaves are used in salad or as is
		Pointed gourd leaves/ *Patolpatra*[54] (*Trichosanthes dioica*)	Reduces pitta, easily digestible, hot in nature, reduces fever	Decoction of leaves of pointed gourd and coriander in *Pitta*-dominant fever.
		Chickpea leaves/ Chana leaves[54] Bengal gram leaves	Increases *Kapha* and *Vata*	Cooked leaves with spices and condiments in the form of salad or as is
		Moringa leaves/ horseradish tree[54] (*Moringa pterygosperma*)	- Pungent - Hot potency - Reduces worms - Reduces *Kapha* and *Vata*.	Fresh leaves are consumed by cooking on steam in the form of salad. Cooked flowers are also used in salad. Crushed skin and leaves are used as external application.
		Karvella (bitter gourd)[54] (*Momordica charantia*)	- Bitter, cold potency, easy to digest, vitiates *Vata*, reduces *Kapha* and *Pitta*	Cooked fruit is used as salad, chips (roasted; not fried).
		Sponge gourd[54] Nenua/*Koshataki/ Luffa aegyptiaca*	- Reduces *Vata* - Moisturizing (*Snigdha*)	Cooked fruit is used as salad, chips (roasted; not fried), *Raita*.
		Ridge gourd/Torai/ *Rajkoshataki*[54] (*Luffa acutangula*)	- Sweet in taste, increases digestive fire, reduces *Pitta*, reduces fever reduces cough	Cooked fruit is used as salad, chips (roasted; not fried), *Raita*.
		Pointed gourd/ *Patola* (*Trichosanthes dioica*)	- Reduces *Kapha* - Little bit hot in potency - Reduces burning	Cooked fruit is used as salad, chips (roasted; not fried), *Raita*.

(*Continued*)

TABLE 4.3 *(Continued)* Nutrition Categories for Inflammatory Disorders

S. NO.	*DIET CATEGORY*	*INGREDIENT NAME* (DRAVYA *NAME*)	*MEDICINAL QUALITY*	*METHOD OF PREPARATIONS*
		Brinjal/*Vruntak*/ *Vartak*/Aubergine (*Solanum melongena*) Indian round gourd[54] (*Cirullus vulgaris*)	- Increases *pitta*, easy to digest, reduces *Kapha*, reduces fat, reduces *Vata* from body	Roasted brinjal on coal – reduces *Kapha* and *Vata* dominant fever - Cooked brinjal is consumed by roasting, cooking with spices - Cooked fruit is used as salad, chips (roasted; not fried)
		Spiny gourd *Karkoti*/Kartoli (*Momordica dioica*)[54]	- Pungent - Increases digestive fire - Reduces fever, cough	- Cooked fruit is used as salad, chips (roasted; not fried)
D.	*Kandavarga*/ Root vegetables/ Tubers	Elephant foot yam/ Suran[54] (*Amorphophallus campanulatus*)	- Astringent - Pungent - Increases digestive fire - Reduces sliminess in body (*Kled shoshak*)	- Used in the form of gravy, vegetable, cooked salad, roasted salad, Burfi, soup, etc. (it must be used along with sour items such as Tamarind or Kokum)
		Radish/Mulak/Mula[54] (*Raphanus sativus*)	- Unripe radish – Pungent - Easy to digest - Stabilizes all three *Doshas* - Using ripe radish in raw form can vitiate all *Doshas*	- If ripped radish is to be used; use it by roasting in oil/ stir-fried - Stir-fried soup - Salad can be consumed
		Giant Taro[54] (*Alocasia indica*)	- Easy to digest, cold potency reduces pitta, increases urine amount	- Dried form of the tuber is used with gruel (manda) of rice.
		Indian Kudzu/Kudzu/ *Vidarikanda*[47] (*Pueraria tuberosa*)	- Sweet in taste - Increases urine output (*mutral*) - Cold potency	- Used in the form of gravy, vegetables, cooked salad, roasted salad, Burfi - soup, etc.
		Carrot/Grunjanak[47] (*Daucus carota*)	- Absorbent (*Grahi*) - Reduces *Vata* and *Kapha*.	- Used in the form of gravy, vegetable, cooked salad, roasted salad, Burfi - soup etc.
E.	Fruits (*Falavarga*)	Raisins/Manuka/ Mrudvika[47] (*Vitis vinifera*)	- Reduces *Vata*, pitta, burning - Sweet in taste, cold potency - Moisturizing	- Consumed directly, by making decoction, by making juice - Used in various delicacies
		Carambola/ Karmarakh/ Karmarang/ Bhavya[47] (*Averrhoa carambola*)	- Absorbent (*Grahi*) - Sweet, sour in taste - Astringent taste - Cold potency	- Consumed directly - By making decoction - By making juice - Used in various delicacies

TABLE 4.3 *(Continued)* Nutrition Categories for Inflammatory Disorders

S. NO.	*DIET CATEGORY*	*INGREDIENT NAME* (DRAVYA *NAME*)	*MEDICINAL QUALITY*	*METHOD OF PREPARATIONS*
		Wood apple/ Elephant apple/ Kapiththa[47] (*Limonia acidissima*)	Absorbent (*Grahi*) - Sweet, sour in taste - Astringent taste	(Only raw fruit is used for inflammatory disorders) - Its pulp is consumed directly, by making decoction, by making juice jam, pickles can be made from it, used in various delicacies
		Pomegranate[47] (*Punica granatum*)	- Absorbent (*Grahi*) - Sweet, sour in taste - Astringent taste	- Consumed directly, by making decoction, by making juice, used in various delicacies, its dried powder is used to sprinkle on foodstuffs.
F.	Meat/ *Mansavarga*	Quail meat[48]	Astringent, sweet, light in digestion, increases digestive fire	Meat soup, curry should be taken.
		Rooster meat[48]	- Lubricating, hot potency - Reduces *Vata*	Meat soup, curry should be taken.
		Crab meat[47]	Lubricating, hot potency reduces *Vata*	Meat soup, curry should be taken.
H.	Milk and milk product group/ *Dugdha* varga	Cow milk[46,51]	- Helps easy movement of bowel - Relives chronic fevers - Relives dysuria	- *Sidhdha Dugdha* Medicated milk should be used. - Milk medicated using black pepper + long pepper + dry ginger powder or by *Trivrutta* means *Operculina turpethum* or by using chitraka means *Plumbago zeylanica*
		Camel Milk[46,51]	- A little bit hot in potency, dry in nature, salty taste, light for digestion, increases digestive fire*	- Having only camel milk in the diet for 7 days or 1 month can relieve inflammation#.
		Curd/*Dadhi*	- Increases inflammation	-
		Buttermilk/*Takra*[46]	- Easy to digest, sour, Astringent, Improves digestion, balances *Kapha* and *Vata*	- Made by churning the curd with water
		Whey/Watery part of curd/Mastu/ Supernatant liquid of curds[46]	- Easy to digest	- Consume only the watery part on top of the curd without disturbing the curd.

(Continued)

TABLE 4.3 *(Continued)* Nutrition Categories for Inflammatory Disorders

S. NO.	*DIET CATEGORY*	*INGREDIENT NAME (DRAVYA NAME)*	*MEDICINAL QUALITY*	*METHOD OF PREPARATIONS*
		Butter/*Navneet*[46]	- Cold potency - Increases digestive fire - Absorbent	- After boiling full-fat milk, the cream/on top of it is collected and curd is made out of it. After churning this curd it turns into buttermilk and butter gets separated from the buttermilk and floats on top of it; it is then collected as butter and consumed.
I.	Sugarcane family (*Ikshuvarga*)	Honey[46,48]	- Dry in nature - Sweet - Astringent - The scrape of the internal lining of tubes in the body (*Lekhan*)	- Should not be used by heating it or by those who are from heat problems, not to be used in the hot season, hot climate, hot country, or with hot potency foodstuffs - Should not be consumed in excessive quantity.
J.	*Aharayogi* varga/ Foodstuffs used in support for making any recipe	Castor oil	- Pungent, sweet, bitter, reduces *Vata*, *Kapha*, promotes nature of movement of body fluids	- Can be used with hot water, with hot milk, in making chapatis/Indian flatbreads especially in RA and other inflammations.
		Dry ginger powder/ *Sunthi*[44]	Little bit moisture containing, increases digestive fire, sweet after digestion, hot in potency, reduces *Vata* and *Kapha*	- It is used for preparation of different decoctions, infused tea, medicated milk.

7. CLINICAL SIGNIFICANCE OF THERAPEUTIC NUTRITION IN INFLAMMATORY DISORDERS

In practice, there can be different permutations and combinations of inflammation, which can be summarized as follows.

Freshly cooked food is essential in any inflammatory disorder. Following are a few freshly prepared recipes from the authors' clinical experiences.

1. Medicated water – water medicated by boiling:
   - For vitiated *Vata Pitta-dominant inflammation*, medicated water with dry ginger
   - For vitiated *Pitta*-dominant inflammation, medicated water with dates and grapes
   - For vitiated *Kapha-dominant* inflammation, medicated water with black pepper powder
2. Buttermilk – buttermilk can be used in various *Santarpanottha* inflammation medications:
   - Freshly prepared fat-removed buttermilk by churning curd should be consumed.

TABLE 4.4 Nutritional Therapy in Different Texts of Ayurveda

VARGA	SHOTHA VYADHI (CHARAK)	SUSHRUTA	ASHTANG HRIDAYA	ASHTANG SANGRAHA
Shukadhanya	*Shashtikshali, Yava, Godhuma*[44]	*Godhuma Anna sevan* (wheat grain food)[55]	Red rice and Linseed[51]	
Shami varga	*Mudga, Kulatha* [44]		*Nishpava*[46]	*Kulatha, Nishapava*[56]
Mansa varga	*Vishkir Pashu – Lava, Tittar* (chicken bird) [44]		Meat soup of ostrich, peacock	
Shaka	*Suvarchala, Parwal, Kakamachi, Nimba*[44,47]	Red drumstick garlic, Putikarnja [43]	*Tulsi, grujannak* [46]	*Tulsi*[56]
Phala	---			
Harit varga	*Mulak, grujjana, Rason*[44]			
Madhya	*Jagal, Pakvarasa Sidhu, Apakvarasa Sidhu*[47]	Linseed *Saktu*, Sugarcane *Sidhu*, *Surasava*[43]	*Sidhu*[28]	*Jagalmedaka* (herbal alcohol), *Sidhu*[56]
Jala varga	---	Low water intake in inflammatory condition[25]		
Dugdha	Milk of camel[44] *Takrapana*[47]	Milk of camel[25]	Camel milk, Buttermilk[46]	Milk of camel, Buttermilk[56]
Ikashu	----			
Kruttana	---		*Mudga* and *kulatha Yusha*[46]	*Yush, Yavagu* [57]
Aharayogi	*Shunthi, Hingu, saindhava, yavakshar*[44]			
Taila			Erand tail[51]	Erand tail[56]
Mutra			Cow urine[46]	All types of Urine mentioned in Samhita
Aushadha				
Takra		Takra[43]		
Ghee		Ghee of camel[43]		
Mudgadi		Linseed[43]		
Matradi				*Dadhi; Shothakrita, Takra Shothahara*[56]

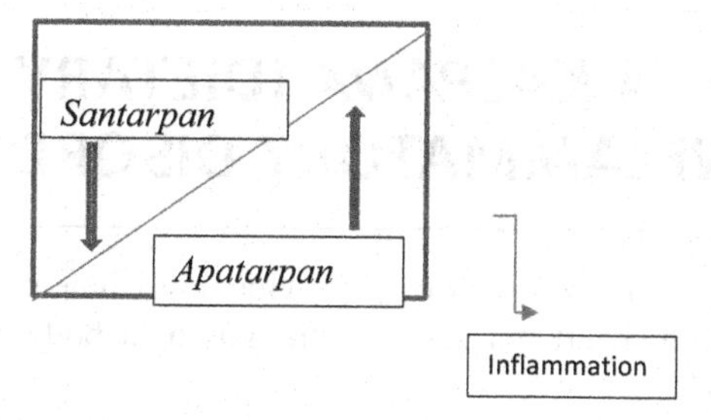

FIGURE 4.5 Inflammation on the basis of *Santarpan*.

- Pinch of black salt can be added to around 50–100 ml of buttermilk.
- Black pepper powder and dry ginger with black salt and honey should be given to the patient, mixed with buttermilk.
- Buttermilk, a watery part of curd useful for inflammatory colitis conditions, relieves constipation and cleanses body channels (*Strotoshuddhikar*).

3. Milk:
   - Cow milk is used in *Nirama* (with no history of indigestion) type of inflammation, in constipation.
   - Camel milk is used in inflammation in piles and ascites.
4. Gruel (*Triticum aestivum*) – This can be used in inflammation with weak digestive power:
   - Gruel made from boiling rice grains with water should be given by sprinkling the cumin powder with kokum or with Indian jujube (Indian berry)
   - Black pepper powder + long pepper powder + dry ginger powder
5. Chicken/mutton soup – mainly in inflammation having *Apatarpan*.
   - Soup made from boiling chicken/mutton in 12 times water should be given by sprinkling the cumin powder (for *Vata-Kapha* dominant inflammation)
   - With *kokum* (*Pitta* dominant)

Other than this, various gruels, roasted grains, soup made from various pulses and lentils, rice cooked from 1- or 2-year-old uncooked rice, and barley should be consumed on a daily basis.

Different food preparations of *Shashtishali* (rice harvested in 60 days) are used in all types of inflammatory joint disorders, rheumatoid arthritis, and colitis.

Food preparations made with barley are used in inflammation where fluid accumulation is more, newly developed inflammation, newly developed fever, fever due to indigestion, fever with rhinitis, and rheumatic arthritis (RA).

Wheat soup is useful in inflammation due to fracture. It is useful to reduce the pain of the fracture.

Foxtail millet is beneficial in fracture inflammation and external application in RA (Poultice).

Great millet is useful in *Vata*-dominant inflammation (inflammation having less fluid accumulation). Bengal gram is used in inflammation without constipation and fever.

Horse gram is useful in inflammations due to indigestion.

Garden purslane can be used where dolor and tumor signs are more in renal and bladder inflammation. Its external application is muscle sprain, burning in palms, and external inflammation.

Bladder dock is useful in colon inflammation, stomach burning, vomiting, and stomach inflammation/gastritis – seeds are used in inflammation due to insect bites.

Bitter gourd is useful in fever where *Pitta* and *Kapha* are dominant, RA, and gout. Sponge gourd is used if dolor sign is more in inflammation. Ridge gourd acts as an anti-inflammatory in *Pitta*-dominant fever. *Brinjal* is used in *Vata*- and *Kapha*-dominant fever. *Suran* and Carrot are useful in inflammation due to RA and hemorrhoids. Indian Kudzu increases urine output and so can be used in renal inflammation and in inflammation when fluid accumulation is more.

8. *AHAREYA KALPANA* (DIETARY RECIPES) IN INFLAMMATORY DISORDERS

Manda increases metabolic power, *Peya* and *Yawagu* provide nourishment and strength, used in diarrhea, and *Vilepi* is beneficial for a healthy heart and provides strength to the body. *Laaja Peya* is made of parched rice and helps relieve fatigue.[46]

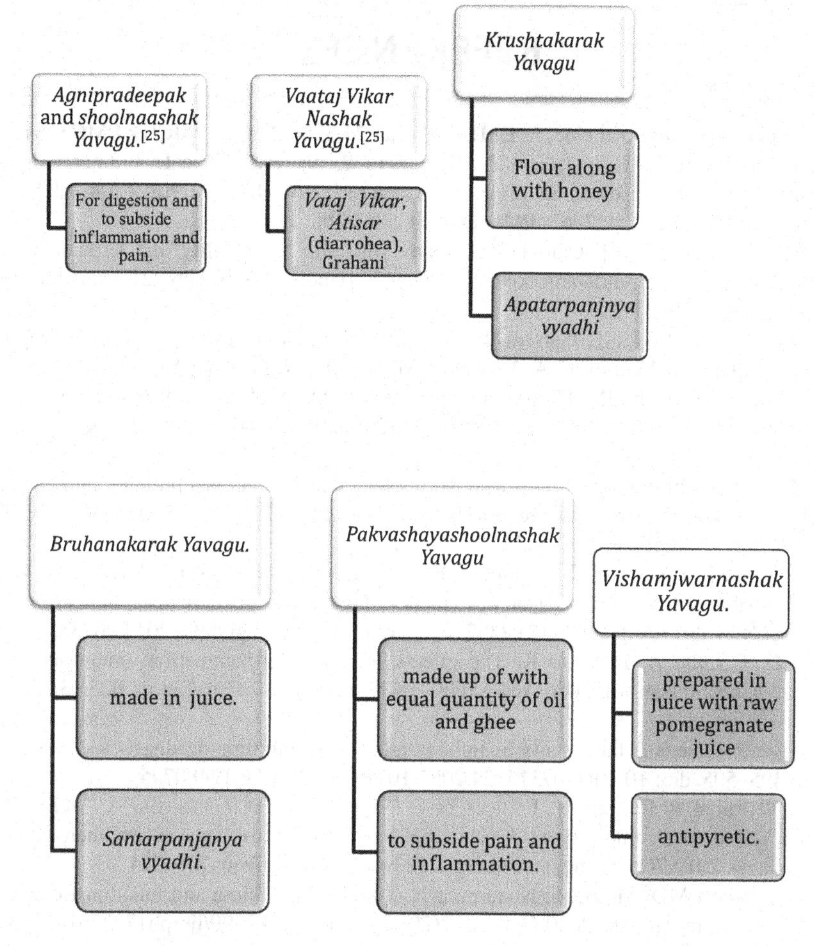

FIGURE 4.6 *Yavagu* which help in different inflammatory condition.

Different types of *Aharaiya Kalpana* are mentioned in Table 4.6. *Charak* in *Sutrasthan* explains 28 types of *Yavagu*.[46]

Following are some *Yavagu* that help in the treatment of different inflammatory conditions.

9. SCOPE FOR FUTURE RESEARCH

The efficacy of functional food may change with risk factors or disease conditions or food–food interaction, food–drug interaction, drug–food timing interaction, and food and circadian rhythm interaction, so studies in this direction are needed.

Nowadays, the topic of gut microbiota is gaining significant importance. In Ayurveda, it is said that the cause of diseases is diminished metabolic energy. Personalized nutrition is an important part in terms of nutrigenomics. In the future, there is a need for research on the evaluation of the clinical impact of genetic testing and personalized nutrition for the welfare of the healthcare system and public health.

REFERENCES

1 Smith R. "Let food be thy medicine...". BMJ. 2004 Jan 24;328(7433):0. PMCID: PMC318470
2 Chen L, Deng H, Cui H, Fang J, Zuo Z, Deng J, Li Y, Wang X, Zhao L. Inflammatory responses and inflammation-associated diseases in organs. Oncotarget. 2017 Dec 14;9(6):7204–7218. doi: 10.18632/oncotarget.23208. PMID: 29467962; PMCID: PMC5805548.
3 InformedHealth.org [Internet]. Cologne, Germany: Institute for Quality and Efficiency in Health Care (IQWiG); 2006. What is an inflammation? 2010 Nov 23 [Updated 2018 Feb 22]. Available from: www.ncbi.nlm.nih.gov/books/NBK279298/
4 Furman D, Campisi J, Verdin E, Carrera-Bastos P, Targ S, Franceschi C, Ferrucci L, Gilroy DW, Fasano A, Miller GW, Miller AH, Mantovani A, Weyand CM, Barzilai N, Goronzy JJ, Rando TA, Effros RB, Lucia A, Kleinstreuer N, Slavich GM. Chronic inflammation in the etiology of disease across the life span. Nat Med. 2019 Dec;25(12):1822–1832. doi: 10.1038/s41591-019-0675-0. Epub 2019 Dec 5. PMID: 31806905; PMCID: PMC7147972
5 Meessen ECE, Warmbrunn MV, Nieuwdorp M, Soeters MR. Human postprandial nutrient metabolism and low-grade inflammation: A narrative review. Nutrients. 2019 Dec 7;11(12):3000. doi: 10.3390/nu11123000. PMID: 31817857; PMCID: PMC6950246.
6 Bordoni A, Danesi F, Dardevet D, Dupont D, Fernandez AS, Gille D, Nunes Dos Santos C, Pinto P, Re R, Rémond D, Shahar DR, Vergères G. Dairy products and inflammation: A review of the clinical evidence. Crit Rev Food Sci Nutr. 2017 Aug 13;57(12):2497–2525. doi: 10.1080/10408398.2014.967385. PMID: 26287637.
7 Giugliano D, Ceriello A, Esposito K. The effects of diet on inflammation: emphasis on the metabolic syndrome. J Am Coll Cardiol. 2006 Aug 15;48(4):677–85. doi: 10.1016/j.jacc.2006.03.052. Epub 2006 Jul 24. PMID: 16904534.
8 Simopoulos AP. Omega-3 fatty acids in inflammation and autoimmune diseases. J Am Coll Nutr. 2002 Dec;21(6):495–505. doi: 10.1080/07315724.2002.10719248. PMID: 12480795.
9 Rinninella E, Raoul P, Cintoni M, Franceschi F, Miggiano GAD, Gasbarrini A, Mele MC. What is the healthy gut microbiota composition? A changing ecosystem across age, environment, diet, and diseases. Microorganisms. 2019;7(1):14. https://doi.org/10.3390/microorganisms7010014
10 Al Bander Z, Nitert MD, Mousa A, Naderpoor N. The gut microbiota and inflammation: An overview. Int J Environ Res Public Health. 2020 Oct 19;17(20):7618. doi: 10.3390/ijerph17207618. PMID: 33086688; PMCID: PMC7589951.
11 De Filippis F, Pellegrini N, Vannini L, Jeffery IB, La Storia A, Laghi L, Serrazanetti DI, Di Cagno R, Ferrocino I, Lazzi C, Turroni S, Cocolin L, Brigidi P, Neviani E, Gobbetti M, O'Toole PW, Ercolini D. High-level adherence to a Mediterranean diet beneficially impacts the gut microbiota and associated metabolome. Gut. 2016 Nov;65(11):1812–1821. doi: 10.1136/gutjnl-2015-309957. Epub 2015 Sep 28. PMID: 26416813
12 Sánchez-Alcoholado L, Ordóñez R, Otero A, Plaza-Andrade I, Laborda-Illanes A, Medina JA, Ramos-Molina B, Gómez-Millán J, Queipo-Ortuño MI. Gut microbiota-mediated inflammation and gut permeability in patients with obesity and colorectal cancer. Int J Mol Sci. 2020 Sep 16;21(18):6782. doi: 10.3390/ijms21186782. PMID: 32947866; PMCID: PMC7555154
13 Banerjee S, Debnath P, Debnath PK. Ayurnutrigenomics: Ayurveda-inspired personalized nutrition from inception to evidence. J Tradit Complement Med. 2015 Mar 24;5(4):228–33. doi: 10.1016/j.jtcme.2014.12.009. PMID: 26587393; PMCID: PMC4624353
14 Adafer R, Messaadi W, Meddahi M, Patey A, Haderbache A, Bayen S, Messaadi N. Food timing, circadian rhythm and chrononutrition: A systematic review of time-restricted eating's effects on human health. Nutrients. 2020 Dec 8;12(12):3770. doi: 10.3390/nu12123770. PMID: 33302500; PMCID: PMC7763532.
15 Hotamisligil GS. Inflammation and metabolic disorders. Nature. 2006 Dec 14;444(7121):860–867. doi: 10.1038/nature05485. PMID: 17167474.
16 Theoduloz C, Alzate-Morales J, Jiménez-Aspee F, Isla MI, Alberto MR, Pertino MW, Schmeda-Hirschmann G. Inhibition of key enzymes in the inflammatory pathway by hybrid molecules of terpenes and synthetic drugs: In vitro and in silico studies. Chem Biol Drug Des. 2019 Mar;93(3):290–299. doi: 10.1111/cbdd.13415. Epub 2018 Oct 30. PMID: 30294891.
17 Payyappallimana U, Venkatasubramanian P. Exploring Ayurvedic knowledge on food and health for providing innovative solutions to contemporary healthcare. Front Public Health. 2016 Mar 31;4:57. doi: 10.3389/fpubh.2016.00057. PMID: 27066472; PMCID: PMC4815005.

18 Jaffey JA, Su D, Monasky R, Hanratty B, Flannery E, Horman M. Effects of a whole food diet on immune function and inflammatory phenotype in healthy dogs: A randomized, open-labeled, cross-over clinical trial. Front Vet Sci. 2022 Aug 23; 9:898056. doi: 10.3389/fvets.2022.898056. PMID: 36082214; PMCID: PMC9447376.

19 Margină D, Ungurianu A, Purdel C, Tsoukalas D, Sarandi E, Thanasoula M, Tekos F, Mesnage R, Kouretas D, Tsatsakis A. Chronic inflammation in the context of everyday life: Dietary changes as mitigating factors. Int J Environ Res Public Health. 2020 Jun 10;17(11):4135. doi: 10.3390/ijerph17114135. PMID: 32531935; PMCID: PMC7312944

20 Lamy PP. Effects of diet and nutrition on drug therapy. J Am Geriatr Soc. 1982 Nov;30(11 Suppl):S99–S112. doi: 10.1111/j.1532-5415. 1982.tb01364.x. PMID: 6752254.

21 Tripathi B, editor. *Charak Samhita; Varanasi.* Chikitsasthan, Grahanichikitsa Addhyaya, chapter 15, verse 9,10. 1st ed. Chaukhamba Surbharti Prakashan; Varanasi, India, 2013.

22 Acharya Jadavaji T, editor. *Charaka Samhita by Agnivesh,* Vimansthana; Rasavimaniya: Chapter 1, Verse 21. 3rd ed. Bombay: Nirnay Sagar Press; 1941, 235

23 Tripathi B, editor. *Charak Samhita; Varanasi.* 1st ed. Vimansthan, Rasviman addhyaya, chapter 1, verse 21. Chaukhamba Surbharti Prakashan; 2013, 664.

24 Kunte AM, Paradkar H. *Ashtang Hridaya* by *Vagbhata* with *Ayurved Dipika commentries* of *Sarvangasundar* by *Arundatta* and *Ayurveda Rasayana* by *Hemadri. Sutrasthana Vartmaroga Vidnyaniya Adhyay.* 8/4-5, 6th ed. Bombay: Nirnaysagar Press; 1939: 148.

25 Sharma S, editor. *Ashtang sangraha Samhita; Varanasi.* 1st ed. Sutrasthan, Annapanviddhi addhyaya, chapter 10, verse 27 Chaukhamba Surbharti Series; 2016.

26 Banerjee S, Debnath P, Debnath PK. Ayurnutrigenomics: Ayurveda-inspired personalized nutrition from inception to evidence. J Tradit Complement Med. 2015 Mar 24;5(4):228–233. doi: 10.1016/j.jtcme.2014.12.009. PMID: 26587393; PMCID: PMC4624353.

27 Lad V. (2003). *Ayurvedic Medicine: The Principles of Traditional Practice.* Churchill Livingstone. London, UK.

28 Singh B, Lohani B. Anti-inflammatory and anti-arthritic activity of Boswellia serrata. J Nat Remedies. 2013;13(1):1–9.

29 Kulkarni SK, Dhir A. Ayurvedic herbs and herbal drugs are used for the treatment of inflammatory disorders. J Tradit Complement Med. 2015;5(3):131–139.

30 Srikumar TS, Ghosh A. Role of Ayurveda in the management of rheumatoid arthritis. J Ayurveda Integr Med. 2013;4(4):231–237.

31 Barrea L, Di Somma C, Muscogiuri G, Tarantino G, Tenore GC, Orio F, Colao A, Savastano S. Nutrition, inflammation and liver-spleen axis. Crit Rev Food Sci Nutr. 2018;58(18):3141–3158. doi: 10.1080/10408398.2017.1353479. Epub 2017 Nov 30. PMID: 28799803.

32 Gabriele M., Pucci L. Diet bioactive compounds: Implications for oxidative stress and inflammation in the vascular system. Endocr Metab Immune Disord Drug Targets. 2017;17:264–275. doi: 10.2174/1871530317666170921142055.

33 Grammatikopoulou MG, Tousinas G, Balodimou C, Anastasilakis DA, Gkiouras K, Dardiotis E, Evangeliou AE, Bogdanos DP, Goulis DG. Ketogenic therapy for Parkinson's disease: A systematic review and synthesis without meta-analysis of animal and human trials. Maturitas. 2022 Sep;163:46–61. doi: 10.1016/j.maturitas.2022.06.001. Epub 2022 Jun 9. PMID: 35714419.

34 Cowan SF, Leeming ER, Sinclair A, Dordevic AL, Truby H, Gibson SJ. Effect of whole foods and dietary patterns on markers of subclinical inflammation in weight-stable overweight and obese adults: A systematic review. Nutr Rev. 2020 Jan 1;78(1):19–38. doi: 10.1093/nutrit/nuz030. PMID: 31429908.

35 Philippou E, Petersson SD, Rodomar C, Nikiphorou E. Rheumatoid arthritis and dietary interventions: Systematic review of clinical trials. Nutr Rev. 2021;79(4):410–428. doi:10.1093/nutrit/nuaa033.

36 Tuttolomondo A, Simonetta I, Daidone M, Mogavero A, Ortello A, Pinto A. Metabolic and vascular effect of the Mediterranean diet. Int J Mol Sci. 2019;20(19):4716. Published 2019 Sep 23. doi:10.3390/ijms20194716.

37 Arouca AB, Santaliestra-Pasías AM, Moreno LA, Marcos A, Widhalm K, Molnár D, Manios Y, Gottrand F, Kafatos A, Kersting M, Sjöström M, Sáinz ÁG, Ferrari M, Huybrechts I, González-Gross M, Forsner M, De Henauw S, Michels N, HELENA study group. Diet as a moderator in the association of sedentary behaviors with inflammatory biomarkers among adolescents in the HELENA study. Eur J Nutr. 2019 Aug;58(5):2051–2065. doi: 10.1007/s00394-018-1764-4. Epub 2018 Jul 4. PMID: 29974229.

38 Iddir M, Brito A, Dingeo G, Fernandez Del Campo SS, Samouda H, La Frano MR, Bohn T. Strengthening the immune system and reducing inflammation and oxidative stress through diet and nutrition: Considerations during the COVID-19 crisis. Nutrients. 2020 May 27;12(6):1562. doi: 10.3390/nu12061562. PMID: 32471251; PMCID: PMC7352291

39 Casas R, Castro-Barquero S, Estruch R, Sacanella E. Nutrition and cardiovascular health. Int J Mol Sci. 2018 Dec 11;19(12):3988. doi: 10.3390/ijms19123988. PMID: 30544955; PMCID: PMC6320919.

40 Grosso G, Laudisio D, Frias-Toral E, Barrea L, Muscogiuri G, Savastano S, Colao A. Anti-inflammatory nutrients and obesity-associated metabolic-Inflammation: State of the Art and Future Direction. Nutrients. 2022 Mar 8;14(6):1137. doi: 10.3390/nu14061137. PMID: 35334794; PMCID: PMC8954840

41 Giugliano D, Ceriello A, Esposito K. The effects of diet on inflammation: emphasis on the metabolic syndrome. J Am Coll Cardiol. 2006 Aug 15;48(4):677–85. doi: 10.1016/j.jacc.2006.03.052. Epub 2006 Jul 24. PMID: 16904534.

42 Sima Balaprasad Jaju, Digambar G. Dipankar, Almas M. Shaikh, Anjali Rajesh Pote, Anupama M. Bathe. Ayurvedic Management of *Vata*rakata (Gout) – A Case Report. Research Journal of Pharmacy and Technology. 2022; 15(11):5026–0. doi: 10.52711/0974-360X.2022.00845.

43 Shastri A. Sushrut Samhita: Sutrasthan: 24/19: 17/7,8,9; 19/32-33; 46/41,199,200,237,239,244,245,278;45/ 45,46,53,84,100,180,181,188,189,218,219,222,223,228: Varanasi: Chaukhamba Sanskrit Sansthan.

44 Tripathi B., editor (1st ed). *Charak Samhita;* Chikitsasthan, 12/ 12,14,24, 25; 30/291,202; 28/37; 12/13,14; 3/ 11,142,160,165,188,189,191,239,291,294: 21 / 108-113.7/82,83;13/96,98;5/110,112,133,134,164,166; 26/18;19/25;15/115;28/184,185 ;3/11,142,160,165,188,189,191,239,291,294: 12/ 60,61, 62,63; 15/53-54:Chaukhamba Surbharti Prakashan; *Varanasi.* 2013

45 Kale V.S. Charak Samhita, Sutrasthan, 27/11,16.19,21,22,23,26, 27,28,29,34,120,125,126,132,135,150,157,158,174, 179,181,184,185; 18/5,6,7,19-24,26,28-31; 27/ 91,181,184,185,220,229; Chaukhamba Sanskrit pratishthan,2014

46 GardeG,editor(1st ed)AshtanghridaySutrasthan5/22,23,26,31,32,33,34,35,36,57,58;14/8,11,21-24,33; 6/ 20,97,108,112,153-157,166; 5/25,34,58,70,71,74,82,83; 27/253; 2/17,18, 19,22,25,26,27,30: Chaukhamba Surbharti Prakashan; *Varanasi.* 2019

47 Kale V.S. Charak Samhita,; Chikitsa sthan 15 Chaukhamba Sanskrit pratishthan,2014

48 Sastri. B. Yogratnakar: Ama*Vata*dhikara: 28/81-82; Shotha chapter,1,2,3,4,5; Jwarchikitsa 1-5Varanasi: Chaukhamba Prakashan, 35,503,148,144,232,235,293,294

49 Aswathy YS, Anandaraman PV. Therapeutic influence of some dietary articles on gut microbiota in the pathogenesis of rheumatoid arthritis (*Amavata*) – A review. Ayu. 2019 Jul–Sep;40(3):147–151. doi: 10.4103/ayu.AYU_192_19. Epub 2020 Aug 8. PMID: 33281390; PMCID: PMC7685262

50 Rajagopala M, Gopinathan G. Ayurvedic management of papilledema. Ayu. 2015 Apr-Jun;36(2):177–9. doi: 10.4103/0974-8520.175545. PMID: 27011720; PMCID: PMC4784129.

51 Garde G, editor (1st ed) Ashtanghriday Chikitsasthan 17/ 9/10; 17/17-20; Chaukhamba Surbharti Prakashan; Varanasi, India. 2019.

52 Shastri A. Sushrut Samhita: Chikitsasthan: 22/65,56; 19/51; 23/12. Varanasi, India: Chaukhamba Sanskrit Sansthan, p. 127, 126.

53 Akhila VG. Management of *Sannipata Jwara* w.s.r to COVID-19 – Case report. J Ayurveda Integr Med. 2022 Jan–Mar;13(1):100416. doi: 10.1016/j.jaim.2021.02.007. Epub 2021 Mar 6. PMID: 33716425; PMCID: PMC7936553.

54 Shastri B, editor (1st ed.). Sarth Bhavprakash; Chaukhamba Sanskrit Bhawan 2018; *Varanasi.* 2019.

55 Shastri A. Sushrut Samhita: Chikitsasthan: 22/65,56; 19/51; 23/12 Varanasi: Chaukhamba Sanskrit Sansthan, p. 127, 126.

56 Sharma S., editor (1st ed). *Ashtang sangraha Samhita;* Sutrasthan: 7//16,17,132,147,195;:6/ 30,38,39,67,85,90,95,105 . Chaukhamba Surbharti Series; Varanasi; 2016; p. 50, 60, 62, 67, 39, 40, 41, 44, 45, 46, 47.

57 Sharma S., editor (1st ed). *Ashtang sangraha Samhita;* Chikitsasthan: 19/17. Chaukhamba Surbharti Series; *Varanasi*; 2016; page no 543.

Therapeutic Nutrition in Ayurveda for Gastrointestinal Disorders

Anagha Ranade and Pankaj Wanjarkhedkar

BACKGROUND

As per the Global Burden of Disease (GBD) study as reported in *Lancet* 2019, there is an urgent need for coordinated efforts in order to improve the dietary habits of people globally.[1]

People from across the world need sensitization for rebalancing their diet and preferring optimal quantities of various foods. The balanced diet concept especially from traditional medicine is reemerging as an essential factor on a large scale. This in fact is the pertinent time to nail the opportunity of creating awareness about therapeutic nutrition described in Ayurveda. The EAT-Lancet Commission has emphasized the consumption of plant-based diets more than animal foods.[2]

Total disease burden in India by cause mentioned in *Lancet* 2019: noncommunicable diseases amount to 5.77%, while 3.95% and 3.49% are contributed by digestive disorders and nutritional diseases, respectively.[3]

CONCEPT OF THERAPEUTIC NUTRITION IN CONVENTIONAL MEDICINE

The concept of therapeutic nutrition is nothing but a modification of the normal diets to one that is best suited per age, viz. infantile diet, toddler's diet, and diet for the adolescent, adult, and geriatric populations.[4] Further, the diet is also modified to make it suitable for patients with specific ailments. It is generally prescribed by physicians and dieticians in particular. The type of modification relates to the condition and need of the individual concerned, viz. liquid diets in the case of infants, bedridden patients, and the elderly are advised for proper digestion. The diet prescribed in the form of therapy is therapeutic nutrition.[5]

DOI: 10.1201/9781003345541-8

There are different categories of therapeutic diets that are often prescribed by dieticians, viz. normal hospital diet, clear liquid diet, full liquid diet, pureed diet, soft fiber restricted diet, connective tissue–containing diet, flavored diet, formula-based diets, etc. The mode of administration, dietary composition, energy intake, calorific value, and dietary component content play a major role in recommendations for a therapeutic diet.[6] Disease-specific diets include diabetic or calorie-controlled diet, low-cholesterol diet, renal diet, low-sodium diet, no-added-salt diet, low-fat diet, and high-fiber diet, which are commonly recommended in respective disorders.[4]

It is observed that modern dietetics lays an emphasis on individual components and energy intake, which is very crucial. Simultaneously, an approach takes into account the actual appetite and interpersonal differences, the role of diurnal variations, seasonal variations, the strength of the patient, and age, which are equally essential while planning the dietary regimen.

THE AYURVEDIC IDEOLOGY OF DIETETICS

In relation to the fundamentals of the physiology of the human body, Ayurveda bestows certain unique principles. Diet (*Ahara*) is an essential component of the three pillars (*Trayopastambha*) quoted by the seers.[7] The classical text, namely Charak Samhita, narrates the significance of health in four separate chapters (*Swasthya Chatushka*).[8] The overall maintenance of health is strongly associated with the quantity (*Matra*) of diet (*Ahara*).[9] An individual possesses sound health in terms of strength (*Bala*), luster, radiance (*varna*), vitality, and immunity (*Oja*) only after optimum consumption of food.[10] The quantity of diet to be consumed is closely dependent upon *Bala* (strength) of *Agni* (metabolism).[11]

The consumption of *Shadrasa* (a diet possessing foods with six tastes, viz. sweet, sour, salty, pungent, bitter, and astringent) has been advised. Any imbalance in the consumption of *Rasa*-dominant *Ahara* leads to pathological conditions and aging.[12]

THERAPEUTIC NUTRITION FOR THE GASTROINTESTINAL SYSTEM – MODERN PERSPECTIVE

Gastrointestinal health has been witnessed to be altered recently due to a transition from traditional dietary practices to modern diet regimens.[13] In order to combat metabolic dysfunction, single-nutrient interventions such as fortifying milk with vitamin D, cereals with iron, and customizing dietary intake of fat, protein, and carbohydrates are the possible solutions. Encouragement has been given toward the consumption of the Mediterranean diet, which comprises a high proportion of fruits and vegetables, legumes, whole grains, fish, and poultry. It also lays an emphasis on monounsaturated fats and antioxidants, unlike the Western-style diet. The latter is characterized by energy-dense foods like butter, high-fat dairy products, refined grains, and processed meat. It lacks fresh vegetables and fruits. The essentiality of the Mediterranean diet has been reported in several epidemiological studies.[14] Dietary fiber plays an important role in maintaining gut health.[15]

Recently, a lot of attention has been paid to the role of gut microbiota in health and disease. The emerging studies have led to the paradigm shift of researchers' interest from the contemporary theories of individual food components to a systems-based approach for decoding gut health.[16] The discovery of numerous data on the role of prebiotics and probiotics has been seen and documented in various studies.[17]

THERAPEUTIC NUTRITION FOR THE GASTROINTESTINAL SYSTEM – AYURVEDA PERSPECTIVE

Ayurveda emphasizes a lot on maintaining gut health, which is the *Mahasrotasa* comprising the alimentary canal.[18] The physiology of digestion has been explained in terms of *Avasthapaka,* which is later followed by actual assimilation and biotransformation. Homeostasis is maintained through adequate ingestion of food and this has been strongly associated with *Matra* (quantity) of *Ahara,* i.e. diet.[19] This appropriate quantity is a further decision based on the strength of *Agni,* i.e. metabolic energy. *Agni* along with *Pitta* is envisaged to be the imperative factor for digestion and metabolism in the body. Seers state that any dysfunction in the metabolic energy is responsible for the exhibition of pathology in the human body due to the production of *Ama* (intermediate residue of assimilation).[20] *Ama* is considered to be the primary etiological factor in exhibiting the symptoms of various disorders, thus referring to the disease as *Amaya,* i.e. originating from *Ama.*[21] Physiologically, *Ama* is daily produced in the body as a part of the waste product/underutilized portion which is otherwise excreted out of the body if digestion and other metabolic processes take place. When there is a malfunction in the whole of this metabolic orchestra, there is a disruption in the equilibrium of metabolic energy, finally leading to pathology.[22] The ill effects are further evident in different systems of the body that are already susceptible, firstly starting from the digestive system, i.e. *Mahasrotasa.*

According to the site of *Ama,* various system-specific, organ-specific symptoms are presented by our body culminating in respective pathologies. Pertaining to the digestive system, a prime cause of *Ama* formation is a poor diet inclusive of overeating; improper timings of food consumption; faulty food combinations; and consumption of excessively fried, heavy-to-digest, sweet-predominant foods coupled with detrimental habits like sedentary lifestyle, emotional causes, and sleep derangement.[23] The diet/food is at the epitome of the normal functioning of *Agni.*

There is a number of ways to aid the human body to digest *Ama* and eliminate it from the body. It supports the body's innate, physiological detoxification process, *Deepana, Pachana* being one of them.[24]

Diet is regarded to be the best medicine by *Kashyap* which itself refers to the origin of the concept of therapeutic nutrition.[25] Managing and establishing equilibrium of this metabolic energy is the ultimate aim in Ayurveda through diet and other therapy.

In the case of gastrointestinal health maintenance, the concept of *Deepana* and *Pachana* is very essential. Seers like *Hemadri* have delineated the clarity in this concept by stating that *Pachana drugs* participate in improving assimilation and rendering digestion of intermediate residues, whereas *Deepana* (appetizers) drugs aid in the separation of *Dosha* from *Dhatu* and thus facilitate tissue metabolism.[26] From a physiological point of view, *Samana Vayu* and *Agni* in communion participate in the *Deepana Pachana* process.[24] Yet, decoding of this process at the biochemical and gut-brain axis level still remains. Efforts have been made to denote the links and implications in the later part of the chapter.

ROLE OF *GRAHANI*

Grahani chapter of *Charak Samhita* focuses on the complete physiology of digestion and pathologies arising due to its malfunction. *Grahani* itself is considered to be a seat of *Pittadhara kala* and possesses a main role in the *Pachana* of *Ahar*. It can be anatomically correlated to the lower portion of the pyloric end of the stomach and the duodenum, which are typically coincidental with essential sites of enzyme production and also the commensal gut microbes.[18]

The digestion process in this chapter has been divided into two types. One is a temporary, reversible process comprising three *Avasthapaka.* This is the site wherein *Sara-Kitta Vibhajan* occurs and the essence of food is further passed on to five *Bhutagnis* (liver first-pass metabolism) and later to *Dhatvagnis* (tissue metabolism), which is a sequential process starting from *rasa dhatu (~ plasma/lymph*)to ultimately *Shukra*(reproductive tissue). Simultaneously, it is quoted that waste products are eliminated through the body in the form of urine, feces, and perspiration. *Bhutagni* functionality deals with selective absorption and assimilation from the gut mucosa. *Avasthapaka* thus can form a part of the mechanical and biochemical digestion process in human physiology. After macrodigestion, *Vipaka role* is pitched in as Chakrapani conspicuously states that this refers to the postdigestion biotransformation process, which is an irreversible metabolism in the gut.[27] The concept exactly coincides with the substantial and specialized role of gut microbiota in metabolizing macronutrients, namely carbohydrates, lipids, and proteins, which is favorable in cellular nutrition.

This depicts the potential role of diet (*ahara*) and its digestion in every bodily function.

Malfunction of *Grahani* typically exhibits loss of its normal function of supporting the partially metabolized food within and keeping a check over proper propulsion and transfer of the metabolized part ahead in the gut. There is a typical term for a malfunction that a person having *Grahani* disease can never fully metabolize the food and successfully excrete normal feces, thus resulting in the production of *Ama* as referred above. [28]

दुर्बलो विदहत्यन्नं तद्यात्यूर्ध्वमधोऽपि वा॥५१॥
अधस्तु पक्वमामं वा प्रवृत्तं ग्रहणीगदः।
उच्यते सर्वमेवान्नं प्रायो ह्यस्य विदह्यते॥५२॥
अतिसृष्टं विबद्धं वा द्रवं तदुपदिश्यते

(Charak Chikitsasthana chapter 15/51-52)

The consistency of stools keeps on changing along with gastric problems -*Pakvapakvam Srujati* followed by *Aasya Vairasya, Arochak* (loss of taste), *Agnimandya,* i.e. loss of appetite, *Aam Gandhi Udgar* (sour belching). Grahani also has *manasik lakshanas.*[29] It coincides with the irritable bowel syndrome , whose management also targets gut health, i.e. modulation of the gut microbial consortium.[30]

Agnimandya, Ajirna, all of these come under the ambit of *Grahani.*[22] *Ama* is considered to be the primary etiological factor in exhibiting the symptoms of these disorders. The ill effects are further evident in different systems of the body, which are already susceptible firstly starting from the digestive system, i.e. *Mahasrotasa.* The verse is summarized as:

दुष्यत्यग्निः, स दुष्टोऽन्नं न तत् पचति लघ्वपि।
अपच्यमानं शुक्तत्वं यात्यन्नं विषरूपताम् ॥
दुर्बलाग्निबलो दुष्टा त्वाममेव विमुञ्चति ।[31]

According to the site of *Ama,* various system-specific, organ-specific symptoms are presented by the body culminating in respective pathologies.[29]

Pertaining to the digestive system, a prime cause of *Ama* formation is a poor diet inclusive of overeating, improper timings of food consumption, faulty food combinations, and consumption of excessively fried, heavy-to-digest, sweet-predominant foods coupled with detrimental habits like sedentary lifestyle, emotional causes, and sleep derangement.

PATHOPHYSIOLOGY OF *AMLA PITTA* (GASTROESOPHAGEAL REFLUX DISEASE [GERD])

Charak has clearly mentioned that *Ajirna* is the root cause of the manifestation of *Amla Pitta,* which typically exhibits symptoms like *Daha* (heart burns), nausea, and relief after emesis. It typically relates to the deranged function of *Pitta* at the *Pitta Dhara Kala* level of *Grahani,* leading to an increase in hyperacidity and acid reflux which further aggravates the integrity of the mucosal lining of the esophagus, causing recurrent symptoms. The disease with a similar presentation, i.e. GERD, also has its roots in the disturbance of gut health.[32] There are reports of change in the gut microbial species in people with GERD comprising viz. *Prevotella, Helicobacter,* and *Moraxella* genera in particular.[33] Same type of symptomatology is associated with *Adhoga Amla pitta,* which comprises of burning sensation, giddiness, and development of skin rashes due to aggravated *pitta.*[34] The treatment protocol completely resides in improving gut health by planning induced emesis, a certain type of oil, and decoction enemas. It directs the potential role of gut microbiota in *Amla Pitta* too.

The information on generalized dietary protocols has been mentioned in all the Ayurveda treatises, namely *Charak Samhita, Sushruta Samhita, Ashtanga hridaya,* and *Ashtanga Sangraha.*

Following these, compendia like *Bhasishyajya ratnavali,*[35] Basavarajiyam,[36] and *Bhavprakash*[37] *Samhita* have separately mentioned disease-specific *Pathya* (dietary recommendations). In the case of Global Disease Burden data of 2019, diarrheal diseases in gastrointestinal disorders have been ranked among the top ten list of leading causes of health loss across the world.[38] The Ayurveda counterparts for the same include *Ajirna* (indigestion), *Agnimandya* (deranged function of metabolic energy), *Atisara* (diarrhea), *Amlapitta* (GERD), and *Grahani* (sprue/malabsorption syndrome). The following sections of this chapter will refer to the dietary recommendations for the above diseases.

APPETITE REGULATION IN MODERN PARLANCE

It is attributed to the hypothalamus, which acts as a guiding center for hunger and satiety. Hormones, namely neuropeptide Y (NPY) and agouti-related protein (AgRP), are involved in appetite stimulation. Neuropeptide Y (NPY) is reported to initiate appetite through G-protein-coupled receptors (GPCRs). AgRP is a potent orexigenic peptide involved in appetite stimulation which is facilitated by Ghrelin, which is secreted by endocrine cells of the gastric mucosa.[39] Hunger and satiety are key contributors to maintaining the metabolic health of an individual; any disruption of appetite control results in the development of various pathologies. It is in line with the Ayurvedic principle of *Agni* and dysfunction contributing to *Agnimandya.*[40] Recently, it has been reported that the gut microbiota possess a role not only in receiving energy from the host for maintaining equilibrium but also in supplying the host with energy via varied metabolites, viz. short-chain fatty acids (SCFAs), amino acids, bile acids, caseinolytic protease B, and lipopolysaccharides (LPSs).[41]

Dipana function is primly attributed to *Samana vata* (neurological factors of the gut) along with the *Agni* component. Thus, *Deepana* activity is probably closely associated with the enhancement of positive signaling of ghrelin and in turn AgRP. A study has been carried out in Japanese medicine named *Rikkunshito,* which exhibited ghrelin signal promotion and secretion effects. This has been considered to be positive in the management of hypophagia in the elderly and cachexia in cancer.[42] The herbs of this formulation include zingiber, glycyrrhiza, and aurantii. A similar *in vitro* study has been carried out to evaluate the orexigenic effect of Japanese medicine.[43] This is an upcoming area wherein some herbs from the *Deepaniya* group can be tested.

Pachana activity may be correlated with the feedback mechanism consisting of regulation of appetite, metabolism, and energy expenditure. Energy metabolism is a key interface for interaction between microbiota and host because the gut microbiota receives energy from the host for normal growth maintenance purposes. They also are reported to participate in supplying the host with energy by releasing a range of enzymes and metabolites, viz. SCFAs, amino acids, bile acids, and LPSs.[44] The role of gut microbiota is crucial in *Pachana Karma* as they serve as an interface to deliver certain metabolites that aid in exerting their effects through direct interaction with receptors present in the gut on enteroendocrine L cells or the Vagus nerve. It also participates in delivering metabolites for translocating through the intestinal epithelium into the peripheral circulation. To date, SCFAs are the most studied components for assessing the gut microbiota and host interactions related to appetite.[45]

Time-restricted eating (TRE) strategy: This is a new area where scientists have observed promising results by adopting a dietary approach limiting the daily eating window, which is very similar to the Ayurvedic concept of *Kalabhojana.*[46] Untimely eating or not abiding by the usual meal timings, i.e. *Atita Kala Bhojana, Samashan,* and *Vishamashan results in manifestation of disease.*[47]

DIETARY RECOMMENDATIONS FOR COMMON GIT DISORDERS

This section will focus on *Ajirna* (indigestion), *Agnimandya* (deranged function of metabolic energy), *Atisara* (diarrhea), *Amlapitta* (GERD), and *Grahani* (sprue/malabsorption syndrome).

1. ***Ajirna*** (indigestion):

Ajirna has been vividly described in compendia like Basavarajiyam,[36] *Bhaishyajya Ratnavali,*[35] and *Bhavprakash.*[37] Dysfunction of *Agni* (metabolic energy) is referred to as the main cause for the development of *Agnimandya* which disturbs the digestion and manifestation into different types of *Ajirna* (indigestion) depending on the involvement of *Dosha.*

a) *Pathya* in *Ajeerna* (dietary recommendations):

SR.NO	*CATEGORY*	*FOOD ITEMS*	*REFERENCE*
1.	**Cereals**	Porridges and soups prepared from harvested old rice, green gram, and red rice have been quoted to be effective in restoring metabolic energy.	Basavarajiyam,[36] Bhaishyajyaratnavali.[35]
2.	**Vegetables**	Preparations of Chenopodium leaves, Tender radish, Garlic, Raw banana, follicles of Drumstick, pointed gourd/parwal, brinjal, lotus root, spine gourd, bitter gourd, the fruit of *Bruhati, Changeri* (wood sorrel) leaves, *Sunishannak* (water clover), fenugreek leaves.	Basavarajiyam, Bhaishajyaratnavali.
3.	**Fruits**	Gooseberry, pomegranate, *Jambira* (lemon variety), Matulunga (citron variety)	Basavarajiyam, Bhaishajyaratnavali.
4.	**Beverages**	Buttermilk, *tushodaka* (fermented gruel made up of Barley), *dhanyamla* (fermented preparation of grains)	Basavarajiyam, Bhaishajyaratnavali.
5.	**Others**	*Navneet* (Butter), *Ghrita* (clarified butter), *Dadhi* (curd), honey, *taptasalila*, *tambula* (betel leaf),	Basavarajiyam, Bhaishajyaratnavali.
6.	**Spices**	- Carom seeds, pepper, coriander seeds, cumin seeds - Ginger and rock salt before meals	Basavarajiyam, Bhaishajyaratnavali.

b) *Apathya* in *Ajeerana* (dietary contraindications):

SR.NO	*CATEGORY*	*FOOD ITEMS*	*REFERENCE*
1.	**Pulses**	Consumption of all preparations of pulses is contraindicated in diet.	Basavarajiyam[36]
2.	**Vegetables and fruits**	*Upodaka*, starchy roots, Jamun fruit,	Basavarajiyam
3.	**Others**	Dairy products like *Kurchika, Kilata,* and *Morat* which are similar to *Chhena*, *paneer*, and contaminated water.	Basavarajiyam

2. *Atisara* (diarrhea):

a) *Pathya* (dietary recommendations):

SR.NO	CATEGORY	FOOD ITEMS	REFERENCE
1.	**Cereals**	Soups and porridges of old harvested rice	Bhaishyajyaratnavali[35]
1.	**Pulses**	*Tuvar daal* soup, *Masoor daal* soup	Basavarajiyam,[36] Bhaishajyaratnavali
2.	**Vegetables**	Pointed gourd, leaves of *Kutaja, Chukrika*, curry leaves, *Karanja* leaves, Chenopodium, flowers of banana	Basavarajiyam, Bhaishajyaratnavali
3.	**Fruits**	Wood apple, *Bael* fruit, banana fruit *Jamun*, Dilenia fruit, *Vikankata, Tinduka, Talaka, Jatiphal*	Basavarajiyam, Bhaishajyaratnavali
4.	**Others**	Curd from sheep milk, oil Goat milk	Basavarajiyam, Bhaishajyaratnavali

b) *Apathya* (dietary contraindications):

SR.NO	CATEGORY	FOOD ITEMS	REFERENCE
1.	**Cereals**	Wheat	Bhaishyajyaratnavali
2.	**Pulses**	*Nishpav*	Basavarajiyam
		Masha, *Yava*	Bhaishajyaratnavali
3.	**Vegetables**	*Kushmanda* (ash gourd), *Karkotaki* (spine gourd), *Upodaki*, *Alabu* (bottle gourd)	Basavarajiyam, Bhaishajyaratnavali
		Chenopodium, *Kakmachi*, drumstick, leafy vegetables	
4.	**Fruits**	Ash gourd, grapes, gooseberry	Bhaishajyaratnavali
5.	**Others**	Meat, jaggery, alcohol, cold infusions, *tambul* (betel leaf), mastu, coconut water	Basavarajiyam

3. *Grahani* (sprue):

a) *Pathya* (dietary recommendations):

SR.NO	CATEGORY	FOOD ITEMS	REFERENCE
1.	**Cereals**	Old harvested rice, gruel of *Shashti* rice variety	Bhaishajyaratnavali
2.	**Pulses**	*Moong dal, Masoor dal, Tuvar dal.*	Bhaishajyaratnavali
3.	**Fruits**	Banana, *Bael* fruit, *Shringataka*, wood apple, *Jatiphal*, Dilenia, pomegranate, Pelavam (apple)	Bhaishajyaratnavali
4.	**Others**	Fish, sesame oil	Bhaishajyaratnavali

b) *Apathya* (dietary contraindications):

SR.NO	CATEGORY	FOOD ITEMS	REFERENCE
1.	**Cereals**	Wheat	Bhaishajyaratnavali
2.	**Pulses**	*Nishpav* (flat beans), *Kalay* (dry peas), *Masha*, *Yava* (barley), *Rajmasha* (kidney beans)	Bhaishajyaratnavali
3.	**Vegetables**	*Chhatrak* (mushroom), *Upodika*, Chenopodium, *Kakmachi*, drumstick, leafy vegetables, *tumbi*, garlic	Bhaishajyaratnavali
4.	**Fruits**	Jujube, ash gourd, *Brihat phal*, grapes, Areca nut, sugarcane	Bhaishajyaratnavali
5.	**Beverages**	*Tushodaka* (fermented gruel made up of Barley), *Dhanyamla* (fermented preparation of grains), *Mastu*, coconut milk.	Bhaishajyaratnavali

4. *Amlapitta* (GERD):

a) *Pathya* (dietary recommendations):

SR.NO	CATEGORY	FOOD ITEMS	REFERENCE
1.	**Cereals**	Wheat	Bhaishajyaratnavali
2.	**Pulses**	Barley, green gram	Bhaishajyaratnavali
3.	**Vegetables**	Banana flower, Chenopodium, spine gourd, bitter gourd, point gourd,	Bhaishajyaratnavali
4.	**Fruits**	Wood apple, pomegranate, gooseberry, ash gourd	Bhaishajyaratnavali

b) *Apathya* (dietary contraindications):

SR.NO	CATEGORY	FOOD ITEMS	REFERENCE
1.	**Cereals**	Newly harvested	Bhaishajyaratnavali
2.	**Pulses**	Masha, Kulathha	Bhaishajyaratnavali
3.	**Others**	Oil, curds, sheep milk, sour substances	Bhaishajyaratnavali

DIETARY IMPLICATIONS IN *AJIRNA* AND ITS POSSIBLE LINKS WITH GUT MICROBIAL FUNCTIONS

The gut microbial dynamics are influenced by lifestyle and dietary factors in human health. The different types of *Ajirna* have been denoted based on dysfunctions of peculiar *Dosha*. It is known that every person possesses functionally different types of metabolic energy, which get hampered due to diet and lifestyle factors aggravating the respective *Dosha*.[48] In *Pathya,* it is observed that certain specific pulses are described. In participants administered with pea protein diet, when analyzed, it was revealed that it results in the rise of gut-commensal *Bifidobacterium* and *Lactobacillus*. Additionally, the reduction has been reported in the pathogenic strains viz. *Bacteroides fragilis* and *Clostridium perfringens*.[49] The intestinal SCFA levels were observed to be elevated, which are considered anti-inflammatory and essential for the maintenance of the mucosal barrier.[50] Short-chain fatty acids are of utmost importance in gut health. Studies have reported that high-protein/low-carb diet intake may aid in weight loss but ultimately is detrimental to health as the butyrate SCFAs are less and microbial communities of *Roseburia* and *Eubacterium rectale* are also decreased, thereby increasing the risk of developing IBD in the future.

The *Pathyas* for GIT, carbohydrate gruels have been prescribed more along with vegetables and fruits, i.e. dietary fiber, which is nutritive for gut microbiota. The plant protein is highly recommended in *Pathya* in classical literature; very meager items fall under the category of *Mamsa Varga.* Starches and sugars, namely glucose, fructose, sucrose, and lactose, are under the category of digestible carbohydrates. Glucose fed in the form of date fruits and grapes (which are in fact cited as *Pathya*) increases the relative abundance of *Bifidobacterium*. The addition of lactose decreases the detrimental clostridia species. It positively adds to the validity of the prescription of certain specific substances in *Dugdha Varga* and *Phala Varga,* wherein rationale can be further sought by planning experiments.

DIETARY IMPLICATIONS IN *ATISARA* (DIARRHEA) AND ITS POSSIBLE LINKS WITH GUT MICROBIAL FUNCTIONS

A poor diet ensues a subclinical chronic inflammatory state in the gut. The gut microbiota plays a major role in maintaining gastrointestinal health as they can resist the colonization of pathogenic bacteria responsible for various diseases including diarrhea. Studies report that proper diet facilitates modulating the commensals, thereby improving their functions and reducing diarrheal diseases.[51] It is observed that the *Pathya* prescribed contains lentils/pulses, which are reported to possess a high concentration of proteins and also micronutrients, viz. folates, iron, zinc, selenium, and carotenoids, which are essentially lost in dehydration due to diarrhea.[52] The lentil varieties, namely *Masur,* are used frequently in the Indian diet and are reported certainly to possess a role as prebiotics.[53] It supports commensal bacterial health. Fruits and vegetables are sources of prebiotics, particularly dietary fibers facilitating SCFA production. Additionally, it helps in replenishing electrolytes and vitamin loss in diarrhea. Still, the evidence must be generated using the specific dietary recommendations mentioned above for validating concepts.

DIETARY IMPLICATIONS IN *GRAHANI* (~IRRITABLE BOWEL SYNDROME) AND ITS POSSIBLE LINKS WITH GUT MICROBIAL FUNCTIONS

The human gastrointestinal tract possesses its own innate immunity.[54] The commensal flora themselves contribute to these defenses by producing antimicrobial factors to limit the entry of pathogenic bacteria. The gut flora also produces SCFAs, thereby reducing the inflammatory markers. The dietary fibers and proteins/peptides have a role in modulating SCFAs, thereby strengthening host response.[55]

DIETARY IMPLICATIONS IN *AMLAPITTA* (GERD) AND ITS POSSIBLE LINKS WITH GUT MICROBIAL FUNCTIONS

People having longstanding dietary intake similar to those in urban areas have been shown to possess chronic inflammation of the esophagus. The highly acidic environment in GERD perpetuates gut dysbiosis, which may further favor the development of pathogenesis. The reason for the induction of gut dysbiosis and GERD is that these patients with low dietary fiber content have a decreased mucus growth rate in the intestinal tract that hampers intestinal permeability.[56] Studies were done on the effect of probiotics in patients who consumed proton pump inhibitors for a very long time.[57] It is already reported that prolonged use of antacids and acid-suppressing medicines leads to hypersensitivity, alkalosis, calculi, and constipation. This is not an apt way to address dyspepsia as it originates due to faulty dietary habits. Proton pump inhibitors are not able to provide the solution as this approach is deficient in managing aggravated *pitta.*[58] Ayurveda believes in refraining from the known etiologies as the first line of treating any pathology. Thus, it is high time to adopt proper dietary recommendations from Ayurveda classics and test them for their effects in arresting GERD.

NUTRAVIGILANCE

Cautious consumption of dietary items in order to safeguard our innate metabolic capacity is the most significant point of vigilance in the classics of Ayurveda, viz. *Sharangdhar Samhita, Bhavprakashasamhita, Madhav Nidana,* and *Yogaratnakara*, which are related to nutrition. *Bhavprakash*[37] *and Bhaishyajyaratnavali*[35] have uniquely mentioned certain substitute dietary items in order to overcome the ill effects of certain foods. As an illustration, in order to overcome indigestion caused by jackfruit and to aid its smooth metabolism, the consumption of bananas has been advocated. Similarly, side effects of the consumption of bananas can be met by administering ghee to the patient and so on. This is also a part of the vigilant approach to dietary recommendations. Classical literature comprises chapters that provide exclusive information on the pros and cons of dietary items that are used in consumption as *Agryadravya*.[59] The lexicons, i.e. nighantus, entail a description of the properties of drugs and food items wherein a vivid knowledge about the harmful effects of excessive consumption has been mentioned, starting from specific cereals, pulses, dairy products, and fruits.[60–62] Pertaining to grains, excerpts suggest that rice harvested in unusual seasons, infested with disease, immature in nature, and in the germinated stage has been mentioned to cause indigestion.[63]

Viruddha Ahara (food incompatibility) is yet another unique concept of Ayurveda which states that regular consumption of such harmful combinations amounts to all the basic pathologies that create *Agnimandya* (diminishing metabolizing power), resulting in *Ama,* thereby culminating in gastrointestinal disorders.[64] These could induce inflammation at a molecular level, disturbing the eicosanoid pathway and creating more arachidonic acid, leading to increased prostaglandin-2 and thromboxane.[65]

Thus, coming to the concluding part, it is clearly evident that the seers of Ayurveda have laid a strong emphasis on the essentiality of apt dietary habits. Adopting these classical recommendations can surely cater to the unmet needs for clinical nutrition in the disorders of the gastrointestinal tract. Generating evidence in this line will surely pave the way for alterations in policy wherein therapeutic nutrition from Ayurveda will emerge as a priority.

REFERENCES

1. Anonymous. The Lancet. "Globally, one in five deaths are associated with poor diet." Science Daily. Science Daily, 3 April 2019. www.sciencedaily.com/releases/2019/04/190403193702.htm.
2. Max Roser, Hannah Ritchie and Fiona Spooner. "Burden of disease". Published online at OurWorldInData.org. 2021. https://ourworldindata.org/burden-of-disease [Online Resource].
3. Willett W, Rockström J, Loken B, Springmann M, Lang T, Vermeulen S, Garnett T, Tilman D, DeClerck F, Wood A and Jonell M. Food in the Anthropocene: the EAT-Lancet Commission on healthy diets from sustainable food systems. The Lancet. 2019 Feb 2; 393(10170): 447–92.
4. Anonymous. California Department of Social Services. In-home Supportive Services. Types of Therapeutic Diets. www.cdss.ca.gov/agedblinddisabled/res/VPTC2/9%20Food%20Nutrition%20and%20Preparation/Types_of_Therapeutic_Diets.pdf.
5. Mchiza ZJ. Diet Therapy and Public Health. Int J Environ Res Public Health. 2022;19(14):8312. doi: 10.3390/ijerph19148312. PMID: 35886174; PMCID: PMC9321782.
6. Mudambi SR, Rajgopala MV. Fundamentals of Food, Nutrition and Diet therapy. Chapter 23, 24. 5th edition. New Age international publishers.2007. pp 257–274.
7. Acharya, J. T. (ed.), Charak Samhita with the Ayurved Dipika commentary, Sutrasthana, Chapter 11, Verse 35, Chaukhamba Krishnadas Academy, Varanasi, 2010, p. 74.
8. Acharya, J. T. (ed.), Charak Samhita with the Ayurved Dipika commentary, Sutrasthana, Chapter 30, Verse 36, Chaukhamba Krishnadas Academy, Varanasi, 2010, p. 190.

9. Acharya, J. T. (ed.), Charak Samhita with the Ayurved Dipika commentary, Vimanasthan, Chapter 2, Verse 4-6, Chaukhamba Krishnadas Academy, Varanasi, 2010, p. 238.
10. Dwivedi, L. (ed.), Charak Samhita, commentary of Chakrapani, Chikitsasthana, Chap 15, verse 5, Chaukhambha Prakashana, Varanasi, 2013, 1st edn, p. 510.
11. Dwivedi, L. (ed.), Charak Samhita, commentary of Chakrapani, Sutrasthana Chap 5, verse 3, Chaukhambha Prakashana, Varanasi, 2013, 1st edn, p. 127.
12. Acharya, J. T. (ed.), Charak Samhita with the Ayurved Dipika commentary, Sutrasthana, Chapter 29, verse 23, 24 and 42, Chaukhamba Krishnadas Academy, Varanasi, 2010, pp. 179–181.
13. Sproesser G, Ruby MB, Arbit N, Akotia CS, Alvarenga MD, Bhangaokar R, Furumitsu I, Hu X, Imada S, Kaptan G, Kaufer-Horwitz M. Understanding traditional and modern eating: the TEP10 framework. BMC Public Health. 2019 Dec;19(1):1–4.
14. De Lorgeril M, Renaud S, Mamelle N, Salen P, Martin JL, Monjaud I, et al. Mediterranean alpha-linolenic acid-rich diet in secondary prevention of coronary heart disease. Lancet. 1994;343(8911):1454–9.
15. Chen Y, Michalak M, Agellon LB. Importance of Nutrients and Nutrient Metabolism on Human Health. Yale J Biol Med. 2018 Jun 28;91(2):95–103. PMID: 29955217; PMCID: PMC6020734.
16. Krumbeck JA, Maldonado-Gomez MX, Ramer-Tait AE, Hutkins RW. Prebiotics and synbiotics: dietary strategies for improving gut health. Curr Opin Gastroenterol. 2016;32:110–9. doi: 10.1097/MOG.0000000000000249.
17. Krumbeck JA, Maldonado-Gomez MX, Martinez I, Frese SA, Burkey TE, Rasineni K. et al. In vivo selection to identify bacterial strains with enhanced ecological performance in synbiotic applications. Appl Environ Microbiol. 2015;81:2455–65. doi: 10.1128/AEM.03903
18. Ranade A, Gayakwad S, Chougule S, Shirolkar A, Gaidhani S, Pawar SD. Gut microbiota: metabolic programmers as a lead for deciphering Ayurvedic pharmacokinetics. Current Science. 2020 Aug 10;119(3):451.
19. Arya Parvathy R, Haritha Chandran, Haroon Irshad, C Ushakumari, Leena P Nair, Utility of Ahara Matra w.r.t. Ashta Ahara Vidhi Vishesha Ayatanani. J Ayu Int Med Sci. 2022;7(1):319–324. https://jaims.in/jaims/article/view/1591
20. Manohar PR. Critical review and validation of the concept of Āma. Anc Sci Life. 2012 Oct;32(2):67–8. doi: 10.4103/0257-7941.118524. PMID: 24167329; PMCID: PMC3807959.
21. Thakur M, Ann Ambitia, Saini S and Manglesh R. Role of Ama in samprapti of various diseases. WJPMR. 2019;5(4):113–115.
22. Anonymous. Charak Samhita with Ayurved Dipika commentary, Chikitsasthana, Chapter 15, verse 38-41, e-Samhita Designed and Developed by National Institute of Indian Medical Heritage, Hyderabad, Central Council for Research in Ayurveda and Siddha (CCRAS), New Delhi. 2010. https://niimh.nic.in/ebooks/ecaraka/.
23. Sorathiya AP, Vyas SN, Bhat PS. A clinical study on the role of ama in relation to Grahani Roga and its management by Kalingadi Ghanavati and Tryushnadi Ghrita. Ayu. 2010 Oct;31(4):451–5. doi: 10.4103/0974-8520.82041. PMID: 22048538; PMCID: PMC3202250.
24. Choudhary K, Gupta N and Mangal G. Therapeutic impact of Deepana-Pachana in Panchakarma: an overview, IRJAY, 2021; 4(1): 252–258. doi: 10.47223/IRJAY.2021.4108.
25. Sharma, H. (ed.), Kashyapa Samhita, Khila Sthana, Chap 4, verse 6, Chaukhamba Sanskrit Sansthana, Varanasi, 2013, p 249.
26. Paradkar H. editor. Aṣṭanghṛdayam with Commentaries Sarvangasundara of Aruṇdatta and Ayurvedarasayana of Hemadri, Chaukhambha orientalia, Varanasi: Reprint 2017; Sutrasthana; chapter 13/ 29, p. 217.
27. Ranade AV, Shirolkar A, Pawar SD. Gut microbiota: One of the new frontiers for elucidating fundamentals of Vipaka in Ayurveda. Ayu. 2019 Apr;40(2):75.
28. Anonymous. Charak Samhita with Ayurved Dipika commentary, Chikitsasthana, Chapter 15, verse 44, e-Samhita Designed and Developed by National Institute of Indian Medical Heritage, Hyderabad, Central Council for Research in Ayurveda and Siddha (CCRAS), New Delhi. 2010. https://niimh.nic.in/ebooks/ecaraka/
29. Anonymous. Charak Samhita with Ayurved Dipika commentary, Chikitsasthana, Chapter 15, verse 45-49, e-Samhita Designed and Developed by National Institute of Indian Medical Heritage, Hyderabad, Central Council for Research in Ayurveda and Siddha (CCRAS), New Delhi. 2010. https://niimh.nic.in/ebooks/ecaraka/
30. Mazzawi T. Gut Microbiota Manipulation in Irritable Bowel Syndrome. Microorganisms. 2022;10(7):1332. doi: 10.3390/microorganisms10071332. PMID: 35889051; PMCID: PMC9319495.

31. Anonymous. Charak Samhita with Ayurved Dipika commentary, Chikitsasthana, Chapter 15, verse 56-57, e-Samhita Designed and Developed by National Institute of Indian Medical Heritage, Hyderabad, Central Council for Research in Ayurveda and Siddha (CCRAS), New Delhi. 2010. https://niimh.nic.in/ebooks/ecaraka/
32. Kawar N, Park SG, Schwartz JL, Callahan N, Obrez A, Yang B, Chen Z, Adami GR. Salivary microbiome with gastroesophageal reflux disease and treatment. Scientific reports. 2021 Jan 8;11(1):1–8.
33. Upadhayay Y. editor, Madhavanidana of Shrimahavakara with Madhukosa Sanskrit Commentary by Sri Vijayaraksita and Srikanthadatta with the Vidyotini Hindi Commentary Part-II, Chaukhambha Prakashan: Reprint 2014; verse 51/3 pp 203.
34. Dr G. Prabhakararao. Bhaisajyaratnavali of Kaviraj Shri Govind Das Sen English Translation and Shri Ramana Prabhakara Commentary, Vol-II, Reprint 2014, Chaukhambha Orientalia, Chapter 56, verse 156-159 pp. 373.
35. Lochan Kanjiv. Editor. Bhaishyajya ratnavali, Vol 1, Reprint 2006 Chaukhambha Sanskrit Sansthan Varanasi, Chap 8/617-625, pp573–574; Chap 10/9-13, pp 632–637; Vol Chap 56/156-161, pp137–138.
36. Narayanan Ala. Editor. Basavarajiyam. First edi 2013, NIIMH, Hyderabad Unit of CCRAS, New Delhi, Chap 10/135-138, pp389; Chap 12/224-236, pp445.
37. Sitaram Bulusu. Editor. Bhavprakash. First edition 2010. Chaukhambha Ayurved Pratisthan, Varanasi. Chap 6/128-145, pp156–157.
38. Owen D, Apfel A. The Lancet: Latest global disease estimates reveal perfect storm of rising chronic diseases and public health failures fuelling COVID-19 pandemic. Institute for Health Metrics and Evaluation. 2020.
39. Austin J, Marks D. Hormonal regulators of appetite. Int J Pediatr Endocrinol. 2009;2009:141753. doi: 10.1155/2009/141753. Epub 2008 Dec 3. PMID: 19946401; PMCID: PMC2777281.
40. Anonymous. Charak Samhita with Ayurved Dipika commentary, Sutrasthana, Chapter 21, verse 5-10, e-Samhita Designed and Developed by National Institute of Indian Medical Heritage, Hyderabad, Central Council for Research in Ayurveda and Siddha (CCRAS), New Delhi. 2010. https://niimh.nic.in/ebooks/ecaraka/
41. Fetissov SO. Role of the gut microbiota in host appetite control: bacterial growth to animal feeding behaviour. Nat Rev Endocrinol. 2017;13:11–25.
42. Yamada C, Hattori T, Ohnishi S, Takeda H. Ghrelin enhancer, the latest evidence of Rikkunshito. Front Nutr. 2021 Dec 9;8:761631. doi: 10.3389/fnut.2021.761631. PMID: 34957179; PMCID: PMC8702727.
43. Miyano K, Ohshima K, Suzuki N, et al. Japanese herbal medicine Ninjinyoeito mediates Its orexigenic properties partially by activating orexin 1 receptors. Front Nutr. 2020;7:5. doi: 10.3389/fnut.2020.00005. PMID: 32175325; PMCID: PMC7056666.
44. Han H, Yi B, Zhong R, Wang M, Zhang S, Ma J, Yin Y, Yin J, Chen L, Zhang H. From gut microbiota to host appetite: gut microbiota-derived metabolites as key regulators. Microbiome. 2021 Jul 20;9(1):162. doi: 10.1186/s40168-021-01093-y. PMID: 34284827; PMCID: PMC8293578.
45. Marcel van de Wouw, Harriët Schellekens, Timothy G Dinan, John F Cryan, Microbiota-Gut-Brain Axis: Modulator of Host Metabolism and Appetite, The Journal of Nutrition, 2017; 147(5): 727–745, https://doi.org/10.3945/jn.116.240481.
46. Anonymous. Charak Samhita with Ayurved Dipika commentary, Sutrasthana, Chapter 25, verse 40, e-Samhita Designed and Developed by National Institute of Indian Medical Heritage, Hyderabad, Central Council for Research in Ayurveda and Siddha (CCRAS), New Delhi. 2010. https://niimh.nic.in/ebooks/ecaraka/
47. Schuppelius B, Peters B, Ottawa A, Pivovarova-Ramich O. Time restricted eating: A dietary strategy to prevent and treat metabolic disturbances. Front Endocrinol (Lausanne). 2021;12:683140. doi: 10.3389/fendo.2021.683140. PMID: 34456861; PMCID: PMC8387818.
48. Saini N, Pal PK, Byadgi PS. Critical analysis of etiological factors of Ajirna (indigestion). Int J Complement Alt Med, 2017; 5(1): 00141. DOI: 10.15406/ijcam.2017.05.00141.
49. Świątecka D, Dominika Ś, Narbad A, Arjan N, Ridgway KP, Karyn RP, et al. The study on the impact of glycated pea proteins on human intestinal bacteria. Int J Food Microbiol. 2011;145:267–72.
50. Kim CH, Park J, Kim M. Gut microbiota-derived short-chain fatty acids, T cells, and inflammation. Immune Netw. 2014;14:277. http://synapse.koreamed.org/DOIx. php?id=10.4110/in.2014.14.6.277.
51. Sonnenburg J. L., Bäckhed F. (2016). Diet-microbiota interactions as moderators of human metabolism. Nature 535 (7610), 56–64. 10.1038/nature18846.

52. Thavarajah D, Thavarajah P, Wejesuriya A, Rutzke M, Glahn RP, Combs GFJr, Vandenberg A. The potential of lentil (*Lens culinaris* L.) as a whole food for increased selenium, iron, and zinc intake: Preliminary results from a 3 year study. Euphytica. 2011; 180 (1), 123–128.
53. Johnson CR, Thavarajah D, Combs Jr. GF, Thavarajah P. Lentil (*Lens culinaris* L.): A prebiotic-rich whole food legume. Food Research International. 2013; 51(1), 107–113.
54. Wu HJ, Wu E. The role of gut microbiota in immune homeostasis and autoimmunity. Gut Microbes. 2012 Jan-Feb;3(1):4–14. doi: 10.4161/gmic.19320. Epub 2012 Jan 1. PMID: 22356853; PMCID: PMC3337124.
55. Koh A, De Vadder F, Kovatcheva-Datchary P, Bäckhed F. From dietary fiber to host physiology: short-chain fatty acids as key bacterial metabolites. Cell. 2016;165(6):1332–45.
56. Okereke I, Hamilton C, Wenholz A, Jala V, Giang T, Reynolds S, Miller A, Pyles R. Associations of the microbiome and esophageal disease. J Thorac Dis. 2019;11(Suppl 12:S1588-S1593. doi: 10.21037/jtd.2019.05.82. PMID: 31489225; PMCID: PMC6702393.
57. Liu W, Xie Y, Li Y, Zheng L, Xiao Q, Zhou X, Li Q, Yang N, Zuo K, Xu T, Lu NH. Protocol of a randomized, double-blind, placebo-controlled study of the effect of probiotics on the gut microbiome of patients with gastro-oesophageal reflux disease treated with rabeprazole. BMC gastroenterology. 2022 May 20;22(1):255.
58. Meenakshi K, Vinteshwari N, Minaxi J, Vartika S. Effectiveness of Ayurveda treatment in Urdhwaga Amlapitta: a clinical evaluation. J Ayurveda Integr Med. 2021;12(1):87–92. doi: 10.1016/j.jaim.2020.12.004. Epub 2021 Feb 3. PMID: 33546994; PMCID: PMC8039346.
59. Athavale, AD. Ed., Ashtangsangraha with shashilekha hindi commentary by Indu, Pune: Atreya Publications, Sutrasthana, chapter 13, 1980, pp. 114–115.
60. Ranade A, Acharya R. Contribution of Dhanwantari Nighantu towards drug safety: a critical review. Global Journal of Research on Medicinal Plants & Indigenous Medicine. 2015 Feb 1;4(2):20.
61. Ranade AV, Acharya R. The pharmacovigilance concern as quoted in various chapters of Madanapala Nighantu. Global Journal of Research on Medicinal Plants & Indigenous Medicine. 2016 Mar 1;5(3):92.
62. Kolhe R, Acharya R. Analyzing the drug safety issue in Bhavaprakasha nighantu–A critical review. Ayurpharm Int J Ayur Alli Sci. 2015;4(10):183–96.
63. P.V. Sharma, Ed., Sushruta samhita. Varanasi: Chaukhambha Vishwabharati Prakashana, Sutrasthana chapter 46/50, 46/210, 46/29, 2008, pp. 473, 503,520.
64. Athavale, AD. Ed., Ashtangsangraha with shashilekha hindi commentary by Indu, Pune: Atreya Publications, Sutrasthana, chapter 9/3, 1980, pp.82.
65. M. Sabnis. "Unwholesome Food Revisited," Nat Ayurvedic Me. 2021;5(1):1–2. https://medwinpublishers.com/JONAM/Unwholesome%20Food%20Revisited.pdf

Therapeutic Nutrition in Ayurveda for Hepatology

Swarupa M. Bhujbal

1. INTRODUCTION

The liver is the largest organ that performs all the vital functions of the body.[1] Carbohydrate-glycogen storage, protein synthesis, lipogenesis, detoxification, and many more are its major functions.[2] The biochemical and enzymatic secretions of the liver depend not only on the quality and quantity of food consumed but also on the method and time of its preparation.[3] The term 'liver diseases' refers to damages to cells and tissue, be it from endogenous causes or exogenous ones, such as viruses, endotoxins, exposure to alcohol or harmful chemicals, over-the-counter dietary supplements, and many more.[4–6]

The global burden of cirrhosis is 2 million deaths annually, out of which 1 million occur due to cirrhosis, hepatitis, and hepatocellular carcinoma.[7] Liver disorder is eleventh most common cause of death affecting 60% male between age group 45 to 60 years and 3.55% of death due to liver cancer worldwide. The global scenario is that 2 billion people consume alcohol, out of which 75 million are at risk of alcoholic liver disease.[8] In the next five decades it is predicted that graph of alcoholic and nonalcoholic liver disease[9] would rise. Owing to this there is a growth in epidemiological awareness in the community regarding the impact of diet on liver cells.

Understanding therapeutic nutrition is the need of the hour in the case of hepatology. It is very important to address the interim stage of NAFLD (nonalcoholic fatty liver disease), missing which would certainly lead to HCC (hepatocellular carcinoma).[10] Though there are many nutritional etiological factors as nonalcoholic contributors to disease progression such as a sedentary lifestyle, disproportionate sleep, high-calorie diet, and tobacco.[11,12] That leads to oxidative stress which needs a comprehensive hepatoprotection regimen. Scientists have evaluated many such antioxidants that are hepatoprotective, but their efficacy in humans is not yet clear. Proteolysis is the outcome of deranged liver function, where protein energy malnourishment is the most common morbid condition of liver cirrhosis, worsening with a reduced rate of survival.[13] Malnourishment is the cause of concern in liver disease for further complications of ascites and hepatorenal syndrome.[14]

Hepatology therapeutics in acute/chronic and critical care are supported with latest nutritional formulae such as albumin, aromatic amino acid (AAA), branched-chain amino acid (BCAA), and L-carnitine though they fulfill the need for nutritional insufficiencies; still malabsorption and maldigestion leads to poor clinical outcome of malnourishment. The refractory stage of the liver disorder requires repeated vigilant management of dietary sources of enteral/parenteral (EN/PN) albumin.[15] Ayurveda offers a unique way of dietary designing on an individual level apart from biochemical parameters, even during its prodromal stage.

Ayurveda food preparation methods are effectively meant for the precise assimilation of nutrients with bio-accessibility and availability. However, we emphasize mainly the *Ayurveda* perspective of liver disease and the interrelation of body constituents with food consumed. Therapeutic nutrition in Ayurveda (TNA) is a composite therapy summarizing from *Ahara-Vidhi* (dietary guidelines),[16] *Ahariya-Hetu*

DOI: 10.1201/9781003345541-9

(dietary causes),[17] and selected *Dwadasha Varga* (12 groups of food matrices)[18] up to efficient functional *Aharia Kalpana*[19,20] (food synergy) meaningful solutions for liver diseases. This chapter focuses on a thorough understanding of the digestibility and absorption of nutrients selected from food matrices/pyramids. Specific food synergies, i.e. multifold dietary strategies, are recommended to regenerate and conserve the affected liver and its functions. This renders nutritional counseling, assessment, and requirements, which are highly endorsed as *Ashta-aharavidhi-Visheshayatana* (eight directives of food ingestion guideline)[21] in established and emerging liver pathogenesis.

2. NUTRITION IN HEPATOLOGY FROM CLASSICAL TEXTS

Charaka unanimously states that *Rakta Dhatu* is the fundamental element and origin of *Yakrut as Raktavaha Srotas.*[22] The *Yakrut* is a vital organ that regulates transformations of metabolites, reflecting an array of functions such as pigmentation of *Rasa Dhatu*[23] (*Ranjana*),[24] *Prinana* (replenishing), *Jivan* (vital for life), *Lepana* (coating), *Snehana* (oleation), *Dharana* (holding), and *Purana* (filling) too. The *liver* plays a critical role in the conversion of nutrients from food ingested to the nutrients needed for vital functions. The liver is an organ that is instrumental in producing building blocks of the body as protein; hence any nutrient that enhances activity will certainly regenerate its efficiency.[26] *Rasayana karma* is such function that performs the above action (immune nutrients).[27]

The etiologies that vitiate *Pitta* and *Rakta Dhatu* are contributors to the pathologic behavior of the liver. This state hampers various functions from *Prinana* (replenishing) to *Purana*[28] (filling), reflected as fatty liver disease (FLD) to HCC. The *Ahara rasa* is the basis of all prospective transformations needed for nourishing body elements. This in turn depends upon food consumed and the dietary signals that are received from various *Prakruti* categories expressed (phenotypic).[29] It makes the entire system a complex conglomeration of *Doshas, Dhatus*, dietary signals depend upon phenotypic variations, and factors of *Ashta-Vidhi-Vishesh Ayatanani* (*ASV*).[21] It provides fundamental dietary guidelines which, if precisely followed, aid in attaining normalcy of *Aharasa, Rasa Dhatu,* and its functions. The *Dwadasha Ahara Varga* (12 categories of food matrices) are simple categorizations of food items by their attributes and benefits on *Doshas* and *Dhatus,* respectively, giving rise to diseases such as *Pandu* (anemia), *Shotha* (edema), *Kamala* (hepatitis), and *Udar* (ascites).

3. LIVER DISEASES AND CONVENTIONAL NUTRITIONAL PRINCIPLES

The liver metabolizes albumin, protein, carbohydrates, short-chain fatty acids, PUFA, and fibers.[30] The insufficiencies and overuse of nutrients along with toxic exposures cause hepatic injuries. This leads to acute, chronic, immunogenic, carcinogenic, and end-stage liver disease.[31]

NASH

The NAFLD, presently MFLD (metabolic dysfunction associated with fatty liver disease), is an acquired metabolic disorder.[12] Insulin resistance (IR) and hepatic dysregulation altering the SREB1c pathway is the outcome of metabolic stress due to excess consumption of obesogenic nutrients and excessive intake of high carbohydrates, that is, soft drinks.[7] This leads to the accumulation of fat because of disturbed pathways of free fatty acid uptake, synthesis, degradation, and secretion. The steatohepatitis condition is the accumulation of fat surrounded by inflammatory cells, which turns into fibrosis if it remains untreated. So, hepatic steatosis is the reversible culmination of fat metabolism.[32]

Autoimmune Concept – Liver Gut Axis

Dietary signaling is the basic unit of hepatic functioning, and gut microbiota displays pathogenesis in terms of dysbiosis, leading to inflammation and immune complex activity.[33] The nutritional therapy design will be facilitated by immunosurveillance[34] of pathogenic molecules and degradation of immunogenic molecules from gut microbiota.

Viral Hepatitis

It is an inflammation mediated by different type of viruses A, B and C. Its virulence and increase in viral load causes complications such as liver cirrhosis and hepatocellular carcinoma.[35]

Ascites

It is decompensated state of liver cirrhosis that consists of a pathological accumulation of fluid within the peritoneal cavity.[36]

4. CLASSIFICATION OF DISEASES IN HEPATOLOGY ACCORDING TO AVASTHA (DISEASE CONDITIONS)

The liver disorders comprise vitiated *Pitta* and *Rakta* causing morbidities such as one of *Pittaj Pandu* (type of anemia),[37] *Shotha* (edema),[38] *Kamala, Kumbha-Kamala* and *Halimaka* (various types of hepatitis),[39] *Udar* (ascites),[40] and *Nanatmaja-Pittaj* and *Kaphaj Vikaras*.[41]

5. GENERAL THERAPEUTIC NUTRITION IN HEPATOLOGY

Therapeutic nutrition in Ayurveda is the consideration of dietary causes/etiologies for the disturbance of liver functions establishing diseases. The treatment principles are contemplated based on the functionality of dietary attributes such as hepatoprotection mainly.

PANDU (ANEMIA)

The *Pandu* is characterized by pallor from varied shades of yellow and white in eyes, lips, and skin as per vitiation of *Doshas*.[29] This is the consequence of unwholesome and antagonistic food. The excessive

TABLE 6.1 Classification and Correlation of Liver Diseases

SR. NO	*CONVENTIONAL*	*STATE/TYPE OF VITIATION*	*POSSIBLE CORRELATION IN AYURVEDA*
1	NASH (Nonalcoholic steatohepatitis)/FLD, Ascites.	*Santarpana (Nourishing)*	*Kaphaj Vikara, Udar*
2	Liver cirrhosis, Infectious hepatitis, Alcoholic hepatitis,	*Aptarpana (Depleting)*	*Kumbh Kamala, Halimaka,* *Rakta Dhatu Vikara*
3	Anemia, Jaundice, Obstructive jaundice, Hepatic encephalopathy,	(*Saama/Nirama*)	*Pandu* *Kamala* *Rudhapatha Kamala* *Mada* *Mruccha* *Sanyasa*
4	Chronic liver cirrhosis, Chronic liver disease,	*Jirna* (Chronic)	*Kumbhakamala* *Halimaka*
5	Autoimmune hepatitis, Portal hypertension,	*Anukta* (Unexplained in Ayurveda classical texts)	*Pitta Vikara*

intake of *Amla* (sour), *Lavan* (saline), and *Kshara* (alkali) leads to vitiation of *Sadhaka Pitta* which resides in the heart. This vitiated *Pitta* is forcefully expelled from the heart by *Vata Dosha* reflecting paleness of skin and tissue throughout body.[42]

Liver diseases – Dietary etiologies, attributes, and therapeutic principles

Kamala (Hepatitis)

The disturbance in the interplay between *Pitta* and *Rakta* vitiates or obstructs the channels, causing hepatitis of varied types such as *Bahupitta* (hepatitis), *Ruddhapath-Kamala* (obstructive hepatitis), *Kumbha-Kamala*, and *Halimaka*[34] (liver cirrhosis).

Bahupitta Kamala (Jaundice)

Constant consumption of *Pitta* aggravating food vitiates *Rakta* and *Mamsa* mainly in anemic individuals, featuring *Pitta*-predominant symptoms as *Klama* (fatigue) and giving reddish-black color to stool and yellow to urine (Table 6.2).

Ruddhapatha/Alpapitta/Shakhashrita Kamala (Obstructive Jaundice)

Excessive cold food having unctuous and sweet properties vitiates *Kapha Dosha,* obstructing the path of *Pitta.* At the same time, *Vata is* also aggravated due to excessive exercise. This consequence of *Pitta* obstruction due to *Vata* and *Kapha* disables its flow to the *Koshta* (gut), leading to *Tilapishta* (steatorrhea) (Table 6.2).

TABLE 6.2 Application of Nutritional Principles for Liver Diseases

DOSHA/DISEASE	*DIETARY ETIOLOGIES/ATTRIBUTES*	*NUTRITIONAL PRINCIPLE*
Vata Dosha	*Shaietya* (cold), *Rooksha* (dry), *Laghu* (light), *Paryushitha* (stale), *Tikta* (bitter), and *Kashaya* (astringent).	Hot, unctuous, fresh food, sweet, sour, salty taste[33]
Pitta Dosha	*Ushna* (hot), *Tikshna* (sharp), *Katu* (Spicy), *Amla* (sour), and *Lavan* (salty).	Cold, unctuous, hydrated, sweet, bitter, and astringent taste[33]
Kapha Dosha	Cold, unctuous, *Guru* (heavy), *Madhur* (sweet), salty, sour.	Hot, dry, light, spicy, bitter, and astringent taste[33]
Pandu (Anemia)	Sour, alkali, excessively hot, antagonistic, black gram, sesame seeds.	*Virechana* cleansing therapy followed by old grains, barely, wheat, gruel prepared from green gram, and lentil[29]
Bahupitta kamala (Hepatitis)	*Ushna* (hot), *Tikshna* (sharp), *Katu* (spicy), *Amla* (sour), *Lavan* (salty).	Dry, sour, and spicy meat, dry radish, gruel of horse gram added with sour lemon, ginger juice, long pepper and after cooling add honey[32]
Rudapatha Kamala (Obstructive jaundice)	Dry, cold, heavy, excessive exercise, suppression of natural urges.	Spicy, *Sukshma* (penetrating), hot food, salty[32]
Kubha-Kamala (Liver cirrhosis)	*Shaietya* (cold), *Rooksha* (dry), *Laghu* (light), *Paryushitha* (stale), *Tikta* (bitter), *Kashaya* (astringent).	*Virechana* followed by appetizers and sweet taste as *Vata* pacifying[32]
Jalodar (Ascites)	Excessive hot, salty, alkali, sour, and toxic exposure to dry and antagonistic food.	*Raktha Shali* (brown rice) barely, green gram, meat from dry region, milk, honey. The soups adding ghee and black pepper are advisable[31]
Tandra and *Murccha* (Syncope)	Excessive meat and alcohol consumption, alkali and spicy food.	Old clarified butter, *Sukshma* food such as powders of ginger, black pepper, piper longum, sour orange[39]

Kumbha-Kamala and *Halimaka* (Liver Cirrhosis and Its Complications)

Conditions of *Kamala* when ignored and untreated progresses in *Kostha* (gut). It further damages the liver, causing irreversible changes, which are reflected as roughness/nodularity to the liver.[43] The *Kumbha-Kamala* displays severe inflammation of the liver, deteriorating *Agni* and leading to pathologic *Trishan* (thirst), *Tandra* (drowsiness), *Moha* (illusion), and red discoloration of eyes, face, urine, and vomitus. However, in *Halimaka* if anemia persists in the above condition there is a further progression of morbidity, causing cardinal symptoms such as loss of vigor and strength,[44] *Trishna* (pathologic thirst),[45] *Tandra* (drowsiness), *Bhrama* (giddiness), and *Angamarda* (body ache), giving greenish-yellow color to the excreted matter.[46,47]

Jalodar (Ascites)

Ascites are clusters of morbidities featuring an enlargement of the abdomen, especially the *Yakrutodar* (hepatomegaly). There are various classifications of *Udar*, but only *Yakrutodar* will be addressed in this chapter. Stages of ascites are categorized according to the quantity of fluid accumulated in the abdomen such as *Ajatodaka* (the stage without fluid), *Jatodaka* (with fluid), and *Pichhavastha.*[48] The etiologies mentioned (Table 6.2) associated with toxic abuse leads to impairment in *Jatharagni,* causing obstruction in perspiration and flatulence, leading to chronic constipation. The vitiated *Dosha* obstructs the channels of *Sweda* (sweat) and *Udaka* (fluid), afflicting functions of *Prana*, *Agni*, and *Apana Vayu*.[49]

Yakrutdalyodar (Hepatomegaly)

Hepatomegaly has been explained in a very unique way by Charaka. The vitiated *Rasa* and *Rakta* due to dietary etiologies and constant physical strain lead to malnourishment. The accumulated *Rasa* and *Rakta* increase the size of not only the liver but also the spleen, causing hepatomegaly and splenomegaly. The main features of *Yakrutodar* are debility, *Klama* (fatigue), *Aruchi* (ageusia), *Ajirna* (indigestion), *Malavabadhaata* (constipation), *Murccha* (darkness in front of the eyes), *Trushna* (excessive thirst), and *Moha* (transient loss of consciousness).[44,50]

Anukta Vikara (Unexplained Disease)

The *Anukta Vikaras* are diseases mentioned in the text with a group of symptoms; however, nonalcoholic steatosis and infectious and autoimmune hepatitis are considered *Anukta-Vikaras. Kaphaj* and *Pittaj Samanyaja Vikara* ensue due to respective etiologies with *Asatmyaindriyarth Sanyog,* leading to a cluster of symptoms.[51]

Kaphaj Vikara (NASH)

The vitiated *Kapha* by etiologies (Table 6.2) with specific attributes of *Sneha, Shaietya,* and *Picchilta* established in the liver. It is correlated with NASH as the symptomatology is similar to etiological components with current lifestyle contributors to disease. The *Kapha* disturbed function in the liver with *Shaietya, Sneha* (oleation), and *Kleda* (secretory) attributes. The symptoms like *Atishthoulya* (overweight), *Tandra* (drowsiness), *Nidradhikya* (sleepiness), *Alasya* (lethargy), and paleness of the body.[52]

6. NUTRITION IN HEPATOLOGY AS PER *DWADASHA AAHAR VARGA* (12 GROUPS OF FOOD MATRICES)

Dwadasha Varga (Table 6.3) is a systemic categorization of food items with properties/attributes according to their origin/species. Thus, food synergies and holistic preparatory methods aid in defining customized food preparation for each patient in an individual way, for instance, if the patient is suffering from *Udar* (ascites) due to liver cirrhosis and has *Alpa-mutrapravartana* (oliguria) and *Malavabadhata* (constipation). The dietary prescription is based on the *Udar Chikista* principle, which is *Virechana* primarily, so the selection of *Gorasa Varga* along with the *Shrushta-Mutrapravartana* (diuretics) and *Vitbhedi* (stool softener) group can be selected for diet designing (Tables 6.3 and 6.4).

Selection of food from *Dwadasaha Ahara Varga*[75] while quantity of dietary prescription of the above food item may differ as per appetite of patient and state of disease mainly.

TABLE 6.3 *Dwadasha Ahara Varga* (Hepatoprotective Food Matrices)

SR. NO	*VARGA (CATEGORY)*	*FOOD ITEMS*
1	*Shuka* (grains)	Rice, wheat, and barley
2	*Shimbi/Shami* (millets and pulses)	Black gram (*Vigna mungo*), Horse gram (*Macrotyloma uniflorum*), Kidney beans (*Phaseolus vulgaris*), Moth beans (*Vigna aconitifolia*), Sesame seeds (*Sesamum indicum*)
3	*Mansa* (meat and fish)	*Jangal* (meat from the arid region)
4	*Shaka* (leafy vegetables)	Leaves of *Kakmachi* (night Shade/*Solanum nigrum*), bottle gourd (*Lagenaria siceraria*), Green sorrel (*Rumex acetosa*), *Upodik* (Malabar spinach/*Basella alba*), *Amaranthus viridis*, *Madukaparni* (penny wart/*Centella asiatica*), bamboo shoots (*Bambusa arundinacea*), bitter gourd, *Gojivha* (aster), *Patol* (silk squash/*Chinese okra*), yam, lotus (*Nelumbo nucifera*), *Shatavari* (*Asparagus racemosus*), sunflower, *Cucumis sativus*, *Kushamanda* (ash gourd/(*Benincasa hispida*).
5	*Phala* (fruits)	*Mrudvika* (raisins), *Khrjur* (dates), ice apple (*Borassus flabellifer*), berry pear, *Kapitha* (wood apple/*Limonia acidissima*), *Bilva* (stone apple/ *Aegel marmelos*), raw mango (*Mangifera indica*), black plump, jujube fruit (*Ziziphus jujuba*), apple (*Malus pumila*), jackfruit (*Artocarpus heterophyllus*), pomegranate (*Punica garnatum)*, malabar tamarind (*Garcinia cambogia*), walnut
6	*Harita-varga* (tuber and spices)	*Aardrak* (ginger/*Zingiber officinalis*), lemon (*Citrus limon*), *Mulak* (radish/ *Raphanus sativum*), *Ajwain* (thymol seeds (*Trachyspermum ammi*), black cumin (*Nigella sativa*), onion (*Allium cepa*), garlic (*Allium sativum*)
7	*Madya* (alcohol)	Red wine prepared with grape, sugar cane, honey
8	*Jala* (water)	Warm or cold water
9	*Gorasa* (dairy)	Milk – Cows, buffalo, camel, goat, skimmed milk, low-fat milk, pasteurized milk, and whole milk Cow's clarified butter and colostrum
10	*Ikshu* (sugar)	Sugarcane, jaggery, sugar candy, honey sugar
11	*Krutanaa* (Sauté food)	*Vilepi, Manda, Lajapeya, Lajamanda, Siddha Laja Manda, Krushara*
12	*Ahar-yoni* (spices and pickles)	*Vesvar (mixed spices), Rasala (hung curd), Raga, and Shadav (pickles)*

7. CONCEPT CONFLUENCE OF TNA IN HEPATOLOGY

Charaka has postulated that dietary guidelines such as dosage of food, method of preparation, and *Agni* (digestion capacity) play a major role in functions such as secretions, storage, transformation, and immunological complex activity.[53]

As per the *Dosha* and *Dhatu's* attributes, similar nutrients can be prescribed to maintain homeostasis and functioning of organ physiology in accordance to *SamanyaVishesh* Siddhanta.[54] However, thoughtful designing of food combination is based upon nutrient attributes such as *Guru* (heavy), *Laghu* (light), *Rooksha* (dry), and *Snigdha* (unctuous)[55,56] focusing its functions *Santarpana* (specific nourishment),

Brunhana (tonicity) and *Langhana* (reducing)[71] which helps to design hepatoprotective regimen to reduce impairment of liver functions and enhances regeneration capacity.

The *Rasayana* (immune nutrients) therapy is nutrient rich and aids in the maintenance of the quality of all *Dhatus* at optimum levels.[57] The hepatoprotection mechanism can be managed with *Rasayana* food synergy.

While treating various liver diseases there is an array of clinical conditions categorized as general and advanced clinical conditions. The diet designed with functional food and their recipes from *Kshemakutuhal* and *Bhojankutuhal* is considered with therapeutic nutritional principles of *Charaka, Ashtangahrudayam,* and *Shushruatasamhita*. The general conditions/symptoms are those such as *Klama, Ajirna, Aruchi, Malavabhtata,* and *Shotha*. These conditions are treated with food groups/recipes which have the potential of *Virya Vardhana, Pachaniya* (appetizer), *Dipaniya* (boost digestive fire), *Vitbhedi, Medopachana,* and *Rasayana*. The diseases progress with advanced symptoms such as *Kamala, Jalodar, Pillavruddhi* (splenomegaly), *Raktaj chhardi* (variceal bleeding), and *Tandra*. The *Chhardighana* (antiemetic), *Shothaghana* (reducing edema), *Pillhanaghan* (reducing splenomegaly), and *Shrushta-mutrapravartana*

TABLE 6.4A Guidelines of Hepato-protective Functional Food

SR. NO	*DIETARY FUNCTIONS*	*GENERAL CLINICAL CONDITIONS*	*CLASSICAL TEXT FOOD ITEMS/RECIPES*
1.	*Virya-vardhana* (strength and vitality)	*Klama (fatigue), Mamsakshya (sarcopenia and frailty) Daurbalya (general debility)*	*Yava saktu* (*Hordeum vulgare*/barely flour), *Shali Saktu* (*Triticum aestivum*/wheat flour), Ragi. *Varaee* (*Panicum miliaceum*/ Proso millet), *Javas* (*Linus usitatissimum*/ flax seeds), black mung (*Vigna mungo*), *Kadalishaka* (banana stem), Mansa *purna, Rajika mamsa, Mansederi, Kshiraamruta, Mamsapurit Vruntaka* (different meat preparation).
2.	*Rasayana* (rejuvenation)	*Klama Mamsakshaya Daurbalya*	*Amala* (*Phyllanthus emblica*/gooseberry), *Kakamachi* (*Solanum nigrum*/black night shed), *Devdali* (gourd), *Krushan chanak* (black horse gram).
3.	*Vitbhedi* (action excreta)	*Malavabadhata* (constipation)	*Adarak* (wet ginger), *Badar* (jujube fruit), *Karkati* (cucumber), gourd, oats, barely
4.	*Jatharagni Deepana* (digestive)	*Aruchi* (apathy) *Agnimandya* (loss of appetite) *Klama*	*Ambastha* (roselle plant), *Saindhav* (pink salt), *Samudralavana* (rock salt), *Amaranthus, Laja* (rice puff), *Amlakushmanda* (sour ash gourd preparation)
5.	*Medhya* (brain)	*Mruccha (giddiness) Tandra (drowsiness)*	*Amala Kushmand (ash gourd), Kakamachi.*
6.	*Medopachana* (lipid and fat lowering)	*Kaphaj Vikara* (FLD)	*Methi-shaka (Trigonella foenum*/fenugreek), *Kusumbha* (*Carthamus tinctorius*/ safflower leaves), drumstick leaves, *Mulak shaka* (*Raphanus Sativus*/radish leaves), *Yav, Kodrava* (Koda millet-*Paspalum scrobiculatum), Adark-Karchari* (wet ginger preparation)
7.	*Pachaniya* (appetizer)	*Aruchi Malavabadhatva*	Pink salt, *Sudbhodak yog* (gooseberry, ghee, and cumin seeds), lemon, ginger

TABLE 6.4B Guidelines of Hepatoprotective Functional Food

SR. NO.	*ADVANCE CLINICAL CONDITION*	*CLASSICAL TEXT FOOD ITEMS/RECIPES*
1.	*Kamala*	Bitter gourd (*Momordica charantia*), pumpkin (*Cucurbita pepo*)
2.	*Pandu*	Bitter gourd, Yam (Dioscora), *Agasthi* (*Sesbania grandiflora/* hummingbird flower)
3.	*Jalodar* (ascites)	Drumstick, Ivy gourd (*Coccinia grandis*), *Asthisamharak* (veld grape/ *Cissus qudrangularis*) cumin seeds (*Cuminum cyminum*), dry ginger (*Zingiber officinalis*), camel milk, horse gram (*Macrotyloma uniflorum*)
4.	*Pliahagna* (splenomegaly)	White goosefoot (*Chenopodium album*), drumstick, Dili, dry ginger, green gram, rice
5.	Inflammation	Amaranthus viridis
6.	*Madya* (alcoholic hepatitis)	*Kasmarda* (*Senna occidentalis*/Nigro coffee), pumpkin leaves, Amaranthus, bitter gourd, *Kartoli* (spiny gourd)
7.	*Sa-rakta Mutra Pravartana* and *Raktaj Chardi* (Hematuria/ hematemesis/variceal bleeding)	Small cake made from ash gourd (*Benincasa hispida*) and *Ragi* (*Eleusine coracana*), *Udumbar* (cluster fig/*Ficus racemose*), *Kadalikanda* (*Musa acuminata*)/banana stem), red Amaranthus, *Sita* (raw sugar), goat milk
8.	*Shrushtamutra-pravartana* (diuretics)	Amaranthus, *Chamkura* (taro leaves/*Colocasia esculenta*), *Trapus, Valuka and Pshatbhujakarkati*) (variety of cucumber), *Demase* (squash grapes), *Tadphal* (palm fruit), jaggery, sugar cane (*Saccharum officinarum*), brinjal (*Solanum melongena*)

(diuresis) functional food is considered for treating advanced conditions. This aids in reducing symptoms and arresting disease progression.

Dietary recipes from *Kshmakutuhal*[58] and *Bhojanakutuhal*[59] using *Dwadasha Ahara Varga* food items.[63]

Nanatmaja Kaphaj Vikara (Fatty Liver Disease)

Nutritional recommendations/suggestion to any of the following liver disease is based upon respective nutritional attributes and its function which is mentioned in (Table 6.5) helping to overcome the liver function impairment.

Nutritional recommendation

Food based on the nutritional attributes and principles mentioned in Tables 6.2 and 6.5 that help in the shedding of fat is recommended in FLD. The principle of fat absorption and regulation is based upon *Laghana* (fasting); *Laghu Apatarpana Ahara* (low-calorie food) leads to the regulation of *Agni* and *Malpravtarna*. The selection *Laghu Apatrapna* food is as *Yav Saktu* (flour barely) and *Bajara* (pearl barley) (Table 6.4B) flour having *Rooksha* (dry), and *Virya-vardhak* (tonicity and strength) quality (Table 6.4A). These flours have dietary soluble fiber, beta-glucans, bran, and phenolic acids,[60] which help in the reducing formation of lipotoxic lipids that contribute to cellular stress. While Roasted rice helps in reducing antinutrients such as phytate and polyphenols.[61,62] Thus, *Haridranna* (Table 6.6) is useful for deranged *Meda* (fat) .Various gruels such as *Karshniya Yavagu* (depleting gruel)[63] prepared from millets corn, *Jawari* (sorghum), and *Bajari* (pearl millets) and when administered with honey reduces fat deposits.[64] The gruel from barley flour and gooseberry will help overcome the sticky nature of fat.[65,66] The beverages

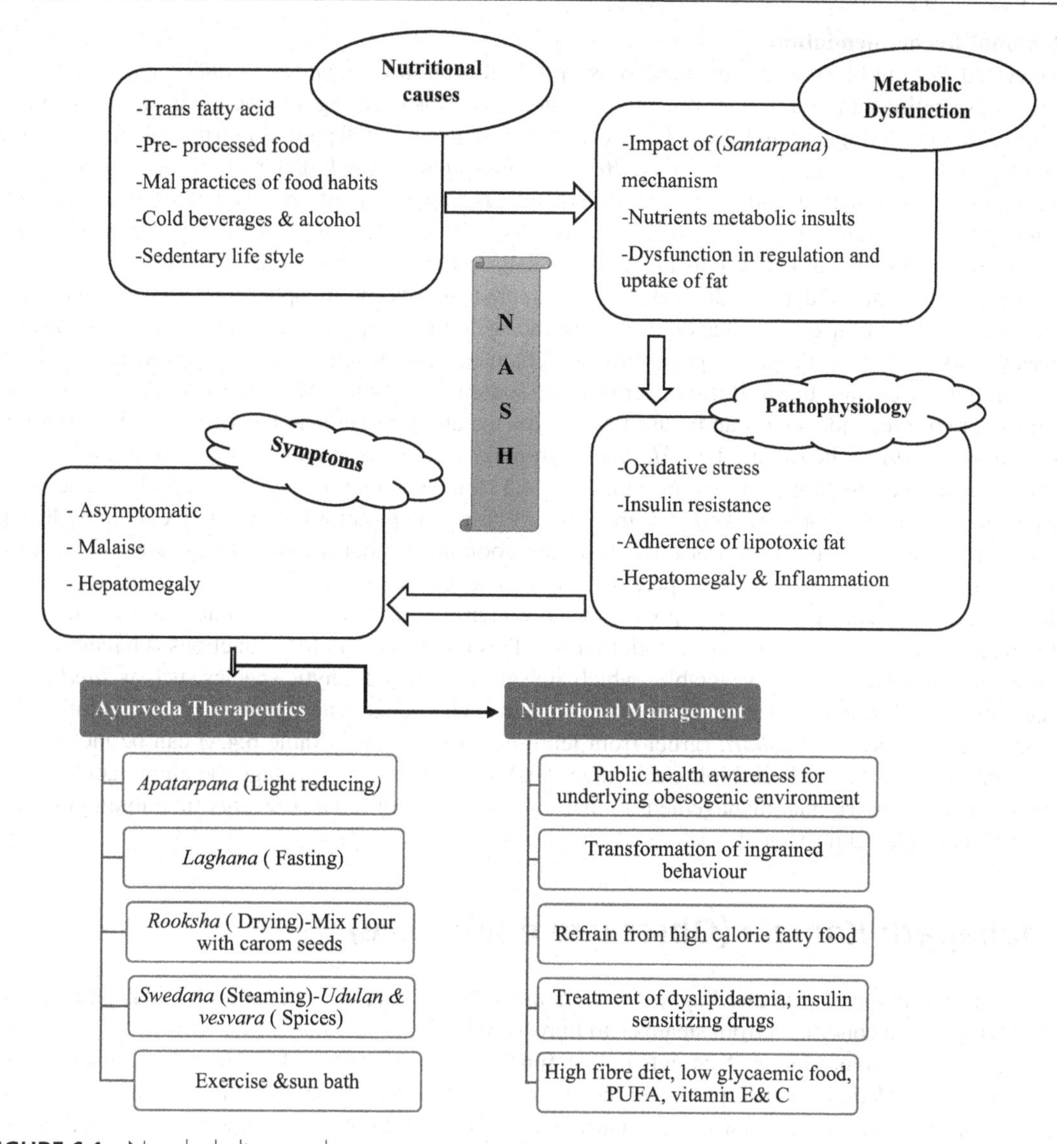

FIGURE 6.1 Nonalcoholic steatohepatitis.

of jujube (Table 6.6) and wood apple having *Kaphanghana* properties with phytochemical agents such as ascorbic acid and triterpenic acid are helpful in hepatoprotection.[67] The health drinks used are buttermilk, *Badar Panak,* and *Nimbu Panak*. The Niger and flax seeds have lignin and dietary fiber. It helps in reducing *Kapha* and *Meda* effectively.[68]

Kamala (Hepatitis)

The stages of hepatitis differ in preicteric, icteric, and convalescent pathogeneses according to standard of care that restricts only to liver metabolism, but *Charaka* elucidates *Kamala*[68] comprehensively where all the stages focus on varied levels of vitiation in *Pitta, Rakta,* and *Mamsa Dhatu.*

Nutritional Recommendation

The elevated *Pitta* in blood and urine is seen as raised bilirubin level manifested due to enzymatic dysfunction in secretion and excretion of bile pigment and salts. This needs aggressive medical management. *Virechana* therapy for aggravated Pitta. This is followed by *Kamala* (Hepatitis) nutritional attributes and Principles (Table 6.5) as *Laghu Santarpna, Bruhana,* and *Viryavardhak* dietary regimen (Table 6.4A). The patient is having malaise and fatigue, predominantly as an outcome of disturbed glucose homeostasis which empties glycogen storage in liver. This needs to be fulfilled with energy source of glucose/*Madhur Rasa.* Health drinks such as decoctions made from raisins (phenolic compound, fructose, and sucrose),[69] pomegranate (linolenic acid and anthocyanin),[70] dates (phytosterols and tocopherols),[71] and rice puffs can be used. This will help in kindling the digestive fire and hydrating transcellular and paracellular pathways permeable to these drinks through aquaporin water channels. The prompt vigilance and management of *Siddha Jala* are necessary to avoid fluid overload and sodium retention.[72] The roasted millets and cereals like barley, ragi, rye, and *Jawar* can be used as *Bhojya* (eatable). Sugarcane has been mentioned in this condition as *Choshya* (chewable). This *Madhur rasa* mainly contains apigenin, which is an inhibitor of lipid peroxidation correcting alanine transaminase (ALT) and aspartate transaminase (AST).[73] The anorexia and nausea can be treated by *Ardrak Karchari* (wet ginger preparation) (Tables 6.4A and 6.6) digestibility enhancer with the absorption of micronutrients, sodium, and potassium.[74] The vitiated pitta needs to be pacified with *Pitta Shamak* food preparation, thus vegetables selected from *Cucurbitaceae* family such as pumpkin, snake gourd, bitter gourd, and ivy gourd[75] (Table 6.3) are Sautee in cows clarified butter, which is a source of soluble fiber with PUFA.[76] This impairment in liver functions is balanced with selection of Amaranthus family vegetables which are salt-tolerant halophytic species such as *Tanduliya*[77] (Amaranthus) and *Lal math* (red Amaranthus) having phytochemical composition of carotenoids, flavonoids, and apigenin. *Krushara* (gruel from lentil and green gram) (Table 6.4A) can be included in meal planning and acts as a high-biological-value (HBV) protein used to manage the amino acid profile. However, hepatitis nutritional management is focused on *Rakta* and *Mamsa Dhatus* functioning majorly with *Prinana* (replenishing) and *Tarpana* (satiation) preparation.

Ruddhapath Kamala (Obstructive Jaundice)

There is an obstruction of *Vata* and *Kapha Dosha* in the path of *Pitta Dosha.* This occurs in hepatic and as well in extrahepatic conditions from steatosis to fibrosis too.

Nutritional recommendation: Nutrients that eliminate the obstruction and enhance *Pitta* function will ultimately help in *Meda* (fat) absorption and thereby rectify stool formation. Condition is preceded by the guidelines mentioned in *Bahupitta Kamala's* nutritional attributes (Table 6.5). For abdominal pain and nausea along with medical management, the food synergy used is wet ginger/gooseberry/ash gourd slice, lemon, and *Saindhav* combination having a sour, salty taste; this helps to fluidize and alkalinize bile (Tables 6.4A and B).[78] The vegetable preparation is mainly with *Vesavar Choorna* (asafetida, ginger, cumin, pepper, turmeric,[79] and coriander, in increasing order) having spices of hot potency and antibacterial origin, which is needed to clear the path of stasis of bile. The *Agni* gradually enhances and nausea reduces in patients; they can be administered as a regimen comprising various probiotics and soluble fiber[80] such as vegetables *Vastuka* and *Upodika* prepared in butter milk. The excretory function of bile is improved with *Vitbhedi Ahara* (Table 6.4A) such as *Devdali* (flavonoids – apigenin), *Badar, Adarak,* oats, and barley (Table 6.3).

Kumbhakamala and *Halimaka* (Liver Cirrhosis and Its Complications)

It is a chronic catabolic stage where a considerable amount of proteolysis, shedding out from *Mansa Dhatu,* causes severe emaciation.[81] The prognosis is worsen as malnourishment.

Nutrition recommendation: Primarily jaundice is the symptom to be addressed with nutritional attributes and principles mentioned in (Table 6.5) *Sneha-sahit Tarpana* and *Mantha Kalpna* enhance appetite and digestive capacity. So, *Tarpana* can be from coconut water added with lemon juice and buttermilk processed with black pepper salt and ghee. *Laja-Tarpana*[82] can be used with ghee and honey. The patients are in need of energy from *Madhur rasa* (sweet), as glucose source is utilized, but glycogen store remains empty, leading to fat oxidation. The repeated servings of *Laghu Santarpana* restore and maintains glucose supply by *Madhur rasa* preparation from ash gourd such as *Khand-kushamnada* (Table 6.6). While soup of ragi, barely, *Saktu Mantha* prepared from chestnut flour, and raw coconut sweet can be used. Further, the main focus is to maintain a negative nitrogen balance with the homeostasis of the amino acid pool in tissues and muscle mass. Thus, *Yavagu* prepared from green gram, lentil, and ragi rock salt sautéed with clarified butter are used. Meat preparation with ghee and grain (Table 6.6) is advised for HBV protein without inducing hepatic encephalopathy. Constipation is major symptom that aggravate poor clinical outcome of ammoniagenic complications which has to be managed immediately with varied combination of buttermilk with vegetable, rice and spices. The soluble fibers are from pectin (fibers), gums (barley, oats, beans, peas), and lentils, while insoluble fibers are cellulose (wheat),[83] hemicellulose (grains),[84] and lignin (green vegetables). The recipes advised are *Vastuka* and *Tanduliya Shaka* (Tables 6.4A and B), representing a combination of probiotics with soluble fiber, reducing ammonia genic effect and facilitating fat-soluble vitamins D and E.

Rakta Dushti Vikara as a Consequence of *Madyapana* (Alcoholic Hepatitis)

The chronic consumption of excessive alcohol and insufficient diet leads to alcoholic hepatitis. There is a severe deficiency of micronutrients with hypermetabolism.[85]

Nutritional recommendation: The dietary strategies based upon nutritional attributes (Table 6.5) are *Laghu Santarpana, Virya-vardhak*, and *Vishghana Ahara* followed by *Bruhan,* mainly hot potency and unctuous with sour taste recipes/preparation (Tables 6.4 A and B). The beverages/*Panak Kalpna* from lemon, berries, and mega lemon can be recommended only based on the physician's advice. Sour taste is beneficial as *Tarpak*. So, preparation of vegetables and Dal(pulses) using sour taste *Amchur* (dry mango powder), *Anardana*, or *Kokam* is advisable. *Meat* balls preparation mentioned in (Table 6.6) with *Vesavar choorna* is helpful to manage *Mamsa dhatu.* Regular use of *Vishanika* (water chestnut)[86] and ash gourd (Table 6.3) is recommended chiefly as it enhances nourishment by fulfilling micronutrient demands thoroughly. Dairy, lean meat, and red meat hold the property of choline,[87] methionine, and L-carnitine to restore hepatic glutathione reserves and attenuate liver injury. Pathologic thirst can be managed with *Mantha Kalpna*. The bamboo and *Asparagus racemosus* shoots are used in meal planning as a rich source of phenols, minerals (iron, copper, manganese, and selenium) and Vitamin (C and E).[88,89]

Anukta Pittaj Vikara (Autoimmune Hepatitis)

It is *Pittaj Vikara* that enables functions of *Grahani*, leading to symptoms of *Anatar-daha* (burning sensation) and *Kamala* (hepatitis)[90] (Table 6.2). The internal environment of gut microbiota forms the essential contributors to the immune complex depending upon the dietary factors. However, asymptomatic patients with ALT and AST cytosol and mitochondrial function disturbance raised >1000 times can be readily advised with *Rakatpittashamak* and *Rasayana* recipes (Tables 4A and B). The *Udumbar* (*Ficus racemosa*) and *Shatavari* (*Asparagus racemosus*) siddha jal (processed water) having lupeol and antioxidant property can be used as an immediate Pitta pacifier.[91] Use of *Dadhi Mastu* can aid in the regulation of gut microbiota,[80] while vegetables prepared in buttermilk help in immune activity of liver-gutaxis.

Ascites – Refractory Stage

Apart from derailed liver functions, there is a critical diminution in protein synthesis,[92] which can be detected as emaciation and edema.

Nutritional recommendations: The diet which reduces *Kleda,* enhances *Agni* and digestion power, reduces flatulence, and loosens the stool. *Virechana*[44] is a guideline to reduce ascitic fluid and bring homeostasis of nutrients and micronutrients.

Dietary tips/advice for lowering ascitic fluid:

- *Ajatodaka avastha* (distension/flatulence): *Dipaniya* and *Pachaniya* (Tables 6.3 and 6.4A) food along with buttermilk
- *Jatodaka* (fluid accumulation): *Virechana* (purgation)
- Cleansing procedure supported: Cow's milk and goat milk[93] (if available)
- Reduced abdominal girth and enhancement in digestibility: *Yush* (green gram, lentil, and horse gram)
- Milk is naturally abundant in an amino acid pool, a rich source of the branched-chain amino acid (BCAA), calcium, phosphorus
- Reduction of edema and urine output enhancement: *Gud* (jaggery)[94] and wet ginger
- Ascites:
  Alcohol – use of vegetables from *Visha* and *Rasayana* group (Tables 6.4A and B)
  Infection – add *Udulan Choorna* (cardamom, clove, pepper, and cinnamon)
  Splenomegaly – Radish leaves and black horse gram
  Malignancy – *Rasayana* and *Viryavardhak* group (Table 6.4)
- Sodium and water restriction – sodium (1 g) and water ~2000 mL, vary per conditions.
- Family counselling and vigilance is priority need for water consumption in recommended quantity.

End-stage Liver Disease (ESLD)

ESLD can be a consequence of any disorder mentioned above, critically affecting the quality of life (QoL) toward terminal sickness.[95] In order to prevent aggravation of *Pitta Dosha* careful recommendations of *Tarpana, Mantha, Yush, Yavagu, and Krushara* (Table 6.4A) are advisable.

Advanced/Critical Conditions

The clinical insight of managing critical symptoms with therapeutic nutrition in *Ayurveda* is to enhance QoL. However, these conditions are critical in nature and often need tube feeding (TF).

Mamsa-Kshaya (Sarcopenia and Frailty)

The chronicity and decompensation of liver disease leads to further *Dhatu* vitiation, mainly *Mamsa* and *Meda,* leading to *Mamsa* and *Medakshaya* sarcopenia and frailty.[96] The condition can be addressed with *Tarpan, Santarpana* (nourishing), and *Bruhana* (tonicity) too.

The Sarcopenia can be compensated/replenished by the use of the *Shaka, Shuka,* and *Mamsa Varga* recipes.[97] Appetizers, vigor and vitality, and *Rasayana* group of food preparation can be useful.

TABLE 6.5 Management of Hepatocellular Disorder – An Integrated Approach

CONDITION/ DISEASE	*EXAMINATION AND ASSESSMENT*	*COUNSELLING*	*ATTRIBUTES-BASED NUTRITIONAL THERAPEUTICS*
Kaphaj Vikara (FLD)	BMI Lipid profile Imaging	Regular exercise, weight reduction, refrain from alcohol.	*Laghana*, *Kapha* and *Meda* reducing diet *Sadya Laghu Aptarpana* Taste – spicy, bitter, and astringent Attributes – hot and dry
Bahupitta kamala (Hepatitis)	Icterus-eyes and skin hue Bilirubin, alanine transaminase (ALT), aspartate transaminase (AST), and serum albumin	Prevention from causative factors (Table 6.1).	*Vitbhedi, Laghu Santarpana, Bruhana* Taste – sweet, bitter, and astringent Attributes – cold, light, and unctuous properties
Rudapatha kamala (Obstructive Jaundice)	Pruritus and grayish stool color Elevated levels of indirect bilirubin, alanine transaminase (ALT), gamma-glutamyl transpeptidase (GGTP), and lipid.	Prompt vigilance and action are needed to avoid portal hypertension associated with ascites.	*Dipanaiya, Pachaniya and Varcho bhedaka* Taste – salty, bitter, and sour Food type – appetizers, digestive, and stool softeners
Kumbha-kamala (Liver cirrhosis)	Abdominal distension, palmar erythema, spider navi Liver function test (LFT), prothrombin time (PT), international normalized ratio (INR), lipoprotein, natremia, ammonia, C-reactive protein (CRP), homocysteine.	Avoid dry *Ahara*, antagonistic food and alcohol consumption.	*Bruhana, Snehasahita Tarpana, and Santarpana* Due to inappropriate hydration in dhatus, there is a need for unctuous food predominant. Taste – sweet, bitter, sour taste.
Progressive condition of LC	Sarcopenia and Muehrcke's nails/lines Reduced level of testosterone	Avoid mix taste food and dry food	*Dipaniya, Pachaka from Harit varga, virya-vardhak, and bruhana* Appetizer, digestive followed by aphrodisiac
Raktaj Vikara (Alcoholic hepatitis)	Puffiness in the face gamma-glutamyl transpeptidase (GGTP), alanine transaminase (ALT), AST, amylase and lipase.	Strictly refrains alcohol	*Rakatpitta shamak* Taste – sweet, bitter Agent – antipurpuric
Jalodar (Ascites)	Increased abdominal girth, caput medusa and edema LFT, albumin, ascitic fluid examination spontaneous bacterial peritonitis (SBP), adenosine deaminase (ADP).	Refrain from sodium and water intake	*Virechana* Taste-sweet

Tandra (Hyponatremia)

The blood sodium level below 126 mEq/L is irritable, rowdy, and sensorium altered.[98] This occurs due to sodium loss because of large-volume paracentesis and diuresis It is advisable to recommend *Shadav* (pickle -rich in salt) for instant relief of hyponatremic symptoms. while salt containing *Panak* and *Siddha jal* can satisfy the need of sodium.

Raktaj Chardi (Hematemesis/Variceal Bleeding)

It is a medical emergency of portal hypertension >13 mmHg leading to variceal bleeding.[99] Treatment includes alpha-adrenergic blockers and endoscopic variceal ligation.[100] *Pachana* is advised for clearing *Pitta* and *Rakta Dushti*. Primarily, after the episode of hematemesis *Laghana* is advisable. Nasogastric tube feeding with *Udumbar Sidhha jal (Ficus racemosa) and Kushamanda (Benincasa hispida)* juice is permissible as *Pitta* pacifier *and Raktasatmbhak* (Table 6.4B). The precise and effective emergency management for raktashamana, raktastambhak and energy replenishing source can be selected from *Shaleshmanatak* (*Cordia dichotoma*)[101] fruits, *Shatavari* (*Asparagus racemosus*) and Bamboo shoots (*Bambusa arundinacea)* soup. Intermittent sips of goat milk and pomegranate juice are advisable in this condition.

Mutravarodha and *Alpta* (Hepatorenal Syndrome)

The chronicity of liver disease leads to renal insufficiency without the involvement of the kidney.[102] Oliguria and edema conditions are advised with selection of food items from *Shrushtamutrapravartna* (diuretics) group mentioned in Table 6.4B. The *Shothaghna* (reduces oedema) and *Kahpaghna Dravyas* such as drumstick leaves, black horse gram, radish leaves, buttermilk, jaggery, and wet ginger, are used (Tables 6.3 and 6.4A and B).

8. *AHAREYA KALPANA* (DIETARY RECIPES) IN HEPATOLOGY

The *Ahariya Kalpna*/food synergy is the selection of *Ahariya Dravyas* from *Dwadasha Ahara* food matrices in a systematic way, benefiting from maintaining wellness, prevention, and therapeutics for liver diseases. In liver disorder, the significant area to work on is liver functions, chronic constipation, malnourishment, and complications with micronutrients mainly.

9. Scope for Further Study

- There is a need to explore Nutritional Medicine through *Ayurveda* principles to reduce burden of morbid poor clinical outcome malnourishment which eventually leads to hypoalbuminemia and hyperammonemia.
- *Ayurveda* Nutrigenetics perspective will prevail over critical conditions such as hepatic encephalopathy and ESLD.

TABLE 6.6 Hepato-protective Functional Food

CATEGORIES	*MAIN INGREDIENT (NAME OF THE FOOD ITEM)*	*PREPARATORY METHODS*	*BENEFITS/BIOACTIVE FEATURE*
Shaka varga (leafy vegetable)	*Tanduliya* (Amaranthus)	Blanching, sauté in cows' ghee/ buttermilk adding asafetida.	A soluble functional fiber with probiotic action. Enhances BCAA level and immunogenic factor. Preserves rheological property.
Phala varga (fruit)	*Khanda-kushamanda* (ash gourd)	Square shape slicing of fruit, boil it in milk, strain it, smash it, sauté with ghee, adding cardamon and black pepper.	Micronutrient and flavonoids rich.
Harita varga (tuber)	*Ardrak karchari* (wet ginger) *Gud Ardrak* (jaggery and ginger)	Wet ginger Sliced, add salt, and sauté in ghee. Paste of jaggery and ginger* (Proportion 4:1).	Digestible soluble fiber. Gingerol- activates heat receptors and terpenoids helps in lipid metabolism.
4. *Mamsa varga* (meat)	*Rajika Mamsa* *Praleha*	Meat, pulses, flour, and buttermilk -Crushed meat with green gram flour add *Vesvar choorna* prepare small ball, steam it followed by frying it in ghee, then dip in buttermilk, curry with cinnamon. *Ghrita, Vesvar Choorna*, coriander powder, asafetida, and curd strain it and add boiled meat to it.	High biological value, protein with probiotics. Comprehensive protein, flavonoids, and probiotics.
5. *Jal Varga* (water)	*Badar Panak* (ripe jujube)	Boil the fruit in water, separate the fruit from the seeds, and add sugar candy, cardamom, and bay leaf.	Antioxidant, micronutrients.
6. *Shuk Varga* (cereals)	*Haridranna* (Rice)	Cook rice with turmeric powder, black pepper, and carom seeds.	Flavonoids. Enhances digestibility.
7. *Gorasa Varga* (dairy)	*Dadimirasa* (buttermilk) *Shirashaka* (paneer)	Sauté pomegranate seeds in clarified butter, strain it, and add in buttermilk. Paneer sauté with ghee, salt.	Probiotics. Rich in Zn. Break the ammonia pathway.

* Food synergy from *kshemakutuhal*.[58,59]

Conclusion

In hepatology, Ayurveda dietary strategies play a pivotal role in treating diseases and managing longer-duration recurrence. Poor outcome of liver diseases is malnourishment, sarcopenia and frailty, which leads to morbidity, can be controlled with TNA dietary strategies reducing recurrence and hospitalization. Liver diseases have high micronutrient demands with HBV and amino acids, which is well structured in *Aharia Kalpana* (food synergy). However, common prevalent NASH is an escalating silent disease in public health because of its ignored symptoms such as abdominal discomfort and malaise. TNA will restrict further progression of NASH with minimal anti-nutrient free dietary guidelines. TNA understanding provides Customized Dietary hepatoprotective regimen to affected individual. *Aharia kalpana* helps to overcome malabsorption and maldigestion of micronutrients and enhances liver functions.

Limitations

TNA lacks in parenteral nutrition (PN), section though high biological value recipes with micronutrients rich sources are available.

10. NUTRAVIGILANCE

Food synergy is a composite concept of food preparation. The *Dosha, Dhatu, Kala* (time), and *Avastha* (condition) of the disease are important when prescribing a diet for a liver disorder. Milk diet according to dosha is one of the key principle. There is need of vigilance in management of ascites through milk in alcoholic and viral hepatitis specifically with severe constipation as it eventually leads to hepatic encephalopathy. Selecting buttermilk along with *Shaka varga* would be more advisable than milk. This gives management for clearing dysbiosis of gut with clearing constipation and automatically reduction in formation of ammonia genic waste to avoid poor clinical outcome of hepatic encephalopathy.[103]

REFERENCES

1 Ikeda Y, Murakami M, Nakagawa Y, Tsuji A, Kitagishi Y, Matsuda S. Diet induces hepatocyte protection in fatty liver disease via modulation of PTEN signalling. Biomed Rep. 2020 Jun;12(6):295–302. Doi: 10.3892/br.2020.1299. Pub 2020 Apr 22. PMID: 32382414; PMCID: PMC7201141

2 Yasutake K, Kohjima M, Nakashima M, Kotoh K, Nakamuta M, Enjoji M. Nutrition therapy for liver diseases based on the status of nutritional intake. Gastroenterol Res Pract. 2012; 2012:859697. Doi: 10.1155/2012/859697. Pub 2012 Nov 14. PMID: 23197979

3 Morales-González Á, García-Luna y González-Rubio M, Aguilar-Faisal JL, Morales-González JA. Review of natural products with hepatoprotective effects. World J Gastroenterol. 2014 Oct 28;20(40):14787–804. doi: 10.3748/wig. v20.i40.14787. PMID: 25356040; PMCID: PMC4209543.

4 Trovato FM, Catalano D, Martines GF, Pace P, Trovato GM. Mediterranean diet and non-alcoholic fatty liver disease: the need of extended and comprehensive interventions. Clin Nutra. 2015 Feb;34(1):86–8. Doi: 10.1016/j.clnu.2014.01.018. Pub 2014 Jan 31. PMID: 24529325.

5 Trovato FM, Catalano D, Martines GF, Pace P, Trovato GM. Mediterranean diet and non-alcoholic fatty liver disease: the need of extended and comprehensive interventions. Clin Nutra. 2015 Feb;34(1):86–8. Doi: 10.1016/j.clnu.2014.01.018. Pub 2014 Jan 31. PMID: 24529325.

6 Osna NA, Donohue TM Jr, Kharbanda KK. Alcoholic liver disease: pathogenesis and current management. Alcohol Res. 2017;38(2):147–161. PMID: 28988570; PMCID: PMC5513682.

7 Ginès P, Krag A, Abraldes JG, Solà E, Fabrellas N, Kamath PS. Liver cirrhosis. 2021 Sept 17; Seminar Volume 398(10308):P1359–1376, October 09, 2021doi: 10.1016/S0140-6736(21)01374-X

8 GBD 2019 Diseases and Injuries Collaborators. Global burden of 369 diseases and injuries in 204 countries and territories, 1990–2019: a systematic analysis for the Global Burden of Disease Study 2019. Lancet. 2020 Oct 17;396(10258):1204–1222. doi: 10.1016/S0140-6736(20)30925-9. Erratum in: Lancet. 2020 Nov 14;396(10262):1562. PMID: 33069326; PMCID: PMC7567026

9 Prof Pere Ginès, MD. Prof Aleksander Krag, MD. Prof Juan G Abraldes, MD. Prof Elsa Solà, MD. Prof Núria Fabrellas, PhD, Prof Patrick S Kamath, MD; Liver cirrhosis; SEMINAR VOLUME 398, ISSUE

10308, P1359-1376, OCTOBER 09, 2021; Published: September 17, 2021. DOI: https://doi.org/10.1016/S0140-6736(21)01374-X

10 Powell EE, Wai-Sun Wong V, Rinella M. Non-alcoholic fatty liver disease. Lancet 2021;397:2212–2.

11 Purkins L, Love ER, Eve MD, Wooldridge CL, Cowan C, Smart TS, Johnson PJ, Rapeport WG. The influence of diet upon liver function tests and serum lipids in healthy male volunteers' resident in a Phase I unit. Br J Clin Pharmacol. 2004 Feb;57(2):199–208. doi: 10.1046/j.1365-2125.2003.01969.x. PMID: 14748819; PMCID: PMC1884438.

12 Osna NA, Donohue TM Jr, Kharbanda KK. Alcoholic liver disease: pathogenesis and current management. Alcohol Res. 2017;38(2):147–161. PMID: 28988570; PMCID: PMC5513682

13 Plauth M et al., ESPEN guideline on clinical nutrition in liver disease, Clinical Nutrition, https://doi.org/10.1016/ j.clnu.2018.12.022

14 Tandon P, Montano-Loza AJ, Lai JC, Dasarathy S, Merli M. Sarcopenia and frailty in decompensated cirrhosis. J Hepatol. 2021 Jul;75 S147–S162. doi: 10.1016/j.jhep.2021.01.025. PMID: 34039486; PMCID: PMC9125684

15 Rehman MT, Khan AU. Understanding the interaction between human serum albumin and anti-bacterial/anti-cancer compounds. Pharm Des. 2015;21(14):1785–99. doi: 10.2174/1381612821666150304161201. PMID: 25738491

16 Aharia vidhi ASV: Dangayach R, Vyas M, Dwivedi RR. Concept of Ahara in relation to Matra, Desha, Kala and their effect on Health. Ayu. 2010 Jan;31(1):101–5. doi: 10.4103/0974-8520.68194. PMID: 22131693; PMCID: PMC3215310

17 Veena, V. V., & Gehlot, S. Historical perspectives of nutrition science: Insights from Ayurveda. J Nat Rem. 2019;19(1):32–42.

18 Agnivesa: (2004): A Textbook entitled Charak Samhita, edited by Acharya J.T, Chaukhamba Publication, Sanskrit, Varanasi. Annaapana vidhiyam Adhyaya, page number 152–352.

19 Sharir, K. An appraisal on Ayurvedic diet and dietary intake considerations in view of nutrition science. Indian J Nutr Diet. 2018;55(1):88.

20 Jacobs, D.R., Tapsell, L.C. & Temple, N.J. Food Synergy: The Key to Balancing the Nutrition Research Effort. Public Health Rev. 2011;33:507–529. doi: 10.1007/BF03391648.

21 Payyappallimana U, Venkatasubramanian P. Exploring Ayurvedic Knowledge on Food and Health for Providing Innovative Solutions to Contemporary Healthcare. Front Public Health. 2016 Mar 31;4:57. doi: 10.3389/fpubh.2016.00057. PMID: 27066472; PMCID: PMC4815005

22 Kasar NV, Deole YS, Tiwari S. Systematic review of the concept of Yakrutotpatti (embryology of liver). Ayu. 2014 Jan;35(1):5–8. doi: 10.4103/0974-8520.141895. PMID: 25364192; PMCID: PMC4213968.

23 Chakrapanidatta;(2019); Ayurveda Deepika Commentary of Charak Samhita, edited by Choukhamba Surabharati Prakashan, Varanasi Grahani 15/8; page number 512.

24 Vagbhata:(2000): A Textbook entitled ASTANGAHRDAYAM, edited by Paradkar H. Krishnadas Academy, Sanskrit Publication, Varanasi. Doshadi Vidnyaniyam Adhyaya,11/4, page number 183.

25 Sharma V, Chaudhary AK. Concepts of Dhatu Siddhanta (theory of tissues formation and differentiation) and Rasayana; probable predecessor of stem cell therapy. Ayu. 2014 Jul-Sep;35(3):231–6. doi: 10.4103/0974-8520.153731. PMID: 26664231; PMCID: PMC4649578

26 De Feo P, Lucidi P. Liver protein synthesis in physiology and in disease states. Care. 2002 Jan;5(1):47–50. Doi: 10.1097/00075197-200201000-00009. PMID: 11790949.

27 Goyal M. *Rasayana* in perspective of the present scenario. Ayu. 2018 Apr-Jun;39(2):63–64. Doi: 10.4103/ayu.AYU_300_18. PMID: 30783358; PMCID: PMC6369608.

28 Sushruta (2002) A text book entitled Sushruta Samhita by Atridev Motilal Banarasidas publication, Delhi, Chikista sthana Anagatbadhapratished adhyam 24/68.

29 Bhalerao S, Deshpande T, Thatte U. Prakriti (Ayurvedic concept of constitution) and variations in platelet aggregation. BMC Complement Altern Med. 2012 Dec 10; 12:248. doi: 10.1186/1472-6882-12-248. PMID: 23228069; PMCID: PMC3562518

30 Boyer JL. Bile formation and secretion. Compr Physiol. 2013 Jul;3(3):1035–78. doi: 10.1002/cphy.c120027. PMID: 23897680; PMCID: PMC4091928.

31 Trefts E, Gannon M, Wasserman DH. The liver. Curr Biol. 2017 Nov 6;27(21):R1147–R1151. doi: 10.1016/j.cub.2017.09.019. PMID: 29112863; PMCID: PMC5897118

32 Ullah R, Rauf N, Nabi G, Ullah H, Shen Y, Zhou YD, Fu J. Role of nutrition in the pathogenesis and prevention of non-alcoholic fatty liver disease: recent updates. Int J Biol Sci. 2019 Jan 1;15(2):265–276. doi: 10.7150/ijbs.30121. PMID: 30745819; PMCID: PMC6367556.
33 Wang R, Tang R, Li B, Ma X, Schnabl B, Tilg H. Gut microbiome, liver immunology, and liver diseases. Cell Mol Immunol. 2021 Jan;18(1):4–17. doi: 10.1038/s41423-020-00592-6. Epub 2020 Dec 14. PMID: 33318628; PMCID: PMC7852541.
34 Robinson MW, Harmon C, O'Farrell C. Liver immunology and its role in inflammation and homeostasis. Cell Mol Immunol. 2016 May;13(3):267–76. doi: 10.1038/cmi.2016.3. 2016 Apr 11. PMID: 27063467; PMCID: PMC4856809
35 Cheng X, Xia Y, Serti E, Block PD, Chung M, Chayama K, Rehermann B, Liang TJ. Hepatitis B virus evades innate immunity of hepatocytes but activates cytokine production by macrophages. Hepatology. 2017 Dec;66(6):1779–1793. Doi: 10.1002/hep.29348. PMID: 28665004; PMCID: PMC5706781
36 Mansour D, McPherson S. Management of decompensated cirrhosis. Clin Med 2018 Apr 1;18(Suppl 2):s60–s65. doi: 10.7861/clinmedicine.18-2-s60. PMID: 29700095; PMCID: PMC6334027.
37 Agnivesa: (2004): A Textbook entitled Charak Samhita, edited by Acharya J.T, Chaukhamba Publication, Sanskrit, Varanasi. Panduroga Chikitsitam Adhyaya.
38 Agnivesa: (2004): A Textbook entitled Charak Samhita, edited by Acharya J.T, Chaukhamba Publication, Sanskrit, Varanasi. Shvayathu chikista.
39 Charak (1983) A Textbook entitled Charak Samhita, edited by Sharma P, Chaukhamba Orientalia Publication, Sanskrit, Delhi. Panduroga Chikitsitam Adhyayam, page number 272-276 verse 16/34,124,133.
40 Charak: (1983): A Textbook entitled Charak Samhita, edited by Sharma P, Chaukhamba Orientalia Publication, Sanskrit, Delhi. Udar Chikitsitam Adhyayam, page number verse 13/22.
41 Agnivesa: (2004): A Textbook entitled Charak Samhita, edited by Acharya J.T, Chaukhamba Publication, Sanskrit, Varanasi. Maharogadhyay Adhyaya, page number 114 verse 20/13-14.
42 Baikampady SV. Dyspnoea on exertion in patients of heart failure as a consequence of obesity: An observational study. Ayu. 2013 Apr;34(2):160–6. doi: 10.4103/0974-8520.119671. PMID: 24250124; PMCID: PMC3821244.
43 Ferrell L. Liver pathology: cirrhosis, hepatitis, and primary liver tumors. Update and diagnostic problems. Mod Pathol. 2000 Jun;13(6):679–704. doi: 10.1038/modpathol.3880119. PMID: 10874674.
44 Hari A. Muscular abnormalities in liver cirrhosis. World J Gastroenterol. 2021 Aug 7;27(29):4862–4878. doi: 10.3748/wjg. v27.i29.4862. PMID: 34447231; PMCID: PMC8371506.
45 John S, Thuluvath PJ. Hyponatremia in cirrhosis: pathophysiology and management. World J Gastroenterol. 2015 Mar 21;21(11):3197–205. doi: 10.3748/wjg. v21.i11.3197. PMID: 25805925; PMCID: PMC4363748
46 Charak: (1983): A Textbook entitled Charak Samhita, edited by Sharma P, Chaukhamba Orientalia Publication, Sanskrit, Delhi. Panduroga Chikitsitam Adhyayam, page number 273 verse 16/11.
47 Atridev (2002) A text book entitled Sushruta Samhita edited by Acharya Dr Ghanekar, Motilal Banarasidas Publication, Varanasi Panduroga chikitsa Adhyaya page number 729.
48 Charak: (1983): A Textbook entitled Charak Samhita, edited by Sharma P, Chaukhamba Orientalia Publication, Sanskrit, Delhi. Udara Chikitsa Adhyayam, page number 210 verse 13/55–58.
49 Agnivesa: (2004): A Textbook entitled Charak Samhita, edited by Acharya J.T, Chaukhamba Publication, Sanskrit, Varanasi. Ashtodariyam Adhyayam, page number 115 verse 20-15/16.
50 Vagbhata: (2000): A Textbook entitled ASTANGAHRDAYAM, edited by Paradkar H. Krishnadas Academy, Sanskrit Publication, Varanasi. Doshabhediyam Adhyaya, page number 206.
51 Vagbhata: (2000): A Textbook entitled ASTANGAHRDAYAM, edited by Paradkar H. Krishnadas Academy, Sanskrit Publication, Varanasi. Doshabhediyam Adhyaya, page number 206.
52 Chakrapani Datta (2005) Charak Samhita edited by Y.G Joshi Vaidya Mitra prakashana Marathi, Pune Maharoga adhayaya 266.
53 Ferenczi P. Hepatic encephalopathy. Gastroenterol 2017 May;5(2):138–147. doi: 10.1093/gastro/gox013. pub 2017 Apr 18. PMID: 28533911; PMCID: PMC5421503.alpa.
54 Chakrapanidatta (2005) Charak Samhita edited by Y.G Joshi Vaidya mitra prakashana Marathi, Pune Dirghajivitiya Adhyaya page 17–18.
55 Agnivesa: (2004): A Textbook entitled Charak Samhita, edited by Acharya J.T, Chaukhamba Publication, Sanskrit, Varanasi. Langhana Bruhaniyam Adhyayam, page number 120.
56 Vagbhata: (2000): A Textbook entitled ASTANGAHRDAYAM, edited by Paradkar H. Krishnadas Academy, Sanskrit Publication, Varanasi. Annaswarupa Vidnyaniyam Adhyayam, page number 84–124.

57 Kuchewar VV, Borkar MA, Nisargandha MA. Evaluation of antioxidant potential of Rasayana drugs in healthy human volunteers. Ayu. 2014 Jan;35(1):46–9. doi: 10.4103/0974-8520.141919. PMID: 25364199; PMCID: PMC4213967.

58 Tripathi (1978) Kshemakutuhal edited by Dr. Gorakhanath Chaturvedi, Chowkhamba vishwa Bharti, Varanasi Page no 1–229.

59 Raghunath (1933) Bhojanakutuhal edited by Dr. Madhav Shastri 8–334.

60 Chen J, Raymond K. Beta-glucans in the treatment of diabetes and associated cardiovascular risks. Vasc Health Risk Manag. 2008;4(6):1265–72. doi: 10.2147/vhrm. s3803. PMID: 19337540; PMCID: PMC2663451.

61 Choi WH, Um MY, Ahn J, Jung CH, Ha TY. Cooked rice inhibits hepatic fat accumulation by regulating lipid metabolism-related gene expression in mice fed a high-fat diet. J Med Food. 2014 Jan;17(1):36–42. doi: 10.1089/jmf.2013.3058. PMID: 24456353.

62 Malik S. Pearl millet-nutritional value and medicinal uses. International Journal of Advance Research and Innovative Ideas in Education. 2015;1(3):414–8.

63 Sharma H, Kumar P, Deshmukh RR, Bishayee A, Kumar S. Pentacyclic triterpenes: New tools to fight metabolic syndrome. Phytomedicine. 2018 Nov 15; 50:166–77.

64 Agnivesa: (2004): A Textbook entitled Charak Samhita, edited by Acharya J.T, Chaukhamba Publication, Sanskrit, Varanasi. Apamargatanduliyam Adhyayam, page number 26.

65 Susruta: (2005): A textbook entitled Susruta Samhita, edited by Acharya J. T., Choukhambha Orientalia publication, Sanskrit, Varanasi. Annapanavidhi Adhyayam, page number 227.

66 Agnivesa: (2004): A Textbook entitled Charak Samhita, edited by Acharya J.T, Chaukhamba Publication, Sanskrit, Varanasi. Santarpana Adhyayam, page number 123.

67 Mbaoji, F., & Nweze, J. A. (2020). Antioxidant and hepatoprotective potentials of active fractions of Lannea barteri Oliv. (Anarcadiaceae) in rats. *Heliyon*, *6*(6), e04099. https://doi.org/10.1016/j.heliyon.2020.e04099

68 Dr. Anna Kunthe; reprint 2002 Astanga Hridaya of Vagbhata with the commentaries: sarvangasundara of arundatta & ayurvedarasayana of hemadri; Choukhamba Surabharati, Varanasi; Nidansthan 13/15-17.

69 Olmo-Cunillera A, Escobar-Avello D, Pérez AJ, Marhuenda-Muñoz M, Lamuela-Raventós RM, Vallverdú-Queralt A. Is Eating Raisins Healthy? Nutrients. 2019 Dec 24;12(1):54. doi: 10.3390/nu12010054. PMID: 31878160; PMCID: PMC7019280.

70 Wong TL, Strandberg KR, Corley CR, Fraser SE, Nagulapalli Venkata KC, Fimognari C, Sethi G, Bishayee A. Pomegranate bioactive constituents target multiple oncogenic and onco suppressive signalling for cancer prevention and intervention. Semin Cancer Biol. 2021 Aug; 73:265–293. doi: 10.1016/j.semcancer.2021.01.006. Epub 2021 Jan 24. PMID: 33503488.

71 Maqsood S, Adiamo O, Ahmad M, Mudgil P. Bioactive compounds from date fruit and seed as potential nutraceutical and functional food ingredients. Food Chem. 2020 Mar 5; 308:125522. doi: 10.1016/j.foodchem.2019.125522. Epub 2019 Sep 30. PMID: 31669945.

72 Prof.K.R.srikanth Murthy; 4th edition 2005; Astanga sangraha of vagbhata (text, English translation, notes, Appendices and index) Vol.I,II,III; Choukhamba Orientalia, Varanasi; Sutrasthan 6/16.

73 Basaranoglu M, Basaranoglu G, Bugianesi E. Carbohydrate intake and non-alcoholic fatty liver disease: fructose as a weapon of mass destruction. Hepatobiliary Surg Nutr. 2015 Apr;4(2):109–16. doi: 10.3978/j.issn.2304-3881.2014.11.05. PMID: 26005677; PMCID: PMC4405421.

74 Mao QQ, Xu XY, Cao SY, Gan RY, Corke H, Beta T, Li HB. Bioactive compounds and bioactivities of ginger (*Zingiber officinale* Roscoe). Foods. 2019 May 30;8(6):185. doi: 10.3390/foods8060185. PMID: 31151279; PMCID: PMC6616534.

75 Kulczyński B, Gramza-Michałowska A. The profile of carotenoids and other bioactive molecules in various pumpkin fruits (*Cucurbita maxima* Duchesne) Cultivars. Molecules. 2019 Sep 4;24(18):3212. doi: 10.3390/molecules24183212. PMID: 31487816; PMCID: PMC6766813.

76 Nagpal R, Behare PV, Kumar M, Mohania D, Yadav M, Jain S, Menon S, Parkash O, Marotta F, Minelli E, Henry CJ, Yadav H. Milk, milk products, and disease-free health: an updated overview. Crit Rev Food Sci Nutr. 2012;52(4):321–33. doi:10.1080/10408398.2010.500231. PMID: 22332596.

77 Sarker, U., Oba, S. Drought stress enhances nutritional and bioactive compounds, phenolic acids and antioxidant capacity of *Amaranthus* leafy vegetable. *BMC Plant Biol* **18**, 258 (2018). https://doi.org/10.1186/s12870-018-1484-1

78 Mahan K. L. Escott-Stump S. & Krause M. V. (2008). *Krause's food and nutrition therapy|krause's food nutrition and diet therapy* (12th ed.). edited L. Kathleen Mahan, Sylvia Escott-Stump by Elsevier Saunders. 713–738.

79 Gul P, Bakht J. Antimicrobial activity of turmeric extract and its potential use in food industry. J Food Sci Technol. 2015 Apr;52(4):2272–2279. doi: 10.1007/s13197-013-11.

80 Slavin J. Fiber and prebiotics: mechanisms and health benefits. Nutrients. 2013 Apr 22;5(4):1417–35. doi: 10.3390/nu5041417. PMID: 23609775; PMCID: PMC3705355.95-4. Epub 2013 Nov 8. PMID: 25829609; PMCID: PMC4375173.
81 Green GR. Mechanism of hypogonadism in cirrhotic males. Gut. 1977 Oct;18(10):843–53. doi: 10.1136/gut.18.10.843. PMID: 590844; PMCID: PMC1411687.
82 Agnivesa: (2004): A Textbook entitled Charak Samhita, edited by Acharya J.T, Chaukhamba Publication, Sanskrit, Varanasi. Santarpaniyam Adhyayam, page number 123
83 Ullah, R., Rauf, N., Nabi, G., Ullah, H., Shen, Y., Zhou, Y., & Fu, J. (2019). Role of nutrition in the pathogenesis and prevention of non-alcoholic fatty liver disease: recent updates. *International Journal of Biological Sciences*, *15*(2), 265–276. https://doi.org/10.7150/ijbs.30121
84 Junfen Fu1 Ross AB, Godin JP, Minehira K, Kirwan JP. Increasing whole grain intake as part of prevention and treatment of non-alcoholic Fatty liver disease. Int J Endocrinol. 2013; 2013:585876. doi: 10.1155/2013/585876. Epub 2013 May 16. PMID: 23762052; PMCID: PMC3670556.
85 Osna NA, Donohue TM Jr, Kharbanda KK. Alcoholic liver disease: pathogenesis and current management. Alcohol Res. 2017;38(2):147–161. PMID: 28988570; PMCID: PMC5513682.
86 Adkar P, Dongare A, Ambavade S, Bhaskar VH. Trapa bispinosa Roxb.: a review on nutritional and pharmacological aspects. Adv Pharmacol Sci. 2014; 2014:959830. doi: 10.1155/2014/959830. Epub 2014 Feb 10. PMID: 24669216; PMCID: PMC3941599.
87 Jill L Sherriff, Therese A O'Sullivan, Catherine Properzi, Josephine-Lee Oddo, Leon A Adams, Choline, Its Potential Role in Non-alcoholic Fatty Liver Disease, and the Case for Human and Bacterial Genes, *Advances in Nutrition*, Volume 7, Issue 1, January 2016, Pages 5–13, https://doi.org/10.3945/an.114.007955
88 Nongdam P, Tikendra L. The nutritional facts of bamboo shoots and their usage as important traditional foods of Northeast India. Int Sch Res Notices. 2014 Jul 20; 2014:679073. doi: 10.1155/2014/679073. PMID: 27433496; PMCID: PMC4897250
89 Alok S, Jain SK, Verma A, Kumar M, Mahor A, Sabharwal M. Plant profile, phytochemistry and pharmacology of *Asparagus racemosus* (Shatavari): A review. Asian Pac J Trop Dis. 2013 Jun;3(3):242–51. doi: 10.1016/S2222-1808(13)60049-3. PMCID: PMC4027291
90 Vagbhata: (2000): A Textbook entitled ASTANGAHRDAYAM, edited by Paradkar H. Krishnadas Academy, Sanskrit Publication, Varanasi. Doshabhediyam Adhyayam,12/51,52, 201.
91 Saleem M. Lupeol, a novel anti-inflammatory and anti-cancer dietary triterpene. Cancer Lett. 2009 Nov 28;285(2):109–15. doi: 10.1016/j.canlet.2009.04.033. Epub 2009 May 22. PMID: 19464787; PMCID: PMC2764818
92 Anand AC. Nutrition and muscle in cirrhosis. J Clin Exp Hepatol. 2017 Dec;7(4):340–357. doi: 10.1016/j.jceh.2017.11.001. Epub 2017 Nov 8. PMID: 29234200; PMCID: PMC5719462
93 Clark S, Mora García MB. A 100-year review: Advances in goat milk research. J Dairy Sci. 2017 Dec;100(12):10026–10044. doi: 10.3168/jds.2017-13287. PMID: 29153153
94 Bhattacharya K. Investigation and management of the hepatic glycogen storage diseases. Transl Pediatr. 2015 Jul;4(3):240–8. doi: 10.3978/j.issn.2224-4336.2015.04.07. PMID: 26835382; PMCID: PMC4729058
95 Bhanji RA, Carey EJ, Watt KD. Review article: maximising quality of life while aspiring for quantity of life in end-stage liver disease. Aliment Pharmocol Ther. 2017 Jul;46(1):16–25. doi: 10.1111/apt.14078. Epub 2017 May 2. PMID: 28464346
96 Tandon P, Montano-Loza AJ, Lai JC, Dasarathy S, Merli M. Sarcopenia and frailty in decompensated cirrhosis. J Hepatol. 2021 Jul;75 Suppl 1(Suppl 1):S147–S162. doi: 10.1016/j.jhep.2021.01.025. PMID: 34039486; PMCID: PMC9125684
97 Ooi, P. H., Hager, A., Mazurak, V. C., Dajani, K., Bhargava, R., Gilmour, S. M., & Mager, D. R. (2019). Sarcopenia in chronic liver disease: impact on outcomes. *Liver Transplantation*, *25*(9), 1422–1438. https://doi.org/10.1002/lt.25591
98 John S, Thuluvath PJ. Hyponatremia in cirrhosis: pathophysiology and management. World J Gastroenterol. 2015 Mar 21;21(11):3197–205. doi: 10.3748/wig. v21.i11.3197. PMID: 25805925; PMCID: PMC4363748
99 Alqahtani, S. A., & Jang, S. (2021). Pathophysiology and management of variceal bleeding. *Drugs*, *81*(6), 647–667. https://doi.org/10.1007/s40265-021-01493-2
100 Pfisterer N, Unger LW, Reiberger T. Clinical algorithms for the prevention of variceal bleeding and rebleeding in patients with liver cirrhosis. World J Hepatol. 2021 Jul 27;13(7):731–746. doi: 10.4254/wjh.v13.i7.731. PMID: 34367495; PMCID: PMC8326161. .

101 Raghuvanshi D, Sharma K, Verma R, Kumar D, Kumar H, Khan A, Valko M, Alomar SY, Alwasel SH, Nepovimova E, Kuca K. Phytochemistry, and pharmacological efficacy of Cordia dichotoma G. Forst. (Lashuda): A therapeutic medicinal plant of Himachal Pradesh. Biomed Pharmacother. 2022 Sep; 153:113400. doi: 10.1016/j.biopha.2022.113400. Epub 2022 Jul 20. PMID: 36076525.

102 Low G, Alexander GJ, Lomas DJ. Hepatorenal syndrome: aetiology, diagnosis, and treatment. Gastroenterol Res Pract. 2015; 2015:207012. doi: 10.1155/2015/207012. Epub 2015 Jan 12. PMID: 25649410; PMCID: PMC4306364

103 Prakash, R., & Mullen, K. D. (2010). Mechanisms, diagnosis and management of hepatic encephalopathy. *Nature Reviews Gastroenterology &Amp; Hepatology*, *7*(9), 515525. https://doi.org/10.1038/nrgastro.2010.116

Therapeutic Nutrition in Ayurveda for Cardiology

Pranesh G. Sanap

INTRODUCTION

Despite advances in modern medicine, cardiovascular disease (CVD) mortality remains a major cause of death in men and women worldwide.[1] Due to the aging of the population in developed countries and the increased prevalence of risk factors and comorbidities like type 2 diabetes mellitus (T2DM) and hypertension (HTN) in developing countries, CVD, especially atherosclerotic CVD (ASCVD), is increasing rapidly.[2]

Prevalence

In 2020 approximately 19 million deaths were attributed to CVD globally, which amounted to an increase of 18.7% from 2010 (American Heart Association), and 75% of CVD deaths occur in low- and middle-income countries.[3] About 85% of all CVD deaths are due to heart attack and stroke.[4]

The world's biggest cause of death is ischemic heart disease (IHD), responsible for 16% of the world's total deaths. Since 2000 the largest increase in deaths has been of IHD, rising by more than 2 million.[5]

The average annual direct and indirect cost of CVD in the United States was an estimated $378 billion from 2017 to 2018.

IHD ranked second in 2019 for a disability-adjusted life year (DALY) loss of 180 million compared to 2000, when it was 144 million. Heart failure (HF) and IHD patients morbidity is increasing because of early therapeutic interventions like thrombolysis and percutaneous angioplasty (PTCA).[2]

UNMET NEED FOR THERAPEUTIC NUTRITION

Suboptimal diet was found to be related to nearly half of the deaths in the United States from heart disease (CVD), stroke, and T2DM in a study conducted in 2012 (Box 7. 1).

DOI: 10.1201/9781003345541-10

BOX 7.1 THE 10 DIETARY FACTORS ASSOCIATED WITH HALF OF THE ESTIMATED CVD DEATHS IN THE US IN 2012(6)

1. High sodium
2. Low nuts and seeds, legumes
3. High-processed meats
4. Low seafood
5. Low vegetables
6. Low fruits
7. High-sugar sweetened beverages
8. Low whole grains
9. Low PUFA replacing carbohydrates or saturated fats
10. High unprocessed red meat

Despite the well-established correlation between suboptimal diet and CVD deaths, clinicians continue to provide rudimentary advice to patients about diet preferences (like eating more fresh fruits and vegetables, cutting down on sweets and processed foods, increasing consumption of fish, nuts, and legumes, etc.). This is because of inadequate nutritional education during medical schooling and for residency doctors. In a recently updated survey, 56% of the senior cardiology fellows reported receiving no nutrition education during their training, and 90% of practicing cardiologists reported receiving no or minimal nutritional education during their fellowship.[7]

So there is a need to fill this gap in therapeutic nutritional education among clinicians and people.

NUTRITION IN CARDIOLOGY FROM CLASSICAL TEXTS

Ayurved believes in holistic health, i.e. prevention is better than cure, and so emphasizes a healthy lifestyle with special chapters for foods and diet patterns (*Annavarga* and *Aharvidhi*) in *CharakSamhita, Ashtang Hriday,* and *Bhavprakash Nighantu.*

Hrudya foods (*Hrudya*) means beneficial for cardiovascular health, described in *Gana*[8] (group of foods/herbs) in Ayurvedic scriptures.

Following are some food categories described as *Hrudya* for cardiac health in Ayurved classical texts:

NUTRITION IN CARDIOLOGY FROM CONVENTIONAL SCIENCE

Cardiovascular diseases are a set of interrelated disorders comprising the following diseases (Box 7.2):

TABLE 7.1 *Hridya* (Cardiotonic) Food in *Dwadash Ahar Kalpana* (12 Food Classes)

CATEGORY	*FOOD*		*DESCRIPTION*
***1. Shuka Dhanya* (Whole Grains)**	Rice unpolished (*Oryza sativa*) Brown rice		Brown rice is recommended for its properties as *Hrudya*.9 A brown rice porridge is *Balya* (nutritious) and *Ruchikar* (appetizer).[10]
***2. Shami Dhanya* (Pulses and legumes)**	Whole green *Moong* (*Vigna radiate*) *Masoor (Lens culinaris)* *Mash* – Black gram (*Vigna mungo*)		Whole green *Moong* (green lentils) (unsprouted) is the healthiest of all pulses; *Madhur Kashay Ras* and *Laghu Guna* (sweet pungent taste and light on digestion) all collectively work well for heart patients.[11] Other lentils include *Masoor* (red lentils) and black gram which can also be used.[11]
***3. Mamsa Varga* (Meat products)**	*Ajabal Mamsa*[12] lean goat meat (*Capra hircus*) *Rohumatsyamamsa*[13] (*Labeo rohita*)		*Ajabalmamsa*[12] (Unprocessed meat), *Mansras*[13] (Meat/Chicken soup*)*, *Ghrutpakvamans*[14] (Meat cooked in cow ghee*)*, and *Rohumatsyamamsa*[15] (*Labeo rohita* – a fish) are described as beneficial for CVD.
***4. Shak Varga* (Vegetables)**	**Leafy vegetables** *Kakmachi* (*Solanum nigrum*) leaves[16] *Chanak shak*[19] (*Cicer arietinum* – Chickpea) leaves *Punanrnava*[21] (*Boerhavia diffusa*) leaves *Shigru*[23] (*Moringa oleifera*) leaves *Dhanyak*[25] (*Coriandrum sativum* – Coriander) leaves *Ambada*[27] (*Hibiscus sabdariffa* – Roselle leaves)	**Fruit vegetables** *Patol*[17,18] (*Trichosanthes dioica*) Pointed gourd *Kushmand*[20] (*Benincasa hispida*) – Ash gourd *Alabu*[22] Lauki (*Lagenaria siceraria*) bottle gourd *Vruntak*[24] (*Solanum melongena* – Eggplant) *Tindis*[26] (*Praecitrullus fistulosus* – Indian round gourd) *Shigru*[28] (*Moringa oleifera*) fruit	Leafy greens and fruit vegetables are generally considered beneficial for heart health.

TABLE 7.1 *(Continued)* *Hridya* (Cardiotonic) Food in *Dwadash Ahar Kalpana* (12 Food Classes)

CATEGORY	*FOOD*	*DESCRIPTION*
***5. Fal Varga* (Fruits and Dried Fruits)**	*Dadim* (*Punica granatum* – Pomegranate), *Rajbadar* (*Ziziphus mauritiana* – berry – Indian jujube), *Badar* (*Ziziphus jujuba* – berry), *Karmard* (*Carissa carandas* – Bengal currant), *Vrikshamla* (*Garcinia indica*), *Amlavetas (Garcinia pedunculata)*, *Mahalunga* (*Citrus medica* Linn. – Huge lemon),[29] *Jambir Nimbu*[30] (*Citrus limon* Linn. Burm. F. – Lemon) *Parushak*[31] (*Grewia asiatica* – Falsa) *Narikeljal*[32] (*Cocos nucifera* – Coconut water), *Khajur*[33] (*Phoenix dactylifera* – Dates), *Akshotak*[34] (*Juglans regia* – Walnut) *Priyal*[35] (*Buchanania Lanzan*) is described as *Hridya*.	
***6. Haritavarga* (herbs/tubers consumed raw)**	*Rason* (*Allium sativum* – Garlic), *Shatpushpa* (*Anethum sowa* – Dill) seeds, *Ajmoda* (*Trachyspermum ammi* – Carom) Seeds,[36] *Shatavari* (*Asparagusracemosus*), *Twakpatra* (*Cinnamomumverum* – Cinnamon leaves),[37] *Tulsi* (*Ocimum sanctum*) leaves, *Sabja* (*Ocimum basilicum*) leaves, Lemongrass (*Cymbopogon citratus*), *Aasuri* (*Brassica juncea* Linn. – Mustard) seeds[38]	Can be used for the enhancing taste of food
***7. Madya Varga* (Fermented products)**	*Tushodak*, *Jirna Madira* (Old wine)[39]	*Tushodak* is a clear liquid from fermented gruel made of barley with husk, *Jirna Madira* (old wine)described as *hrudya*[39]
***8. Jala Varga* (Water)**	*Antariksh Jala* (rainwater before touching earth) – Mountain water	*Antariksh Jala* is pure, clean, and light on digestion. Its properties change according to land receiving it. Water from rivers originating from high mountains is considered better[40]
***9. Goras Varga* (Milk and milk products)**	Cow curd *Takrapind* (unsalted yogurt), *Piyush* (colostrums), *Kilat* (thickened soured milk by adding lemon/curd when boiling), *Ksheer shaka*(unboiled soured milk)[42] *Takra*[42](Buttermilk), *Navneet* (Unsalted cow butter),[43] Cow ghee – clarified butter	Cow curd mixed with jaggery *Navneet* (Unsalted cow butter)[43] and cow ghee are preferred to cook food
***10. Ikshu Varga* Sugarcane and products; honey**	Old jaggery (*Saccharum officinarum*) (older than 1 year)	Old jaggery (older than 1 year) is described as *Hridya*[44]
***11. Krutanna Varga* Processed food**	*Vilepi* (thick soup)[45] and *Lajamanda* (Popped rice soup)[46]	Food made with Wheat is stated as *Hridya*, *Vilepi* (thick soup)[45] and *Lajmanda* (Popped rice soup) are beneficial for heart disease[46]

(*Continued*)

TABLE 7.1 (Continued) *Hridya* (Cardiotonic) Food in *Dwadash Ahar Kalpana* (12 Food Classes)

CATEGORY	*FOOD*	*DESCRIPTION*
12. Aharyoni Varga Vegetable oils, Salt, *Kshar* (alkali), and Spices like dry ginger, Pepper, asafetida, long pepper	Oil – *Erand tailam* – Castor oil (*Ricinus communis*) Salt – *Saindhav* (Rock salt) *Sauvarchal* (Black salt) Potassiumnitrate Spices – *Shunthi* (*Zingiber officinale*)	*Erand tel* (Castor oil) is *Hrudrujanashak* (effective for pain due to heart disease) and effective for heart diseases[47,48] *Saindhav* or Rock salt is healthier form to consume. It is *Hridya;* *Sauvarchal* (Black salt) is stated as *Hridya*[49] and so can be used in the first place for salt. *Shunthi*[50] – Dry Ginger
13. Pushpa Varga **Flowers**	*Shatapatri* – Rose (*Rosa rubiginosa*)	Rose flower, described as *Hrudya*,[51] can be consumed in the form of rose petal powder or rose petal, a flavor enhancer in foods like vegetables, rice, and soup

BOX 7.2 CARDIOVASCULAR DISEASES (CVDS) CLASSIFICATION

Congenital Heart Diseases – **CHD**
Cyanotic CHD, CHD in adults
Valvular Heart Disease – **VHD**
Rheumatic Heart Disease
Nonrheumatic VHD
Cardiac Arrhythmia
Bradyarrhythmias, Tachyarrhythmias
Supraventricular, Ventricular Arrhythmia – **AF, SVT, VT, VF**
Atherosclerotic CVD – **ASCVD**
Coronary Artery Disease – **CAD**
Ischemic Heart Disease – **IHD**; Myocardial Infarction – **MI**; Acute Coronary Syndrome – **ACS**
Heart Failure – **HF**
  Heart Failure with Preserved Ejection Fraction – **HFpEF**
  Heart Failure with Reduced Ejection Fraction – **HfrEF**
Hypertension – **HT**
Primary/essential hypertension
  Secondary hypertension
Pulmonary Hypertension – **PHT**
Diseases of Myocardium
  Cardiomyopathies (**CMP**), Dilated CMP
  Myocarditis and Pericardial effusion, Constrictive pericarditis

ATHEROSCLEROTIC CARDIOVASCULAR DISEASE (ASCVD)

ASCVD involves the narrowing of small blood vessels due to plaque (built-up lesion). The plaque known as atheroma can rupture, leading to the formation of a blood clot that blocks the artery resulting in myocardial ischemia and infarction (heart attack).

RISK FACTORS FOR ASCVD

Modifiable risk factors are physical inactivity, abdominal obesity, cigarette smoking, consumption of too few fruits and vegetables, consumption of too much alcohol, psychological factors like stress and type-A personality and hypercholesterolemia, i.e. increased lower density lipoprotein (LDL), and decreased high-density lipoprotein (HDL).[52] Metabolic syndrome (MetS)[52] has also emerged as a new modifiable risk factor for ASCVD (Box 7.3).

Unmodifiable risk factors are increasing age, male sex, and genetic predisposition (presence of ASCVD before age of 50 in first-degree relatives).

More than 50% of heart attacks occur in individuals with normal serum cholesterol, which has led to research on novel risk factors like inflammatory markers (hs-CRP), homocysteine, and TMAO (trimethylamine-N-oxide), which is a gut biota–dependent metabolite that contributes to CVD.[53]

HEART FAILURE

Heart failure (HF) is the disease of aging. The prevalence of HF rises from less than 1% in individuals below 60 years to nearly 10% in those over 80 years of age.[54] HF incidence and prevalence is increased in recent years due to increased comorbid populations like T2DM, HTN, IHD, VHD, and CKD patients survival.

Heart failure is classified as heart failure with preserved ejection fraction HFpEF and with reduced ejection fraction HFrEF, where left ventricular EF is reduced to normal. Clinical differentiation is

BOX 7.3 METABOLIC SYNDROME (METS) DEFINITION

Abdominal obesity – large waist circumference (WC) ≥90 cm in men and ≥80 cm in women independent of **BMI**

Dyslipidemia – increased triglycerides, decreased **HDL**

Increases fasting glucose **FBG** >110 mg/dL

Hypertension

Presence of acanthosis nigricans

difficult in both forms as exertional dyspnea, orthopnea, and easy fatigability are present in both types. Echocardiography plays an important role in accessing left ventricular function and thus estimating ejection fraction LV EF.

Rehospitalization (within 3 to 6 months) rate is as high as 30 to 50% in HF patients.[54] So a multidisciplinary approach including intense diet modification is important in advanced HF patients.

LIFESTYLE MODIFICATION IN CVD

As diet plays an important role in altering modifiable risk factors[55] for ASCVD, various diet patterns like the Mediterranean diet (MeD),[56] a dietary approach to stop hypertension (DASH) diet,[56] Ornish diet,[57] and vegan diet[58] are endorsed by organizations like American Heart Association (AHA) and National Cholesterol Education Program (NCEP) (Table 7.2).

Regular exercise reduces CVD risk in healthy individuals, specifically in those who are prone to ASCVD.

AYURVEDA PERSPECTIVE ON CVD

The heart is described as an important organ in the human body, *Hrudaya*, which is a place of *Ojas*[61] (essence of vitality).

All three *Doshas* (*Vata, Pitta,* and *Kapha*)at equilibrium maintain and regulate the healthy condition of the heart and vessels (Table 7.3).

Disease (*roga*) is a result of the derangement of three *Doshas* called a *Dosh prakop*[62] (Figure 7.1).

Dushya is a place or organ system where the disease occurs, like the cardiovascular system.

Dosh-Dushya sammurchana is the pathogenesis of the disease.

Ayurveda has described *Hrudrog,* which describes pathological conditions of the heart and main vessels.[63]

Hetu (general causative factors) of *Hrudroga*

Prolonged ingestion of spicy and salty food, excessive exertion, psychological stress, *Aamdosha* (toxins generated from improper food habits), trauma, and extensive dietary changes are some of the causative factors besides age, sex, and genetic predisposition.[63]

***Hrudroga* (CVD), *Hetu, and Samprapti* (Etiopathogenesis)**[63]

All three *Doshas* vitiated causes *Sannipataj Hrudrog,* also called *Krimij Hrudrog* in some scriptures.

THERAPEUTIC NUTRITION IN AYURVEDA FOR CVD *(AHAR IN AYURVEDA)*

Nutrition: General Considerations

Ayurveda believes in holistic health, i.e. prevention is better than cure, and so emphasizes a healthy lifestyle with special chapters for foods and diet patterns. *Ahar* (food) can work as *Aushadhi* (medicine) if taken in a proper way. Food is responsible to preserve health and soul.[64]

TABLE 7.2 AHAs Nutritional Guidelines for ASCVD[59,60]

FOOD CATEGORY	*GUIDELINES*	*SERVING SIZE/DAY*
Fruits and vegetables	1. Plenty of vegetables and fruits except starchy vegetables like white potatoes 2. Deeply colored like leafy greens 3. Whole fruits should be preferred as they provide more fiber and satiety rather than juice	5 servings of vegetables (2.5 cups) 4 servings of fruits (2 cups)
Whole Grains Barley, Brown rice, Oatmeal, Popcorn, and Whole grain bread	1. Whole grains have been reported to improve cardiovascular risk 2. Products made with at least 51% whole grain are classified as whole grain 3. Whole grains are rich in fiber and beneficial for laxation and gut microbiota	3–6 servings (3–6 ounces)
Protein Legumes, nuts, nonfried fish and seafood, lean meat, skinless poultry, eggs, and seeds	Healthy source of protein 1. Protein, preferably plantsourced (legumes and nuts) 2. Fish and seafood – AHA recommends eating fatty fish (high in omega-3 fatty acids) per week 3. Lean meat if desired	1–2 servings (5.5 ounces)
Oils	Use of liquid plant oils rich in unsaturated (polyunsaturated and monounsaturated) fats Nontropical vegetable oils good for a heart like olive, peanut, sunflower, safflower oil Tropical oils like palm and coconut oil are found to increase LDL cholesterol	3 tablespoons of Nontropical oil
Processed Foods Beveragesand foods with added sugar	**Unprocessed or Minimally Processed Food** Avoid intake of beverages and foods with high sugar, like soft drinks	
Salt (Sodium)	Little or no added salt is preferable in food preparations	Moderate salt restriction, i.e. 5–6 g of salt per day, is advised for all HF patients[14]
Alcohol	1. Drinking wine or any other alcohol for unproven health benefits is discouraged 2. Intake should be limited, if at all 3. Not more than 1 drink a day for women and 2 drinks a day for men	1 drink is -12 ounces of beer -5 ounces of wine -1.5 ounces of 80proof spirit (40% ABV), e.g. most of rums, tequilas, gins, whiskeys, vodkas
Caffeine from coffee, tea, soft drinks, chocolate, and some nuts	High intake related to increased risk of Coronary Artery Disease is still under study.	Moderate coffee drinking(1-2 cups a day) doesn't seem to be harmful 1–2 cups of coffee or tea a day
Dairy	Low-fat or no-fat dairy products over full-fat dairy products are advised	3 servings (3 cups)

(*Continued*)

TABLE 7.2 (Continued) AHAs Nutritional Guidelines for ASCVD

FOOD CATEGORY	*GUIDELINES*	*SERVING SIZE/DAY*
Fats	AHA recommends replacing saturated fats with unsaturated fats (PUFA, MUFA) 1. Avoid saturated fats that do rise LDL cholesterol 2. Avoid trans fats (partially hydrogenated oil is no longer generally recognized as safe [GRAS] in human food, for example, fried foods, baked goods like akes, pie crusts, biscuits, frozen pizza, cookies, crackers, margarine, and other spread)	
Nuts	Healthier nuts like Almond, hazelnut, peanuts, pecans, pistachios, and walnuts Walnuts are high in omega-3 fatty acids	1.5 ounces of the whole nut
Omega-3 fatty acids	High omega-3 fatty acid supplements had a beneficial effect on the incidence of HF and hospitalization in patients with T2DM, but not in those without T2DM This benefit was stronger in Asians and African patients with T2DM	
Water	Sufficient water intake is associated with a lower risk of CVD. The atherosclerosis risk in communities (ARIC) study concluded that there is an association between higher sodium levels and elevated risk for heart failure (March 2022). Decreased body water content is the most common factor that increases serum sodium; therefore, results suggest that staying well hydrated may slow down the aging process and prevent or delay chronic disease.[63]	

BOX 7.4 REGULAR PHYSICAL ACTIVITY TO MAINTAIN ENERGY (CALORIE) INTAKE AND EXPENDITURE[59]

150 minutes of moderate physical activity OR

75 minutes of vigorous physical activity OR

Combination of 50 minutes each of moderate and vigorous physical activity per week

Activity should be spread throughout the week

TABLE 7.3 Functions of *Doshas* at Normal Pace

KAPHA DOSHA	*PITTA DOSHA*	*VATA DOSHA*
1. Holds and maintains structures (Myocardium) 2. Normal endocardial, endothelial,and pericardial functions.	1. Responsible for normal coronary circulation 2. Digests and eliminates undesired products and protects against inflammatory conditions	1. Important for *Spandan* (Sinoatrial [SA]and Atrioventricular [AV] nodal activities) 2. Normal movements of valves and cardiac muscle

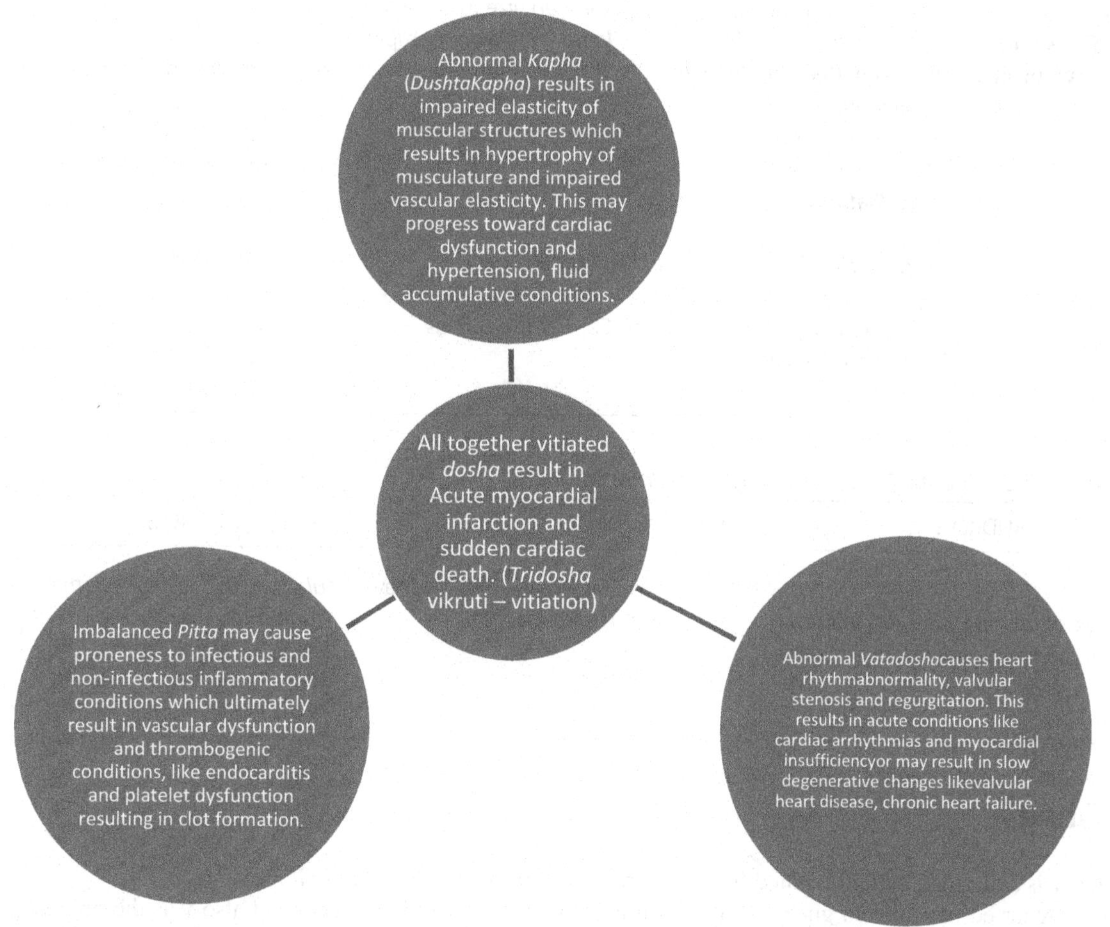

FIGURE 7.1 *Dosha Prakop* (Dosha vitiation).

Hridya means beneficial for cardiovascular health. *Hridya* foods are described in *Gana*(8)(group of foods/herbs) in *Ayurvedic* scriptures.

Hridya Gana includes foods that are sour and sweet in taste, as *Amla* (sour) is the taste for heart health.[66] A separate *Hridya Gana* is mentioned in scriptures.[8]

Also, the attributed chapter describes *Hitakar ahar*[67] (diet to be practiced) for cardiac patients. Food guidelines specific to types of heart disease are described in the literature (Table 7.5).

TABLE 7.4 *Hrudrog* Etiopathogenesis

Hetu (**Causes**)	Sedentary lifestyle, Intake of sweet, fatty food, *Santarpanjanya* (Overnutrition) ↓	Vigourous exercises, psychological stress, *Aptarpanjanya* (Undernutrition) ↓	Salty, spicy food habits, *Amadosha*, alcohol addiction, mental irritability ↓
Samprapti (**Process**)	*Prakupit* (Vitiated) ***Kapha*** predominance ↓	*Prakupit* (Vitiated) ***Vata*** predominance ↓	*Prakupit* (Vitiated) ***Pitta*** Predominance ↓
Lakshan (**Signs and symptoms**)	*Stambha* (heaviness), *Stimit* (chest discomfort), *Bharikam* (feeling of stone pressed) *Aruchi* (Anorexia) *Alasya* (heaviness) *Avrodhatmak* ↓	*Kampan* (irregular beats) *Jadta* (inefficient pumping) *Stambha* (numbness) *Ruja* (throbbing pain) *Murchha* (Giddiness) ↓	*Hriddaha* (chest burning) *Trushna* (thirst) *Amlodgar* (dyspepsia) *Sweda* (sweating) *Bhrama* (giddiness) *Ruja* (pain) ↓
Vyadhi (**Disease**)	***Kaphaj Hrudrog*** Hypertension, Pulmonary Hypertension, Ischemic Heart disease	***Vataj Hrudrog*** Chronic Heart Failure, Cardiac arrythmias	***Pittaj Hrudrog*** Rheumatic Heart Failure, Infective and noninfective carditis, Acute Heart Failure

TABLE 7.5 Specific Foods per Category of *Hrudroga*

VATAJ HRUDROG	PITTAJ HRUDROG	KAPHAJ HRUDROG
Milk, curd, cow ghee, *Jalchar-mamsa* (fish and seafood) *Anoopmans* (animals in or near water bodies)	Grape juice, Sugarcane juice, Honey *Parushak* (*Grewia asiatica*)	*Kulattha yush*, (*Macrotyloma uniflorum* soup) *Jangal-mamsa soup* *Madya* (wine) *Jav* (*Hordeum vulgare*)

Ghee

Ghee is a clarified butter obtained from cow milk or buffalo milk known as *Ghrita.*

Ayurved has utilized ghee as a therapeutic agent for thousands of years and also was the preferred cooking oil in ancient India.

Ghee nourishes *Ojas*, pacifies *Vata* and *Pitta,* and is acceptable for *Kapha* in moderation.[68]

A study invitro showed that consumption of ghee in the amount of 10% of total fat may show a rise in triglyceride levels but does not increase lipid peroxidation processes that are related to increased CVD risk.[70]

Another study conducted on the Indian rural population showed a lower prevalence of coronary heart disease in men with higher consumption of ghee (approx. 50% of total fat).[71]

Previous research and data available in the literature do not support a conclusion or harmful effects of the moderate consumption of ghee in the general population.

When ghee was used as the sole source of fat at a 10% level, there was a large increase in oleic acid levels and a large decrease in arachidonic acid levels in serum lipids.[72] Thus cow ghee can be used as a food ingredient or medicated ghee in cardiology patients safely.[73]

TABLE 7.6 Fatty Acid Profile of Ghee[69]

TOTAL FAT CONTENT	*98%*
Saturated fatty acids	Above 55% of total
Unsaturated fatty acids	Cow ghee – 28.31% Buffalo ghee – 25.19%
Monounsaturated fatty acids (MUFA)	Cow ghee – 24.28% Buffalo ghee – 22.74%
Polyunsaturated fatty acids (PUFA)	Cow ghee – 4.02% Buffalo ghee – 2.45%

AHAREYA KALPANA (DIETARY RECIPES) IN CARDIOLOGY

Ayurved cookbook recipes for Cardiac patients (Box 7.5):

BOX 7.5 RECIPES – *HRIDYA AHAREYA KALPANA*

Whole Grains

1. *Raktashali Vilepi* – brown rice cooked in four times (volume:volume) of water with black salt added for taste, tampered with cumin seeds, coriander seeds, turmeric, and ginger in ghee.[9]
2. *Lajamanda* – Popped rice boiled four times (volume:volume) of water for 5 to 6 minutes, with pink salt added (at medium flame of gas stove; temperature approximately 170°C open vessel).

Legumes

Soup made with green lentils, red grams, and horse gram in eight times (volume:volume) of water mixed with salt to taste, seasoned with ginger in ghee.[74]

Lean Meat and Fish

1. Tender lean mutton pieces cooked in ghee, pink salt, and suggested spices[75]
2. *Rohu* fish cooked in mustard oil with rocksalt and spices[76]

Vegetables

1. Eggplant Bharit[77] – Smashed and tossed eggplant with lemon juice, pink salt, ginger, and asafetida.
2. Leafy vegetables are first steamed and then cooked with a mixture of chickpea flour and buttermilk, spice, and salt added accordingly.[78]
3. Ash gourd (*Kaddu*) fruit – cooked, smashed, and seasoned with ghee, salt, and cumin seeds.[79]

Fruits

1. A simple lemonade or *Karwand* (*Carissa carandas*) juice is described.[80]
2. A mixed fruit drink made with pomegranate, *falsa* (*Grewia asiatica*), and *Jamun* (Indian blackberry) is called *Rag-shadava*ndis known to be *Hridya.*[81]
3. *Amraras* (mango juice) with jaggery is also mentioned as *Hridya.*[82]
4. *Amla* (*Emblica officinalis*) boiled and then smashed and cooked with asafoetida, pink salt, and ghee.[83]

CLINICAL SIGNIFICANCE OF THERAPEUTIC NUTRITION IN CARDIOLOGY TODAY

Dietary preventive interventions for cardiovascular diseases can be classified as[84]:

A) Primary Preventive

This applies to the healthy population that is prone to atherosclerotic cardiovascular diseases, like people with a positive family history of CVD. Patients with risk factors like obesity, existing diabetes, and hypertension are also to be considered for primary prevention.

AHA suggests that primary prevention of CVD should begin in children older than 2 years age.[85]

B) Secondary Preventive

Patients who already have suffered one or more episodes of ASCVD need intensive dietary changes and lifestyle modifications, which can be outlined as shown in Table 7.8.

DISCUSSION

The suboptimal diet was found to be related to nearly half of the deaths in the United States from heart disease (CVD), stroke, and T2DM in a study conducted in 2012.

TABLE 7.7 A Sample Diet Chart as Primary Prevention

CLASS	RECOMMENDATION
Emphasizes	Whole grains – Brown rice, whole wheat, *Jawar*, *Bajra*
Should include in daily food regime, minimum servings 4–6 per day	Legumes – Green Moong, Red lentil, Black gram, Chickpea, Horse gram
	Vegetables – Rich in fiber: leafy greens like leaves of Moringa, Chickpea, Coriander, Roselle, *Punarnava*
	Fruit vegetables like Pointed gourd, Ash gourd, Bottle gourd, Eggplant, Indian round gourd, Moringa
	Fruits – Pomegranate, all types of Berries, Citrus fruits, Coconut water
Includes	Dairy products preferably from cow milk – Low fat
Include in daily plate, serving size 2–3 per day	Milk, buttermilk, curd, cottage cheese
	Fish, Poultry, lean meat
	Nontropical oil – Sesame oil, olive, rice bran, Peanut oil and Cow ghee
	Dry nuts – Walnut, Almond, *Chironji*
Limits	Highly processed food- canned food
Try to avoid in daily regime	Bakery products – cake, biscuits, confectionaries
Avoid	High sugar-containing beverages – soft drinks
Should avoid	Alcohol

TABLE 7.8 Sample Diet Plan for Secondary Prevention

MEAL	*FOOD PREPARATIONS*	*FREQUENCY*
Breakfast	a. Boiled green lentils/Boiled chickpea/porridge	Twice weekly
	b. Ragi(*Eleusine coracana*) dosa/rice and black gram dosa, upma	Twice weekly
	c. *Thaleepith* (multigrain bread made with green vegetable)	Twice weekly
	d. Vegetable (carrot/pumpkin/bottle gourd/fenugreek leaves, etc.) paratha with less oil.	Twice weekly
	e. Boiled egg white	Twice weekly
	f. Any seasonal local fruit and dry nuts	Daily
Lunch	a. *Roti*/flatbread of either: Wheat (*Triticum vulgare*) or *Ragi* (*Eleusine coracana*) or Jowar (*Sorghum bicolor*) With green leaf or fruit vegetable (like pumpkin, eggplant, bottle gourd, ash gourd, pointed gird) with masoor or moong dal, rice (preferably brown or unpolished), salad of cucumber, radish, tomato, asparagus	Daily
	b. Lean meat	Weekly
Snacks	a. A seasonal fruit of choice/dates – Nuts, raisins, makhanas b. Popped amaranth/puffed rice/puffed jowar c. Beaten curd with honey/Buttermilk, d. Vegetable soup/*Ragi* porridge e. Multi-whole grain soup f. *Panak* (lemon/kokum/green mango)/Coconut water	Combination of any two daily
Dinner	a. *Ragi/Jowar roti*, vegetable Dal	Daily
	b. Fish curry/Grilled fish	Twice weekly

Ayurveda describes CVD as *Hrudroga* in detail, with special emphasis on diet and lifestyle management. Food categories like plant-based food sources like whole grains (*Shukdhanya*), legumes, green vegetables and fruits (*Shaka*), medicated alcohol (*Asava*, *Arishta*), fruits with a sour taste (*Amlaras fal*) – berries and dry fruits – spices, and herbs are described as *Hridya* (beneficial for heart health). Also, eating a wide variety of fruits and vegetables every day helps ensure an adequate intake of many vitamins, minerals, phytochemicals, and fiber. Studies have shown that eating this food is protective against CVD and cancer.

Previous research and data available in the literature do not support a conclusion about the harmful effects of the moderate consumption of ghee in the general population. AHA also advises replacing saturated fat with more healthy fats such as nontropical oils and unsaturated fats rich in MUFA, PUFA, omega-3 FA, and linoleic and oleic acids, which are beneficial for the heart.

Ayurveda emphasizes plant-based local food obtained from fresh sources prepared with the least refinement and processes, which is composed of fruits, vegetables, whole grains, nuts, and spices. Animal products should be used of native and open-farmed domestic animals, which comprises cow milk and milk products and poultry products.

With the advances in the food industry, packaged food and staple food are flooding the market. Those should be better avoided or selected wisely with the above guidelines.

Further study on the nutritional value of cultural food preferences (native food) with behavioral factors is necessary.

TABLE 7.9 What to Be Looked for on Food Packets/Labels

Sugar	High-fructose corn syrup, corn syrup, agave barley, malt syrup, dehydrated cane juice
Sodium	Salt, sodium benzoate, disodium/monosodium glutamate (MSG), sodium nitrite (present in most of packaged food/ tinned food as a preservative)
Trans fats	Partially hydrogenated oil and hydrogenated oil, soft margarine

Modern agricultural practices like genetically modified crops, hybrid crops, and indiscriminate use of chemical fertilizers, insecticides, and pesticides have tremendously reduced the nutritive value of all food sources. Therefore, traditional organic methods should be promoted for agricultural practices as food is a therapy for a disease-free life.

NUTRAVIGILANCE

Knowing the food and making a better choice for heart health is an important step toward protection from CVD. Food picked from the grocery shop or ready-to-eat food packets or that delivered via online should be looked for ingredients. Reading nutritional information or facts on food labels and the ingredients listed on them is good practice. Generally, ingredients are listed in order of quantity, but the main culprit ingredients like sodium, added sugar, and saturated and trans fats may be listed at last or somewhere in the middle with several other names we do know (Table 7.9).

The American Heart Association (AHA) has endorsed the iconic **Heart-Check Mark** on food packets in grocery stores since 1990. More efforts from governing bodies like spreading awareness and promoting nutritional vigilance among people and clinicians are important, as diet can heal and suboptimal diet can slowly kill.

REFERENCES

1. World Health Organization. (2019). Cardiovascular diseases. Retrieved June 11, 2019, from www.who.int/health-topics/cardiovascular-diseases
2. Papadakis, M. (Ed.). (2022). Current Medical Diagnosis & Treatment 2022 (59th ed.). McGraw Hill Lange Publications. p. 354.
3. World Health Organization. (n.d.). Cardiovascular diseases. Retrieved from www.who.int/health-topics/cardiovascular-diseases#tab=tab_1
4. World Health Organization. (2021). Cardiovascular diseases (CVDs). Retrieved from www.who.int/news-room/fact-sheets/detail/cardiovascular-diseases-(cvds)
5. World Health Organization. (n.d.). Global health estimates: Leading causes of death. Retrieved from www.who.int/data/gho/data/themes/mortality-and-global-health-estimates/ghe-leading-causes-of-death
6. MichaR, PenalvoJL, CudheaF, et al. Association between dietary factor and mortality from heart disease, stroke and T2DM in the US. JAMA. 2017;317(9):912–924.
7. DevriesS, Agatston A, AgarwalM, et al. A deficiency of nutritional education and practice in cardiology. Am J Med. 2017;130:1298–1305.
8. CharakSamhita Pratham Khand. Hindi translation by Kaviraj Shri Atridevji Gup. Bhargav Pustakalay Publication 2nd edition., 2000 . Sutrasthan, Adhyay 4, verse 10, page 51.
9. Dhanyavarga, verse 3-9. In: Kaiyadeva Nighantu [Internet]. National Institute of Indian Medical Heritage; [cited 2023 Mar 4]. Retrieved from https://niimh.nic.in/ebooks/e-Nighantu/Kaiyadeva Nighantu/?mod=read.

10. CharakSamhita PrathamKhand. Hindi translation by Kaviraj Shri Atridevji Gup. Bhargav Pustakalay Publication, 2nd edition., 2000. Sutrasthan, Adhyay 27, verse 250, page 359.
11. KrutannaVarga, verse 62-73. In: Kaiyadeva Nighantu [Internet]. National Institute of Indian Medical Heritage; [cited 2023 Mar 4]. Retrieved from https://niimh.nic.in/ebooks/e-Nighantu/Kaiyadeva Nighantu/?mod=read.
12. Kshem kutuhal Manjula-Hindi VyakhyaVibhushitam by Dr IndradevTripathi. Shashthotsav, verse 82-83.
13. Charak Samhita Pratham Khand. Hindi translation by Kaviraj Shri AtridevjiGup. BhargavPustakalay Publication, 2nd edition., 2000. Sutrasthan, Adhyay 27, verse 274-276, page 362.
14. Kshem kutuhal Manjula-Hindi Vyakhya Vibhushitam by Dr Indradev Tripathi, Chaukhambha Orientalia publication 1978. Shashthotsav, verse 73.
15. Kshem kutuhal Manjula-Hindi VyakhyaVibhushitam by Dr IndradevTripathi, Chaukhambha Orientalia publication 1978. Saptamotsav, verse 19.
16. Kshem kutuhal Manjula-Hindi VyakhyaVibhushitam, by Dr IndradevTripathi, Chaukhambha Orientalia publication 1978. Ashtamotsav. 1978;126.
17. Kaiyadeva. Krutanna varga, verse 69-70. In: National Institute of Indian Medical Heritage. e-Nighantu. Retrieved from https://niimh.nic.in/ebooks/e-Nighantu/Kaiyadeva Nighantu/?mod=read.
18. VagbhatAHS. Ashtang Hriday–Sartha Vagbhat – by Lt.Dr.Ganesh Krushna Garde. Anmol Prakashan Edi 1891; reprinted 2003. Sutrasthan. Adhyay 6,verse 77.
19. Tripathi I. Kshemkutuhal Manjula-Hindi Vyakhya Vibhushitam, Chaukhambha Orientalia publication 1978. Ashtamotsav.1978;136.
20. Kaiyadeva. Aushadhivarga, verse 526-527. In: National Institute of Indian Medical Heritage. e-Nighantu. Retrieved from https://niimh.nic.in/ebooks/e-Nighantu/Kaiyadeva Nighantu/?mod=read.
21. Tripathi I. Kshem kutuhal Manjula-Hindi VyakhyaVibhushitam, Chaukhambha Orientalia publication 1978. Ashtamotsav. 1978;152.
22. Chunekar KC, Pandey GD, eds. Bhavprakash Nighantu Hindi Vyakhya (Commentary) by Chaukhambha Bharti Akadami reprint 2015. Shakvarga, verse 58-59.
23. Vagbhat AHS. Ashtang Hriday–Sartha Vagbhat – by Lt.Dr.Ganesh Krushna Garde. Anmol Prakashan Edi 1891; reprinted 2003. Sutrasthan. Adhyay 15, verse 59.
24. Vagbhat AHS. Ashtang Hriday–Sartha Vagbhat – by Lt.Dr.GaneshKrushnaGarde. AnmolPrakashan Edi 1891; reprinted 2003. Sutrasthan. Adhyay 6, verse 79-80.
25. Dhanvantari. Shatpushpadivarga, verse 66-68. In: National Institute of Indian Medical Heritage. e-Nighantu. Retrieved from https://niimh.nic.in/ebooks/e-Nighantu/dhanvantarinighantu/?mod=read.
26. Tripathi I. Kshem kutuhal Manjula-Hindi VyakhyaVibhushitam. Ashtamotsav. 1978;47.
27. Charak. Charak Samhita Pratham Khand. Hindi translation by Kaviraj Shri AtridevjiGup. Bhargav Pustakalay Publication, 2nd edition, 2000. Sutrasthan, Adhyay 4, verse 10.
28. Chunekar KC, Pandey GD, eds. Bhavprakash Nighantu Hindi Vyakhya by (Commentary). Chaukhambha Bharti Akadami reprint 2015. Guduchyadivarga, verse 104–106.
29. Charak Samhita Pratham Khand, Hindi translation by Kaviraj Shri AtridevjiGup, BhargavPustakalay Publication, 2nd edition, 2000. Sutrasthan, Adhyay 4, verse 10.
30. Ashtang Hriday–Sartha Vagbhat – by Lt. Dr. Ganesh KrushnaGarde by Anmol Prakashan Edi 1891; reprinted 2003, Sutrasthan. Adhyay 6.
31. Bhavprakash Nighantu Hindi Vyakhya (Commentary) by Prof. K.C. Chunekar edited by Lt. Dr. Gangasahay Pandey by Chaukhambha Bharti Akadami reprint 2015, Amradifalvarga, verse 98 and 99.
32. Bhavprakash Nighantu Hindi Vyakhya (Commentary) by Prof. K.C. Chunekar edited by Lt. Dr. Gangasahay Pandey by Chaukhambha Bharti Akadami reprint 2015, Amradifalvarga, verse 40-41.
33. Bhavprakash Nighantu Hindi Vyakhya (Commentary) by Prof. K.C. Chunekar edited by Lt. Dr. Gangasahay Pandey by Chaukhambha Bharti Akadami reprint 2015, Amradifalvarga, verse 115-118.
34. National Institute of Indian Medical Heritage. Kaiyadeva Nighantu [Internet]. Aushadhivarga; c2017 [cited 2023 Mar 4]. Retrieved from https://niimh.nic.in/ebooks/e-Nighantu/Kaiyadeva Nighantu/?mod=read, Aushadhivarga, 374-376.
35. Bhavprakash Nighantu Hindi Vyakhya (Commentary) by Prof. K.C. Chunekar edited by Lt. Dr. Gangasahay Pandey by Chaukhambha Bharti Akadami reprint 2015, Amradifalvarga, verse 85.
36. Bhavprakash Nighantu Hindi Vyakhya (Commentary) by Prof. K.C. Chunekar edited by Lt. Dr. Gangasahay Pandey by Chaukhambha Bharti Akadami reprint 2015, Haritakyadi varga, verse 221-223, 89-92, 79.
37. Bhavprakash Nighantu Hindi Vyakhya (Commentary) by Prof. K.C. Chunekar edited by Lt. Dr. Gangasahay Pandey by Chaukhambha Bharti Akadami reprint 2015, Karpuradi varga, verse 87, 65.

38. Ashtang Hriday–Sartha Vagbhat – by Lt. Dr. GaneshKrushnaGarde by Anmol Prakashan Edi 1891; reprinted 2003, Sutrasthan. 6, verse 104-105.
39. Bhavprakash Nighantu Hindi Vyakhya (Commentary) by Prof. K.C. Chunekar edited by Lt. Dr. Gangasahay Pandey by Chaukhambha Bharti Akadami reprint 2015, Sandhanvarga, verse 6-31.
40. CharakSamhita Pratham Khand, Hindi translation by Kaviraj Shri AtridevjiGup, BhargavPustakalay Publication, 2nd edition, 2000. Sutrasthan, Adhyay 27, verse 194–209.
41. Bhavprakash Nighantu Hindi Vyakhya (Commentary) by Prof. K.C. Chunekar edited by Lt. Dr. Gangasahay Pandey by Chaukhambha Bharti Akadami reprint 2015, Digdhavarga, verse 30-33.
42. KaiyadevaNighantu, Dravvarga, verse 230-231. Retrieved from https://niimh.nic.in/ebooks/e-Nighantu/Kaiyadeva Nighantu/?mod=read.
43. KaiyadevaNighantu, Dravvarga, verse 253. Retrieved from https://niimh.nic.in/ebooks/e-Nighantu/Kaiyadeva Nighantu/?mod=read.
44. KaiyadevaNighantu, Aushadhivarga, verse 168-169. Retrieved from https://niimh.nic.in/ebooks/e-Nighantu/Kaiyadeva Nighantu/?mod=read.
45. CharakSamhita Pratham Khand, Hindi translation by Kaviraj Shri AtridevjiGup, Bhargav Pustakalay Publication, 2nd edition, 2000. Sutrasthan, Adhyay 27, verse 270.
46. KaiyadevaNighantu, Krutannavarga, verse 55–56. Retrieved from https://niimh.nic.in/ebooks/e-Nighantu/Kaiyadeva Nighantu/?mod=read.
47. Bhavprakash Nighantu Hindi Vyakhya (Commentary) by Prof. K.C. Chunekar edited by Lt. Dr. Gangasahay Pandey by Chaukhambha Bharti Akadami reprint 2015, Tailvarga, verse 22–24.
48. Charak Samhita Pratham Khand, Hindi translation by Kaviraj Shri Atridevji Gupt, Bhargav Pustakalay Publication, 2nd edition, 2000. Sutrasthan, Adhyay 27, verse 287.
49. Ashtang Hriday–Sartha Vagbhat – by Lt. Dr. Ganesh Krushna Garde by Anmol Prakashan Edi 1891; reprinted 2003, Sutrasthan, Adhyay 6, verse 142-143.
50. Charak Samhita Pratham Khand, Hindi translation by Kaviraj Shri Atridevji Gupt, Bhargav Pustakalay Publication, 2nd edition, 2000. Sutrasthan, Adhyay 27, verse 284-285.
51. Chunekar KC, Pandey G (ed). Bhavprakash Nighantu Hindi Vyakhya. Chaukhambha Bharti Akadami; reprint 2015. Guduchyadivarga, verse 23–24.
52. Basnet P, Hussain H, Thapa P, et al. Traditional herbal medicine use among people living with HIV/AIDS in Nepal. BMC Complement Altern Med. 2018;18(1):111. doi: 10.1186/s12906-018-2195-6.
53. Saggam A, Kim DY, Akter R, et al. Systematic review on medicinal plants used for management of diabetes mellitus in South Asia. Front Pharmacol. 2020;11:1062. doi: 10.3389/fphar.2020.01062.
54. Papadakis MA (ed). Current Medical Diagnosis & Treatment 2022. McGraw Hill Lange Publications; 2022:404–412.
55. Narasimhan S, Balasubramanian N, Kumar DS. Antimicrobial activity of some important medicinal plants against plant and human pathogens. J Med Plants Stud. 2017;5(4):157–161.
56. National Center for Biotechnology Information. (n.d.). NCBI–WWW Error Blocked Diagnostic. Retrieved from www.ncbi.nlm.nih.gov/pmc/articles/PMC6413235/
57. Kandaswami C, Lee LT, Lee PP, et al. The antitumor activities of flavonoids. In: Huang MT, Osawa T, Ho CT, Rosen RT (eds). Food Phytochemicals for Cancer Prevention I. Fruits and Vegetables. American Chemical Society; 1994:87–98. doi: 10.1021/bk-1994-0546.ch006.
58. Narasimhan S, Balasubramanian N, Kumar DS. Antimicrobial activity of some important medicinal plant against plant and human pathogens. J Med Plants Stud. 2017;5(4):157–161.
59. Eckel RH, Jakicic JM, ArdJD, et al. 2013AHA/ACC guideline on lifestyle management to reduce cardiovascular risk: a report of the American College of Cardiology/American Heart Association Task Force on Practice Guidelines. Circulation. 2014;129(25 suppl 2):S76–S99. doi: 10.1161/01.cir.0000437740.48606.d1.
60. American Heart Association. AHA diet and lifestyle recommendations. Retrieved from www.heart.org/en/healthy-living/healthy-eating/eat-smart/nutrition-basics/aha-diet-and-lifestyle-recommendations
61. GardeG.Ashtang Hriday–Sartha Vagbhat. AnmolPrakashan; reprint 2003. Sutrasthan. Adhyay 6; verse 37.
62. GardeG. Ashtang Hriday–Sartha Vagbhat. AnmolPrakashan; reprint 2003. Sutrasthan. Adhyay 1, verse 5–6.
63. Tripthi B (ed). CharakSamhitaVol II with Charaka-Chandrika Hindi Commentary. Chaukhmabha Surbharti Prakashan; Chikitsasthan, Adhyay 26, verse 77-80.
64. CharakSamhita PrathamKhand, Hindi translation by Kaviraj Shri AtridevjiGup, Bhargav Pustakalay Publication, 2nd edition, 2000. Sutrasthan, Adhyay 27, verse 1.

65. Chunekar KC, Pandey GS, editors. Bhavprakash Nighantu Hindi Vyakhya by (Commentary) by Prof. K.C. Chunekar edited by Lt. Dr. Gangasahay Pandey by Chaukhambha Bharti Akadami reprint2015.
66. CharakSamhita PrathamKhand, Hindi translation by Kaviraj Shri AtridevjiGupt, Bhargav Pustakalay Publication, 2nd edition, 2000. Sutrasthan, Adhyay 26, verse 40/2.
67. Garde GK. Ashtang Hriday–Sartha Vagbhat – by Lt. Dr. Ganesh Krushna Garde by Anmol Prakashan Edi 1891;reprinted 2003, Sutrasthan. Adhyay 6, verse 40,42,43, 48.
68. Lad V. The Complete Book of Ayurvedic Home Remedies. Harmony Books; 1998.
69. Sharma A, Choudhary P, Sharma SD. Traditional Indian fermented foods: a rich source of beneficial microorganisms. Crit Rev Food Sci Nutr. 2019;59(17):2780–2796. doi: 10.1080/10408398.2018.1488962.
70. Acharya JD, Acharya D, Rudraprayag IC. Role of Ayurveda in the management of obesity: a review. J Evid Based Complementary Altern Med. 2013;18(4):237–242. doi: 10.1177/2156587213491732.
71. Sharma A, Bhatnagar S, Mishra S. A comparative study of the hypolipidemic activity of vedic guard and a marketed formulation of Ayurvedic medicine in albino rats. J Ethnopharmacol. 1997;56(3):205–210. doi: 10.1016/S0378-8741(97)00023-9.
72. Kumar MV, Sambaiah K, Mangalgi SG, Murthy NA, Lokesh BR. Effect of medicated ghee on serum lipid levels in psoriasis patients. Indian J Dairy Biosci. 1999;10:20–23.
73. Sharma H, Zhang X, Dwivedi C. The effect of ghee (clarified butter) on serum lipid levels and microsomal lipid peroxidation. AYU. 2010;31(2):134. doi: 10.4103/0974-8520.72361.
74. Kaiyadeva. Krutannavarga, verse 62-73. Retrieved from: https://niimh.nic.in/ebooks/e-Nighantu/Kaiyadeva Nighantu/?mod=read.
75. Tripathi I. Kshem kutuhal Manjula-Hindi VyakhyaVibhushitam by Dr Indradev Tripathi, Chaukhambha Orientalia publication 1978. Shashthotsav, verse 86-87.
76. Tripathi I. Kshem kutuhal Manjula-Hindi Vyakhya Vibhushitam, Chaukhambha Orientalia publication 1978. Saptamotsav, verse 16-17; 1978.
77. Tripathi I. Kshem kutuhal Manjula-Hindi Vyakhya Vibhushitam, Chaukhambha Orientalia publication 1978. Ashtmotsav, verse 27, 28; 1978.
78. Tripathi I. Kshem kutuhal Manjula-Hindi Vyakhya Vibhushitam, Chaukhambha Orientalia publication 1978. Ashtmotsav, verse 103; 1978.
79. Tripathi I. Kshem kutuhal Manjula-Hindi Vyakhya Vibhushitam, Chaukhambha Orientalia publication 1978. Ashtmotsav, verse 50; 1978.
80. Tripathi I. Kshem kutuhal Manjula-Hindi Vyakhya Vibhushitam, Chaukhambha Orientalia publication 1978. Dwadashotsav, verse 168; 1978.
81. AtridevjiGupt K. Charak Samhita Pratham Khand. Bhargav Pustakalay Publication. Sutrasthan, Adhyay 27, verse 279.
82. Atridevji GuptK. Charak Samhita Pratham Khand. Bhargav Pustakalay Publication. Sutrasthan, Adhyay 27, verse 280.
83. TripathiI. Kshem kutuhal Manjula-Hindi Vyakhya Vibhushitam, Chaukhambha Orientalia publication 1978. Ashtamotsav, verse 64; 1978.
84. NCBI–WWW Error Blocked Diagnostic. (n.d.). Retrieved from www.ncbi.nlm.nih.gov/pmc/articles/PMC 6100800/
85. Gidding SS, Dennison BA, Birch LL, et al. Implementing American Heart Association pediatric and adult nutrition guidelines: a scientific statement from American Heart Association Nutrition Committee on the Council of the Nutrition, Physical Activity and Metabolism, Council of Cardiovascular Diseases in the Young, Council on Atherosclerosis, Thrombosis and Vascular Biology, Council on Cardiovascular Nursing, Council on Epidemiology and Prevention, and Council for High Blood Pressure Research. Circulation. 2009;119(8):1161.

Therapeutic Nutrition in Ayurveda for Dermatology

Pallatheri Nambi Namboodiri

INTRODUCTION

Skin is the largest connective tissue in the body. It is an external tissue and covers every part of the body. That itself makes it a very important organ. The skeleton provides the foundation on which the rest of the tissues are built. The skin is the tissue that wraps it up and gives the body the shape we see.

The skin is where we first look for most health and ill-health conditions. Good and healthy skin is lustrous and blemish free. It is moist and glowing, while unhealthy skin is dry, chapped, and irregular. Conditions like fever and inflammation are noticed in the skin. Poisoning and allergies are also exhibited on the skin, making it a very important organ for diagnosis and treatment. Maintaining healthy skin is very much necessary for health. Good nutritious food keeps the body healthy and the same reflects on the skin. A balanced healthy diet is therefore necessary to keep the skin healthy.

CLASSIFICATION OF SKIN DISEASES

Skin diseases have been broadly divided into two groups:

1. *Mahakushta* (severe skin issues)[1]
2. *Kshudra Kushta* (mild skin issues)[2]

Maha Kushtas are seven in number (shown in Table 8.1) and are not easy to cure once they become chronic.

By their symptoms, these seven can be understood and classified into the following skin disorders used in conventional medicine.

Kshudra Kushta are 11 in number (shown in Table 8.2) and are easier to manage and don't lead to any other complications. *Kshudra Kushta* can be understood with symptoms in the following manner and according to conventional medicine.

Conditions like leucoderma have not been classified under the topic of skin diseases. In spite of not being mentioned, the treatment and approach are the same for *Mahakushta* and *Kshudrakushta.*

DOI: 10.1201/9781003345541-11

TABLE 8.1 *Maha Kushta* Classification[3]

	AYURVEDIC NAME	CONVENTIONAL MEDICINE EQUIVALENT
1	*Kapalam*	Scleroderma
2	*Oudumbaram*	Boil/Furuncle
3	*Mandalam*	Psoriasis
4	*Rikshajihva*	Lichen Planus
5	*Pundareeka*	Urticaria
6	*Sidhma*	Pityriasis Versicolor
7	*Kakanaka*	Squamous cell carcinoma

TABLE 8.2 *Kshudra Kushta* Classification[4]

	AYURVEDIC NAME	CONVENTIONAL MEDICINE EQUIVALENT
1	*EkaKushta*	Ichthyosis Vulgaris
2	*Charmakhya*	Lichen simplex chronicus
3	*Kitibha*	Psoriasis
4	*Vipadika*	Cracked feet
5	*Alasaka*	Prurigo nodularis
6	*Dadru*	Tinea infection
7	*Charmadala*	Impetigo
8	*Pama*	Scabies
9	*Visphota*	Superficial folliculitis
10	*Shataru*	Pyoderma gangrenosum
11	*Vicharchika*	Eczema

All skin diseases are due to *Raktadosha* (impurity and/or imbalance in the blood).[5] Therefore during the management of all skin diseases, special care has to be taken with respect to the quality of blood. Food or activity which can vitiate the blood has to be completely avoided.

There are multiple factors that lead to skin diseases. *Nidansya cha varjanam* (abstaining from the causative factor)[6]: the food which caused the skin disease has to be avoided, and compatible food is introduced during the management of the skin disease.

GENERAL THERAPEUTIC NUTRITION IN DERMATOLOGY

The foremost and primary drink during a skin disease is warm water.[7] Avoid water that is cold or very hot. Cold water blocks all the channels and reduces metabolism, while extremely hot water delays the healing of the skin disease.

Even though curd has been contraindicated in all types of skin diseases, one can consume curd made out of camel's milk, but in small quantities.[8] Camel's milk and curd are hot in potency and are not as thick as other curds. It does not block the channels if taken in small quantities.

For all skin conditions, *Purana Ghritha* (ghee which is 10 years old) is advisable.[9] Ghee does not get spoilt for more than 10 years when stored properly. The odor and taste might change, but the efficacy improves with age.[10]

Sorting food with neem oil[11] or with mustard oil[12] is recommended for all skin conditions. Neem oil is extremely bitter and sometimes not palatable and is recommended for infective skin conditions. Neem oil has very good antifungal, antibacterial, and wound-healing properties. The use of mustard oil reduces itching and urticaria.[13]

Kakamachi (*Solanum nigrum*) is a very healthy vegetable to be consumed.[14] *Kakamachi* is very good for all skin issues.[15]

Uncooked food has not been recommended during the treatment. Salads and other fruits also can be avoided. Uncooked food is heavy and takes a long time to digest, which will affect the metabolism. Fresh garlic can be taken in small quantities.[16] Garlic is a good blood purifier and also prevents indigestion.

Changeri or *Oxalis corniculata*[17] can be ground, or one can make a drink out of these by boiling them in water and being consumed warm. *Oxalis* leaves are slightly sour and are used in the kitchen instead of tamarind.

Patola – Trichosanthes cucumerina – is a vegetable advisable for all skin diseases.[18] It has to be steam-cooked and served. Another healthy fruit is *Shigruphala – Moringa olifera.*[19] Red rice, barley, wheat, and millet are the recommended foods to be taken.[20]

Green gram (*Vigna radiata*), *Masoor Dal* (*Lens culinaris*), and *Tuvar Dal* (*Cajanus cajan*) also can be consumed.[20] These pulses are easily digestible.

All vegetables which are bitter in taste are good. Bitter-tasting foods are easier to digest and purify the blood.

The meat of animals living in dry lands (*Jangalamamsa*) – deer and sheep are advisable, while others like pork and beef have to be avoided. The former help in promoting strength while they are not fattening, while the latter takes a long time to digest.

Buttermilk is advisable for all skin diseases. Being sour and bitter in taste and easy to digest, it opens up all the pores.[21]

Food that is salt, spice, and sour predominant is to be avoided because such foods increase the dryness in the body, which may lead to further itching.

Curd, milk, jaggery, sesame, and meat of animals living in marshy lands (*Anupadesha*) – cow, buffalo, yak, etc. – are some examples of animals living in marshy lands. These meats have to be avoided as they promote itching and sliminess in the system.

Sesame and black gram have to be completely avoided. Sesame is very hot, while black gram is heavy to digest and too sticky, thereby interfering with the metabolism.

There is no conclusive evidence to prove the efficacy of fermented milk products, like curd, on skin diseases.[22] Overall, there is early and limited evidence that fermented dairy products, used both topically and orally, provide benefits for skin health.[22]

MANAGEMENT OF DIFFERENT SKIN DISEASES

There are many different kinds of skin diseases. Skin diseases can be broadly classified as infective, metabolic, or autoimmune. Almost all skin diseases have the same management as far as the diet goes.

Skin diseases including superficial fungal infections like ringworm or athlete's foot also have to be managed with diet restrictions. Food that is heavy and difficult to digest can reduce metabolisms like cold water or curd and have to be completely avoided.

GENERAL WHOLESOME AND UNWHOLESOME FOODS FOR ALL SKIN DISEASES

For all skin diseases, avoid fresh rice (harvested less than a year ago),[23] all nonvegetarian food including eggs (all poultry like chicken, turkey, or duck),[24] jaggery, pulses (except green gram and masoor dal), black gram (*Vigna mungo*), and milk and milk products.[25]

All these foods take a longer time to digest. In the process, they block all channels too.[26]

Studies have shown that a low-protein (protein-restricted) diet will improve response during psoriasis treatment when compared to the Western diet, predominantly in meat, saturated fats, and sugar.[27]

If the skin disease is in an acute and aggressive state, the patient is not allowed to drink too much water. Too much water reduces the metabolism, thereby slowing down healing. Advised to consume warm water in moderation.[28]

Patients suffering from skin disorders can consume wheat or millets like ragi or jowar. The cereals should be properly cooked, preferably open cooking. Open cooking is a procedure where the cereal doubles the amount of water added and the vessel is kept open and cooked on medium heat. This cooking makes food lighter and easier to digest.

All vegetables which grow on creepers (Cucurbitaceae family) like bitter gourd (*Momordica charantia*), snake gourd (*Trichosanthes cucumerina*), bottle gourd (*Lagenaria siceraria*), ash pumpkin (*Benincasa hispida*), and yellow pumpkin (*Cucurbita pepo*) are predominantly bitter and easy to digest. All the vegetables are to be steam-cooked and garnished with spices, like dry ginger or pepper. These spices help the vegetables digest faster.[29] Adding turmeric powder into each and every dish helps the skin disease heal as turmeric is also antibacterial and antifungal.[30] The preferred cooking medium is ghee. Consuming food cooked with neem oil or mustard oil is not palatable to all due to its typical bitter taste. Neem and mustard oil have antibacterial and antifungal properties and are best indicated for all skin diseases, while ghee is light and strength promoting.

Drinking water is usually processed by boiling *Khadira* (*Acacia catechu*) in the ratio of 3 g to a liter of water and consumed warmly. *Khadira* is a good blood purifier.[31]

Bread, *Idli,* and *Dosa* are to be avoided as they are fermented. Fermented food clogs all the microchannels of the body.[32] The alternative to these is freshly cooked foods like *Khichadi, Idiyappam, Upma,* or *Puttu*. *Dosa* made of wheat also is preferable. Brinjal which is very heat generating and sticky is one vegetable to be avoided for all skin issues. Underground vegetables are other major vegetables to be avoided as they are all heavy, predominantly sweet, and cause indigestion. Cut down on spicy, sour, and deep-fried food. Pepper is a better choice for spice. Green or red chilies and bell pepper increase the heat in the body and are not preferable while healing is happening.

THERAPEUTIC NUTRITION IN SPECIFIC SKIN DISEASES

Every skin disease, irrespective of whatever nature it belongs to, infective, metabolic, or autoimmune, passes through three stages:

a. The beginning stage is dominated by *Kapha*
b. The middle stage is dominated by *Pitta* and
c. The last stage is dominated by *Vata*.

The main treatment objective is to bring the skin condition to the third stage as quickly as possible. When the skin disease is brought to this stage and managed, the chances of recurrence are very low.

INFECTIVE SKIN DISEASES

In all infective conditions like scabies, athlete's foot, or molluscum contagiosum this dietary regimen has to be followed.

UNWHOLESOME FOOD

1. Yoghurt and black gram (*V. mungo*). Black gram is the main ingredient in *Idli* and *Dosa*. These are thick, sticky, and difficult to digest and interfere with metabolism.[25]
2. Chickpea (*Cicer arietinum*). This interferes with proper digestion and increases flatulence.
3. Red meat (beef, lamb, mutton, pork, etc.) has to be completely avoided. This is heavy to digest.[24]
4. All milks, including plant based, have to be avoided. This is also because milk and milk products are heavy to digest.[25]
5. Eggplant (*Solanum melongena*), spice in excess, sour in excess, deep fry (French fries, chips, etc.).

WHOLESOME FOOD

1. Tea without sugar and milk. It helps in reducing the ooze in skin disease.
2. Wheat and millets (fox tail [*Setaria italica*] and ragi [*Eleusine coracana*]) can be consumed.
3. Rice that is older than a year. This is easy to digest.
4. Green gram (*V. radiata*) is the preferred pulse to be consumed. It is easily digestible.
5. Drinking water is boiled with dry ginger (3 g to a liter of water). This prevents indigestion.

AFTER THE INFECTIVE STATE

1. Avoid heavy foods, both plant and animal, that take a long time to digest. This promotes indigestion.
2. Can introduce 1:1 water-diluted cow's milk into the diet. Diluted milk promotes strength and does not interfere with digestion.
3. Ghee is preferred along with food. The first morsel of rice is mixed with ghee and a little bit of salt and taken.
4. Add turmeric to food while cooking. Fastens healing of the skin disease.

METABOLIC SKIN DISEASES

Many skin diseases like eczema and atopic dermatitis can become severe if the diet is not compatible. The main reason for skin diseases itself is incompatible food.[24]

Metabolic skin diseases are also understood in three stages (*Kapha*, *Pitta,* and *Vata*) like infective skin diseases. Metabolic skin diseases can be further again classified into two major heads:

- Wet
- Dry

Food restrictions for both wet and dry are slightly different.

WET ECZEMA

Unwholesome Food

Food that adds to the increase of the ooze should be avoided.

1. Avoid yogurt and other milk products which are difficult to digest.[24]
2. All kinds of sweets (especially milk) are to be avoided. This is difficult to digest and prolongs the healing time. Anything which takes a long time to digest increases *Kapha,* which will in turn increase the ooze.
3. Jaggery has to be avoided as this increases the ooze.[32]
4. Eggs (chicken, turkey, and duck) are to be avoided. This also increases the itching and ooze.
5. Black gram (*V. mungo*) has to be avoided as this is very sticky and increases the healing time. It also contributes to the ooze.
6. Avoid all cold substances including cold water to drink. Warm water promotes healing.

Wholesome Food

1. Millets like *Ragi* (*Eleusine coracana*) and *Jowar* (*Sorghum bicolor*) are very good to be consumed. This is drying in nature and thereby fastens healing.[33]
2. All vegetables which grow on creepers are bitter in taste and are indicated in skin diseases. They are light and easy to digest. They also have a natural blood-purifying action because they are bitter in taste.[34]
3. Water boiled with *Neem* flowers (*Azadirachta indica*) is good for consumption; 3 g to a liter of water.
4. Consuming 5 g of ground fresh curry leaves (*Murraya koenigii*)[35] and 3 g of turmeric paste on empty stomach. This helps in quick healing and is also antibacterial/antifungal.[30]
5. Steam-cooked leafy vegetables like moringa, palak (*Spinacia oleracea*), and red *Palak* (*Amaranthus cruentus*) are indicated for skin diseases.
6. Boil water with dry ginger for drinking; 3 g to a liter. This water promotes digestion.

DRY ECZEMA

Unwholesome Food

1. Avoid jaggery as it increases the itching.[33]
2. Sesame oil and mustard oil[36] intake has to be avoided. Not good for any skin diseases as it is hot in nature.
3. Unripe banana increases dryness and so do potatoes.
4. Avoid unprocessed buttermilk. It increases dryness. Buttermilk is *Ruksha.*[37]
5. Avoid yogurt as it promotes itching.
6. Buffalo milk and buffalo ghee are to be avoided. It is very difficult to digest and may lead to itching.
7. No fermented food like *Idli, Dosa,* or bread. These are too sticky and clog all the channels of the body. They also increase itching.[32]

Wholesome Food

1. All foods are to be cooked in ghee. Ghee is the best to heal dryness.
2. Diluted cow's milk or camel's milk can be given. This provides nourishment and doesn't increase itching.
3. Vegetables growing on creepers are bitter. They are light and easy to digest and promote fast healing.
4. One-year-old rice is good (freshly harvested rice is heavy).[23] It is nourishing and provides relief to dryness.
5. For children, diluted milk can be given. Ghee is very good and should be added well with all kinds of food. This helps in healing and improving the immunity in kids.[38]
6. Buttermilk processed with turmeric and curry leaves (*Murraya koenigii*) is good for mixing with rice and having; 3 g of turmeric and 10 g of curry leaves are boiled with 1 liter of buttermilk. Salt added to taste.

Autoimmune Skin Diseases

The most common autoimmune skin disorder is psoriasis. The diet restriction and indications for autoimmune and metabolic skin disorders are the same. For autoimmune, apart from the restrictions for metabolic disorders, there are a few other dietary restrictions to be followed. Studies have not found a very clear correlation between diet and psoriasis. LCD (low-calorie diet) helps in psoriasis. Dietary changes alone do not cause a large effect on psoriasis but may become an important adjunct to current first-line treatments.[39]

Unwholesome Food

1. Avoid all pulses except green gram (*V. radiata*) and orange lentil (*L. culinaris*).
2. All kinds of meat (red and white) and sea foods to be avoided. These take a long time to digest, are irritants, and promote pruritus.

3. Eggs (turkey, chicken, and duck) and other heavy proteins are to be avoided. They block all channels and moreover increase the itch.

Wholesome Food

1. Consume cumin-boiled drinking water. ½ teaspoon of cumin to 1 liter of water. This prevents indigestion and gastric issues.[40]
2. For conditions with severe itching, have water boiled with neem leaves; 5 g to 1 liter of water.

Preparation of Selective Local Foods

1. *Khichadi* is made by cooking rice and green gram together. A little ghee is added to soften it.
2. *Puttu* is steamed rice powder. Wet the rice powder and add salt to taste. Add coconut gratings and steam cook.
3. *Idiyappam* is made by adding more water to the rice powder to make it a paste. Add salt to taste and squeeze this out like noodles and steam cook.
4. Wheat *Dosa* is made by thoroughly mixing wheat powder and water to get a semisolid consistency. Salt is added to taste and cooked over a heated pan.

Modern science as well as Ayurveda describes seven skin layers.[41] The difficulty in managing and curing a skin disease depends upon the layer where the disease is affected. The first layer of diseases like fungal infections doesn't need too many food restrictions, while autoimmune conditions which can be classified under the *Maha Kusta* are in the fifth layer and below[42] and need strong dietary restrictions and take a longer period of time to get cured.

CONCLUSION

Skin disease is the most common disease which arises due to wrong dietary habits.[24] Apart from the treatment being provided for skin disorders, a strict diet has to be followed.

TABLE 8.3 Preferable Dietary Options for Infective Skin Diseases

CATEGORY	*WHOLESOME*	*UNWHOLESOME*
Grains	Rice which is older than a year Wheat and millets (fox tail and ragi)	Newly harvested rice
Legumes	Green gram	Black gram, chickpea
Dairy	Buttermilk	Milk from all sources, Yoghurt
Vegetables	On creepers	Eggplant
Fish and meat		Red meat (beef, lamb, mutton, pork, etc.)
Spices and condiments	Pepper, dry ginger	All spices in excess
Nuts and seeds		All
Oils	Ghee	Sesame
Beverages	Tea without sugar and milk Drinking water is boiled with dry ginger	Cold water

TABLE 8.4 Preferable Dietary Options for Metabolic Skin Diseases

CATEGORY	*WHOLESOME*	*UNWHOLESOME*
Grains	Rice which is older than a year Wheat and millets (jowar and ragi)	Newly harvested rice
Legumes	Green gram	Black gram, chickpea
Dairy	Processed buttermilk, diluted cow or camel milk	Milk from all sources, Yoghurt
Vegetables	On creepers	Eggplant
Fish and Meat		Red meat (beef, lamb, mutton, pork, etc.)
Spices and condiments	Turmeric, pepper, dry ginger	All spices in excess
Oils	Ghee	Buffalo ghee, sesame oil, mustard oil
Beverages	Tea without sugar and milk Drinking water is boiled with dry ginger	Cold water

TABLE 8.5 Preferable Dietary Options for Autoimmune Skin Diseases

CATEGORY	*WHOLESOME*	*UNWHOLESOME*
Grains	Rice which is older than a year Wheat and millets (Foxtail and ragi)	Newly harvested rice
Legumes	Green gram	Black gram, chickpea
Dairy	Buttermilk	Milk from all sources, Yoghurt
Vegetables	On creepers	Eggplant
Fish and Meat		Red meat (beef, lamb, mutton, pork, etc.)
Spices and condiments	Turmeric, pepper	All spices in excess
Oils	Ghee	
Beverages	Tea without sugar and milk Drinking water boiled with cumin or neem leaves	Cold water

The skin has seven layers and each layer has been attributed to the seven tissues of the body. The deeper and more chronic the disease is, the more difficult it gets to cure the disease.[42] The skin is the place where the underlying disorder is being exhibited.

Very few studies are there wherein the effect of food on skin has been discussed. The effect of fermented food on skin diseases is a very important topic as Ayurveda says ground and fermented foods like *Idli*, *Dosa*, and bread have to be avoided. Further studies on this topic would help the larger population that depends on *Idli* and *Dosa* as their staple food.

Since skin is the *Upadhatu* (subtissue) of *Mamsadhatu* (muscle)[43] any dietary habit that can deplete or aggravate the muscle tissue can have a direct effect on the skin. Excess protein (could be plant or animal protein) or proteins which are difficult to digest (like paneer or eggs), even in small quantities, have to be avoided.

Gluten is a naturally occurring protein and studies on its effect on psoriasis and other skin diseases have been done. A GF (gluten-free) diet showed improvement in the condition of patients suffering from psoriasis.[44]

Fermented foods (*Idli*, bread, or curd), which are difficult to digest, also aggravate skin diseases. Ground and fermented food (*Pistanna*),[45] which are difficult to digest, will block all the microchannels in the body, which aggravate skin disease. While fermented drinks like buttermilk are advisable as they are light, digest fast, and also help open up the channels. Dishes like *Khichadi, Puttu*, or *Idiyappam*, which are prepared fresh, keep the body healthier.

REFERENCES

1. Charaka Samhita, Chikitsa Sthana, 7/14-20, Sharma RK, Dash VB, Choukhambha Sanskrit Studies, Varanasi.
2. Charaka Samhita, Chikitsa Sthana, 7/21-26, Sharma RK, Dash VB, Choukhambha Sanskrit Studies, Varanasi.
3. Das P. A comparative study of Mahakushta with Modern medicine. Int Ayurvedic Med J. 2017;5(9):3917.
4. Chalapathi RS. Critical evaluation of Kshudrakushtas of Charaka Samhita: An Ayurvedic treatise in light of modern medicine. Int J Res Ayurveda Pharm. 2016;7(Suppl 1):16.
5. Charaka Samhita, Sutra Sthana, 28/11, Sharma RK, Dash VB, Choukhambha Sanskrit Studies, Varanasi.
6. Charaka Samhita, Vimana Sthana, 7/32, Sharma RK, Dash VB, Choukhambha Sanskrit Studies, Varanasi.
7. Susrutha Samhita Sutra Stanam Chapter 45/ shloka 46, Srikantha Murthy KR, Chaukhambha Orientalia, Varanasi.
8. Susrutha Samhita Sutra Stanam Chapter 45/ shloka 70, Srikantha Murthy KR, Chaukhambha Orientalia, Varanasi.
9. Susrutha Samhita Sutra Stanam Chapter 45/ shloka 108, Srikantha Murthy KR, Chaukhambha Orientalia, Varanasi.
10. Astanga Hridayam Sutra Sthana Chapter 5- shloka 40-41, Govindan Vaidyar, Devi Book Stall, Kodungalloor, Kerala.
11. Srikantha Murthy KR. Susrutha Samhita Sutra Stanam Chapter 45/ shloka 115. Varanasi: Chaukhambha Orientalia; Year not available.
12. Srikantha Murthy KR. Susrutha Samhita Sutra Stanam Chapter 45/ shloka 117. Varanasi: Chaukhambha Orientalia; Year not available.
13. Sharma RK, Dash VB. Charaka Samhita, Sutra Sthana, Chapter 27/ Shloka 290. Varanasi: Choukhambha Sanskrit Studies; Year not available.
14. Srikantha Murthy KR. Susrutha Samhita Sutra Stanam Chapter 46/ shloka 266. Varanasi: Chaukhambha Orientalia; Year not available.
15. Sharma RK, Dash VB. Charaka Samhita, Sutra Sthana, Chapter 27/ Shloka 89. Varanasi: Choukhambha Sanskrit Studies; Year not available.
16. Sharma RK, Dash VB. Charaka Samhita, Sutra Sthana, Chapter 27/ Shloka 176. Varanasi: Choukhambha Sanskrit Studies; Year not available.
17. Chunekar KC. BhavaprakashaNighantu. SaakaVarga 15 commentary. Varanasi: Choukhambha Bharati Academy; Year not available.
18. Chunekar KC. BhavaprakashaNighantu. SaakaVarga 47 commentary. Varanasi: Choukhambha Bharati Academy; Year not available.
19. Chunekar KC. BhavaprakashaNighantu. SaakaVarga 78 commentary. Varanasi: Choukhambha Bharati Academy; Year not available.
20. Vaidyar G. AstangaHridayam Sutra Sthana Chapter 7- shloka 25-27. Kodungalloor, Kerala: Devi Book Stall; Year not available.
21. GovindanVaidyar. AstangaHridayam Sutra Sthana Chapter 5- shloka 38. Kodungalloor, Kerala: Devi Book Stall; [date unknown].
22. Vaughn AR, Branum A, Sivamani RK. Effects of fermented dairy products on skin: a systematic review. J Altern Complement Med. 2015 Jul;21(7):380–5. doi: 10.1089/acm.2014.0142. PMID: 26154186.
23. Srikantha Murthy KR. Susrutha Samhita Sutra Stanam Chapter 46/ shloka 51. Varanasi: Chaukhambha Orientalia; [date unknown].
24. Srikantha Murthy KR. Susrutha Samhita Nidana Stanam Chapter 5/ shloka 3. Varanasi: Chaukhambha Orientalia; [date unknown].
25. Srikantha Murthy KR. Susrutha Samhita Chikitsa Stanam Chapter 9/ shloka 4. Varanasi: Chaukhambha Orientalia; [date unknown].
26. Sharma RK, Dash VB. Charaka Samhita, Sutra Sthana, Chapter 23/ Shloka 7. Varanasi: Choukhambha Sanskrit Studies; [date unknown].

27. Musumeci ML, Nasca MR, Boscaglia S, Micali G. The role of lifestyle and nutrition in psoriasis: current status of knowledge and interventions. Dermatol Ther. 2022 Sep;35(9):e15685. doi: 10.1111/dth.15685. Epub 2022 Jul 18. PMID: 35790061; PMCID: PMC9541512.
28. GovindanVaidyar. AstangaHridayam Sutra Sthana Chapter 5- shloka 13. Kodungalloor, Kerala: Devi Book Stall; [date unknown].
29. GovindanVaidyar. AstangaHridayam Sutra Sthana Chapter 6- shloka 160-166. Kodungalloor, Kerala: Devi Book Stall; [date unknown].
30. Chunekar KC. BhavaprakashaNighantu. HareetakyadiVarga 119 commentary. Varanasi: ChoukhambhaBharatu Academy; [date unknown].
31. Chunekar KC. BhavaprakashaNighantu. Vatadi Varga 31-32 commentary. Varanasi: ChoukhambhaBharatu Academy; [date unknown].
32. Sharma RK, Dash VB. Charaka Samhita, Chikitsa Sthana, Chapter 7/ Shloka 7. Varanasi: Choukhambha Sanskrit Studies; [date unknown].
33. GovindanVaidyar. AstangaHridayam Chikitsa Sthana Chapter 19- shloka 25-27. Kodungalloor, Kerala: Devi Book Stall; [date unknown].
34. GovindanVaidyar. AstangaHridayam Sutra Sthana Chapter 10- shloka 14-15. Kodungalloor, Kerala: Devi Book Stall; [date unknown].
35. Apte VG. Rajanighantu Gudoochyadi prathama varga -49). Pune: Anandashrama Publication; [date unknown].
36. GovindanVaidyar. AstangaHridayam Sutra Sthana Chapter 5- shloka 61. Kodungalloor, Kerala: Devi Book Stall.
37. GovindanVaidyar. AstangaHridayam Sutra Sthana Chapter 5- shloka 38. Kodungalloor, Kerala: Devi Book Stall.
38. GovindanVaidyar. AstangaHridayam Sutra Sthana Chapter 5- shloka 42-45. Kodungalloor, Kerala: Devi Book Stall.
39. Pona A, Haidari W, Kolli SS, Feldman SR. Diet and psoriasis. Dermatol Online J. 2019 Feb 15;25(2):13030/qt1p37435s. PMID: 30865402.
40. Dr KC Chunekar (commentator). BhavaprakashaNighantu. HareetakyadiVarga 81. Varanasi: Choukhambha Bharatu Academy.
41. KR Srikantha Murthy. Susrutha Samhita Sareera Stanam Chapter 4/ shloka 4. Varanasi: Chaukhambha Orientalia.
42. GovindanVaidyar. AstangaHridayam Nidana Sthana Chapter 14- shloka 32. Kodungalloor, Kerala: Devi Book Stall.
43. Dr. Ram Karan Sharma and Vaidya Bhagwan Dash. Charaka Samhita, Chikitsa Sthana, Chapter 15/ Shloka 17. Varanasi: Choukhambha Sanskrit Studies.
44. Bell KA, Pourang A, Mesinkovska NA, Cardis MA. The effect of gluten on skin and hair: a systematic review. Dermatol Online J. 2021 Apr 15;27(4):13030/qt2qz916r0. PMID: 33999573.
45. Shastri A (commentator). Ayurved Tattva Sandipikahindi commentary on Susruta samhita of Maharshi Sushruta, Sutrasthan; Chapter 42., Verse 499. Varanasi: Chaukhambha Orientalia, 2016; 286.

Therapeutic Nutrition in Ayurveda for Oncology

Pankaj Wanjarkhedkar

1. CURRENT SCENARIO OF CANCER AND NUTRITION

Cancer remains the first or second most common contributor to premature mortality across the globe. It has been predicted that the 2020 incidence of cancer will be doubled by 2070.[1]

Indian population–based registry suggests that the incidence of cancer cases is estimated to increase by 12.8 percent in 2025 as compared to 2020.[2]

Proactive nutritional interventions shall be an integral part of cancer treatments as malnutrition and weight loss are commonly observed due to different mechanisms involving the tumor, the host response to the tumor, and anticancer therapies, therefore improving clinical outcomes and quality of life of the patients.[3]

A clinical nutrition study reported that malnutrition is a risk factor for mortality in older cancer patients, especially with solid tumors.[4]

Nutrition in oncology and updated practical and concise recommendations for cancer patients receiving chemoradiotherapy play an important role in the clinical outcome of the treatment.[5]

Cancer metabolism is influenced by the availability of nutrients in the microenvironment, which target oncogenic metabolism, underlining the role of dietary modifications in cancer treatments.[6]

2. THERAPEUTIC NUTRITION FROM AYURVEDA CLASSICAL BOOKS

The nutrition-related principle, as discussed by Vaidya Jeevan, states the importance of an advisable and avoidable diet for any disease. The verse states that following recommended nutrition is as useful as medicine; at the same time, medicine will not be useful if nutrition is improper or not as recommended.

पथ्ये सति गदार्तस्य किमऔषधनिषेवणे: I

पथ्येऽसति गदार्तस्य किमऔषधनिषेवणे: II

Ayurveda classical books do not directly correlate the disease condition with cancer, but tumors have been discussed in diagnosis and treatment sections.[7]

DOI: 10.1201/9781003345541-12

Langhana (therapeutic fasting) has been suggested as first line of treatment, to begin with the therapeutic diet for *Arbuda* (tumors).[8] Types of *Langhana* are *Anashana Langhana* which is complete abstinence from diet and *Alpashana Langhana* which is partial abstinence from the diet. In cancer, the *Alpashana Langhana* is preferred over *Anashana Langhana* given the general condition of a patient who undergoes multiple treatments at respective stages.[9]

According to the principle of detoxification or purification to restore biology toward normalcy, four treatments are suggested[10]; therapeutic fasting is one of them, which is preferred in cancer patients depending upon the *Agni* (metabolic energy) of the patient.

Purana Ghrita (aged ghee), *Jeerna Rakta-shali* (aged red), *Mudga* (*Phaseolus aureus* Roxb.), *Yava* (barley), *Patola* (pointed gourd; *Trichosanthes dioica*), *Rakta shigru* (red variety of Moringa), and *Kathillaka* (*Boerhavia diffusa*) have been suggested for diet in treatment plan of *Arbuda*.[11]

Cancer-related fatigue (CRF), which is usually experienced by maximum patients before, during, or after undergoing surgery, chemotherapy, radiation therapy, or immunotherapy, results in depleted energy levels.[12]

The recommended nutritional guidelines from *Charaka-Samhita* for such *Ksheena* patients (patients with severe weakness/fatigue) are immediate replenishment on the lines of *Sadya-santarpana,* by administering gruel, meat soups, milk, and ghee given biology, *Agni*, quantity, and time of nutrient administration, along with *Basti* (per rectal administration of nutrition/medicines) and *Abhyanga* (whole body oleation) as and when indicated.[13]

2.1 General Guideline for *Apathya* (Contraindications) in Cancer

Milk, milk products , products of sugarcane (like white sugars), meat of *Anupa* animals and region (marshy land regions), fine flour (for example, bakery products prepared from fine flour, without whole grains), excess sweet, *Guru* (heavier to digest), and *Abhshyandi* (food causing and retaining secretions in the body) food items shall be avoided.[16]

In the present era, the scope of *Anupa* animals may be extended to not only those from marshy lands but also meat from poultries, raw fish, and fish as a yield of aqua farming, by extrapolating the underlined etiological impact on *Tridoshas*.

3. CONTEMPORARY ASPECTS OF NUTRITION IN CANCER

Nutrition improves nutritive values, body composition, symptoms, quality of life, and ultimately survival in cancer patients. The involvement of a nutritionist is recommended with/without oral nutritional supplements (ONSs). The escalation of nutrition measures is decided based on factors including: (1) 50% of intake vs. requirements for more than 1–2 weeks; (2) if it is anticipated that undernourished patients will not eat and/or absorb nutrients for a long period; and (3) if the tumor itself impairs oral intake.[17]

3.1 Dietary Epigenetics in Cancer

The epigenetic aberrations can be corrected, with one of the known approaches in cancer being dietary control. The epigenetic alterations can be achieved through the quality as well as the number of calories consumed.[18]

TABLE 9.1 Essentials of *Dwadasha aahar varga* in *Arbuda* (Tumor) Treatments[14]

DIETARY ITEM	*QUALITIES*	*CLASSICAL INDICATIONS*	*CONTRAINDICATION*
Water	*Jeevaniya, Tarpaneeya, Hridyam,*	To reduce Fatigue, Replenishing	Contraindicated in persons with *Agni* dysfunction
Lukewarm to Warm Water	*Deepana, Pachana, Laghu, Bastishodhana*	Bloating, Dysuria, Fever, Rhino-sinusitis, Dry cough, Indigestion	
Coconut Water	*Snigdha, Vrishya, Laghu, Deepana, Bastishodhana*	Excessive Thirst, Peptic diseases, Dysuria	
Milk	*Snigdha, Dhatuvardhana, Vrishyam*	Mitigating Vata and Pitta disorders	
Ksheerapaka (Milk and water cooked in a specific way)	*Laghutara* – Easily assimilable		
Unboiled Milk	*Abhishyandi, Guru*		Contraindicated in *Kaphaja* disorders
Cow Milk	*Jeevaneeya, Rasayana, Balya, Medhya*	Emaciation, Fatigue, Constipation, Dysuria, Excess hunger	
Buffalo Milk		Insomnia, Sleep disorders	
Goat Milk	*Laghu*	Emaciation, Asthma, Diarrhea, GI bleeding disorders	
Camel Milk	*Ushnam, Ruksham, Lavana, Laghu*	Inflammatory disorders, Ascites, Hemorrhoids	
Lamb Milk	*Ahridyam, Guru, Snigdha*	Indicated only in *Vataja shwasa*	
Curds	*Grahi, Guru, Ushna*	Improves taste, Dysuria nourishes fatty tissue	Contraindicated to consumption at night
Buttermilk	*Laghu, Deepana,*	Improves taste, Hemorrhoids, Anemia, oliguria, and Ascites	
Freshly Churned Unsalted Butter	*Agnikruta, Vrishya*	Improves complexion, anemia, Hemorrhoids, improves digestion	
Ghee	*Best Snehana, Sheeta*	Brain vitalizer, Antiaging, improves memory	
	Balya, Guru, Vishtambhi	Improves sleep	
Sugarcane	*Guru, Snigdha, Bruhana, Vrishya*	Diuresis, alleviate gastritis, Nourishing	
Jaggery		Diuresis, Improves bowel movements, Nourishing the muscle mass	
Honey	*Ruskha, Kashaya, Madhura*	Controls diarrhea, thirst, hiccups, and gastritis; improves healing	
Edible Oils	*Tikshna, Vyavayi, Sukshma*	Nourishment, Constipation, indicated for selective use in Obesity	
Castor Oil	*Ushna, Guru, Saram*	Constipation, Ascites, Bloating, Backache	
	Teekshanam, Laghu	Increases *Pitta* and *Rakta*, Indicated for skin diseases	
Linseed (Flax seed) Oil	*Ushanam*	Indicated in Skin diseases, increases *Kapha* and Pitta	
Coconut[15] Oil	*Bruhana*	*Balavardhana*	

In cancer care, epigenetic modifications can play a significant role in occurrence and pathogenesis. The most common epigenetic mechanisms are DNA methylation and chromatin remodeling. A range of bioactive food compounds like gingerol ginger have a proven role in cancer prevention through an epigenetic mechanism.[19]

The epigenetic diet is the dietary factors that act to modify the epigenome. Bioactive nutritional components of an epigenetic diet can be used therapeutically for medicinal or chemopreventive purposes.[20]

3.2 Diet – Microbiome in Cancer

The drug metabolism, immune activation, and response to immunotherapy significantly vary with the microbiome. One of the critical factors affecting the microbiome structure and function is diet; together they affect the immune response.[21]

The optimum intestinal microbiota structure can modulate the function of the human immune system, which affects anticancer response. A strong association between diet, gut microbiome composition, and the outcome of immunotherapy has been reported in a comprehensive review of diet, microbiome, and immunotherapy.[22]

4. CATEGORIES OF CANCERS BASED ON THERAPEUTIC NUTRITION

The cancers are categorized as per the respective organ and site (*Adhishthana*). The therapeutic nutrition from Ayurveda perspective based on Ayurveda *Samprapti* (etiopathogenesis of cancer as per Ayurveda fundamentals) of individual cancers or based on histology type is yet to evolve and get validated. Therefore, conventionally accepted and widely known categories of cancer have been considered for discussion in this chapter.

Diseases are classified based on etiopathogenesis by *Charak Samhita* as *Santarpaottha* and *Apatarpanottha*,[23] which can be extrapolated for oncological concepts given current understanding in contemporary science.

While designing therapeutic nutrition with applied principles of Ayurveda to carcinogenesis, the proposed dietary guidelines can be divided into two main categories as mentioned by *Sushruta Samhita*: *Guru Vipaka* and *Laghu Vipaka* food items.[24]

Therefore, dietary items with predominant *Madhura, Amla,* and *Lavana Rasa* will be restricted in *Santarpanottha,* and *Katu, Tikta* and *Kashaya Rasa* will be advised to avoid in *Apatarpanottha* category of cancers.

5. THERAPEUTIC NUTRITION IN CANCER SUBSITES AND GROUPS

Nutritional treatments vary as per the *Sapta-dhatus* (seven biophysiological tissues) and their respective *Agni* (metabolic energy of the tissue). The gross categorization of cancers is discussed briefly in Figure 9.1 and Figure 9.2.

Additionally, hematological cancers have been discussed as per therapeutic dietary studies and clinical experience.

5.1 Site-Specific Nutritional Patterns in Cancer

TABLE 9.2 A Systemic Review of Dietary Patterns and Sites of Cancer[25]

SITE OF CANCER	*REDUCED RISK*	*ADDED RISK*
Breast Cancer Postmenopausal	A diet rich in vegetables, fruits, and whole grains, and lower in animal-source foods and refined carbohydrates	
Colorectal Cancer	Higher in vegetables, fruits, legumes, whole grains, lean meats and seafood, and low-fat dairy – associated with lower risk of colon and rectal cancer	Higher in red and processed meats, French fries, potatoes, and sources of sugars (e.g. sugar-sweetened beverages, sweets, and dessert foods) – associated with a greater colon and rectal cancer risk
Lung Cancer	Frequent servings of vegetables, fruits, seafood, grains and cereals, legumes and lean fat meats, and lower-fat or nonfat dairy products – are associated with a lower risk of lung cancer, primarily among former smokers and current smokers	

5.2 Need for Specific Categorization for Cancers in Gynecology

Discussion on therapeutically important nutrition among gynecological cancers can be narrowed down to breast cancers, ovarian cancers, and endometrium cancers, which predominantly fall under the *Santarpanottha* category.

5.3 Cancer Sites of Microbiome Significance

The data regarding the effects of several common dietary items on intestinal microbiota suggests that consuming a particular type of food produces predictable shifts in existing host bacterial genera, which can further affect host immune and metabolic parameters with broad implications on human health.[26]

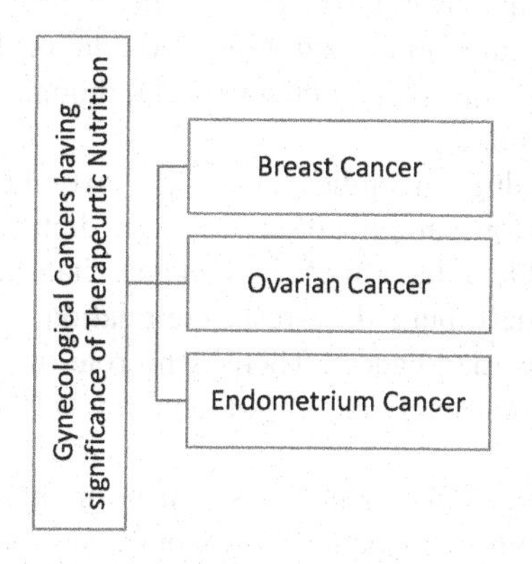

FIGURE 9.1 Systemwise significance of therapeutic nutrition.

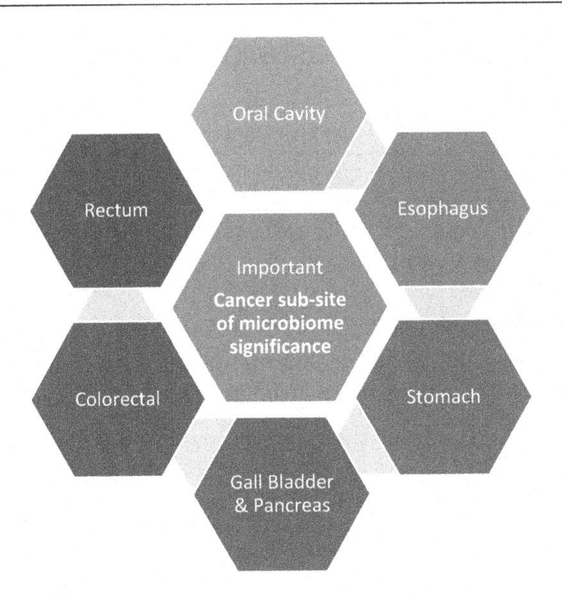

FIGURE 9.2 Cancer and microbiome.

The cancers which can lead to postulate a role of microbiome as one of the etiopathogenetic factor and need to address with an understanding of dietary epigenetics while planning therapeutic nutrition in Ayurveda, mainly include cancers from the Oral cavity to the rectum; have been sketched in Figure 9.2.

6. APPLIED STUDIES IN THERAPEUTIC NUTRITION CLASSIFIED AS PER *AHARA-VARGA*

Gorasa-varga/Dugdhavarga

A metaanalysis of 34 prospective cohort studies on dairy consumption and total cancer, and cancer-specific mortality suggests that high milk consumption (with high fat) was associated with higher mortality, while fermented milk consumption was associated with low cancer mortality. The cancer-specific mortality was increased for liver, ovarian, and prostate cancers. High cheese consumption was mainly associated with higher colorectal cancer mortality than overall cancer mortality.[27]

The results from epidemiological studies examining the relationship between intake of dairy foods and breast or ovarian cancer risk are mixed and not consistent, though animal studies have suggested that galactose may be toxic to ovarian cells.[28]

A metaanalysis of 29 epidemiological studies has shown that whole milk intake might contribute to higher ovarian cancer risk, whereas low-fat milk, dietary Ca, and dietary vitamin D might reduce the risk.[29]

There are debates on the possible reduction of calcium and vitamin D in those ovarian cancer patients with restricted milk and milk products, but a dose-response relation metaanalysis concluded that vitamin D could not decrease the risk of ovarian cancer. There is no role of calcium intake in reducing ovarian cancer risk. Besides, the combination use of calcium and vitamin D has not shown any additional benefits for ovarian cancer prevention.[30]

A systemic review and dose-response metaanalysis of hormone-positive breast cancer prospective studies have concluded that a nonlinear association was observed only for milk, such that higher milk intake increased the risk, while no association was observed for a lower intake of milk. The study also concluded that high intakes of vegetables, fruit, cheese, and soya products and low intakes of red meat and processed meat were associated with lower risks of breast cancer.[31]

A Mendelian randomization study about the correlation between sugar-sweetened and sugary beverages and colorectal cancers has reported a causal association with colonic malignant neoplasms risk but did not support such a relationship with the rectal malignant neoplasms.[32]

Vidahi is the characteristic term that indicates a proinflammatory phenomenon, which may hypothesize the conflagrate of the underlined *Dhatu* (biophysiological tissue) and its metabolism may lead to protumorogenesis if the *Vidahi* phase gets prolonged or repeated.

Dietary advanced glycation end products (dAGEs) can be formed during the heating/reheating of the diet (Maillard reaction). The biological effects of dAGEs can cause proinflammatory effects influencing digestion, absorption, formation, and degradation of the gastrointestinal tract.[33] The concept of *Vidaha* may be correlated with the glyceraldehyde-derived advanced glycation process supporting tumor growth.

The data derived from clinical and experimental studies have demonstrated that the receptor of advanced glycation end products (RAGE)/AGEs axis plays an important role in the onset of a crucial and long-lasting inflammatory milieu, thus supporting tumor growth and development.

Once AGEs bind to RAGE, the secretion of several proinflammatory cytokines is also involved in the promotion of gallbladder cancer invasion and migration; association between high dietary AGEs intakes and high risk for gallbladder cancer has been observed, which is further supported by emerging dietary intervention data to reduce gallbladder cancer risk.[34]

Mamsavarga

In the metaanalysis of prospective observational studies, high processed meat consumption was associated with increased breast cancer risk, particularly in postmenopausal women, while no significance has been reported per hormone-positive or hormone-negative breast cancers.[35]

A metaanalysis summarizing food groups and risk of colorectal cancers suggested that a diet plan with a high intake of whole grains, vegetables, fruit, and dairy products and low amounts of red meat and processed meat was associated with a lower risk of colorectal cancers (CRCs).[36]

A systemic, comprehensive review of 148 published articles on red meat and its association with increased risk of cancer revealed that high red meat intake was positively associated with the risk of breast cancer, endometrial cancer, colorectal cancer, colon cancer, rectal cancer, lung cancer, and hepatocellular carcinoma, and high processed meat intake was positively associated with risk of breast, colorectal, colon, rectal, and lung cancers.[37]

An association of risk as per histological type has been evaluated in esophagus cancer, and it was concluded that a high intake of red meat and a low intake of poultry is associated with an increased risk of esophageal squamous cell carcinoma. High intake of meat, especially processed meat, is likely to increase esophageal adenocarcinoma risk, while fish consumption may not be associated with the incidence of esophageal cancer.[38]

Phala Varga

A Japanese cohort study suggested that total fruit intake and total vegetable intake had inverse and positive associations, respectively, with pancreatic cancer risk.[39]

The frequent consumption of organic food was associated with a reduced risk of cancer.[40]

Ikshuvarga

Antiproliferative effects were shown to be mediated via alteration in cytokines, VEGF-1, and NF-κB expression in a cell line study of sugarcane extract.[41]

Madya varga – Liquor Consumption in the Present Era

Acetaldehyde is a metabolite of ethanol, which can cause DNA damage and block DNA synthesis and repair, while both ethanol and acetaldehyde can disrupt DNA methylation. Further DNA damage and lipid peroxidation can result due to ethanol-induced inflammation.[42]

A high vegetable and fruit intake before diagnosis had an inverse association with overall mortality in survivors of head and neck and ovarian cancer.[43]

Dugdha varga

The large population-based cohort study data in women suggests that consumption of unpasteurized milk does not increase the risk of cancer.[44]

Taila varga

Coconut oils are available as virgin coconut oil (VCO), crude coconut oil (ECO), and refined coconut oil (RCO). Consumption of VCO during chemotherapy helped improve the functional status and global QOL of breast cancer patients, while in other cancers it reduces the cisplatin-induced myelosuppression and hepatotoxicity. The renal histology revealed reduced glomerular/tubular congestion and necrosis.[45,46]

Shuka-dhanya varga

A systematic review of metaanalysis of cohort and case-control studies consistently demonstrates that whole grain intake is associated with a lower risk of total and site-specific cancer and supports current dietary recommendations to increase whole grain consumption, while the relationship between refined grain intake and cancer risk remained inconclusive.[47]

Shimbi varga, particularly *Mudga* (*Phaseolus mungo*) being *Laghu* (early to digest) from Ayurveda Nutrition, has a preventive and protective role against carcinogenesis. The mung bean has been documented to prevent cancer as well as possess hepatoprotective and immunomodulatory activities.[48]

A group of pulses (beans, peas, chickpeas, pigeon peas, lentils, groundnut) may play an important role in reducing the risk of cancer occurrence and its progression by reducing inflammation and cancer cell proliferation/metastasis by inducing apoptosis in cancer cells.[49]

An overview suggested cancer inhibitory activity of high legumes (peas, beans, lentils) especially common beans (*Phaseolus* species) and their nondigestible fractions that modulate genes and proteins are associated with reduced cancer risk in human populations, while common beans have shown its antiproliferative activity and induce apoptosis in preclinical studies.[50]

7. CLINICAL EXPERIENCE IN THERAPEUTIC NUTRITION AS PER AYURVEDA GUIDELINES

7.1 Ayurveda Nutrition Practice and Contemporary Evidence

The comprehensive treatment of cancer includes guidelines for the patients to follow as per Ayurveda classical treatment principles, which are supported by contemporary research data, unraveling the spectrum of therapeutic nutrition in Ayurveda.

7.1.1. Twaksidhha *Jala (Cinnamon Water) Prescription*

There was a study conducted on spices and their effect on inhibition of advanced glycation end product formation that suggested the rank of AGE inhibition ability to be as follows, arranged high to low: Lavang (cloves) > Maricha (black pepper) > Jeeraka (cumin) > Shweta maricha (white pepper) > Twak/Dalchini (cinnamon) > Ela (cardamom) > Ardrak (ginger) > Jatiphala (nutmeg).[51]

7.1.2 Red Meat Restriction in Non-Hodgkin's Lymphoma (NHL)

A population-based case-control study to test the hypothesis that meat consumption correlates with NHL provided evidence that red meat consumption is associated with an increase in NHL risk,[52] which supports the principle of Ayurveda nutrition to avoid meat consumption including frozen meat being *Vidahi* in blood disorders.

7.1.3 Sugar and Sugar Products Restriction in Ovarian Cancer

The assessment to study the association between sugar intake and cancer risk was done using Cox proportional hazard models adjusted for known risk factors (sociodemographic, anthropometric, lifestyle, medical history, and nutritional factors), which suggested that sugars may play a role in cancer etiology as a modifiable risk factor for cancer prevention.[53]

A population-based case-control study aiming at assessing the correlation between glycemic loss, glycemic index, and ovarian cancer revealed that diets with a high glycemic load may increase the risk of ovarian cancer, particularly among overweight/obese women.[54]

An age-stratified subcohort study among Canadian women suggested that relatively high sugar-containing beverages intake was associated with a higher risk of type I endometrial cancers and ovarian cancers but not of breast or colorectal cancers.[55]

A systemic review and metaanalysis conducted to study the correlation between insulin resistance and ovarian cancer revealed that levels of IGF-1 and IGFBP-3 are lower, while higher levels of IGBP-2 and IGBP-1 are found in patients with ovarian cancer.[56]

A strong direct relationship between circulating insulin-like growth factor-I (IGF-I) levels and the risk of developing ovarian cancer before age 55 has been reported in a study correlating it with the risk of ovarian cancer.[57]

IGF-1 is overexpressed in low-grade serous ovarian cancers when compared with serous borderline ovarian tumors and high-grade serous ovarian cancers.[58]

7.1.4 Jangal-mamsa *in Upper Gastrointestinal (GI) Tract*

Ayurveda nutritional chapters from classical texts have suggested taking meat that is *Laghu* (light/early digestion) and categorizing it as per the habitat of the animals; they broadly classified *Jangal* white meat as *Laghu* while red meat as Guru (heavy/delayed digestion), and that is why the integrative cancer care nutrition plan suggests restriction of red meat, stale meat or frozen meat in gastric and upper GI cancer patients.

The data published after an overall and dose-response metaanalysis supports the clinical practice reporting that an increase in white meat consumption may reduce the risk of gastric cancer, while red or processed meat may increase the risk of gastric cancer.[59]

Honey is used as an adjunct to Ayurveda cancer therapy. It is an important vehicle for Ayurveda medicines as well as a diet supplement. The role of honey in targeted key hallmarks of carcinogenesis, including uncontrolled proliferation, apoptosis evasion, angiogenesis, growth factor signaling, invasion, and inflammation, has been studied.[60]

7.1.5 Emphasis on Boiled and Cooked Vegetables vs. Raw Vegetables

Shakah vardhante roga is one of the nutritional guidelines in Ayurveda; therefore the advice from integrative cancer care, to prefer boiled and cooked vegetables over raw vegetables is usually recommended.

The data from the case-control study revealed that consumption of cruciferous vegetables, particularly raw cruciferous vegetables, is a modifiable lifestyle behavior that may be inversely associated with pancreatic cancer.[61]

7.1.6 *Advice on the Use of Dietary Herbs as per Guidelines in Ayurveda Classical Books*

Turmeric vs curcumin

Ayurveda classic books mainly advise turmeric as a whole to be used for dietary consumption. In recent days the chemotherapeutic effect of curcumin, one of three major curcuminoids derived from turmeric, has been reported. The nutritional or therapeutic effects of more analogs from complex turmeric extracts remained unexplored, as well as the relative importance of the three curcuminoids and their metabolites as anticancer agents. The pharmacodynamic effects of curcuminoids on human breast cancer cell growth and tumor cell secretion of parathyroid hormone-related protein (PTHrP), an important driver of cancer bone metastasis, have been reported.[62]

7.1.7 *Advice on* Langhana *(Therapeutic Fasting)*

Langhana is an important treatment among the main categories of treatment; the other one is *Bruhana*. Therapeutic fasting, usually a partial type, is part of treatment in patients who have completed the primary treatment like surgery/chemotherapy/radiotherapy and are in the disease-free/progression-free phase. To enhance the *Agni* (metabolic energy), which is the basis to develop innate immunity *Langhana,* plays a key role, though it's one of the nonpharmacological treatments in Ayurveda.[63]

Fasting or fasting-mimicking diets (FMDs) lead to wide alterations in growth factors and metabolite levels, reducing the capability of cancer cells to adapt and survive, thus improving the effects of cancer therapies. Additionally, fasting or FMDs increase resistance to chemotherapy in normal but not cancer cells and promote regeneration in normal tissues, which may mitigate the side effects of chemotherapy. Fasting is hardly tolerated by patients; the cycles of low-calorie FMDs are feasible and overall safe.[64]

7.2 *Annaraksha* AHS 7/2

7.2.1 *Cooking at Optimum Temperature*

The Maillard reaction is a simple but ubiquitous reaction that occurs in preclinical studies during the cooking or processing of foods under high-temperature conditions, such as baking, frying, or grilling. The glycation of proteins is a posttranslational modification that forms temporary adducts, which, on further crosslinking and rearrangement, form permanent residues known as advanced glycation end products (AGEs). Cooking at high temperatures results in various food products having high levels of AGEs.[65]

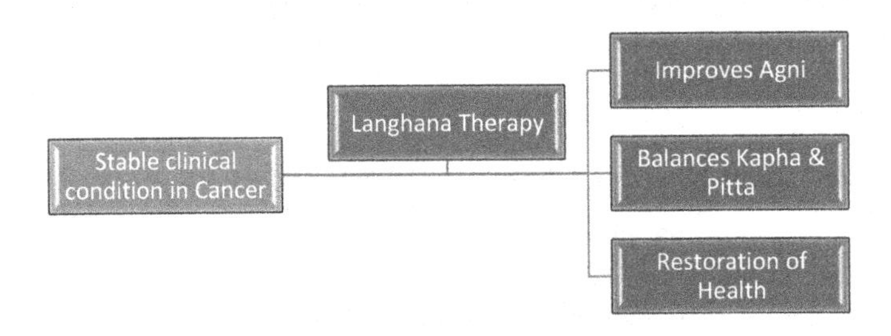

FIGURE 9.3 Therapeutic *Langhana*: (*Ashtang Sangraha Chikitsasthana 1/3*).

7.2.2 Reheating Food

The extensive consumption of repeatedly heated cooking oil has been associated with an increased risk of cancers, including lung, colorectal, breast, and prostate cancers.[66]

7.3 *Anukta* Dietary Characters

The category which is not mentioned in Ayurveda classical texts but in routine dietary practice in India or the rest of the world is termed ***Anukta Ahara Dravya***.

7.3.1 Olive Oil

Highest olive oil consumption was associated with a 31% lower likelihood of any cancer, breast, gastrointestinal, upper aerodigestive, and urinary tract cancer.[67]

7.3.2 Cranberry

Cranberry preclinical study has indicated the inhibitory action toward cancers of the esophagus, stomach, colon, bladder, prostate, glioblastoma, and lymphoma, with mechanisms of inhibiting cellular death induction via apoptosis, necrosis, and autophagy; reduction of cellular proliferation; alterations in reactive oxygen species; and modification of cytokine and signal transduction pathways.[68]

7.3.3 Strawberry

Strawberry is one of the berries from western India, individual components of which have demonstrated anticancer activity by blocking the initiation of carcinogenesis and suppressing the progression and proliferation of tumors.[69]

7.3.4 Milk of Vegetable Origin

An example is soy milk; the isoflavones in soy products are inversely associated with the risk of cancer.[70]

7.3.5 Nuts

A study of 97 breast cancer patients suggested that the high consumption of peanuts, walnuts, or almonds reduced the risk for breast cancer significantly, almost by 2–3 times.[71]

7.3.6 Coconut Yogurt

The available whole-fat coconut yogurt and reduced-fat coconut yogurt may be possible alternatives, where dairy-based milk products are not recommended. The medium- and long-chain fatty acid in coconut has health benefits as they act metabolically differently through a source of saturated fats.[72]

7.4 Route of Administration of Therapeutic Nutrition

Peroral is the most commonly practiced and preferred route for nutrition. In certain clinical conditions, after oral or upper GI surgeries, head and neck cancer patients on chemoradiation or in critically ill patients where feeding through nasogastric (NG) tube or percutaneous endoscopic gastrostomy (PEG) feeding tube insertion may be recommended, given possible chances of aspiration or microaspiration.

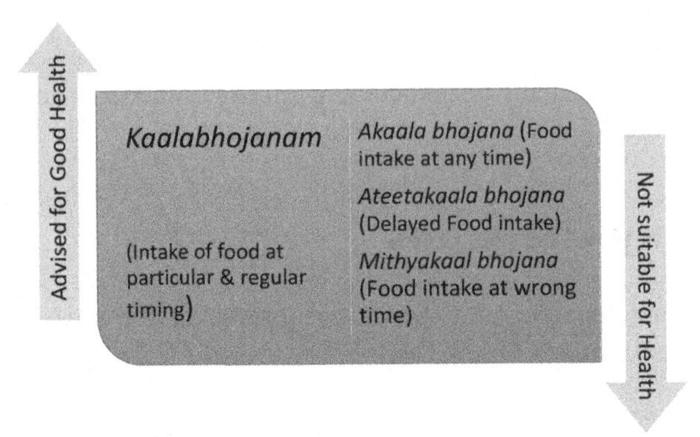

FIGURE 9.4 Principles of diet intake.

Routine administration of therapeutic nutrition can be achieved through an NG tube or prepyloric PEG tube, while it needs meticulous selection of *Aahareeya Kalpana* like *Manda*, *Mantha, Yavagu, Yusha,* and *Ksheerpaka* through the PEG tube at postpyloric junction where it gets inserted in duodenum or jejunum bypassing the stomach.

7.5 Education on Food Intake

Observing particular and regular timing of food intake is an equally important aspect of Ayurveda therapeutic nutrition for the maintenance of a healthy state of the body. The advice on food intake is described below in Figure 9.3.

8. *AHAREYA KALPANA*

The *Aaahareeya Kalpana* (dietary recipes), which can be used in cancer treatments as per the symptoms during or after treatment, has been discussed.

In cancer care, the main support of dietary recipes is needed during the chemotherapy while patients have anorexia, dysgeusia, nausea, odynophagia, or during radiotherapy, which can cause similar side effects in addition to dysphagia to solids and xerostomia.

The recipes tabulated below are mainly advised for symptom care during and after chemotherapy and radiotherapy.

9. NUTRAVIGILANCE

9.1 Heating Honey

Heating honey directly or indirectly leads to the formation of dangerous compounds like 5-hydroxymethyl furfural (HMF), which is not naturally present in honey and is hazardous to human health. HMF is a potential carcinogenic, mutagenic, and cytotoxic agent.[74]

TABLE 9.3 *Aahareeya Kalpana* in Clinical Practice

AAHARA-KALPANA	*KEY INGREDIENTS*	*THERAPEUTIC ACTION*
Sharakarodaka	*Ela, Lavang, Marich*	Reduces Vomiting, Burning
Prapanakam	*Aam-amram, Sita, Karpoor, Marich*	Improves taste, Replenishing
Amlika-phala Panakam	*Amlika, Marich Sita*	Improves taste, Enhances Metabolism
Nimbu-phala Panakam	*Nimbuphala, Lavang Marich*	Improves taste, Enhances Metabolism
Dhanyak Panakam	*Sharkara, Dhanyak, Karpoor*	Alleviates *Pitta*
Takrapanakam	Curd, *Cumin, Hingu, Lavangam*	Improves taste, prodigestion
Tandul-Payasam[73]	Rice, Milk, Water	Balyaam, Pittaghna, Bruhanama, Saram, Dhatupushtipradam

9.2 *Viruddha anna* (Incompatible Diet)

There is a concept of an incompatible diet in Ayurveda. Following is a study documenting a similar principle of incompatibility in diet processing.

High consumption of mustard oil was observed to have an association with gall bladder cancer risk. An increased risk of gall bladder cancer was observed with deep frying of fresh fish in mustard oil.[75]

9.3 Repeated Heating of Edible Oils

The repeated heating of edible vegetable oils is a commonly observed practice; a study on the use of repeatedly heated coconut oil in vivo indicates that it can be genotoxic and may cause preneoplastic changes in the liver.[76]

9.4 Ultraprocessed Food

A large prospective cohort study with more than 2000 subjects concluded that a 10% increase in consumption of ultraprocessed food in diet can lead to a 10% increase in overall cancers and breast cancer.[77]

10. SCOPE FOR FURTHER STUDY

Therapeutic nutrition in Ayurveda is clinically important while treating diseases like cancer; at times nutrition awareness, education, and execution discussion are more important between the treating physician and the caregiver in the presence of the patient.

Further studies on establishing an association between the therapeutic diet and treatment modalities in cancer together or individually shall be conducted at multiple cancer centers.

An important aspect of cancer treatment is to delay or prevent recurrence; the food, which is *Satmya,* that is most suitable for that individual shall play a vital role. The nutritional advice shall be offered considering the patient's food habits and pattern of eating.

Nutrition assessment studies have their own limitations as there can be multiple confounders and even with adjusted ratios, there are chances that the data retrieved and analyzed may not answer the hypothesis. Even though the data remains inconclusive, further studies shall be continued, unless the results indicate a negative correlation.

In the present chapter, the discussion is more inclined toward solid tumors in adults; there is potential to explore hematological cancers equally and pediatric cancers as it can be challenging to plan a therapeutic diet in children.

REFERENCES

1 Soerjomataram I, Bray F. Planning for tomorrow: global cancer incidence and the role of prevention 2020–2070. *Nat Rev Clin Oncol* 18, 663–672 (2021). https://doi.org/10.1038/s41571-021-00514-z
2 Sathishkumar K, Chaturvedi M, Das P, Stephen S, Mathur P. Cancer incidence estimates for 2022 & projection for 2025: Result from National Cancer Registry Programme, India. Indian J Med Res. 2022 Dec 13. doi: 10.4103/ijmr.ijmr_1821_22. Epub ahead of print. PMID: 36510887.
3 von Meyenfeldt M. Cancer-associated malnutrition: an introduction. Eur J Oncol Nurs. 2005;9 Suppl 2:S35–8. doi: 10.1016/j.ejon.2005.09.001. PMID: 16437756.
4 Zhang X, Pang L, Sharma SV, Li R, Nyitray AG, Edwards BJ. Malnutrition and overall survival in older patients with cancer. Clin Nutr. 2021 Mar;40(3):966–977. doi: 10.1016/j.clnu.2020.06.026. Epub 2020 Jul 1. PMID: 32665101.
5 Cotogni P, Pedrazzoli P, De Waele E, Aprile G, Farina G, Stragliotto S, De Lorenzo F, Caccialanza R. Nutritional therapy in cancer patients receiving chemoradiotherapy: Should we need stronger recommendations to act for improving outcomes? J Cancer. 2019 Jul 10;10(18):4318–4325. doi: 10.7150/jca.31611. PMID: 31413751; PMCID: PMC6691712.
6 Morita M, Kudo K, Shima H, Tanuma N. Dietary intervention as a therapeutic for cancer. Cancer Sci. 2021 Feb;112(2):498–504. doi: 10.1111/cas.14777. Epub 2020 Dec 31. PMID: 33340176; PMCID: PMC7893991.
7 Yadavji Trikamji Acharya, Sushruta Samhita Dalhana Commentary Choukhamba Surabharati Publication Nidanasthana Chapter 11 verse 13–15 page 310.
8 Prem Kishor, Padhi MM., Vishwanath Sen's *Pathyapathay Vinishchaya,* Central Council of Research in Ayurveda & Siddha (CCRAS) India, 1999, verse 332, Page 74.
9 Vaidya Yadavji Trikamji Acharya, The Charaka Samhita of Agnivesa, Ayurveda Dipika commentary of Chakrapanidatta, 5th edition, 1992, Sutra sthana, Chapter 22 verse 19–22, page 121.
10 Vaidya Yadavji Trikamji Acharya, The Charaka Samhita of Agnivesa, Ayurveda Dipika commentary of Chakrapanidatta, 5th edition, 1992, Sutra sthana, Chapter 22 verse 18, page 121.
11 Prem Kishor, Padhi MM, Vishwanath Sen's *Pathyapathay Vinishchaya*, Central Council of Research in Ayurveda & Siddha (CCRAS) India, 1999, Page 74.
12 Savina S, Zaydiner B. Cancer-related fatigue: Some clinical aspects. Asia Pac J Oncol Nurs. 2019 Jan-Mar;6(1):7–9. doi: 10.4103/apjon.apjon_45_18. PMID: 30599009; PMCID: PMC6287376.
13 Vaidya Yadavji Trikamji Acharya, The Charaka Samhita od Agnivesa, Ayurveda Dipika commentary of Chakrapanidatta, 5th edition, 1992, Sutra sthana, Chapter 23 verse 31–22, page 123.
14 Vagbhata, Ashtanghridayam; Arundatta & Hemadri commentary; Choukhamba Orientalia, 9th Edition, 2005, Sutrastahana Chapter 5 verse 1–2, 13–14, 19–25, 29–31, 33,35,39,41–42, 47, 51–52, 55–59, 61; Page 61–79.
15 Raghunath Suri, Bhojankutuhala, University Press, Trivendrum 1955 print ed, Parichheda 1, Tailavarga, Page 154.
16 Prem Kishor, Padhi MM, Vishwanath Sen's *Pathyapathay Vinishchaya,* Central Council of Research in Ayurveda & Siddha (CCRAS) India, 1999, verse 337–338, Page 74–75.
17 Ravasco P. Nutrition in cancer patients. J Clin Med. 2019 Aug 14;8(8):1211. doi: 10.3390/jcm8081211. PMID: 31416154; PMCID: PMC6723589.
18 Tollefsbol TO. Dietary epigenetics in cancer and aging. Cancer Treat Res. 2014;159:257–267. doi: 10.1007/978-3-642-38007-5_15. PMID: 24114485; PMCID: PMC3875399.
19 Nasir A, Bullo MMH, Ahmed Z, Imtiaz A, Yaqoob E, Jadoon M, Ahmed H, Afreen A, Yaqoob S. Nutrigenomics: Epigenetics and cancer prevention: A comprehensive review. Crit Rev Food Sci Nutr. 2020;60(8):1375–1387. doi: 10.1080/10408398.2019.1571480. Epub 2019 Feb 7. PMID: 30729798.
20 Hardy TM, Tollefsbol TO. Epigenetic diet: impact on the epigenome and cancer. Epigenomics. 2011 Aug;3(4):503–18. doi: 10.2217/epi.11.71. PMID: 22022340; PMCID: PMC3197720.

21 Greathouse KL, Wyatt M, Johnson AJ, Toy EP, Khan JM, Dunn K, Clegg DJ, Reddy S. Diet-microbiome interactions in cancer treatment: Opportunities and challenges for precision nutrition in cancer. Neoplasia. 2022 Jul;29:100800. doi: 10.1016/j.neo.2022.100800. Epub 2022 Apr 29. PMID: 35500546; PMCID: PMC9065883.

22 Szczyrek M, Bitkowska P, Chunowski P, Czuchryta P, Krawczyk P, Milanowski J. Diet, microbiome, and cancer immunotherapy-A comprehensive review. Nutrients. 2021 Jun 28;13(7):2217. doi: 10.3390/nu13072217. PMID: 34203292; PMCID: PMC8308287.

23 Vaidya Yadavji Trikamji Acharya, The Charaksamhita of Agnivesha, the Ayurveda-Deepika commentary, 5th edition 1992, *Sutrasthana* Chapter 23, Verse 5, Page 122–123.

24 Yadavji Trikamji Acharya, Sushruta Samhita Dalhana Commentary Choukhamba Surabharati Publication Sutrasthana Chapter 40 verse 10 page 179.

25 Boushey C, Ard J, Bazzano L, Heymsfield S, Mayer-Davis E, Sabaté J, Snetselaar L, Van Horn L, Schneeman B, English LK, Bates M, Callahan E, Butera G, Terry N, Obbagy J. Dietary Patterns and Breast, Colorectal, Lung, and Prostate Cancer: A Systematic Review [Internet]. Alexandria (VA): USDA Nutrition Evidence Systematic Review; 2020 Jul. PMID: 35129907.

26 Singh RK, Chang HW, Yan D, Lee KM, Ucmak D, Wong K, Abrouk M, Farahnik B, Nakamura M, Zhu TH, Bhutani T, Liao W. Influence of diet on the gut microbiome and implications for human health. J Transl Med. 2017 Apr 8;15(1):73. doi: 10.1186/s12967-017-1175-y. PMID: 28388917; PMCID: PMC5385025.

27 Jin S, Je Y. Dairy consumption and total cancer and cancer-specific mortality: A meta-analysis of prospective cohort studies. Adv Nutr. 2022 Aug 1;13(4):1063–1082. doi: 10.1093/advances/nmab135. PMID: 34788365; PMCID: PMC9340963.

28 Rock CL. Milk and the risk and progression of cancer. Nestle Nutr Workshop Ser Pediatr Program. 2011;67:173–185. doi: 10.1159/000325583. Epub 2011 Feb 16. PMID: 21335998.

29 Liao MQ, Gao XP, Yu XX, Zeng YF, Li SN, Naicker N, Joseph T, Cao WT, Liu YH, Zhu S, Chen QS, Yang ZC, Zeng FF. Effects of dairy products, calcium and vitamin D on ovarian cancer risk: a meta-analysis of twenty-nine epidemiological studies. Br J Nutr. 2020 Nov 28;124(10):1001–1012. doi: 10.1017/S0007114520001075. Epub 2020 Mar 19. PMID: 32189606.

30 Xu J, Chen K, Zhao F, Huang D, Zhang H, Fu Z, Xu J, Wu Y, Lin H, Zhou Y, Lu W, Wu Y, Xia D. Association between vitamin D/calcium intake and 25-hydroxyvitamin D and risk of ovarian cancer: a dose-response relationship meta-analysis. Eur J Clin Nutr. 2021 Mar;75(3):417–429. doi: 10.1038/s41430-020-00724-1. Epub 2020 Aug 19. PMID: 32814859.

31 Kazemi A, Barati-Boldaji R, Soltani S, Mohammadipoor N, Esmaeilinezhad Z, Clark CCT, Babajafari S, Akbarzadeh M. Intake of various food groups and risk of breast cancer: A systematic review and dose-response meta-analysis of prospective studies. Adv Nutr. 2021 Jun 1;12(3):809–849. doi: 10.1093/advances/nmaa147. PMID: 33271590; PMCID: PMC8166564.

32 Liu C, Zheng S, Gao H, Yuan X, Zhang Z, Xie J, Yu C, Xu L. Causal relationship of sugar-sweetened and sweet beverages with colorectal cancer: A Mendelian randomization study. Eur J Nutr. 2022 Aug 30. doi: 10.1007/s00394-022-02993-x. Epub ahead of print. PMID: 36040623.

33 van der Lugt T, Opperhuizen A, Bast A, Vrolijk MF. Dietary advanced glycation endproducts and the gastrointestinal tract. Nutrients. 2020 Sep 14;12(9):2814. doi: 10.3390/nu12092814. PMID: 32937858; PMCID: PMC7551018.

34 Rojas A, Lindner C, Schneider I, Gonzàlez I, Morales MA. Receptor of advanced glycation end-products axis and gallbladder cancer: A forgotten connection that we should reconsider. World J Gastroenterol. 2022 Oct 21;28(39):5679–5690. doi: 10.3748/wjg.v28.i39.5679. PMID: 36338887; PMCID: PMC9627425.

35 Farvid MS, Stern MC, Norat T, Sasazuki S, Vineis P, Weijenberg MP, Wolk A, Wu K, Stewart BW, Cho E. Consumption of red and processed meat and breast cancer incidence: A systematic review and meta-analysis of prospective studies. Int J Cancer. 2018 Dec 1;143(11):2787–2799. doi: 10.1002/ijc.31848. Epub 2018 Oct 3. PMID: 30183083; PMCID: PMC8985652.

36 Schwingshackl L, Schwedhelm C, Hoffmann G, Knüppel S, Laure Preterre A, Iqbal K, Bechthold A, De Henauw S, Michels N, Devleesschauwer B, Boeing H, Schlesinger S. Food groups and risk of colorectal cancer. Int J Cancer. 2018 May 1;142(9):1748–1758. doi: 10.1002/ijc.31198. Epub 2017 Dec 14. PMID: 29210053.

37 Farvid MS, Sidahmed E, Spence ND, Mante Angua K, Rosner BA, Barnett JB. Consumption of red meat and processed meat and cancer incidence: a systematic review and meta-analysis of prospective studies. Eur J Epidemiol. 2021 Sep;36(9):937–951. doi: 10.1007/s10654-021-00741-9. Epub 2021 Aug 29. PMID: 34455534.

38 Zhu HC, Yang X, Xu LP, Zhao LJ, Tao GZ, Zhang C, Qin Q, Cai J, Ma JX, Mao WD, Zhang XZ, Cheng HY, Sun XC. Meat consumption is associated with esophageal cancer risk in a meat- and cancer-histological-type dependent manner. Dig Dis Sci. 2014 Mar;59(3):664–73. doi: 10.1007/s10620-013-2928-y. Epub 2014 Jan 7. PMID: 24395380.

39 Yamagiwa Y, Sawada N, Shimazu T, Yamaji T, Goto A, Takachi R, Ishihara J, Iwasaki M, Inoue M, Tsugane S. Fruit and vegetable intake and pancreatic cancer risk in a population-based cohort study in Japan. Int J Cancer. 2019 Apr 15;144(8):1858–1866. doi: 10.1002/ijc.31894. Epub 2018 Oct 30. PMID: 30255932.

40 Baudry J, Assmann KE, Touvier M, Allès B, Seconda L, Latino-Martel P, Ezzedine K, Galan P, Hercberg S, Lairon D, Kesse-Guyot E. Association of frequency of organic food consumption with cancer risk: Findings from the NutriNet-Santé prospective cohort study. JAMA Intern Med. 2018 Dec 1;178(12):1597–1606. doi: 10.1001/jamainternmed.2018.4357. Erratum in: JAMA Intern Med. 2018 Dec 1;178(12):1732. PMID: 30422212; PMCID: PMC6583612.

41 Prakash MD, Stojanovska L, Feehan J, Nurgali K, Donald EL, Plebanski M, Flavel M, Kitchen B, Apostolopoulos V. Anti-cancer effects of polyphenol-rich sugarcane extract. PLoS One. 2021 Mar 10;16(3):e0247492. doi: 10.1371/journal.pone.0247492. PMID: 33690618; PMCID: PMC7946306.

42 Rumgay H, Murphy N, Ferrari P, Soerjomataram I. Alcohol and cancer: Epidemiology and biological mechanisms. Nutrients. 2021 Sep 11;13(9):3173. doi: 10.3390/nu13093173. PMID: 34579050; PMCID: PMC8470184.

43 Hurtado-Barroso S, Trius-Soler M, Lamuela-Raventós RM, Zamora-Ros R. Vegetable and fruit consumption and prognosis among cancer survivors: A systematic review and meta-analysis of cohort studies. Adv Nutr. 2020 Nov 16;11(6):1569–1582. doi: 10.1093/advances/nmaa082. PMID: 32717747; PMCID: PMC7666913.

44 Sellers TA, Vierkant RA, Djeu J, Celis E, Wang AH, Kumar N, Cerhan JR. Unpasteurized milk consumption and subsequent risk of cancer. Cancer Causes Control. 2008 Oct;19(8):805–11. doi: 10.1007/s10552-008-9143-8. Epub 2008 Mar 15. PMID: 18344007; PMCID: PMC2575230.

45 Law KS, Azman N, Omar EA, Musa MY, Yusoff NM, Sulaiman SA, Hussain NH. The effects of virgin coconut oil (VCO) as supplementation on quality of life (QOL) among breast cancer patients. Lipids Health Dis. 2014 Aug 27;13:139. doi: 10.1186/1476-511X-13-139. PMID: 25163649; PMCID: PMC4176590.

46 Narayanankutty A, Illam SP, Rao V, Shehabudheen S , Raghavamenon AC. Hot-processed virgin coconut oil abrogates cisplatin-induced nephrotoxicity by restoring redox balance in rats compared to fermentation-processed virgin coconut oil. Drug Chem Toxicol. 2022 May;45(3):1373–1382. doi: 10.1080/01480545.2020.1831525. Epub 2020 Oct 15. PMID: 33059468

47 Gaesser GA. Whole grains, refined grains, and cancer risk: A systematic review of meta-analyses of observational studies. Nutrients. 2020 Dec 7;12(12):3756. doi: 10.3390/nu12123756. PMID: 33297391; PMCID: PMC7762239.

48 Hou D, Yousaf L, Xue Y, Hu J, Wu J, Hu X, Feng N, Shen Q. Mung bean (*Vigna radiata* L.): Bioactive polyphenols, polysaccharides, peptides, and health benefits. Nutrients. 2019 May 31;11(6):1238. doi: 10.3390/nu11061238. PMID: 31159173; PMCID: PMC6627095.

49 Rao S, Chinkwo KA, Santhakumar AB, Blanchard CL. Inhibitory effects of pulse bioactive compounds on cancer development pathways. Diseases. 2018 Aug 3;6(3):72. doi: 10.3390/diseases6030072. PMID: 30081504; PMCID: PMC6163461.

50 Campos-Vega R, Oomah BD, Loarca-Piña G, Vergara-Castañeda HA. Common beans and their non-digestible fraction: Cancer inhibitory activity-An overview. Foods. 2013 Aug 2;2(3):374–392. doi: 10.3390/foods2030374. PMID: 28239123; PMCID: PMC5302293.

51 Starowicz M, Zieliński H. Inhibition of advanced glycation end-product formation by high antioxidant-leveled spices commonly used in European cuisine. Antioxidants (Basel). 2019 Apr 15;8(4):100. doi: 10.3390/antiox8040100. PMID: 30991695; PMCID: PMC6523868.

52 Aschebrook-Kilfoy B, Ollberding NJ, Kolar C, Lawson TA, Smith SM, Weisenburger DD, Chiu BC. Meat intake and risk of non-Hodgkin lymphoma. Cancer Causes Control. 2012 Oct;23(10):1681–92. doi: 10.1007/s10552-012-0047-2. Epub 2012 Aug 14. PMID: 22890783.

53 Debras C, Chazelas E, Srour B, Kesse-Guyot E, Julia C, Zelek L, Agaësse C, Druesne-Pecollo N, Galan P, Hercberg S, Latino-Martel P, Deschasaux M, Touvier M. Total and added sugar intakes, sugar types, and cancer risk: results from the prospective NutriNet-Santé cohort. Am J Clin Nutr. 2020 Nov 11;112(5):1267–1279. doi: 10.1093/ajcn/nqaa246. PMID: 32936868.

54 Nagle CM, Kolahdooz F, Ibiebele TI, Olsen CM, Lahmann PH, Green AC, Webb PM; Australian Cancer Study (Ovarian Cancer) and the Australian Ovarian Cancer Study Group. Carbohydrate intake, glycemic

load, glycemic index, and risk of ovarian cancer. Ann Oncol. 2011 Jun;22(6):1332–1338. doi: 10.1093/annonc/mdq595. Epub 2010 Dec 3. PMID: 21131370.

55 Arthur RS, Kirsh VA, Mossavar-Rahmani Y, Xue X, Rohan TE. Sugar-containing beverages and their association with risk of breast, endometrial, ovarian and colorectal cancers among Canadian women. Cancer Epidemiol. 2021 Feb;70:101855. doi: 10.1016/j.canep.2020.101855. Epub 2020 Nov 18. PMID: 33220638.

56 Gianuzzi X, Palma-Ardiles G, Hernandez-Fernandez W, Pasupuleti V, Hernandez AV, Perez-Lopez FR. Insulin growth factor (IGF) 1, IGF-binding proteins and ovarian cancer risk: A systematic review and meta-analysis. Maturitas. 2016 Dec;94:22–29. doi: 10.1016/j.maturitas.2016.08.012. Epub 2016 Aug 20. PMID: 27823741.

57 Lukanova A, Lundin E, Toniolo P, Micheli A, Akhmedkhanov A, Rinaldi S, Muti P, Lenner P, Biessy C, Krogh V, Zeleniuch-Jacquotte A, Berrino F, Hallmans G, Riboli E, Kaaks R. Circulating levels of insulin-like growth factor-I and risk of ovarian cancer. Int J Cancer. 2002 Oct 20;101(6):549–554. doi: 10.1002/ijc.10613. PMID: 12237896.

58 King ER, Zu Z, Tsang YT, Deavers MT, Malpica A, Mok SC, Gershenson DM, Wong KK. The insulin-like growth factor 1 pathway is a potential therapeutic target for low-grade serous ovarian carcinoma. Gynecol Oncol. 2011 Oct;123(1):13–18. doi: 10.1016/j.ygyno.2011.06.016. Epub 2011 Jul 2. PMID: 21726895; PMCID: PMC3171566.

59 Kim SR, Kim K, Lee SA, Kwon SO, Lee JK, Keum N, Park SM. Effect of red, processed, and white meat consumption on the risk of gastric cancer: An overall and dose-response meta-analysis. Nutrients. 2019 Apr 11;11(4):826. doi: 10.3390/nu11040826. PMID: 30979076; PMCID: PMC6520977.

60 Porcza LM, Simms C, Chopra M. Honey and cancer: Current status and future directions. Diseases. 2016 Sep 30;4(4):30. doi: 10.3390/diseases4040030. PMID: 28933410; PMCID: PMC5456322.

61 Morrison MEW, Hobika EG, Joseph JM, Stenzel AE, Mongiovi JM, Tang L, McCann SE, Marshall J, Fountzilas C, Moysich KB. Cruciferous vegetable consumption and pancreatic cancer: A case-control study. Cancer Epidemiol. 2021 Jun;72:101924. doi: 10.1016/j.canep.2021.101924. Epub 2021 Mar 11. PMID: 33714902; PMCID: PMC8278290.

62 Wright LE, Frye JB, Gorti B, Timmermann BN, Funk JL. Bioactivity of turmeric-derived curcuminoids and related metabolites in breast cancer. Curr Pharm Des. 2013;19(34):6218–6225. doi: 10.2174/1381612811319340013. PMID: 23448448; PMCID: PMC3883055

63 Qian J, Fang Y, Yuan N, Gao X, Lv Y, Zhao C, Zhang S, Li Q, Li L, Xu L, Wei W, Wang J. Innate immune remodeling by short-term intensive fasting. Aging Cell. 2021 Nov;20(11):e13507. doi: 10.1111/acel.13507. Epub 2021 Oct 27. PMID: 34705313; PMCID: PMC8590100.

64 Nencioni A, Caffa I, Cortellino S, Longo VD. Fasting and cancer: molecular mechanisms and clinical application. Nat Rev Cancer. 2018 Nov;18(11):707–719. doi: 10.1038/s41568-018-0061-0. PMID: 30327499; PMCID: PMC6938162.

65 Gill V, Kumar V, Singh K, Kumar A, Kim JJ. Advanced glycation end products (AGEs) may be a striking link between modern diet and health. Biomolecules. 2019 Dec 17;9(12):888. doi: 10.3390/biom9120888. PMID: 31861217; PMCID: PMC6995512.

66 Ganesan K, Sukalingam K, Xu B. Impact of consumption of repeatedly heated cooking oils on the incidence of various cancers – A critical review. Crit Rev Food Sci Nutr. 2019;59(3):488–505. doi: 10.1080/10408398.2017.1379470. Epub 2017 Oct 20. PMID: 28925728.

67 Markellos C, Ourailidou ME, Gavriatopoulou M, Halvatsiotis P, Sergentanis TN, Psaltopoulou T. Olive oil intake and cancer risk: A systematic review and meta-analysis. PLoS One. 2022 Jan 11;17(1):e0261649. doi: 10.1371/journal.pone.0261649. PMID: 35015763; PMCID: PMC8751986.

68 Weh KM, Clarke J, Kresty LA. Cranberries and cancer: An update of preclinical studies evaluating the cancer inhibitory potential of cranberry and cranberry derived constituents. *Antioxidants*. 2016; 5(3):27. https://doi.org/10.3390/antiox5030027

69 Hannum SM. Potential impact of strawberries on human health: a review of the science. Crit Rev Food Sci Nutr. 2004;44(1):1–17. doi: 10.1080/10408690490263756. PMID: 15077879.

70 Fan Y, Wang M, Li Z, Jiang H, Shi J, Shi X, Liu S, Zhao J, Kong L, Zhang W, Ma L. Intake of soy, soy isoflavones and soy protein and risk of cancer incidence and mortality. Front Nutr. 2022 Mar 4;9:847421. doi: 10.3389/fnut.2022.847421. PMID: 35308286; PMCID: PMC8931954.

71 Soriano-Hernandez AD, Madrigal-Perez DG, Galvan-Salazar HR, Arreola-Cruz A, Briseño-Gomez L, Guzmán-Esquivel J, Dobrovinskaya O, Lara-Esqueda A, Rodríguez-Sanchez IP, Baltazar-Rodriguez LM, Espinoza-Gomez F, Martinez-Fierro ML, de-Leon-Zaragoza L, Olmedo-Buenrostro BA, Delgado-Enciso

I. The protective effect of peanut, walnut, and almond consumption on the development of breast cancer. Gynecol Obstet Invest. 2015;80(2):89–92. doi: 10.1159/000369997. Epub 2015 Jul 10. PMID: 26183374.

72 Hewlings S. Coconuts and health: Different chain lengths of saturated fats require different consideration. J Cardiovasc Dev Dis. 2020 Dec 17;7(4):59. doi: 10.3390/jcdd7040059. PMID: 33348586; PMCID: PMC7766932.

73 Yogaratnakara – Payasam, Pt. Sadashiv Shatri Joshi, Togaratnakara, 1996, The Choukhamba Sanskrit Series, 1996 edition Payasam, verse 1 Page 35.

74 Afroz R, Tanvir EM, Zheng W, Little PJ (2016) Molecular pharmacology of honey. Clin Exp Pharmacol. 6: 212. doi: 10.4172/2161-1459.1000212

75 Mhatre S, Rajaraman P, Chatterjee N, Bray F, Goel M, Patkar S, Ostwal V, Patil P, Manjrekar A, Shrikhande SV, Badwe R, Dikshit R. Mustard oil consumption, cooking method, diet and gallbladder cancer risk in high- and low-risk regions of India. Int J Cancer. 2020 Sep 15;147(6):1621–1628. doi: 10.1002/ijc.32952. Epub 2020 Mar 30. PMID: 32142159

76 Srivastava S, Singh M, George J, Bhui K, Murari Saxena A, Shukla Y. Genotoxic and carcinogenic risks associated with the dietary consumption of repeatedly heated coconut oil. Br J Nutr. 2010 Nov;104(9):1343–1352. doi: 10.1017/S0007114510002229. Epub 2010 Aug 6. PMID: 20687968.

77 Fiolet T, Srour B, Sellem L, Kesse-Guyot E, Allès B, Méjean C, Deschasaux M, Fassier P, Latino-Martel P, Beslay M, Hercberg S, Lavalette C, Monteiro CA, Julia C, Touvier M. Consumption of ultra-processed foods and cancer risk: results from NutriNet-Santé prospective cohort. BMJ. 2018 Feb 14;360:k322. doi: 10.1136/bmj.k322. PMID: 29444771; PMCID: PMC5811844.

Therapeutic Nutrition in Ayurveda for Ophthalmology

Pravin M. Bhat

1. INTRODUCTION

The eye is a one-of-a-kind sensory organ with intricate anatomical and physiological characteristics. The rays of light fall on the cornea are followed by the lens, which leads to the retina, which is subsequently followed by the optic nerve. The activation of photoreceptor cells allows for signal transmission to the brain and the production of an image. The eyes had direct interaction with the external environment such as toxins, ionizing radiation, smoking, and various air pollutants are all examples of contaminants. The retina has the highest metabolic rate among human bodily tissue. It requires the most oxygen and is exposed to light constantly. As a result, there is a considerable risk of oxidative damage caused by light.[1] Fortunately, our eyes have evolved to be extremely sensitive to nutritional variables that protect us from stress.[2]

Carotenoids, organic pigments obtained from plant sources that protect the eye from free radical damage, are the most essential of these protective elements for the eye. Blue light in the 400 to 500 nm range is powerful enough to convert oxygen into reactive oxygen, causing retinal damage.

There are two types of blindness classified as reversible blindness and irreversible blindness found in the population.[3] Reversible blindness is also termed avoidable blindness, which can be treated significantly with the help of medicinal and surgical aid.

Approximately 253 million people around the world experience varied degrees of visual loss.[4] Leading causes include a number of eye diseases that are examined, including cataracts, ARMD, glaucoma, and diabetic retinopathy.[5] With an aging population, it is expected that the number of people affected by these disorders would rise exponentially. These conditions disproportionately impact older adults.[6]

Although the causes of age-related eye illness are complicated and diverse, oxidative stress has been identified as a major contributing factor. The eye is especially prone to oxidative stress because of its high oxygen consumption, high content of polyunsaturated fatty acids, and cumulative exposure to high-energy visual light. Reactive oxygen species (ROS) are produced due to these interrelated variables, which can cause oxidative damage to the eye's tissues.[7]

DOI: 10.1201/9781003345541-13

TABLE 10.1 *Chakshyushya* Dietary Items Recommended and Not Recommended in Different Classics

AYURVEDIC CLASSIC	*RECOMMENDED*	*NOT RECOMMENDED*
Yogaratnakar	Ghee, Milk, Rock Salt, Honey, Brown-Colored Sugar, *Patola* (Snake Gourd), Red Rice, Green Gram, Wheat, *Triphala*, Carrots, *Lodhra*, Leafy Vegetables, *Jeevanti, Matsyakshi, Punarnava, Meghanada*, rainwater	*Urad Dal*, Excessive Hot, Sour Heavy Foods, Tobacco, Selected Leafy Vegetables, Curd, Selected Sea Fishes, Alcohols, Dry Meat, Sour Food Items.
Vaghbhata	*Ghee*, Milk, *Triphala*, Pomegranate, Black Grapes, Gooseberry, Meat of *Tittar* Bird	Cold Food (after cooking, kept for more time), Food Which Is Infected, Dry, Rough Food, Stale food (food prepared at night and eaten next day)
Sushruta Samhita	*Purana Ghrita*, *Triphala* Asparagus, Barley, Snake Gourd, Small Radish, Small Brinjal, Bitter gourd, Meat of Home Grown Poultry, Leaf of Drumstick	
Bhavaprakasha	*Ghee*, Butter Milk, Urine, Mustard Oils, Honey, Black Grapes, Sandal Jasmin Oil, Banana, *Kataka* Fruit	Pineapple, Sour Mango
Sargadhara Samhita	*Ghrita*	
Raja Nighantu	Cow's Urine, Intake of Minerals *Prapaundarika*	

2. NUTRITION IN OPHTHALMOLOGY FROM CLASSICAL TEXT

According to *Ayurveda, Ahara* is one of the important pillars of life along with sleep and regulated sex.[8] The definition of *Swasthya* is "to be established in one's self or own natural state. These are mainly focused on nourishing the tissue of the eye and protecting the eye from the diminution of vision. The *Chakshyushya* dietary items mentioned in different classics are shown in Table 10.1.

3. NUTRITION IN OPHTHALMOLOGY FROM CONVENTIONAL SCIENCE

Vision impairment or diminution of vision is a major concern in ophthalmology which can be classified into two categories: distant visual impairment and near visual impairment. The leading causes of visual impairment or blindness are cataracts, glaucoma, diabetic retinopathy, age-related macular degeneration, and uncorrected refractive errors.[9]

The global burden was estimated on the basis of the measurement of disability-adjusted life years (DALYs) and classified into three groups: Group I (trachoma, vitamin A deficiency), Group II (cataract, diabetic retinopathy, glaucoma, and macular degeneration), and Group III (other eye diseases including ocular trauma). World Bank and other classifications also suggest these diseases as a global burden.[10] Cataract and refractive errors are the major contributors to the global burden of eye diseases.

3.1 Cataract

Any discernible opacity within the normally clear crystalline lens of the eye is referred to as a cataract. Depending on the anatomical location of the opacity, cataract can also be categorized as cortical, nuclear, or posterior subcapsular. Over 60 million people around the world suffer from cataract-related visual impairment; however, the prevalence of cataract-associated blindness varies significantly by region, making up less than 22% of blindness in high-income nations and more than 44% in South East Asia.[11] The biggest risk factor for cataractogenesis is age.

High concentrations of polyunsaturated fatty acids, cumulative exposure to high-energy visible light, and high oxygen consumption are the mainstream factors that cause oxidative stress in the eye. ROS are produced as a result of this interaction of variables, which can cause oxidative damage to ocular tissues.[12] According to observational research, consuming adequate amounts of the carotenoids, vitamin A, lutein and zeaxanthin, vitamins C and E, beta-carotene, α-carotene, lycopene, cryptoxanthin, and total carotenoids and taking daily multivitamin supplements can all lower the risk of developing cataracts. However, with carbohydrates, which are given in the high quantity in the American diet, protein glycation weakens cellular defenses and is also significantly related to cataract risk.[13] The nucleus part of the lens is very sensitive to dietary factors and nuclear cataracts have an association with vitamin A, niacin, thiamin, and riboflavin.[14] Linxian Eye Study shows that vitamins and minerals supplements decrease the risk of cataract and emphasizes the importance of nutrition in cataract.[15] This is the only completed randomized trial for nutritional intervention for the prevention of cataract.

3.2 Refractive Errors

A refractive error results from an ocular anatomical abnormality that prevents images of objects from focusing on the retina. Despite spectacle use, refractive errors that go uncorrected are a major contributor to impaired vision globally.[16]

However, according to *Ayurveda* refractive errors are taken as *Timira* and it is a disease of *Drishti,* i.e. vision.

Clinical signs of vision impairments are only observed in *Drishtigata Rogas* in *Ayurveda*. Therefore, the *Timira-Kacha-Linganasha* (refractive errors, partial blindness/diminished vision, cataract) complex can be used as a general term to describe all instances of visual abnormalities. The most significant refractive error, myopia, can be connected with the clinical characteristics of *Timira* (first and second *Patala*).

3.3 Glaucoma

Glaucoma is the second leading cause of blindness worldwide.[17] According to recently published evidence-based preclinical and clinical studies, the etiology of glaucoma may be significantly influenced by metabolic deficits and abnormalities. In addition to lowering intraocular pressure, individuals can modify their diet and exercise regimens, which may be beneficial for retinal ganglion cells in glaucoma.[18] Evidence-based observational studies also emphasized the impact of diet on the progression of glaucoma and elevation of IOP.[19] This suggests developing a therapeutic nutrition alternative from the *Ayurveda* stream in the management of glaucoma.

3.4 Age-Related Macular Degeneration

ARMD is the leading cause of irreversible blindness in Asian and Occidental populations.[20] A choroidal neovascular membrane (CNVM) that causes bleeding, exudation, the elevation of the macular retina, and scarring is linked to visual loss in the wet type of ARMD, while the one related to the atrophy of retinal pigmentary epithelium (RPE) is linked to visual loss in the dry type of ARMD. More than half of the eyes with a CNVM deteriorate to a visual acuity below 20/200. The neovascular variant of ARMD affects the majority of people who become legally blind as a result of the condition.[21]

Dietary elements, especially the carotenoids found in leafy dark green vegetables, have shown significant promise. While making lifestyle changes and changes in the dietary pattern may lower the risk of developing ARMD, it's possible that certain dietary components can further lower the risk.[22] In *Ayurveda* classics, a dedicated *Shaka Varga* has been described which has medicinal properties.[23] That classical knowledge can also be utilized to design therapeutic nutrition in ARMD and different retinal degenerations.

3.5 Diabetes Retinopathy

Diabetes is also termed a chronic metabolic disorder having impaired insulin secretion and its action. The decrease in proinflammatory cytokines level can be achieved through a dietary modification which is ultimately effective in ceasing apoptosis and the inflammatory process.[24] Hyperlipidemia and hyperglycemia are related to being the causative factor for diabetic retinopathy in type 2 diabetes.[25] According to *Ayurveda*, *Agnimandya*, *Aamdosha,* and oxidative stress are important factors in the pathogenesis of diabetic retinopathy.[26] Hence the improvement in *Agni* (~digestive fire) will lead to cessation of *Aam* and relieves oxidative stress. So therapeutic nutrition which is having *Agnivardhak* line of treatment can be useful.

4. CLASSIFICATION OF DISEASES IN OPHTHALMOLOGY ACCORDING TO *AVASTHA* (DISEASE CONDITIONS)

Ayurveda has classified eye diseases according to the involved *Dosha* (~body humor). So broadly all 76 eye diseases are classified as *Vata* predominant, *Pitta* predominant, *Kapha* predominant, *Rakta* predominant, and *Sannipatik* (involving all body humor).[27] *Acharya Sushruta* divided these diseases into *Saam* (~associated with endotoxins) and *Niraam* (~not associated with endotoxins) eye diseases.[28] Considering the classification of eye diseases the abovementioned diseases can be classified according to *Dosha* predominance and *Saam* and *Nirama* conditions. However, the classification according to the treatment includes *Shastrasadhya* (surgical treatment) and *Audhadhisadhya* (medical treatment). These need therapeutic nutrition as a second line of treatment. Many a time, ocular manifestations are also seen in some systemic diseases, so therapeutic nutrition can be considered for these systemic diseases to control the ocular manifestations.[29]

The prodromal phase i.e. *Pooravarupa* condition of eye diseases is considered to be the *Saamavastha.*[30] *Langhana* (synonym – *Upavasa)* i.e. fasting is one of the measures for the management of *Agnimandya* and *Saam* condition of diseases. It is also the first line of treatment for *Santarpanoth* (due to over nourishment) diseases as well. All eye diseases in the prodromal phase need *Langhana,* which helps to alleviate disease in five days.[31]

5. GENERAL THERAPEUTIC NUTRITION IN OPHTHALMOLOGY

5.1 Cataract

Since oxidative stress and degeneration are the main factors of the disease, we need to think of a diet which is having a rich source of antioxidants.[32] According to *Ayurveda,* cataract is termed *Linganasha,* i.e. the disease in which vision becomes impaired and can be restored with *Shastrakarma* (surgical procedures), especially in *Kapahaja Linganasha.* Ayurveda advocated the *Rasayana* (rejuvenation and revitalization therapy) and *Chakshyushya* (beneficial to the eyes as a sense organ) diet for many eye diseases. The consumption of *Achakshyushya* (nonbeneficial to eyes) diet leads to the pathogenesis of these eye diseases.

Ghrita is regarded as a healthy source of food that maintains the body's epithelial tissues and outer lining in the absence of vitamin A. It keeps the eyeball moist and guards against blindness.[33] It helps to prevent degenerative disorders, such as age-related eye disease and oxidative stress to tissues. Ayurvedic classics stressed the role of *Ghrita* as a *Rasayana* and it can be consumed with milk in patients with senile and degenerative eye diseases.[34] This is stated as *Agrya* (having prime importance) *Dravya* and should be consumed daily.

Green grams (*Phaseolus aureus* Roxb.) contain phenolic acid, which has a slow effect in the reduction of blood sugar levels and helps in maintaining the patient from diabetic complications such as cataract and diabetic retinopathy.[35] A cup of green grams contains about 14 mg of vitamin C, which is about 25% of the daily need for vitamin C. Vitamin C and other sources in green gram reduce the risk of cataract and improve capillary function in the eye. The recipes of green gram like *Mudga Yusha* (green gram soup), *Saptamushtik Yusha* (soup having seven contents), and *Mudga Modak* (*Laddu* of green gram) made with cow ghee are some of the *Kritanna Kalpana* which can be helpful as a therapeutic nutrition in senile eye diseases like cataract and ARMD.

Madhu (honey) is one of the most important drugs and natural medicine in *Ayurveda.* Several medicines are given with honey as a coadminister. It also possesses anti-inflammatory and immunomodulatory activity, which can be utilized for inflammatory eye diseases.[36]

Yava (barley, *Hordeum vulgare* Linn.) contains tocopherols (vitamin E), which have been shown in recent research to postpone the development of cataracts and age-related macular degeneration (ARMD). *Yava Rotika* (*chapatis* made from barley) is a preparation that can reduce the progression of cataract if used in a regular diet. *Yavasaktu* (roasted barley flour) and *Chanak Yavasaktu* (roasted horse gram + barley flour) are the preparations that can be used in proliferative diabetic retinopathy and hypertensive retinopathy and included as postsurgical diet.[37]

Amalaki (*Emblica officinalis* Gaertn.) is an important dietary source of amino acids, minerals, and vitamin C which has highly nutritious properties. It possesses an excellent antioxidant property which reduces oxidative stress and thereby free radical scavenging activity. In order to treat nearsightedness, cataracts, and glaucoma, fresh *Amalaki* juice or dried *Amalaki* powder is a helpful supplement. Its *Rechak* (purgative) effect regulates intraocular tension. Due to the antioxidants present, it improves eyesight. *Amalaki* pacifies *Pitta Dosha.*[38] So preparations like *Moravala* (formulation prepared from *Amalaki* fruit mixed with sugar syrup), *Amala* candy, and freshly extracted *Amalaki* juice can be useful in eye diseases like cataract and retinal vascular diseases.

Shatavari (*Asparagus racemosus*) is a traditional herbal medicinal plant that contains many bioactive metabolites. The nutritional properties of food/diet can be enhanced with asparagus powder or extract.[39] *Shatavari Payasa* (pudding made with asparagus) is a useful preparation and dietary item in cataract, glaucoma, and ARMD.

5.2 Refractive Errors

Refractive errors fall within the Ayurvedic category of eye disease *Timira*, which is specifically associated with visual symptoms.[40] *Vata Dosha* is the causative factor for refractive errors and *Pitta* and *Kapha* are associated *Dosha.* Considering these facts, a diet rich with *Snigdha* (unctuous) quality and *Madhur Dravya* (sweet in taste) is essential to alleviate vitiated *Vata Dosha.* Steamed rice flour rolls blended with *Triphala,* desserts, and cakes made with *Triphala* as an ingredient are some of the recipes beneficial for refractive errors.

Daily intake of freshly prepared juice of Indian gooseberry, i.e. *Amalaki,* help to reduce or prevent refractive errors.[41]

5.3 Glaucoma

Diet plays an important role in glaucoma. The neurodegenerative nature of the disease demands diet/nutrition that can prevent/halt the progression of neuronal loss.[42] The *Jangala Mamsa Rasa* and *Anupa Mamsa* along with cow *ghee* are advised. Milk processed with *Vatashamak* (having *Vata* alleviating properties) drugs and *ghee* processed with *Triphala* or *Purana Ghrita* (old ghee) can be taken after a meal.[43] *Yavagu* (gruel) and *Peya* (thin gruel prepared with brown rice) mixed with cow *ghee Agnideepak* (appetizer), *Balya* (gives strength), *Tarpak* (tissue rejuvenation), and *Dhatupushtikar* (enhances tissue metabolism) may help in combating the degenerative diseases of eyes such as glaucoma and ARMD. The basic pathophysiology of eye diseases starts in *Amashaya* (stomach),[44] so these preparations are light in digestion and helpful in correcting *Agni* (digestive metabolic fire), thereby halting the progression of the disease.

It is necessary to eat grains like barley, wheat, rice, and *Jawar* by light roasting on a flat griddle to make them light and digestible before processing them as a food item.[45] *Mudga Modaka* (*Laddu* made with green gram) is beneficial for eyes and tissue rejuvenation property helps in retinal eye diseases. *Saindhav* (rock salt) is the best among all the varieties of salt and its use in daily diet is beneficial for the eyes.[46] Freshly prepared *Navaneet* (butter) and its processed food items are useful, whereas *Vata-Pitta* vitiation is a prime factor in genesis of eye diseases. The unctuous nature of *Navanit* helps in tissue rejuvenation and nourishment.[47] The *Agni* should be good while eating food when it is processed with *Navaneet.*

Narikel Jala (coconut water) is diuretic in nature which is helpful for controlling intraocular pressure. Rice processed with coconut and *Krushara* (*Khichadi*) mixed with cow *ghee* are *Dhatuposhak* (nourish metabolic tissue) in nature, so these can be utilized as therapeutic nutrition in glaucoma.

5.4 Age-Related Macular degeneration

Senile age and malnutrition confounded by alcoholism and smoking are the causative factors in ARMD. The Age-Related Eye Disease Study (AREDS) highlighted the importance of a nutritious diet and dietary components in ARMD.[48] The *Chakshyushya* property of some of the vegetables and fruits described in the classical text can be utilized to overcome oxidative stress–related retinal degenerative diseases like ARMD. The *Chakshyushya Varga* described above is a very good option as a therapeutic nutrition in ARMD.[49] The different preparations of green gram, boiled cow milk mixed with ghee, *Jivanti Shaka* (*Leptadenia reticulata*), preparations of Gotu Kola (*Centella asiatica*),[50] *Kusumbha Shaka* (saffron, *Crocus sativus* L.), *Draksha* (black grapes, *Vitis vinifera*), especially dry ones, cow *ghee, Saindhav* (rock salt), *Shigru* (drumsticks, *Moringa oleifera*), funnel cakes (*Jalebi*) made from pure cow ghee, *Jangal,* and *Anupa Mansa Rasa,* preparations of *Raktashali* (brown rice), *Narikela Khandapaka* (preparation of coconut), fresh juice of Indian gooseberry, fresh juice of *Dadima* (pomegranate, *Punica granatum*), *Kakamachi* (black nightshade, *Solanum nigrum*), *Navanit* (butter), and *Kushmanda* (winter melon, *Benincasa hispida*) preparations like *Petha* (a translucent soft candy made from winter melon) contain the essential nutrients that are beneficial in retinal degenerative diseases like ARMD. Since there are two types of ARMD (dry type and wet type), the diet may vary according to type as the dry type has *Vata* vitiation, so the diet needed should be *Snigdha* and *Vatashamaka,* while in the wet type, *Vata-Pitta* vitiation is seen, so diet needed should be *Deepan-Pachaniya* and should not contain *ghee* in the initial phase.

Some Dietary Recommendations in Retinal Degenerative Diseases

Purana Ghrita, Triphala, Shatavari, Patola, Mudga, Amalaki, Yava, Lohita Shali, whole grains like *Yava* and *Godhuma* (wheat), cooked vegetables of *Jivanti, Sunishannaka, Tanduleeya, Vastuka, Mulaka,* the meat of birds and wild animals, *Karkotaka, Karavella, Vartaka, Karira, Shigru,* and *Tarkari.*[51]

Peya (thin gruel), *Vilepi* (thick gruel), *Tikta* (bitter) and *Laghu Ahara* (light digestible diet), *Shalitandula* (brown rice), *Godhuma, Saindhava, Goghrita, Gopaya* (cow milk), *Sita* (sugar cubes), *Madhu* (honey), *Draksha* (black grapes), *Kustumburu, Surana* (elephant yam, *Amorphophallus paeoniifolius*), *Naveena Mocha, Matsyakshi* (*Alternanthera sessilis*), and *Punarnava* (*Boerhavia diffusa*).[52]

5.5 Diabetic Retinopathy

The main etiological factor for diabetes is *Kapha Dosha.* Eyes very easily get afflicted by *Kapha Dosha.*[53] The etiological factors include *Achakshushya* diet (Nonbeneficial dietary items to eyes), excessive intake of sweet and sour food items, *Masha* (black grams), baked, overcooked, and fried food items (fish), unwholesome diet (milk with fish, milk with fruits), daytime sleeping, overeating, excessive intake of milk products, etc. This type of *Kaphakara Ahara* (*Kapha* vitiating diet) will lead to the derangement of *Dosha* and start the pathogenesis of *Shirobhishyanda* and *Netrabhishyanda* (trickling of *Dosha* in the head region and eyes), which lead to the formation of diabetic complications in eyes as diabetic retinopathy.[54] So avoiding *Abhishyandi* diet produces more moisture in the tissue and causes obstruction of various channels.

Although according to modern science, there are proliferative and nonproliferative types of diabetic retinopathy, the vascular endothelial growth factor plays an important role in the formation of the stages of this disease.[55]

Curcumin is a biologically active compound found in turmeric, also known as Haridra in Ayurvedic medicine. Curcumin is known for its anti-inflammatory and antioxidant properties. Kitchen turmeric, or boiled turmeric, is typically used as a spice in cooking and is not commonly used as a source of curcumin extract.[56]

While Kitchen *Haridra* is not a rich source of curcumin, it still contains some amount of curcuminoids and has many benefits when consumed as a part of diet. It has anti-inflammatory properties which can help reduce pain and inflammation. It has antioxidant properties that can help to protect the body from harmful molecules called free radicals. It also has been used in Ayurvedic medicine for centuries to help improve digestion, reduce gas and bloating, and improve the overall health of the skin.[57]

It is important to note that consuming turmeric in food amounts, as a spice, is considered safe for most people. However, consuming large amounts of turmeric or taking turmeric supplements should be done under the guidance of a healthcare professional.[58]

The active compound of turmeric, i.e. curcumin, was found to have potential therapeutic benefits in the prevention of diabetic retinopathy.[59] Curcumin may help improve tear production and reduce inflammation in the eyes. Oral supplementation of curcumin can improve symptoms of age-related macular degeneration (AMD), a leading cause of blindness in older adults. It may help to reduce inflammation and oxidative stress in the eyes, ultimately slowing the progression of AMD.[60,61]

Recommended Dietary Items in Diabetic Retinopathy

Wheat, *Sattu* (roasted gram flour or Bengal gram flour), Ragi (finger millet), flatbreads of *Yavanaal* (*Jwari,* sorghum, *Sorghum vulgare*), drumsticks, bitter gourd, *Methika* (fenugreek), *Patola* (*Trichosanthes dioica*), *Masoor* (lentil pulse), *Mudga,* the meat of chicken, Tittar bird, goat, and other *Jangala Mamsa, Jambu* (*Eugenia jambolana*), *Kharjur* (dates), buttermilk, and hot water.

6. NUTRITION IN OPHTHALMOLOGY AS PER *DWADASHA AAHAR VARGA*

TABLE 10.2 The *Dwadasha Ahara Varga* (12 Classes of Food) and *Siddhanna* (Food Preparations) in the Context of Ophthalmic Diseases and Their Preparations in Brief According to the Classical Text

SR. NO	DIETARY ITEM	ENGLISH NAME	*KRITANNA KALPANA*[62]	PROPERTIES[63]	INDICATIONS
	Shali Dhanya **(should be used roasted)**				
1.	*Rakta Shali*	Brown rice	*Peya* (thin gruel), *Yavagu* (thick gruel), *Vilepi* (thick gruel), *Manda* (gruel water), *Odana* (rice), *Krushara* (*Khichadi*), *Shalisaktu*. It should be used after roasting before making food items	*Chakshyushya*, *Santarpana* (rejuvenates and nourishes body tissue), increases *Kapha-Pitta*	Degenerative eye diseases, ARMD, squint, facial and eyelid palsies, refractive errors, Dry eye
	Shooka Dhanya **(corn with bristles) (should be used roasted)**				
2.	*Yava*	Barley	*Yava Rotika, Yava Saktu*	Alleviates *Kapha-Pitta*	Cataract, Diabetic retinopathy, Hypertensive retinopathy, ARMD, postsurgical diet
3.	*Yavanaal* (*Jwari*)	Sorghum	Flatbreads, *Yavagu*	It alleviates all three *Dosha* and is light in digestion	Diabetic retinopathy, retinal vascular diseases, prodromal phase of all eye diseases, postsurgical diet
4.	*Mahayaavanaal*	Corn	Corn flakes made the traditional way, *Makki Rotika, Yusha* prepared from *Makki*	Santarpak	Cataract, ARMD, Retinitis pigmentosa, Diabetic retinopathy
	Shimbi Dhanya **(should be used roasted)**				
5.	*Mudga*	Green gram	*Mudga Yusha* (soup of green gram), *Saptamushtik Yusha, Krushara, Mudga Modaka* (*Laddu*), *Mudgendari*	*Chakshyushya*, *Santarpaka*	ARMD, Glaucoma, Diabetic retinopathy, Vitamin A deficiency–related eye diseases, Cataract, all types of retinal degenerative and vascular diseases, Dry eye

TABLE 10.2 *(Continued)* The *Dwadasha Ahara Varga* (12 Classes of Food) and *Siddhanna* (Food Preparations) in the Context of Ophthalmic Diseases and Their Preparations in Brief According to the Classical Text

SR. NO	DIETARY ITEM	ENGLISH NAME	*KRITANNA KALPANA*[62]	PROPERTIES[63]	INDICATIONS
6.	*Masha*	Black grams	*Alikamaccha* (recipe prepared with black gram pasted betel leaves with steamed and fried process), *Masha Vetika, Masha Soup*	*Brihan*	Facial palsies, ptosis, lid dropping, lagophthalmos, neurophthalmic diseases
7.	*Chanaka*	Horse gram	*Chanaka Rotika, Chanaka Yava Saktu, Vedhanika* (funnel cake), *Chanaka Yusha,*	*Apatarpan,* alleviates *Kapha-Pitta*	Diabetic retinopathy
8.	*Adhaki*	Red gram	*Adhaki Yusha*	Alleviates all three *Dosha,* improves digestive fire	Diabetic retinopathy
	***Truna Dhanya* (should be used roasted)**				
9	*Ragi*	Finger millet	*Ragi Yusha, Yavagu,* flatbreads made of *Ragi*	Alleviates *Rakta* and *Pitta*, *Balya*	Proliferative diabetic retinopathy, retinal vascular diseases, Retinitis pigmentosa, orbital fractures
	Sattu				
10	*Yava Sattu*	Roasted barley flour	*Yava Sattu*	*Agnideepan*, light in digestion, alleviates *Kapha-Pitta*	Hypertensive retinopathy, Diabetic retinopathy, postsurgical diet, burning sensation in eyes
11.	*Chanak Yava Sattu*	Roasted horse gram + barley flour	*Chanak Yava Sattu* with *ghee* and sugar	*Agnideepak*	Cataract, specially indicated for consumption in the summer season
	Soup				
12	*Kulattha Yusha*	Soup made from *Dolichos biflorus* L.	*Kulattha Yusha*	Alleviates *Kapha*	Allergic eye diseases. If taken in excess can cause eye diseases
13	*Awal vade*	Amalaki + rice + rock salt + asafoetida	*Awal vade*	Alleviates all three Dosha, *Chakshyushya,*	Diabetic retinopathy, ARMD

(*Continued*)

TABLE 10.2 *(Continued)* The *Dwadasha Ahara Varga* (12 Classes of Food) and *Siddhanna* (Food Preparations) in the Context of Ophthalmic Diseases and Their Preparations in Brief According to the Classical Text

SR. NO	DIETARY ITEM	ENGLISH NAME	*KRITANNA KALPANA*[62]	PROPERTIES[63]	INDICATIONS
	***Shaka Varga* (group of vegetables)**				
	***Patra Shaka* (leafy vegetables)**				
14	*Baalmulak*	Raw daikon radish	Soup made from daikon radish, *Saptamushtik Yusha*	Alleviates *Vata* and *Kapha*	Painful and inflammatory conditions of eyes
15	*Rason*	Garlic	Used in tempering or *Chaunk* while making food, especially vegetables, *Rason Siddha Kshirpaka* (medicated garlic milk)	Alleviates *Vata*	Beneficial in *Vata* predominant eye diseases
16	*Kusumbha Shaka*	*Carthamus tinctorius* Linn.		Balances *Kapha*	Nourishes vision, refractive errors
17	*Kanta Shaka*	-	-	Alleviates *Kapha*	Beneficial for eyes
18	*Kakamachi Shaka*	Black nightshade	Food preparation	*Chakshyushya*	Degenerative eye diseases
19	*Punarnava*	*Boerhavia diffusa*	Food preparation	*Chakshyushya*	Inflammatory eye diseases, macular edema, Diabetic retinopathy, retinal vascular diseases
20	*Jeevanti Shaka*	*Leptadenia reticulate*	Steamed leaf boiled with buttermilk, tempered with asafoetida, oil, and buttermilk	*Chakshyushya*, alleviates all three *Dosha*, *Rasayana,*	All degenerative conditions of eyes, all types of eye diseases
21	***Phala Shaka* (fruit vegetables)**				
22	*Shigru*	Drumstick	*Shigru Yusha,*	Alleviates *Kapha*	Retinitis pigmentosa, Vitamin A deficiency diseases, Diabetic retinopathy, cystoid macular edema, Dry eye
23	*Agastya Pushpa*	*Sesbania grandiflora*	Cooked *Agastya Pushpa* vegetable, *Yusha*	Alleviates *Pitta* and *Kapha*	Night blindness, Retinitis pigmentosa, ARMD, wounds in eye, headache
24	*Kharjur*	Dates	*Kharjuradi Mantha*	Alleviates *Vata*, *Pitta* and *Rakta*	Retinal vascular diseases, hypertensive retinopathy, ARMD, Glaucoma

TABLE 10.2 *(Continued)* The *Dwadasha Ahara Varga* (12 Classes of Food) and *Siddhanna* (Food Preparations) in the Context of Ophthalmic Diseases and Their Preparations in Brief According to the Classical Text

SR. NO	DIETARY ITEM	ENGLISH NAME	*KRITANNA KALPANA*[62]	PROPERTIES[63]	INDICATIONS
25	*Draksha*	Black grapes	*Manuka* infusion	Alleviates *Pitta*	Retinal vascular diseases, Dry eye
26	*Kushmanda*	Benincasa hispida	*Petha* (winter melon candy)	Alleviates *Vata* and *Pitta*	ARMD, Optic neuropathies, retinal degeneration, Dry eye
27	*Udumbar Phala*	Cluster fig, *Ficus racemosa*	Medicated water with fruits of cluster fig	Alleviates *Pitta* and thereby purifies blood	Retinal vascular diseases, choroidal diseases, wet ARMD, retinal neovascularization
28	*Badarbeeja*	Indian jujube seed, *Ziziphus mauritiana*	Powdered form	Alleviates *Vata* and *Kapha*	Inflammatory eye diseases
29	*Karkat*	Cucumber, *Cucumis sativus*	Salad	Alleviates *Kapha* and *Pitta*	ARMD, Cataract, Dry eye
	***Pushpa Shaka* (leafy vegetables)**				
30	*Shatavha*	*Anethum graveolens*, Dill seed	*Yusha*	Alleviates *Kapha*	Wound in eyes, Diabetic retinopathy
	***Lavan Varga* (group of salts)**				
31	*Saindhav*	Rock salt	Used in all food preparations for taste	Alleviates all three *Dosha*, *Chakshyushya*	Refractive errors, ARMD, cataract, Dry eye
	***Dugdha Varga* (group of milk) and its products**				
32	*Godugdha*	Cow's milk	Cow's milk + *ghee*	*Rasayana*, alleviates *Vata Pitta*	All degenerative and senile eye diseases, optic atrophy, Dry eye
33	*Hastini Dugdha*	Elephant milk		*Rasayana*	Nystagmus, refractive errors
	Naristanya	Human breast milk	External application in the form of instillation in eyes, nose	Alleviates *Vata-Pitta, Rakta*	Traumatic eye diseases, ophthalmia neonatorum
34	*Navaneet*	Butter	Can be used for tempering during preparation of vegetables	Alleviates *Vata*	ARMD, retinal atrophies, optic atrophy

(*Continued*)

TABLE 10.2 *(Continued)* The *Dwadasha Ahara Varga* (12 Classes of Food) and *Siddhanna* (Food Preparations) in the Context of Ophthalmic Diseases and Their Preparations in Brief According to the Classical Text

SR. NO	DIETARY ITEM	ENGLISH NAME	*KRITANNA KALPANA*[62]	PROPERTIES[63]	INDICATIONS
	***Dadhi Varga* (group of curds)**				
35	*Aja Dadhi*	Curd prepared from goat milk		*Laghu* (easily digestible), *Agneedeepak*, alleviates all three Dosha, *Chakshyushya*	ARMD, Retinal degeneration, tubercular eye diseases
	***Ghrita Varga* (group of ghee)**				
36	*Goghrita*	Cow's *ghee*	Used in food preparations, can be used for tempering, Cow's milk with ghee	*Rasayana, Chakshyushya*	All types of eye diseases, can be used internally as well as external application, Dry eye
37	*Ajaghrita*	*Ghee* extracted from goat milk	Used in food preparations, can be used for tempering	*Chakshyushya*	Refractive errors, Optic atrophy
38	*Mahish Ghrita*	Buffalo's ghee	Used in food preparations, can be used for tempering	*Chakshyushya*	Refractive errors, Optic atrophy, degenerative eye diseases, Vitamin deficiency–related eye diseases
39	*Purana Ghrita*	Old preserved ghee for 10, 20, or 100 years	Used for internal use	*Chakshyushya, Vranashodhak* (wound healing property), alleviates *Vata*	Glaucoma, ARMD, Diabetic eye diseases, refractive errors, optic atrophy
	***Taila Varga* (group of oils) Not beneficial for eyes if used internally**				
40	*Karanja Taila*	Oil extracted from Indian beech (*Pongamia pinnata*)	External application	Alleviates *Vata*	Eczematous eyelid diseases
41	*Dhanya Taila*	Oil extracted from Jowar, rice bran, wheat	External application	Alleviates all three *Dosha*	External eyelid diseases

TABLE 10.2 *(Continued)* The *Dwadasha Ahara Varga* (12 Classes of Food) and *Siddhanna* (Food Preparations) in the Context of Ophthalmic Diseases and Their Preparations in Brief According to the Classical Text

SR. NO	DIETARY ITEM	ENGLISH NAME	*KRITANNA KALPANA*[62]	PROPERTIES[63]	INDICATIONS
	Mamsa Varga **(group of meat)**				
42	*Aja Mamsa*	Meat of goat	Soup, different food preparations, *Rasodana* (rice with meat soup)	*Rasayana*, *Drishti Prasadak* (improves vision), Balya (strengthens the tissue)	ARMD, degenerative eye diseases, Vitamin deficiency–related eye diseases, Optic atrophy, glaucoma
43	*Varaha Mamsa*	Meat of pig	Soup, different food preparations, *Rasodana* (rice with meat soup)	*Chakshyushya*	ARMD, degenerative eye diseases, Vitamin deficiency–related eye diseases, Optic atrophy
	Madhu Varga **(group of honey)**				
44	*Madhu*	Honey	Used as single drug, can be mixed or spread over *Yava Rotica*, wheat breads, consumed with hot water	*Chakshyushya*	Cataract, ARMD (wet), glaucoma, retinal tessellations

7. CLINICAL SIGNIFICANCE OF THERAPEUTIC NUTRITION IN OPHTHALMOLOGY

7.1 Dry Eye

Dry eye is a chronic inflammatory ocular surface disorder caused due to longer screen time on electronic gadgets like laptops and mobiles, due to hormonal changes in menopausal age in females,[64] vitamin deficiency, and malnutrition. A 39-year-old female patient with a 1-year history of dry, gritty eyes visited the OPD. She reported a normal menstrual history, but her dietary history revealed low intake of oil or ghee. She works in IT and has used a laptop for 8–10 hours daily for 15 years. She had moderate dry eye according to Schirmer's test readings and had been using tear supplements for 3 years without improvement. She was advised to continue tear supplements, make lifestyle and dietary changes, and take *Shatavari Ghrita* and *Laja Manda* daily. After 3 months of this treatment, her symptoms improved and Schirmer's reading increased to 18 mm. This approach may be useful in treating ocular surface disorders.

7.2 Optic Neuropathy

The author treated a case of traumatic optic neuropathy with *Ayurvedic* medicines and *Panchakarma*. The patient showed an improvement in vision and electrophysiological examination (VEP). In this case, the author used *Mudga Yusha,* a dietary preparation as a coadminister for *Snehapana* (drinking medicated oil) considering the *Vata* vitiation. The dietary properties of *Mudga* were used with the medication to get the desired therapeutic effect in this case.[65] Such preparations of *Mudga* described in this chapter can be used in different retinal disorders.

7.3 Headache Associated with Refractive Error

Considering the anatomical and structural changes in refractive errors it creates symptoms of asthenopia many a time associated with headache. In such situations according to *Ayurveda, Vata Dosha* is vitiated. Understanding this etiopathogenesis, winter melon candies are advised daily in the morning and 10 mL of cow *ghee* daily with hot water. This gave relief to the symptoms of headache and asthenopia associated with refractive errors.

7.4 Diabetic Retinopathy

A 61-year-old male patient with diabetes complained of vision problems for 2 months. He was diagnosed with nonproliferative diabetic retinopathy in both eyes. He was treated with Ayurvedic medication, including *Amalaki Swaras* and *Haridra* powder. After 1 month, his vision improved and ophthalmoscopy examination showed a reduction in exudates and hemorrhages. The antioxidant effect of *Amalaki* and *Haridra* were effective in improving vision and disease pathology.

In the hemorrhagic stage, *Pitta* alleviating diet; in the exudative stage *Kapha* alleviating diet; and in the apoptotic stage of the optic nerve involvement in DM retinopathy *Vata-Pitta* alleviating diet are important.

8. *AHAREYA KALPANA* (DIETARY RECIPES) IN OPHTHALMOLOGY

8.1 *Mudga Modaka*

Dried flour, obtained by grinding skinless, wet green gram and mixing it with a sufficient amount of water, was transformed into a semiliquid paste. This paste was then placed over a pan containing cow's *ghee*, using a sieve. The paste droplets were cooked in the *ghee* until they turned golden brown, and subsequently, the small fried balls were extracted. These fried balls were then fashioned into *Laddu* by incorporating sugar syrup. These are called *Mudga Modaka.* These are light to digest, rejuvenate tissue, alleviate all three *Dosha,* increase taste sensation, and are beneficial for the eyes.

9. SCOPE FOR FURTHER RESEARCH

- Diet has long been a major focus in discussions of health.
- Noncommunicable diseases can be prevented with a healthy and nutritious diet.
- Ayurveda recognizes the relationship between food, health, and the mind.
- There is a need for proper methodology to assess the therapeutic effect of diet on various eye diseases.
- Evidence-based data is lacking in the system of dietetics.
- Therapeutic nutrition is often considered a secondary aspect of treatment.

- There is a need to mainstream diet as a primary focus in treating eye diseases.
- Ayurveda has 12 classes of food that can be applied to every system.
- A tailor-made diet can be designed using the standard nutrition value and caloric value of food preparations.
- Ayurveda has the potential to develop system-specific guidelines with therapeutic and wellness approaches.
- Ancient Ayurvedic classics can be used to develop an evidence-based therapeutic approach in ophthalmology.
- A restaurant following Ayurvedic principles with a scientific approach has the potential to enlighten individuals about the significance of traditional cuisine in maintaining good health.

10. NUTRAVIGILANCE

In many places during a literature search, nonbeneficial food items for the eyes were mentioned in *Ayurvedic* classics. It means a nutravigilance approach was already incorporated in *Ayurveda* by ancient *Acharya.* In 12 groups of food, *Taila Varga* (group of oils) has been mentioned as *Achakshyushya* (nonbeneficial for eyes). It should not be consumed orally, especially sesame oil. *Sarshap Taila* (mustard oil) is used in many places across the world. However, it vitiates *Pitta* and *Rakta,* which are the major factors in the etiopathogenesis of eye diseases. Hence the use of *Sarshap Taila* should be avoided in eye diseases, especially in vascular diseases.

The *Shaka Varga* (group of leafy vegetables) is said to be nonbeneficial for the eyes. However, there is a practice of eating green leafy vegetables for eye diseases. *Ayurveda* mentioned the use of cooked food items and not the raw ones. In *Kshemkutuhal,* first the process of cooking is described and its attributes are presented, so these *Shaka,* especially *Patola, Vastuka, Kakamachi, Punarnava,* and *Jivanti,* must be processed, not consumed raw.

Eating food items like milk with fruits and milk with *Khichadi* comes under the incompatible diet category which creates *Aam* (endotoxins).[66] The classics mentioned that an incompetent diet creates diseases like deafness, blindness, and other dreadful eye diseases. So these types of food items should be avoided considering the eye health.[67] The *Achakshyushya* diet as per classics and *Anukta Ahar* (diet not mentioned in classics) should be documented with evidence to create awareness among nutravigilance in ophthalmology is a need of hour.

11. DO'S AND DON'TS IN EYE DISEASES[68]

TABLE 10.3 Beneficial and nonbeneficial categorization of dietary regime

BENEFICIAL DIETARY REGIME	*LATIN NAMES*	*NONBENEFICIAL FOR EYES*
Surana (Elephant foot yam)	*Amorphophallus campanulatus*	Sour-tasting food items like Curd, pickle, etc.
Patola (Snake gourd)	*Trichosanthes dioeca*	Baked food items
Vartaka (Brinjal)		Overcooked food items
Karvellaka	*Momordica charantia* L.	Fried food items
Unripe Banana		Milk with fish
Mulaka	*Raphanaus raphanistrum subsp. Sativus* L.	Milk with fruit

(*Continued*)

TABLE 10.3 *(Continued)* Beneficial and nonbeneficial categorization of dietary regime

BENEFICIAL DIETARY REGIME	*LATIN NAMES*	*NONBENEFICIAL FOR EYES*
Kumari	*Aloe vera* L.	Cooled and boiled items together
Punarnava	*Boerhavia diffusa* L.	Overeating
Kakamachi	*Solanum americanum Mill.*	Unfamiliar food items
Dhanyaka	*Coriandrum sativum* L.	Chilly
Dadima (Pomegranate)	*Punica granatum*	Sea fishes
Draksha (Dry black grapes)	*Vitis vinifera* L.	Prawns
Saindhava (Rock salt)		Daily hot water head bath
Mudga	*Vigna radiate* L.	Daytime sleeping and night awakening
Shatavari	*Asparagus racemosus willd.*	Smoking, Tobacco
Ghee	Butter oil	Suppression of natural urges
Jeevanti	*Holostemma ada-kodien Schult.*	Reading/sleeping in wrong position
Vastuka	*Dysphania ambrosioides* L.	Looking too small or too distant objects continuously
Mastyakshi		hypercaloric food items
Meghanada (Tandulaja)		Refined Carbohydrates
Triphala	Combination of *Terminalia chebula, Terminalia bellerica, Emblica officinalis*	*Masha* (*Vigna mungo* L.)
Puana Ghee	Old butter oil	Sour gruel
Cow's Milk		Sprouted pulses
Drinking rainwater		Green leafy vegetables
Godhuma	Wheat	Alcohol
Kulattha	*Macrotyloma uniflorum Lam.*	Meat of domestic animals
Drumsticks	*Moringa oleifera Lam.*	Excessive indulgence in sex
Barley		
Katak	*Strychnos potatorum* L.	Sour Mango
Indian gooseberry	*Emblica officinalis*	Pork meat
Green gram		
Brown rice		

REFERENCES

1. Kaur C, Foulds WS, Ling EA. Hypoxia-ischemia and retinal ganglion cell damage. Clin Ophthalmol. 2008 Dec;2(4):879–89. doi: 10.2147/opth.s3361. PMID: 19668442; PMCID: PMC2699791.
2. Lien EL, Hammond BR. Nutritional influences on visual development and function. Prog Retin Eye Res. 2011 May;30(3):188–203. doi: 10.1016/j.preteyeres.2011.01.001. Epub 2011 Feb 4. PMID: 21296184.
3. *Vision impairment and blindness*. (2022, October 13). www.who.int/news-room/fact-sheets/detail/blindness-and-visual-impairment
4. Ackland P, Resnikoff S, Bourne R. World blindness and visual impairment: despite many successes, the problem is growing. Community Eye Health. 2017;30(100):71–73. PMID: 29483748; PMCID: PMC5820628.
5. *Vision impairment and blindness*. (2022, October 13). www.who.int/news-room/fact-sheets/detail/blindness-and-visual-impairment
6. Lawrenson JG, Downie LE. Nutrition and Eye Health. Nutrients. 2019 Sep 6;11(9):2123. doi: 10.3390/nu11092123. PMID: 31489894; PMCID: PMC6771137.
7. Saccà SC, Cutolo CA, Ferrari D, Corazza P, Traverso CE. The Eye, Oxidative Damage and Polyunsaturated Fatty Acids. Nutrients. 2018 May 24;10(6):668. doi: 10.3390/nu10060668. PMID: 29795004; PMCID: PMC6024720.

8. Acharya JT. *Charaka Samhita* by *Agnivesha* with *Ayurved Dipika commentary* of *Chakrapanidatta. Sutrasthan Tisraiyashaniya Adhyay*. 11/35, 3rd ed. Bombay: Nirnaysagar Press; 1941. p. 74.
9. Vision Impairment and blindness [Internet]. World Health Organization. World Health Organization; [cited 2022Nov21]. Available from: www.who.int/news-room/fact-sheets/detail/blindness-and-visual impairment#:~:text=This%201%20billion%20people%20includes,well%20as%20near%20vision%20 impairment
10. Ono K, Hiratsuka Y, Murakami A. Global inequality in eye health: country-level analysis from the Global Burden of Disease Study. Am J Public Health. 2010 Sep;100(9):1784–8. doi: 10.2105/AJPH.2009.187930. Epub 2010 Jul 15. PMID: 20634443; PMCID: PMC2920965.
11. Flaxman SR, Bourne RRA, Resnikoff S, Ackland P, Braithwaite T, Cicinelli MV, Das A, Jonas JB, Keeffe J, Kempen JH, Leasher J, Limburg H, Naidoo K, Pesudovs K, Silvester A, Stevens GA, Tahhan N, Wong TY, Taylor HR; Vision Loss Expert Group of the Global Burden of Disease Study. Global causes of blindness and distance vision impairment 1990-2020: a systematic review and meta-analysis. Lancet Glob Health. 2017 Dec;5(12):e1221–e1234. doi: 10.1016/S2214-109X(17)30393-5. Epub 2017 Oct 11. PMID: 29032195.
12. Lawrenson JG, Downie LE. Nutrition and Eye Health. Nutrients. 2019 Sep 6;11(9):2123. doi: 10.3390/nu11092123. PMID: 31489894; PMCID: PMC6771137.
13. Weikel KA, Garber C, Baburins A, Taylor A. Nutritional modulation of cataract. Nutr Rev. 2014 Jan;72(1):30–47. doi: 10.1111/nure.12077. Epub 2013 Nov 26. PMID: 24279748; PMCID: PMC4097885.
14. Cumming RG, Mitchell P, Smith W. Diet and cataract: the Blue Mountains Eye Study. Ophthalmology. 2000 Mar;107(3):450–6. doi: 10.1016/s0161-6420(99)00024-x. PMID: 10711880.
15. Sperduto RD, Hu TS, Milton RC, Zhao JL, Everett DF, Cheng QF, Blot WJ, Bing L, Taylor PR, Li JY, et al. The Linxian cataract studies. Two nutrition intervention trials. Arch Ophthalmol. 1993 Sep;111(9):1246–53. doi: 10.1001/archopht.1993.01090090098027. PMID: 8363468.
16. Wedner S, Dineen B. Refractive errors. Trop Doct. 2003 Oct;33(4):207–9. doi: 10.1177/004947550303300406. PMID: 14620422.
17. Kingman S. Glaucoma is second leading cause of blindness globally. Bull World Health Organ. 2004 Nov;82(11):887–8. Epub 2004 Dec 14. PMID: 15640929; PMCID: PMC2623060.
18. Tribble JR, Hui F, Jöe M, Bell K, Chrysostomou V, Crowston JG, Williams PA. Targeting Diet and Exercise for Neuroprotection and Neurorecovery in Glaucoma. Cells. 2021 Feb 1;10(2):295. doi: 10.3390/cells10020295. PMID: 33535578; PMCID: PMC7912764.
19. Al Owaifeer AM, Al Taisan AA. The Role of Diet in Glaucoma: A Review of the Current Evidence. Ophthalmol Ther. 2018 Jun;7(1):19–31. doi: 10.1007/s40123-018-0120-3. Epub 2018 Feb 8. PMID: 29423897; PMCID: PMC5997592.
20. Jin G, Zou M, Chen A, Zhang Y, Young CA, Wang SB, Zheng D. Prevalence of age-related macular degeneration in Chinese populations worldwide: A systematic review and meta-analysis. Clin Exp Ophthalmol. 2019 Nov;47(8):1019–1027. doi: 10.1111/ceo.13580. Epub 2019 Jul 22. PMID: 31268226.
21. Berson EL. Nutrition and retinal degenerations. Int Ophthalmol Clin. 2000 Fall;40(4):93–111. doi: 10.1097/00004397-200010000-00008. PMID: 11064860.
22. Pratt S. Dietary prevention of age-related macular degeneration. J Am Optom Assoc. 1999 Jan;70(1):39–47. PMID: 10457680.
23. Kunte AM, Paradkar H. *Ashtang Hridaya* by *Vagbhata* with *Ayurved Dipika commentaries* of *Sarvangasundar* by *Arundatta* and *Ayurveda Rasayana* by *Hemadri. Sutrasthan Annaswarup Vidnyaniya Adhyay*. 6/72, 6th ed. Bombay: Nirnaysagar Press; 1939. p. 101.
24. Oza MJ, Laddha AP, Gaikwad AB, Mulay SR, Kulkarni YA. Role of dietary modifications in the management of type 2 diabetic complications. Pharmacol Res. 2021 Jun;168:105602. doi: 10.1016/j.phrs.2021.105602. Epub 2021 Apr 8. PMID: 33838293.
25. Kowluru RA, Mishra M, Kowluru A, Kumar B. Hyperlipidemia and the development of diabetic retinopathy: Comparison between type 1 and type 2 animal models. Metabolism. 2016 Oct;65(10):1570–1581. doi: 10.1016/j.metabol.2016.07.012. Epub 2016 Jul 30. PMID: 27621192; PMCID: PMC5023070.
26. Sahoo PK, Fiaz S. Conceptual analysis of diabetic retinopathy in Ayurveda. J Ayurveda Integr Med. 2017 Apr-Jun;8(2):122–131. doi: 10.1016/j.jaim.2016.12.003. Epub 2017 May 16. PMID: 28526441; PMCID: PMC5496992.
27. Acharya JT. *Sushrut Samhita* of *Sushruta* with *Nibandha Sangraha commentary* of *Shree Dalhanacharya. Uttaratantra Aupadravika Adhyay*. 1/28, Revised 2nd ed. Bombay: Nirnaysagar Press; 1941. p. 539.

28. Acharya JT. *Sushrut Samhita* of *Sushruta* with *Nibandha Sangraha commentary* of *Shree Dalhanacharya. Uttaratantra Aupadravika Adhyay*. 1/28, Revised 2nd ed. Bombay: Nirnaysagar Press; 1941. p. 539.
29. Bhat Pravin M. Ocular manifestations in systemic diseases – an Ayurvedic perspective. World Journal of Pharmaceutical Research. Vol.6, Issue.6, 2017; p.1571–1582.
30. Acharya JT. *Sushrut Samhita* of *Sushruta* with *Nibandha Sangraha commentary* of *Shree Dalhanacharya. Uttaratantra Aupadravika Adhyay*. 1/28, Revised 2nd ed. Bombay: Nirnaysagar Press; 1941. p. 539.
31. Cakrapāṇidatta, Tripāṭhī Indradeva. Chakradatta. 59/3, 1st ed. Varanasi: Chaukhambha Sanskrit Sansthan; 1991, p. 347.
32. Lobo V, Patil A, Phatak A, Chandra N. Free radicals, antioxidants and functional foods: Impact on human health. Pharmacogn Rev. 2010 Jul;4(8):118–26. doi: 10.4103/0973-7847.70902. PMID: 22228951; PMCID: PMC3249911.
33. V. Anuja Singh., & Gowda, S. T. (2017). Critical Analysis on Chakshushya Varga. *Journal of Ayurveda and Integrated Medical Sciences (JAIMS)*, *2*(4). https://doi.org/10.21760/jaims.v2i4.9339
34. Acharya JT. *Charaka Samhita* by *Agnivesha* with *Ayurved Dipika commentary* of *Chakrapanidatta. Sutrasthan Yajjapurushiya Adhyay*. 25/40, 3rd ed. Bombay: Nirnaysagar Press; 1941. p. 131.
35. Lin D, Xiao M, Zhao J, Li Z, Xing B, Li X, Kong M, Li L, Zhang Q, Liu Y, Chen H, Qin W, Wu H, Chen S. An Overview of Plant Phenolic Compounds and Their Importance in Human Nutrition and Management of Type 2 Diabetes. Molecules. 2016 Oct 15;21(10):1374. doi: 10.3390/molecules21101374. PMID: 27754463; PMCID: PMC6274266.
36. Samarghandian S, Farkhondeh T, Samini F. Honey and Health: A Review of Recent Clinical Research. Pharmacognosy Res. 2017 Apr-Jun;9(2):121–127. doi: 10.4103/0974-8490.204647. PMID: 28539734; PMCID: PMC5424551.
37. Bhojan Kutuhal P50.
38. Pravin M Bhat, Hari Umale, Madhukar Lahankar. Amalaki: A review on functional and pharmacological properties. J Pharmacogn Phytochem 2019; 8(3):4378–4381
39. Kohli D, Champawat PS, Mudgal VD. Asparagus (Asparagus racemosus L.) roots: nutritional profile, medicinal profile, preservation, and value addition. J Sci Food Agric. 2022 Nov 25. doi: 10.1002/jsfa.12358. Epub ahead of print. PMID: 36433663.
40. Gopinathan G, Dhiman KS, Manjusha R. A clinical study to evaluate the efficacy of Trataka Yoga Kriya and eye exercises (non-pharmocological methods) in the management of Timira (Ammetropia and Presbyopia). Ayu. 2012 Oct;33(4):543–6. doi: 10.4103/0974-8520.110534. PMID: 23723673; PMCID: PMC3665208.
41. Jain SK, Khurdiya DS. Vitamin C enrichment of fruit juice based ready-to-serve beverages through blending of Indian gooseberry (Emblica officinalis Gaertn.) juice. Plant Foods Hum Nutr. 2004 Spring;59(2):63–6. doi: 10.1007/s11130-004-0019-0. PMID: 15678753.
42. Tribble JR, Hui F, Jöe M, Bell K, Chrysostomou V, Crowston JG, Williams PA. Targeting Diet and Exercise for Neuroprotection and Neurorecovery in Glaucoma. Cells. 2021 Feb 1;10(2):295. doi: 10.3390/cells10020295. PMID: 33535578; PMCID: PMC7912764.
43. Acharya JT. *Sushrut Samhita* of *Sushruta* with *Nibandha Sangraha commentary* of *Shree Dalhanacharya. Uttaratantra Vatabhishyanda Pratishedha Adhyay*. 9/3,8-9, Revised 2nd ed. Bombay: Nirnaysagar Press; 1941. p. 553.
44. Kunte AM, Paradkar H. *Ashtang Hridaya* by *Vagbhata* with *Ayurved Dipika commentries* of *Sarvangasundar* by *Arundatta* and *Ayurveda Rasayana* by *Hemadri. Uttaratantra Vartmaroga Vidnyaniya Adhyay*. 8/1-2, 6th ed. Bombay: Nirnaysagar Press; 1939. p. 804.
45. Rajanighantu. [cited 2023Jan22]. Available from: https://niimh.nic.in/ebooks/e-Nighantu/rajanighantu/?mod=read
46. Mooss NS. Salt in Ayurveda I. Anc Sci Life. 1987 Apr;6(4):217–37. PMID: 22557573; PMCID: PMC3331422.
47. Kunte AM, Paradkar H. *Ashtang Hridaya* by *Vagbhata* with *Ayurved Dipika commentries* of *Sarvangasundar* by *Arundatta* and *Ayurveda Rasayana* by *Hemadri. Sutrasthana Dravadravya Vidnyaniya Adhyay*. 5/35, 6th ed. Bombay: Nirnaysagar Press; 1939. p. 73.
48. Weikel KA, Chiu CJ, Taylor A. Nutritional modulation of age-related macular degeneration. Mol Aspects Med. 2012 Aug;33(4):318–75. doi: 10.1016/j.mam.2012.03.005. Epub 2012 Apr 6. PMID: 22503690; PMCID: PMC3392439.
49. Rajitha S., & Bhat, S. (2020). A Review On *Chakshushya Varga* & *Anjana* As Cosmeceuticals. International Ayurvedic Medical Journal, 8(9), 4472–4477. https://doi.org/10.46607/iamj2808092020

50. Chandrika UG, Prasad Kumarab PA. Gotu Kola (Centella asiatica): Nutritional Properties and Plausible Health Benefits. Adv Food Nutr Res. 2015;76:125–57. doi: 10.1016/bs.afnr.2015.08.001. Epub 2015 Oct 1. PMID: 26602573.
51. Acharya JT. *Sushrut Samhita* of Sushruta with Nibandha Sangraha commentary of *Shree Dalhanacharya. Uttaratantra Drishtigatavyadhi Pratishedha Adhyay*. 17/50-51, Revised 2nd ed. Bombay: Nirnaysagar Press; 1941. p. 572.
52. Yoga Ratnakara Netra Roga Chikitsa Sl 1-4, Yogaratnakara – with vidyotini Hindi Commentary by Vaidya Lakshmipati Sastri, Edited by Bhisagratna Brahamasankar Sastri. 5th ed. Varanasi: Chaukhamba Sanskrit Sansthan; 1993. 504 pp. P 395.
53. Kunte AM, Paradkar H. *Ashtang Hridaya* by *Vagbhata* with *Ayurved Dipika commentries* of *Sarvangasundar* by *Arundatta* and *Ayurveda Rasayana* by *Hemadri. Sutrasthana Dinacharya Adhyay*. 2/5, 6th ed. Bombay: Nirnaysagar Press; 1939. p. 25.
54. Shanthakumari P. *Textbook of Ophthalmology in Ayurveda*. Chapter 2. 1st ed. Trivandrum; 2002. p 53–55.
55. Aiello, L. P., & Wong, J. S. (2000). Role of vascular endothelial growth factor in diabetic vascular complications. *Kidney International*, *58*, S113–S119. https://doi.org/10.1046/j.1523-1755.2000.07718.x
56. Sharifi-Rad J, Rayess YE, Rizk AA, Sadaka C, Zgheib R, Zam W, Sestito S, Rapposelli S, Neffe-Skocińska K, Zielińska D, Salehi B, Setzer WN, Dosoky NS, Taheri Y, El Beyrouthy M, Martorell M, Ostrander EA, Suleria HAR, Cho WC, Maroyi A, Martins N. Turmeric and Its Major Compound Curcumin on Health: Bioactive Effects and Safety Profiles for Food, Pharmaceutical, Biotechnological and Medicinal Applications. Front Pharmacol. 2020 Sep 15;11:01021. doi: 10.3389/fphar.2020.01021. PMID: 33041781; PMCID: PMC7522354.
57. Hewlings SJ, Kalman DS. Curcumin: A Review of Its Effects on Human Health. Foods. 2017 Oct 22;6(10):92. doi: 10.3390/foods6100092. PMID: 29065496; PMCID: PMC5664031.
58. *Turmeric, National Center for Complementary and Integrative Health*. U.S. Department of Health and Human Services. Available at: www.nccih.nih.gov/health/turmeric (Accessed: January 22, 2023).
59. Agrawal R, Agrawal P, Saxena R, Srivastava S. Curcumin prevents experimental diabetic retinopathy in rats through its hypoglycemic, antioxidant, and anti-inflammatory mechanisms. J Ocul Pharmacol Ther. 2011 Apr;27(2):123–30. doi: 10.1089/jop.2010.0123. Epub 2011 Feb 12. PMID: 21314438.
60. Radomska-Leśniewska DM, Osiecka-Iwan A, Hyc A, Góźdź A, Dąbrowska AM, Skopiński P. Therapeutic potential of curcumin in eye diseases. Cent Eur J Immunol. 2019;44(2):181–189. doi: 10.5114/ceji.2019.87070. Epub 2019 Jul 30. PMID: 31530988; PMCID: PMC6745545.
61. Pescosolido N, Giannotti R, Plateroti AM, Pascarella A, Nebbioso M. Curcumin: therapeutical potential in ophthalmology. Planta Med. 2014 Mar;80(4):249–54. doi: 10.1055/s-0033-1351074. Epub 2013 Dec 9. PMID: 24323538.
62. Pandey G. Kshemakutuhalam. Varanasi: Chaukhamba Krishnadas Academy; 2014.ISBN: 978-81-218-0350-2.
63. Balkrishna A. Bhojankutuhal. Haridwar, Uttarakhanda: Divya Prakshan, Patanjali Yogapeeth; 2013. ISBN: 81-89235-90-7.
64. Peck T, Olsakovsky L, Aggarwal S. Dry Eye Syndrome in Menopause and Perimenopausal Age Group. J Midlife Health. 2017 Apr-Jun;8(2):51–54. doi: 10.4103/jmh.JMH_41_17. PMID: 28706404; PMCID: PMC5496280.
65. Bhat PM. Traumatic Optic Neuropathy (TON) and Ayurveda – A case report. J Ayurveda Integr Med. 2022 Jan-Mar;13(1):100494. doi: 10.1016/j.jaim.2021.07.010. Epub 2021 Nov 27. PMID: 34844840; PMCID: PMC8728071.
66. Sabnis M. Viruddha Ahara: A critical view. Ayu. 2012 Jul;33(3):332–6. doi: 10.4103/0974-8520.108817. PMID: 23723637; PMCID: PMC3665091.
67. Acharya JT. *Sushrut Samhita* of *Sushruta* with *Nibandha Sangraha commentary* of *Shree Dalhanacharya. Uttaratantra Pratishyay Pratishedha Adhyay*. 24/17, Revised 2nd ed. Bombay: Nirnaysagar Press; 1941. p. 594.
68. Lokhande, J. N., & Pathak, Y. V. (2021). *Nutraceuticals for Aging and Anti-Aging: Basic Understanding and Clinical Evidence* (1st ed.). CRC Press.

Therapeutic Nutrition in Ayurveda for Orthopedic Conditions

Mangesh Deshpande

1. INTRODUCTION

Orthopedics is a branch of medicine concerned with correcting or preventing deformities, disorders, or injuries of the skeleton and associated structures (such as tendons and ligaments).[1] It includes osteoporosis, osteoarthritis, tendinopathy, and ligament injuries.

According to *Ayurved*, *AsthiwahaSrotas* is the *Sthan* (site) of *Asthidhatu.*[2]

अस्थिवहस्रोतांसिऽपिअस्थ्नांस्थानम्।

The Nature of *Asthi Dhatu*

शरीरगतस्य कठीनतमस्य धातोः अस्थि इति नाम।

According to *Shushrut Samhita,* the hardest *Dhatu* is the *Asthidhatu.*[3]

Causes of Damaging *Asthiwaha Srotas*

Excessive exercise, trauma, and food that vitiates *Vata* are the main causes of damaging the *Asthiwaha Srotas.* [4]

Causes of Vitiation of *Vata*

Excessive exercise, malnutrition, trauma, fractures, sleeping late at night, suppression of natural urges, too much grief, excessive cold, and consuming astringent, bitter, spicy, dry, and pungent food can lead to vitiation of *Vata.*[5]

DOI: 10.1201/9781003345541-14

Composition of *Asthi Dhatu*

Asthi is made up of *Prithvi* (the earth) and *Anil* (the air).[6]

Ayurveda has its approach to planning a daily diet for an individual. It is not based on conventional chemistry; rather, it is based on classical *Panchmahabhautic* chemistry and *Tridoshic* functions. As such, *Ayurveda* considers the planning of a biobalancing diet rather than a balanced diet.[7]

Correlation of *Purishdhara Kala* and *Asthidhara Kala*

The lower one-third of the large intestine is *Purishdhara Kala*. *Purishdhara kala* has a direct correlation to *Asthidhara Kala*; hence anything affecting *Purishdhara Kala* also affects *Asthidhara Kala*.[8] Thus gut health and bone health are interlinked.

Diet is an important controlling factor concerning indigenous microbiotic activities. The gut microflora contains pathogenic, benign, and beneficial microbial species. A predominance of the former can lead to gut upset, which can be both acute (e.g., gastroenteritis) and chronic (e.g., inflammatory bowel disease). Foods that pass through or get close to the gut affect the composition of activities aimed at achieving a more positive metabolism.[9]

Vitamin B6 plays a significant role in regulating many metabolic pathways of neuronal function including the production of neurotransmitters and the conversion of food into energy. The synthesis of it is regulated in the gut. One study examined the effects of vitamin B6 supplements in patients with mild to moderate carpel tunnel syndrome. The results of the study suggest that vitamin B6 supplements were effective in improving nocturnal pain severity, nocturnal awakening frequency due to pain and hand numbness, frequency and persistence of daily pain, hand numbness and weakness, hand tingling, and clumsiness when handling objects.[10]

Functions of Asthi Dhatu in the Body

Meda nurtures *Asthi Dhatu*. *Asthi* nurtures *Majja Dhatu*. *Asthi* is the basic framework on which the body is formed.[11]

There are diet-related conditions that can harm tendons.[12,13]

Hypercholesterolemia is a risk factor for tendinopathy. Excessive intake of cholesterol results in the accumulation of oxidized low-density lipoproteins in the load-bearing region of the tendon, where it may impair type I collagen production and reduce tendon strength and energy-storing capacity. As *Meda Dhatu* has an impact on orthopedic clinical conditions, we should consider foods nurturing *Meda Dhatu* while deciding orthopedic therapeutic nutritional diet.

Vitamin D is specifically said to have a role in neuroprotection and neurotropism, which reduce neurological injuries and neurotoxicity and improve myelination and recovery after nervous system injuries.[14] A diet that nurtures *Majja Dhatu* is equally important in an orthopedic therapeutic nutritional diet.

Conclusion (Unmet Need)

1) The food that nurtures *Asthi* should be a combination of *Prithvi + Anil*.
2) As the nutrition of *Asthi* depends on *Meda*, a diet that nurtures *Meda Dhatu* is equally important.
3) The diet that nurtures *Majja Dhatu* is equally important.
4) The food which has an impact on *Purishdhara Kala* has an effect on *Asthidhara Kala*. Gut health is directly interlinked with bone health.
5) *Vata* vitiating diet is not at all recommended in *Asthi*-related clinical conditions.
6) *Ayurvedic* physicians practice with a thorough understanding of the sources, classification, nutritional merits, adverse effects, and therapeutic indications of food.[15]

2. NUTRITION IN ORTHOPEDICS FROM CLASSICAL TEXTS

1) ***Chaturbeej Churn***:
   Methika (fenugreek seeds), *Chandrashur* (garden cress seeds), *Kalajaje* (black cumin seeds), *Yavani* (cardamom seeds).[16] It is used in back pain and lumbar disc syndrome.
2) *Rason* (garlic) + milk or *Tila* (sesame seeds) + oil/*Ghee* (clarified butter)/*Mamsa Ras* (meat soup)/rice porridge. It is useful for pain in upper limbs (frozen shoulder – adhesive capsulitis).[17]
3) ***Laddu***:
   Chandrasur (garden cress seeds), *Khus Khus* (poppy seeds), *Kharjoor* (black dates), *Methika* (fenugreek seeds), *Tila* (sesame seeds), *Jeerak* (cumin seeds), *Bhallatak* (*Semecarpus anacardium* seeds), *Vatam* (almond) + *Babool* (acacia gum)+ *Guda* (jaggery) +desiccated coconut + *Ghee* (clarified butter).
   It is effective in low back pain and lumbar disc syndrome.[18]
4) The root, fruit, and flower of the drumstick are cooked in oil and boiled in water with the addition of powdered rock salt and asafetida.
   This preparation stimulates the digestive fire. It is mainly used in tendinopathy.[19]
5) ***Taradvati***:
   After kneading *Maida* flour (refined wheat flour) with *Ghee* (clarified butter), one has to render it foamy by adding curds. After preparing *Vadas* (unsweetened doughnuts) from it, one should cook them in *Ghee* (clarified butter), on low flame. These *Vadas* (unsweetened doughnuts) are then soaked in sugar syrup and scented with camphor. With the addition of black pepper, the delicacy is called "*Taradvati.*"
   It is used in the realignment of connective tissues like tendons, ligaments, and bones.[20]
6) ***Karpura-nalika*** (camphor tubes):
   Maida flour (refined wheat flour), mixed with *Ghee* (clarified butter), is kneaded into balls with water. It is then spread out as a square on a raised platform and cooked in the manner of *Polika* (tortilla).On either side of this spread-out *Polika* (tortilla), a sugar candy four fingers in length is placed and wrapped by the *Polika* (tortilla). The two tube-like structures are cooked in *Ghee* (clarified butter) with the addition of the necessary ingredients. The two tubular sweets are then removed and filled with camphor, *Ghee* (clarified butter), and sugar. This preparation is called "*Karpura-nalika*" (camphor tubes).
   This is indicated in the healing of injuries of connective tissues like tendons and ligaments.[21]
7) ***Sevika*** or Semolina:
   After kneading *Maida* (refined wheat flour) into small pieces that measure a *Yava* (1/6 or 1/8 of an *Angula*/finger), the pieces are dried and then cooked like rice. They are eaten with *Ghee* (clarified butter) and sugar.
   This is indicated in the healing of injuries of connective tissues like tendons and ligaments.[22]
8) ***Kheer*** (sweet rice pudding):
   Mix granulated wheat and milk. Make *Kheer* (sweet rice pudding). It enhances osteoblastic activity to heal fractures.[23]
9) ***Ksheerpaka*** (Latte):
   Laksha (*Lacciferlacca*) + Milk: Enhances the healing of fractures.[24]

3. NUTRITION IN ORTHOPEDIC FROM CONVENTIONAL SCIENCE

Fundamentally, food and nutrition serve three basic functions in the body[25]:

- As a source of energy for day-to-day bodily functions
- As the source of biomaterials for the growth and repair of daily wear and tear
- To assist in certain vital functions of the body.

The gross essential components of food are:

- Carbohydrates
- Fat
- Protein
- Vitamins
- Minerals
- Water.

Carbohydrates and fat provide energy to the body. Proteins are the building blocks for growth and repair, while vitamins and minerals assist in a range of vital functions in the body. Water is an important component of food and is essential for hydration and circulatory functions. Depending on a person's age and a range of functional conditions, different individuals require different proportions of these food components. A proper combination and proportion of food components is called a balanced diet.

TABLE 11.1 Essential Components of Food and Its Role in Orthopedics

	ROLE IN ORTHOPEDICS
Carbohydrates	Research suggests that calcium ingested in dietary form along with carbohydrates, lipids, and other nutrients present in food enhances calcium absorption[26]
Fat/Lipids	**Probiotics** are live bacteria/yeast classified as "good" bacteria that live in our body to help promote better digestion and overall gut health. These include foods such as yogurt and cheeses. Fats like nuts, avocados, and olive oil are another dietary recommendation. Extra virgin olive oil is believed to contain a chemical similar to ibuprofen and so may be a helpful addition to a diet.[27] Excessive intake of cholesterol results in the accumulation of oxidized low-density lipoproteins in the load-bearing region of the tendon, where it may impair type I collagen production and reduce tendon strength and energy storing capacity.[28,29] Triglycerides help the body produce and regulate hormones. In the reproductive system, fatty acids are required for proper reproductive health. Women who lack proper amounts may stop menstruating and become infertile. Omega-3 and omega-6 essential fatty acids help regulate cholesterol and blood clotting and control inflammation in the joints, tissues, and bloodstream.[30]

(*Continued*)

TABLE 11.1 *(Continued)* Essential Components of Food and Its Role in Orthopedics

	ROLE IN ORTHOPEDICS
Protein	Malnutrition, which is very prevalent in geriatric populations, is one of the main risk factors for the onset of frailty. A good nutritional status and, wherever necessary, supplementation with macronutrients and micronutrients reduce the risk of developing frailty.[31] Recommending protein supplementation as a stand-alone intervention for healthy older individuals seems ineffective in improving muscle mass and strength.[32] Evidence suggests that manipulating protein intake via dietary protein or free amino acid–based supplementation diminishes muscle atrophy and/or preserves muscle function in experimental models of disuse (i.e., immobilization and bed rest in healthy populations).[33] At low protein intakes insulin-like growth factor production is reduced, which in turn harms calcium and phosphate metabolism, bone formation, and muscle cell synthesis.[34]
Water	The prevalence of sarcopenia in the elderly population was related to inadequate dietary water intake after adjusting for covariates. Adequate water intake in the elderly should be recommended to prevent dehydration-related complications, including sarcopenia (characterized by progressive and generalized loss of skeletal muscle mass and strength).[35]
Vitamin A	Higher bone mineral density (BMD) and lower fracture risk have been reported in individuals with higher vitamin A intake.[36,37] Low vitamin A, C, and E concentrations are associated with an increased risk of hip fracture, possibly mediated through bone turnover mechanisms.[38] **Vitamin A** deficiency resulting from unbalanced dietary habits is associated with exacerbation of male early-onset of ossification of the posterior longitudinal ligament (OPLL).[39]
Vitamin B6	This plays a significant role in regulating many metabolic pathways of neuronal function including the production of neurotransmitters and the conversion of food into energy. The results of a study suggested that vitamin B6 supplements were effective in improving nocturnal pain severity, nocturnal awakening frequency due to pain and hand numbness, frequency and persistence of daily pain, hand numbness and weakness, hand tingling, and clumsiness when handling objects.[40]
Vitamin C	A diet that is high in fruit (e.g., strawberries, blueberries, oranges) has also been associated with decreasing inflammation.[41] Vitamin C supplementation was shown to accelerate tendon healing in rats with Achilles tendon rupture.[42] Low concentrations of leucocyte vitamin C appear to be associated with subsequent development of pressure sores in elderly patients with femoral neck fractures.[43] Vitamin C deficiency is known to cause scurvy and also to contribute to osteoporosis and fragility fractures.[44] Collagen is the primary protein of bone matrix and vitamin C is also involved in the hydroxylation of lysine, the critical first step for collagen biosynthesis.[45]

TABLE 11.1 *(Continued)* Essential Components of Food and Its Role in Orthopedics

	ROLE IN ORTHOPEDICS
Vitamin D	Vitamin D is known to be a neuroactive steroid that suppresses these VEGFs (vascular endothelial growth factors, which are associated with increased vascular proliferation and inflammatory synovial fibrosis that may trigger the onset of the syndrome) and induces nerve growth factors that can help with the prevention of neurological deficits. This vitamin is specifically said to have a role in neuroprotection and neurotrophism, which reduce neurological injuries and neurotoxicity and improve myelination and recovery after nervous system injuries.[46] A study by Angeline et al. (2013) examined the impact of vitamin D deficiency after surgical reattachment of the supraspinatus tendon in rats. The biomechanical and histological data suggested low vitamin D levels may also negatively affect early healing at the rotator cuff repair site.[47] The results of a randomized double-blind placebo-controlled study linked vitamin D to the reduction of "cytokine storms," thus reducing musculoskeletal pain.[48]
Vitamin E	Current literature suggests that vitamin E molecules (α-, β-, γ-, δ-tocopherols and the corresponding tocotrienols) with their antioxidant and anti-inflammatory capabilities may mitigate age-associated skeletal dysfunction and enhance muscle regeneration. Preclinical and human experimental studies show that vitamin E benefits myoblast proliferation, differentiation, survival, membrane repair, mitochondrial efficiency, muscle mass, muscle contractile properties, and exercise capacity.[49] Low vitamin A, C, and E concentrations are associated with an increased risk of hip fracture, possibly mediated through bone turnover mechanisms.[50]
Vitamin K	In a study of postmenopausal women, supplementation of vitamin K1 with vitamin D, calcium, magnesium, and zinc resulted in a significant reduction in femoral neck bone loss compared with a placebo.[51] A recent metaanalysis shows a beneficial effect of the pharmacologic dose of vitamin K2 on reducing vertebral and hip fractures.[51] Low vitamin K contributes to undercarboxylated osteocalcin, leading to low bone mass density.[51]
Omega-3 fatty acids	Omega-3 fatty acids found in fish oil supplements and in cold water–dwelling fish (e.g., salmon, mackerel) have been shown to reduce inflammation.[52] A diet with foods that have less inflammation-causing prostaglandins (PGE2), and more foods with anti-inflammatory prostaglandins (PGE1, PGE3) could be helpful in osteoarthritis[53]
Calcium	Low calcium intake is associated with increased osteoporosis and fracture risk.[54] Research suggests that calcium ingested in dietary form along with carbohydrates, lipids, and other nutrients present in food enhances calcium absorption.[55] Ca administration inhibits postmenopausal osteopenia and there is epidemiological evidence that a liberal Ca intake reduces bone loss in middle adulthood.[56]
Potassium	Potassium is needed for muscle contraction, communication between muscles and nerves, and overall muscular function. Since muscles are found throughout your body, including your arms, legs, and respiratory and digestive tracts, a diet low in potassium can contribute to fatigue and digestive troubles.[57] Observational studies suggest that increased consumption of potassium from fruits and vegetables is associated with increased bone mineral density.[58] A 3-year study looking at a diet rich in potassium, such as fruits and vegetables, as well as a reduced acid load, resulted in the preservation of muscle mass in older men and women.[59]

(*Continued*)

TABLE 11.1 *(Continued)* Essential Components of Food and Its Role in Orthopedics

	ROLE IN ORTHOPEDICS
Phosphorus	The Ca:P ratio for the average diet consumed in these countries (about 1:1.6) appears to be satisfactory; a low intake of dairy foods, coupled with a high intake of other foods rich in natural and added phosphorus, may raise the ratio above 1:2, a value beyond which animal studies indicate that there is a risk of increased bone loss.[60]
Magnesium	A study suggests that a combination of magnesium along with vitamin C could inhibit bone spur formation and reduce inflammation in the knee synovium.[61] A significant decrease in the maximum three-point bend strength of the femurs of Mg-deficient rats was observed. These data support the hypothesis that short-term Mg deficiency affects the pattern of bone mineral formation.[62]
Curcumin (Turmeric)	An antioxidant with positive effects on cell regeneration, wound healing, and other factors related to tendinopathies.[28] A study in 2016 showed that curcumin supplementation in mice with patellar tendon injury had overall better outcomes.[63] These outcomes included the better organization of collagen fibers, improved biomechanical properties, improved healing properties, and increased MnSOD activity.

CLASSIFICATION OF ORTHOPEDIC DISORDERS[64]

Most orthopedic disorders fall within the following groups:

TABLE 11.2 Classification of Diseases in Orthopedics

Deformities	Congenital deformities Acquired deformities
General affections of the skeleton	Bone dysplasia Inborn errors of metabolism Metabolic bone disease Endocrine disorders
Infections of bone and joints	Infections of bone Joint infections
Bone tumors and other local conditions	Tumors of bone Osteochondritis Cystic change
Soft tissue tumors and other diseases	Tumors of soft tissue Inflammatory lesions of soft tissue
Arthritis and other joint disorders	Arthritis Dislocation and subluxation Internal derangements
Neurological disorders	Cerebral palsy Spina bifida Poliomyelitis Peripheral nerve lesions

TABLE 11.3 Various Classification of Orthopedic Diseases According to Ayurveda

ACCORDING TO CHRONICITY[66]	*ACCORDING TO THE STATE OF DIGESTION/ METABOLISM*[67]	*ACCORDING TO CAUSES*[68]	*ACCORDING TO COMPLICATIONS OF DISEASES*[69]
Acute (injury 0–4 days) *Subacute* (5–14 days) *Chronic* (above 14 days)	• *Aam* • *Pachyamanawastha* • *Niraam*	1. *Santarpan* (caused by consumption of regularly heavy-to-digest food like *Paneer*[cottage cheese], milk products, wheat, etc.) 2. *Apatarpan* (caused by consumption of regularly lighter-to-digest food like roasted rice, green gram, etc.)	1. *Uttan* 2. *Gambhir*

Deformity Arising at a Joint[65]

Causes:

The causes of deformity arising at a joint may be summarized under the following headings:

1. Dislocation or subluxation
2. Muscle imbalance
3. Tethering or contracture of muscles or tendons
4. Contracture of soft tissues
5. Arthritis
6. Prolonged abnormal posture
7. Unknown causes.

4. CLASSIFICATION OF DISEASES IN ORTHOPEDIC ACCORDING TO AVASTHA (DISEASE CONDITIONS)

1) **Osteoarthritis[70]:**
Osteoarthritis, sometimes called OA, is a type of arthritis that only affects the joints, usually in the hands, knees, hips, neck, and lower back. It's the most common type of arthritis.[1]
Common Symptoms:
Pain during movements, stiffness, swelling when overused, restriction of movements.
Common Causes:Aging, being overweight, H/o injury/surgery, overuse, family history.
2) **Osteoporosis[71]:**
Osteoporosis is a disease in which bones become fragile and more likely to break (fracture).
Common Causes:
A decrease in estrogen in women at the time of menopause, bedridden patients, family history, alcohol, smoking, low intake of calcium and vitamin D.
3) **Fracture[72]:**
A fracture is a break, usually in a bone.
Common Causes:
Car accidents, falls, or sports injuries are the main causes of fractures. Other causes are low bone density and osteoporosis.
4) **Tendinitis[73]:**
Tendons are the fibrous structures that join muscles to bones. When these tendons become swollen or inflamed, it is called tendinitis. In many cases, tendinosis (tendon degeneration) is also present.

TABLE 11.4 Therapeutic Nutrition Indicated for Diseases in Orthopedics as per *Dwadash Ahar Varga*

S. NO.	*CLASSIFICATION*	*EXAMPLES*	*PROPERTIES*
1.	*Shukadhanya* (Corn)	*Godhum* (Wheat)	*Bhagnasandhankar*[86] (Eases fracture healing)
2.	*Shamidhanya* (Pulses)	*Priyangu* (Indian Millet) *Yava*	*Bhagnasandhankar*[87] (Eases fracture healing) *Urustambhahar*[88]
3.	*Shaka Varga* (Vegetables)	*Shigru* (Drumstick/Moringa) *Lasun* (Garlic)	Diseases of *Snayu*[89] *Bhagnasandhankar*[90] (Eases fracture healing)
4.	*Phala Varga* (Fruits)	*Kharjoor* *Amalaki* Banana floral stem	*Abhigataj Daha*[91] (Posttraumatic burning) *Asthisandhankari*[92] (Eases fracture healing) *Asthisrava*[93] (Used in oozing discharge from bone – in osteomyelitis and postoperative drainage)
5.	*Gorasa Varga* (Milk and milk products)	Mare milk	Effective in arthritis[94]
6.	*Taila Varg* (Oil)	Castor oil	Pain and swelling in lumbosacral region[95]

5) **Ligament Injuries[74]:**
A sprain is a stretched or torn ligament. Falling, twisting, or getting hit can all cause a sprain. According to *Ayurved*, the main diseases of *Asthiwaha Srotas* are
   1) *Asthigat Vata* (pain in carpal and tarsal joints, muscle wasting, insomnia)[75]
   2) *Sandhigat Vata* (swollen joints, pain during joint movements, crepitus during movements)[76]
   3) *Asthikshaya* (sluggish body movements, reduced strength and energy, fat loss, tremors, sudden outburst)[77]
   4) *Avabahuk* (pain in the shoulder joint, muscle wasting, or flattening of shoulder contour)[78]
   5) *Parshnishool* (stiffness and pain in ankle joint)[79]
   6) *Vataakantak* (sprain in ankle joint)[80]
   7) *Kroshtuk Shirsha* (excruciating pain in knee joints)[81]
   8) *Bhagna* (fractures)[82,83]
   9) *Katigraha* (low back pain)[84]
   10) *Grudhrasi* (referred pain from buttocks to heel, stiffness, and numbness in affected lower limb)[85]

5. CLINICAL SIGNIFICANCE IN ORTHOPEDIC

What Type of Breakfast Is to Be Consumed?

Breakfast – Raw rice, *Ragi* (finger millet) can be used that is easy to digest.

Use of type and quantity of *Agni* (fire) while cooking is important for digestion assimilation. Here is an example with the same grain, same quantity, and same nutritional value but different forms of fuel changing the food property from heavy to digest (*Guru*)to easy to digest (*Laghu*). Heavy to digest is used in the degenerative stage of disease, while easy to digest is used in the inflammatory stage of disease.

TABLE 11.5 Orthopedic Nutrition According to Seasons

TYPES OF FOOD ↓→ *SEASON*	*BHOJYA*	BHAKSHYA	CHARVYA	LEHYA	CHOSHYA	PEYA
Winter	Old Brown Rice, Red Gram, Green Gram, Moth Bean Curry. Snake Gourd and Ridge Gourd Vegetables. *Dal* (Pulse Soup), Rice, Cauliflower, Cabbage, Potatoes, Fenugreek Leaves, *Parathas* (Stuffed Flat Bread), Mutton	Fenugreek *Laddu* (Spherical Sweet), *Roti* Prepared From Millet (Flat Bread). Boiled Egg and Pepper *PooranPoli* (Stuffed Sweet Flat Bread), *Bhakri* (Rice Flat Bread), *DinkLaddu* (Edible Gum Spherical Sweet), Carrot *Halwa* (Carrot Pudding)	*Chana* (Chickpeas), *Papad* (Thin Indian Flavored Wafer), *Chikki* (Brittle) Made With Sesame, Walnut. Carrot*Raita* (Seasoned Yogurt Dip or Sauce) Jaggery and Groundnut, Dry Fruit *Chikki* (Brittle), Carrot and Beetroot *Raita* (Seasoned Yogurt Dip or Sauce).	*Chyawanprash* (Traditionally Indian Supplement), Garlic *Chutney* (Cold Sauce/ Spread) , Turmeric Pickle *Shrikhand* (Strained Yoghurt Sweet), *Brahma Rasayan*	Ginger *Barfi* (Milk-Based Sweet), Gooseberry Candy	Ginger Water, Turmeric Milk, and Buttermilk prepared from Asafoetida, Ginger, and Cumin Seeds *Tadka* (Tempering). Green Gram *Yusha* (Soup), *Paya* (Goat Leg) Soup, *Basundi* (Flavored Dense Sweetened Milk), Turmeric Milk, Cinnamon Water, Mutton Soup, Masala Buttermilk
Summer	Green Gram *Khichadi* (Salty Porridge), Bitter Gourd Vegetable, Horse Gram *Dal* (Gram Soup) Green Gram *Dal* (Green Gram Soup) With Rice, Bottle Gourd Vegetable	Puffed Rice *Laddu* (Spherical Sweet). Sorghum *Roti* (Unleavened Flat Bread), *RagiRoti* (Finger Millet Unleavened Flat Bread), *SujiLaddu* (Semolina Spherical Sweet), Coriander *Vadi* (Coriander Leaves Fitters)	Popped Rice (*Lahya*) *Mirgund* of Rice Flakes, *Rajgira Chikki* (Amaranth Brittle Sweet), Radish *Raita* (Radish Yogurt Dip or Sauce) Coconut *Chikki* (Coconut Brittle Sweet), Rice Flake *Papad* (Thin Indian Rice Wafer), Cucumber *Raita* (Cucumber Yogurt Dip or Sauce)	Wood Apple (*Kawath*) *Chutney* (Cold Sauce/Spread), Gooseberry Pickle Coconut *Chutney* (Cold Sauce/ Spread), Rose Petal Jam	Pomegranate, Orange Sugarcane, Mango	Holy Basil Water, Ginger Water, and Buttermilk Prepared from *Trikatu* *KhusSiddha* Water, *Kokam* Juice (*Garcinia indica* Juice), *Kharjuradi Mantha*, Saffron Milk, Buttermilk Prepared From Cumin

(*Continued*)

TABLE 11.5 *(Continued)* Orthopedic Nutrition According to Seasons

TYPES OF FOOD ↓→ *SEASON*	*BHOJYA*	BHAKSHYA	CHARVYA	LEHYA	CHOSHYA	PEYA
Rainy	Old Brown Rice, Lentil *Dal* (Lentil Soup), Ridge Gourd, Lady's Finger, *BhajaniThalipeeth* (Roasted Multigrain Flatbread)	Wheat *Chapati* (Phulka) (Wheat Flat Bread)	Sorghum *Lahya* (Popped Rice), Garlic *Papad* (Garlic-Flavored Wafer), *Moong Papad* (Green Gram–Flavored Wafer), Green Gram *Laddu* (Green Gram Spherical Sweet), *Rajgira Chikki* (Amaranth Brittle Sweet), Carrot *Raita* (Carrot Yogurt Dip or Sauce)	Black Currant *Chutney* (Cold Sauce/Spread), Garlic *Chutney* (Cold Sauce/ Spread)	Garlic, Sweet Lime, Gooseberry, and *Murabba* (Marmalade)	Garlic Milk, Ginger Water, and Buttermilk With Rock Salt

TABLE 11.6 Do's and Don'ts According to the Stage of the Disease

CONDITIONS OF THE DISEASE	*AYURVEDIC APPROACH*	*PATHYA (WHAT TO DO?)*	*APATHYA (WHAT NOT TO DO?)*
Inflammatory conditions	*Vatarakta* (Connective tissue disorders)	*Draksha* (black raisins) Gourd (Snake gourd, bottle gourd, etc.)	*Vidahi* (creates epigastric pain and burning) *Abhishyandi, Guru* (heavy to digest) *Ushna* (which vitiates pith, e.g., spices) *Amla* *Lavana* predominant (sour, salty, e.g., pickles)
Degenerative conditions	*Vatavyadhi*	*Panchamoolashrutaksheeram (Panchamoola – Kantakari, Bruhati, Shalaparni, Prushniparni, Gokshur)* *Mansa rasa + Amla Ras* *Yoosh*	*Amla* (sour) *Lavana* (salty), Pungent, Bitter Taste predominant food

TABLE 11.7 *Ayurvedic* Principles of Selection of Food According to Stages of the Diseases

	STAGE OF DISEASE	*DOSHA PREDOMINANCE (SIGNS AND SYMPTOMS)*	*THE BASIC PRINCIPLE OF DIET SELECTION*
Stage 1	Trauma (*Agantuj*)	*Vatavruddhi* (Vitiated *Vata*) Raktadushti (Vitiated *Rakta*)	*Raktaprasadak* (Improves blood quality) *Vataashamak* (*Vata* pacifier)
Stage 2	Inflammation	*Kapha, Pittavruddhi* (Vitiate *Kapha* and *Pitta*)	*Kapha Pitta Shamak* (*Kapha and Pitta* pacifier) *Aamapachan* (*Aam* digestive) *Deepan* (Appetizer)
Stage 3	Degeneration	*Vataprakop* (Extreme vitiation of *Vata*)	*Vataashamak,* (*Vata* pacifier) *Vataanuloman* (Correcting *Vata* functions), *Snehan* (Oelation)
Stage 4	With complications	Irreversible anatomical changes occur	*Brumhan* (improves nourishment of *Dhatus*) + *Vataahar* (*Vata* pacifier)

COOKED VEGETABLES

TABLE 11.8 Cooking Methods According to Types of Cooking

COOKING METHOD	*TYPE OF COOKING*	*EXAMPLES*	*GURU TO LAGHU*
Raw food		Salads	*Guru ++*
Kukoolapachitham	Steaming	*Idli, Modak, Patoli*	*Guru +*
Koorpparapachitham	Hot plate	Sizzlers	*Guru*
Brashtapachitaham	Frying	*Vada, Bhaji*	*Laghu*
Kandhupachitham	Furnace	*Tandori* dishes	*Laghu +*
Angarapachitham	Burning charcoal	Barbeque	Laghu ++

TABLE 11.9 Properties According to Cooking Methods

COOKING METHOD	*GURU TO LAGHU*
Vegetables processed with *Ghee* (clarified butter) + cumin seeds or oil + mustard seeds + Asafoetida (Hinga) Along with curry leaves + salt	*Guru +*
Vegetables processed with *Ghee* (clarified butter) + cumin seeds or oil + mustard seeds + Asafoetida Along with curry leaves + onion + tomato + potato + salt	*Guru*
Vegetables processed with *Ghee* (clarified butter) + cumin seeds or oil + mustard seeds + Asafoetida Along with curry leaves + onion + tomato + potato + salt Completely smashed	*Laghu*
Vegetables processed with *Ghee* (clarified butter) + cumin seeds or oil + mustard seeds + Asafoetida Along with curry leaves + onion + tomato + potato + spices + salt Completed smashed	*Laghu +*

CURRY (ONE OF THE MAIN INGREDIENTS IN THE MAIN COURSE)

Main Ingredients – Split black gram (*Udid/Maash daal*), red kidney beans (*Rajma*), split Bengal gram (*Chana daal*), black-eyed beans (*Chawli/Lobia*), white and green peas (*Vatana*), dew gram beans (*Matki/Moth*), split red lentils (*Masoor daal*), red gram (Tur/*Arhardaal*), green gram (*Moong daal*).

Additional Ingredients – Spices + lipids (to enhance the bioavailability of the main ingredients, to amplify the taste of the delicacy)

Quantity of Water – The quantity of water used to prepare the delicacy decides its digestion time. Less quantity of water makes the substance heavy to digest (*Guruta*), whereas more quantity of water makes it easy to digest (*Laghuta*).

TIME CONSUMED FOR COOKING

The amount of external fuel used in preparing an item decides the time taken for its digestion. An item cooked for more time makes it easier to digest (used in the inflammatory stage), whereas an item cooked for lesser time is heavier to digest (used in the degenerative stage). This rule is applicable in the case of cooking only. However, one should avoid regular reheating of food as this leads to oxidation and thereby increases its toxicity.

Why Know All These Things?

Easy-to-digest food is mainly used in acute trauma, the inflammatory stage of the disease, for example, fractures, immediate postoperative stage, *Rakta,* and *kapha* predominant phases, whereas in case of chronicity of the disease, heavier-to-digest food is used for strengthening the muscles, bones, etc.

TABLE 11.10 *Dosha* Predominance and Spices

SPICES	*LIPIDS*	*DOSHA PREDOMINANCE*	*DISEASE CONDITION*
Asafoetida, bay leaf, cinnamon, garlic	Sesame oil, coconut oil, olive oil, groundnut oil	*Vata shaman* (*Vata* pacifier)	Degenerative condition (*VataPittaDushti*– Vitiates *Vata and Pitta*)
Cumin seeds, curry leaves, turmeric, clove, cardamom, coriander seeds	*Ghee* (clarified butter), butter, rice barn oil	*Pitta Shamak* (*Pitha* pacifier)	Inflammatory condition (*Pitta KaphaDushti*– Vitiates *Pitta and Kapha*)
Ginger, black pepper, mustard seeds, red chili, green chili, fenugreek seeds	Mustard oil, safflower oil	*Kapha Shamak* (*Kapha* pacifier)	Traumatic condition (*Rakta Pitta Dushti*– Vitiates *Rakta* and *Pitta*)

Recipes

1) Soup of lady's finger:
In degenerative conditions like Perthe's disease, avascular necrosis of the head of the femur (AVN), and fractures in diabetic patients. It enhances the flow and quantity of synovial fluid. Take 2/3 lady's finger + 4 cups of water, boil, reduce it to 1 cup, and temper it with *Ghee* (clarified butter) and cumin seeds.
2) Healthy Soup
10 g wheat, 5 g black gram, 10 g green gram, and a pinch of fenugreek seeds + 4 cups water; boil, reduce it to 1 cup, and temper it with *Ghee* (clarified butter) and cumin seeds. In osteoporosis, degenerative diseases of vertebrae, and lumbar disc syndrome, this type of soup is recommended.
3) Moringa leaves/drumstick pods + rice + 12 cups of water – Boil, reduce it to 2 cups, and temper it with *Ghee* (clarified butter) and cumin seeds. This soup is mainly used in the inflammatory condition of bone and joint disorders. In osteoporosis, rickets, and scurvy this soup is highly beneficial.
4) *Kheer*:
20 g garden cress seeds/*Chandrashur* + 2 cups milk –Boil, reduce to 1 cup, and add 5 g dry dates powder + 2 g grated coconut. Used in postnatal arthropathy. It enhances the formation and repair of bones.
5) Chicken/mutton/black gram soup:
This is mainly used in meniscus injuries as well as in muscle wasting as *Brumhan* for nurturing the injured meniscus/muscle.
There are two main causes of osteoporosis:
a) Nutritional deficiency
b) Obstruction in bone formation
In both cases, mutton soup is highly recommended.
6) Milk + *Ghee* (clarified butter) + herbs of bitter taste having affiliation towards bones are to be used meticulously.
This combination helps in preventing osteoporosis at all stages. It is useful in rickets also.
7) 20 g chicken/mutton/black gram soup + 12 cups of water – boil, reduce it to 4 cups, and temper it with *Ghee* (clarified butter) and cumin seeds.
This is useful mainly in osteoarthritis with muscle wasting.

Evening Snacks

1) Steamed nuts (peanuts, almonds, cashew nuts, pistachio)
2) Raw salads

3) Fruit salads (seasonally available fruits)
4) Steamed corn
5) Green gram split
6) *Dosas* are made up of mixed dal, only green gram

These delicacies are used for *Brumhan*, strengthening *Dhatu*, mainly *Asthi Dhatu*, and increasing the *Oja* of the body.

Coconut

Best for *Maans, Meda, Asthi,* and *Majja Dhatu*. It can be used in various preparations as grated coconut or coconut milk. It is to be avoided in patients having constipation or it has to be processed with cumin seeds, curry leaves, and *Kokum* to enhance its bioavailability/*Brumhan* property.

Pumpkin Soup

Mainly used in inflammatory conditions of bones and vertebrae. If mixed with milk or *Ghee* (clarified butter) it is useful in degenerative conditions. It acts as *Brumhan* mainly in osteoporosis with senile dementia.

Raw Banana

Slightly heavier to digest. Mainly used as *Brumhan*. Useful in bone formation. Contraindicated in patients suffering from hemorrhagic piles.

Onion/Shallots/*Palandu*

Onions are the best *Vata Shamak* (reduces *Vata*)

Snake/Bottle/Ridge/Ivy/Bitter Gourd

Easy to digest + *Pitta Kapha Shamak.*

Used mainly in all clinical conditions related to bones and joints.

Long Beans + Moringa Leaves + Drumstick Pods

The above combination is the best in all inflammatory conditions of bone- and joint-related pathologies.

Beetroot + Buttermilk + Black Pepper

Easy to digest, improves the quality of *Rakta Dhatu.*
Pomegranate
Mainly used in body aches and joint pain associated with *Pandu* (anemic conditions).
Black raisins
To be avoided in distention of abdomen. Mainly used in OA with muscle wasting.

Papaya and mangoes
Both papaya and mangoes have multiminerals and micronutrients.
Payas + Laksha + Madhur Ras/Bala mainly act in degenerative conditions.
In *Bhagna* (fractures) the principle behind the dietetics is:

अल्पाशिनोऽनात्मवतो जन्तोर्वातात्मकस्य च।
उपद्रवैर्वा जुष्टस्य भग्नं कृच्छ्रेण सिध्यति।।३।।[96]

सुश्रुत चिकित्सा स्थान ३/३

Fractures fail to heal, when:

1. The patient doesn't consume nutritious food.
2. The patient is cowardly and timid.
3. The patient consumes food that vitiates *Vata*.
4. The patient has fracture with complications like chronic renal failure and diabetes, etc.

शालिर्मांसरसः क्षीरं सर्पिर्यूषः सतीनजः।
बृंहणं चान्नपानं स्याद्देयं भग्नाय जानता।।५।।[97]

सुश्रुत चिकित्सा स्थान ३/५

Avoid the consumption of red rice, cereals, millet, mutton, or chicken soups, as well as milk (as commonly practiced), as these can exacerbate pus formation in wounds. However, in the context of fractures, the consumption of *Ghee* (clarified butter) and green pea soup (despite its *Vata*-aggravating nature), comes highly recommended as they promote healing. These foods stimulate osteoblastic activities, contributing significantly to the healing of fractures.

What to Avoid in Fractures?

Salty, pungent, spicy, alkaline food, sour food, and oil-free food are to be avoided. It creates vitiation of *Pitta* and *Rakta*. So it is contraindicated in fractures. We have to follow diet rules for fractures for 1 month in the case of children, 2 months for youngsters, and 3 months for aged people. Red chilies, brinjal, pickles, tamarind, and vinegar are to be avoided[98].

6. *AHAREYA KALPANA* (DIETARY RECIPES) IN ORTHOPEDICS

How to Prepare *Yusha*?

Yusha (soup) being one among them is widely discussed in every work of literature on *Ayurveda* and in most diseases as a therapeutic diet. *Yusha* is prepared by taking one part of grain other than paddy preferably legumes and cooking in 14 parts of water.[99]

Yush Kalpana is easier to digest, improves appetite, and facilitates the assimilation of nutrients.

When consumed freshly prepared, combined with *Ghee* (clarified butter) infused with *Tikta Ras*-dominant medicinal herbs, it contributes to enhanced post-trauma osteoblastic activity.

Yusha of *Punarnava, Rasna, Changeri, Bala Dahi, Ghruta* helps in pacifying *Vata Dosha* is known to reduce stiffness in inflammatory arthritis.

Yusha of *Bilva Patra* + *Shigru* + *Eranda* + *Balaa* + *Rasna* + *Trikatu* + *Amra Patra –Patra Yusha)* resolves aggravated *Vata Dosha,* helping in combating degenerative articular disorders.

How to Prepare *Ksheerpaka?*

Ksheerpaka may be prescribed as a medicament. It has nourishing properties and thus is expected to show immunomodulation. These benefits of *Ksheerpaka* make it the preferred choice of drug delivery system in aged people and children who have a low tolerance to various dosage forms and need added nourishment. *Ksheerpaka Kalpana* of *Rason* (garlic) helps to prevent recurrent attacks of stiffness and pain in rheumatoid arthritis, sciatica, and low back pain.

Ksheerpaka Kalpana of *Arjun* (Terminalia *Arjuna*) helps to enhance osteoblastic activity and is hence used in osteoporosis and rickets (a disease found in kids because of malnutrition of bones, lack of Ca in food that weakens bones).

Ksheerpaka Preparation

One part of Churna + 8 parts milk and 32 parts water. Boil and reduce it to 8 parts.[100]

How to prepare *Laddu?*

Calcium and bone beneficial minerals–enhancing *Laddu* (a delicacy used in postnatal conditions)

Wheat flour along with dry fruits, dry coconut powder, *Khus Khus* (poppy seeds), fenugreek seeds, *Chandrasur* (garden cress) seeds, and edible gum is mixed with a little *Ghee* (clarified butter) and fried slightly. Sugar syrup along with cardamom, almond powder, and saffron is then added. It is made into sweet meat. It is given a spherical shape by pressing it with the fist. One has to smoothen it with oily hands.[101]

This Indian delicacy can be taken with a warm glass of milk for breakfast during winter. Kids can enjoy them in their snack box. They have a good shelf life and remain at room temperature for about a month. If prepared without coconut, the shelf life increases to 3 to 4 months. These *Laddus* are given to nursing mothers as it helps to increase milk production and also provide the new mothers with the necessary energy and heat to recover their health. It strengthens the reproductive system of females. It is used in postnatal arthropathies after the fever vanishes.

Scope for Further Research and Nutravigilance

Building upon the previously mentioned information, it's important to take into account factors that delay healing and essential dietary principles in *Bhagna* (fractures):

Oxidative stress affects calcium metabolism and thus delays healing bone fractures.[102]

Studies for evaluating the earlier mentioned foods such as Ghee (clarified butter), soup of green peas, etc. and their role in delaying fracture healing are needed.

In osteoarthritis, after *Ampachan*, *Agnivardhan Chikitsa*, *Brumhan*, and *Dhatupariposhan* are required. *Maansa Samhanan, Vardhan* + *Medakshayakari* (to reduce obesity) treatment is essential.

Regular exercise, aged rice, and grains are of great importance in obesity management.

व्यायामनित्यो जीर्णाशी यवगोधूमभोजनः।

सन्तर्पणकृतैर्दोषैः स्थौल्यं मुक्त्वा विमुच्यते॥२५॥[104]

चरक सूत्रस्थान २३/२५

When a sedentary lifestyle aligns with a Santarpanottha diet and the presence of Rasawaha Srotodushtijanya Hetu, the diet outlined above is implemented for a duration spanning 30 to 90 days. It reduces weight, and the patient feels light (*Laaghaw*).

This is the first stage of osteoarthritis in which *Asthigat Vata* is to be treated along with correction of *Mamsa Balakshaya* (muscle wasting). For every kilogram you weigh, your knees receive four times the amount of stress; 1 kg of weight loss equals a 4 kg reduction in load through the knee joint.

The most important modifiable risk factor for the development and progression of knee osteoarthritis is obesity. Weight loss reduces the risk of symptomatic knee osteoarthritis and is recommended by medical guidelines worldwide.

According to *Ayurved*, a substance containing properties of *Snigdha* (smooth, lubricated) + *Shoshan* (property of drying) + *Khara* (coarse) is *Asthivardhak* (bone enhancing) but there is no single drug present on earth containing these properties in combination. So *Ayurved* suggests *Tikta* (bitter)+ *Ghruta* (clarified butter) +*Ksheera* (milk) to be processed and to be administered for Asthi *Bruhman* (nourishing *Asthi Dhatu*).[105] Further study regarding the same is needed.

REFERENCES

1. "Orthopedics." Merriam-Webster.com Dictionary, Merriam-Webster, www.merriam-webster.com/dictionary/orthopedics. Accessed 2 Jan. 2023.
2. S. G. Vartaka, March 1962, Doshadhatumal Vidnyaniyam. Asthidhatuhu Nakharomach, pg.209.
3. Sushrut sutrasthan 35/16 Maharsi Susruta. Susruta Samhita. Edited by Dr. Anant Ram Sharma. Varanasi: Chaukhamba Surbharati Prakashan;2004. pg.273.
4. Charak Vimana sthan 5/17, Vaidya Y. G. Joshi, Charak Samhita. Part 1: Vaidyamitra Publications, Pune:2017. pg.544.
5. Madhav Nidanam 1/5 Madhukosh Vyakhya Aacharya Vijayrakshita Shrikanth Dutta, Pub. Motilal Banarasidas, Delhi pg.17.
6. Sushrut sutrasthan 15/8 Bhanumati Teeka Maharsi Susruta. Susruta Samhita. Edited by Dr. Anant Ram Sharma. Varanasi: Chaukhamba Surbharati Prakashan;2004. pg.117.
7. Sanjeev Rastogi (2014) *Ayurvedic* Science of Food and Nutrition. The Basic Tenets of *Ayurvedic* Dietetics and Nutrition pg.33.
8. Sushrut kalpasthan 4/45 Dalhan Teeka Maharsi Susruta. Susruta Samhita. Edited by Dr. Anant Ram Sharma. Varanasi: Chaukhamba Surbharati Prakashan;2004. pg.546.
9. Gibson GR (2008) Prebiotics as gut microflora management tools. J Clin Gastroenterol 42(Supp 2):S75–S79.
10. Talebi M, Andalib S, Bakhti S, Ayromlou H, Aghili A, Talebi A. Effect of vitamin B6 on clinical symptoms and electrodiagnostic results of patients with carpal tunnel syndrome. Adv Pharm Bull. 2013;3(2):283–288. doi:10.5681/apb.2013.046
11. Shushrut sutrasthan15/5 Maharsi Susruta. Susruta Samhita. Edited by Dr. Anant Ram Sharma. Varanasi: Chaukhamba Surbharati Prakashan;2004. pg.115.
12. Loiacono C, Palermi S, Massa B, et al. Tendinopathy: Pathophysiology, therapeutic options, and role of nutraceuticals. A Narrative Literature Review. *Medicina (Kaunas)*. 2019;55(8):447. Published 2019 Aug 7. doi:10.3390/medicina55080447
13. Scott A, Backman LJ, Speed C. Tendinopathy: Update on Pathophysiology. JOSPT. 2015 Nov;45(11):833–841.
14. Nageeb R S, Shehta N, Nageeb GS, Omran AA. Body mass index and vitamin D level in carpal tunnel syndrome patients. The Egyptian Journal of Neurology, Psychiatry, and Neurosurgery. 2018;54(1):1–7.
15. Sanjeev Rastogi (2014) *Ayurvedic* Science of Food and Nutrition. The Concept of Diet in *Ayurveda* and Its Implications for the Modern World, pg.36.
16. Nighantu Ratnakar, Original version 1789, Gunadosha Prakaran, Vaidyavarya Vishnu Vasudev Godbole, Nirnay Sagar Press Owner-Pandurang Javji, pg.81.
17. Nighantu Ratnakar, Original version 1789, Nidanasah Chikitsa, Anugraha Vichar, Urdhwavata Vichar, Vaidyavarya Vishnu Vasudev Godbole, Nirnay Sagar Press Owner-Pandurang Javji, pg.417.

18. Nighantu Ratnakar, Original version 1789, Vaidyavarya Vishnu Vasudev Godbole, Nirnay Sagar Press Owner-Pandurang Javji, pg.423.
19. Ksemasarma, Ksemakutuhalam: Indian Institute of Ayurveda & Integrative Medicine, Bangalore: 2009.The Eight utasava pg.241.
20. Ksemasarma, Ksemakutuhalam: Indian Institute of Ayurveda & Integrative Medicine, Bangalore: 2009. The Eight utasava pg.214, 215.
21. Ksemasarma, Ksemakutuhalam: Indian Institute of Ayurveda & Integrative Medicine, Bangalore: 2009. The Eight utasava pg.321, 322.
22. Ksemasarma, Ksemakutuhalam: Indian Institute of Ayurveda & Integrative Medicine, Bangalore: 2009. The Eight utasava pg.323.
23. Nighantu Ratnakar. Edited by Krisnasastri R. Navre, Pandurang Jawaji of Nirnaya-Sagar Press, Bombay:1934. pg.38.
24. Sushrut ChikitsaSthan 3/14 MaharsiSusruta. Susruta Samhita. Edited by Dr. Anant Ram Sharma. Varanasi: ChaukhambaSurbharati Prakashan;2004. pg.191.
25. Sanjeev Rastogi (2014) *Ayurvedic* Science of Food and Nutrition. The Basic Tenets of *Ayurvedic*. Dietetics and Nutrition pg.33.
26. Emkey RD, Emkey GR. Calcium metabolism and correcting calcium deficiencies. Endocrinol Metab Clin North Am. 2012;41:527–56.
27. Eustice C. Is chondroitin effective in treating arthritis?Verywellhealth.com. Accessed April 27, 2021.www.verywellhealth.com/chondroitin-information-189549
28. Loiacono C, Palermi S, Massa B, et al. Tendinopathy: Pathophysiology, therapeutic options, and role of nutraceuticals. A Narrative Literature Review. Medicina (Kaunas). 2019;55(8):447. Published 2019 Aug 7. doi:10.3390/medicina55080447.
29. Scott A, Backman LJ, Speed C. Tendinopathy: Update on Pathophysiology. JOSPT. 2015Nov;45(11):833–841.
30. Human Nutrition. 2018. The role of lipids in food. http://pressbooks.oer.hawaii.edu/humannutrition/chapter/the-functions-of-lipids-in-the-body/
31. Artaza-Artabe I, Sáez-López P, Sánchez-Hernández N, Fernández-Gutierrez N, Malafarina V. The relationship between nutrition and frailty: Effects of protein intake, nutritional supplementation, vitamin D and exercise on muscle metabolism in the elderly. A systematic review. Maturitas. 2016 Nov;93:89–99.doi: 10.1016/j.maturitas.2016.04.009. Epub 2016 Apr 14. PMID: 27125943.
32. Mertz KH, Reitelseder S, Bechshoeft R, Bulow J, Højfeldt G, Jensen M, Schacht SR, Lind MV, Rasmussen MA, Mikkelsen UR, Tetens I, Engelsen SB, Nielsen DS, Jespersen AP, Holm L. The effect of daily protein supplementation, with or without resistance training for 1 year, on muscle size, strength, and function in healthy older adults: A randomized controlled trial. Am J Clin Nutr. 2021 Apr 6;113(4):790–800. doi: 10.1093/ajcn/nqaa372. PMID: 33564844.
33. Howard EE, Pasiakos SM, Fussell MA, Rodriguez NR. Skeletal Muscle Disuse Atrophy and the Rehabilitative Role of Protein in Recovery from Musculoskeletal Injury. Adv Nutr. 2020 Jul 1;11(4):989–1001. doi 10.1093/advances/nmaa015. PMID: 32167129; PMCID: PMC7360452.
34. Ginty F. Dietary protein and bone health. Proc Nutr Soc. 2003 Nov;62(4):867–76. doi 10.1079/PNS2003307. PMID: 15018487.
35. Yoo JI, Choi H, Song SY, Park KS, Lee DH, Ha YC. Relationship between water intake and skeletal muscle mass in elderly Koreans: A nationwide population-based study. Nutrition. 2018 Sep;53:38–42. doi: 10.1016/j.nut.2018.01.010. Epub 2018 Feb 12. PMID: 29655775.
36. de Jonge EA, Kiefte-de Jong JC, Campos-Obando N, Booij L, Franco OH, Hofman A, Uitterlinden AG, Rivadeneira F, Zillikens MC. Dietary vitamin A intake and bone health in the elderly: the Rotterdam Study. Eur J Clin Nutr. 2015 Dec;69(12):1360–8. doi: 10.1038/ejcn.2015.154. Epub 2015 Sep 16. Erratum in: Eur J Clin Nutr. 2015 Dec;69(12):1375. PMID: 26373964.
37. Joo NS, Yang SW, Song BC, Yeum KJ. Vitamin A intake, serum vitamin D and bone mineral density: analysis of the Korea National Health and Nutrition Examination Survey (KNHANES, 2008-2011). Nutrients. 2015 Mar 10;7(3):1716–27. doi: 10.3390/nu7031716. PMID: 25763530; PMCID: PMC4377877.
38. Torbergsen AC, Watne LO, Wyller TB, Frihagen F, Strømsøe K, Bøhmer T, Mowe M. Micronutrients and the risk of hip fracture: Case-control study. Clin Nutr. 2017 Apr;36(2):438–443. doi: 10.1016/j.clnu.2015.12.014. Epub 2015 Dec 23. PMID: 26795217.
39. Endo T, Imagama S, Kato S, Kaito T, Sakai H, Ikegawa S, Kawaguchi Y, Kanayama M, Hisada Y, Koike Y, Ando K, Kobayashi K, Oda I, Okada K, Takagi R, Iwasaki N, Takahata M. Association between vitamin

A intake and disease severity in early-onset heterotopic ossification of the posterior longitudinal ligament of the spine. Global Spine J. 2022 Oct;12(8):1770–1780. doi 10.1177/2192568221989300. Epub 2021 Jan 25. PMID: 33487053; PMCID: PMC9609524.

40. Talebi M, Andalib S, Bakhti S, Ayromlou H, Aghili A, Talebi A. Effect of vitamin B6 on clinical symptoms and electrodiagnostic results of patients with carpal tunnel syndrome. Adv Pharm Bull. 2013;3(2):283–288. doi:10.5681/apb.2013.046.
41. Foods that fight inflammation. Harvard Health Publishing, Harvard Medical School.
42. Ömeroğlu S, Peker T, Türközkan N, Ömeroğlu H. High-dose vitamin C supplementation accelerates the Achilles tendon healing in healthy rats. Archives of Orthopaedic and Trauma Surgery. 2008;129(2):281–286. doi:10.1007/s00402-008-0603-0
43. Goode HF, Burns E, Walker BE. Vitamin C depletion and pressure sores in elderly patients with a femoral neck fracture. BMJ. 1992 Oct 17;305(6859):925–7. doi 10.1136/bmj.305.6859.925. PMID: 1458073; PMCID: PMC1883546.
44. Sahni S, Hannan MT, Gagnon D, et al. Protective effect of total and supplemental vitamin C intake on the risk of hip fracture–a 17-year follow-up from the Framingham osteoporosis study. Osteoporos Int. 2009;20:1853–61.
45. Munday K. Vitamin C and bone markers: investigations in a Gambian population. Proc Nutr Soc. 2003 May;62(2):429–36. doi 10.1079/pns2003247. PMID: 14506891.
46. Nageeb RS, Shehta N, Nageeb GS, Omran AA. Body mass index and vitamin D level in carpal tunnel syndrome patients. The Egyptian Journal of Neurology, Psychiatry, and Neurosurgery. 2018;54(1):1–7.
47. Angeline ME, Ma R, Pascual-Garrido C, et al. Effect of diet-induced vitamin D deficiency on rotator cuff healing in a rat model. The American Journal of Sports Medicine. 2013;42(1):27–34. doi:10.1177/0363546513505421
48. Gendelman O, Itzhaki D, Makarov S, Bennun M, Amital H. A randomized double-blind placebo-controlled study adding high dose vitamin D to analgesic regimens in patients with musculoskeletal pain. Lupus. 2015 Apr;24(4-5):483–489.
49. Chung E, Mo H, Wang S, Zu Y, Elfakhani M, Rios SR, Chyu MC, Yang RS, Shen CL. Potential roles of vitamin E in age-related changes in skeletal muscle health. Nutr Res. 2018 Jan;49:23–36. doi:10.1016/j.nutres.2017.09.005. Epub 2017 Sep 21. PMID: 29420990.
50. Torbergsen AC, Watne LO, Wyller TB, Frihagen F, Strømsøe K, Bøhmer T, Mowe M. Micronutrients and the risk of hip fracture: Case-control study. Clin Nutr. 2017 Apr;36(2):438–443. doi: 10.1016/j.clnu.2015.12.014. Epub 2015 Dec 23. PMID: 26795217.
51. Lanham-New SA. Importance of calcium, vitamin D and vitamin K for osteoporosis prevention and treatment. Proc Nutr Soc. 2008;67:163–76.
52. Tipton KD. Nutritional support for exercise-induced injuries. Sports Med. 2015;45 Suppl 1:S93–S104. doi:10.1007/s40279-015-0398-4
53. Frozen shoulder – can your diet make a difference? Nielasher.com. Accessed April 27, 2021. www.nielasher.com/blogs/video-blog/116257029-frozen-shoulder-can-your-diet-make-a-difference
54. Chapuy MC, Arlot ME, Duboeuf F, et al. Vitamin D3 and calcium to prevent hip fractures in elderly women. N Engl J Med. 1992;327:1637–42.
55. Emkey RD, Emkey GR. Calcium metabolism and correcting calcium deficiencies. Endocrinol Metab Clin North Am. 2012;41:527–56.
56. Draper HH, Scythes CA. Calcium, phosphorus, and osteoporosis. Fed Proc. 1981 Jul;40(9):2434–8. PMID: 7250388.
57. Potassium is good for the heart, bones, and muscles. Accessed January 6, 2023.https://source.colostate.edu/potassium-good-for-heart-bones-and-muscles/
58. Hanley DA, Whiting SJ. Does a high dietary acid content cause bone loss, and can bone loss be prevented with an alkaline diet? J Clin Densitom 2013;16:420–425. [PubMed abstract]
59. Dawson-Hughes B, Harris SS, Ceglia L. Alkaline diets favor lean tissue mass in older adults. American Journal of Clinical Nutrition. 2008;87(3):662–665.
60. Draper HH, Scythes CA. Calcium, phosphorus, and osteoporosis. Fed Proc. 1981 Jul;40(9):2434–8. PMID: 7250388.
61. Yao H, Xu J, Wang J, Zhang Y, Zheng N, Yue J, Mi J, Zheng L, Dai B, Huang W, Yung S, Hu P, Ruan Y, Xue Q, Ho K, Qin L. Combination of magnesium ions and vitamin C alleviates synovitis and osteophyte formation in osteoarthritis of mice. Bioact Mater. 2020 Nov 10;6(5):1341–1352. doi: 10.1016/j.bioactmat.2020.10.016. PMID: 33210027; PMCID: PMC7658330.

62. Boskey AL, Rimnac CM, Bansal M, Federman M, Lian J, Boyan BD. Effect of short-term hypomagnesemia on the chemical and mechanical properties of rat bone. J Orthop Res. 1992 Nov;10(6):774–83. doi 10.1002/jor.1100100605. PMID: 1403290.
63. Jiang D, Gao P, Lin H, Geng H. Curcumin improves tendon healing in rats: a histological, biochemical, and functional evaluation. Connect Tissue Res. 2016;57(1):20–27. doi:10.3109/03008207.2015.1087517.
64. David L Hamblen, A Hamish R W Simpson (2010) Outline of Orthopaedics. General Survey of Orthopaedic Disorders, pg.64.
65. David L Hamblen, A Hamish R W Simpson (2010) Outline of Orthopaedics. Deformities and Congenital Disorders, pg.69.
66. Kenneth L Knight. More precise classification of orthopaedic injury types and treatment will improve patient care. Nata Journals. Athl Train. 2008 Mar–Apr; 43(2): 117–118. doi 10.4085/1062-6050-43.2.117. PMCID: PMC2267322. PMID: 18345334.
67. Madhavnidanam Madhukoshtika /63-65, Aacharya Vijayrakshita Shrikanth Dutta, Pub. Motilal Banarasidas, Delhi, pg.143.
68. Ashtang Hruday Sutrasthan. 14/1 Acharya Vagbhat. Ashtang Hruday, Dwiwidhopakramaneeyam. Edited by Dr. G. K Garde. Pune: Rajesh Publication; pg.63.
69. Charak Chikitsasthan. 29/19 Vaidya Y. G. Joshi, Charak Samhita. Part 2: Vaidyamitra Publications, Pune:2016. pg.659.
70. "Osteoarthritis" MedlinePlus.gov. Accessed January 8, 2023, https://medlineplus.gov/osteoarthritis.html
71. "Osteoporosis" MedlinePlus.gov. Accessed January 8, 2023, https://medlineplus.gov/ency/article/000360.htm
72. "Fractures" MedlinePlus.gov. Accessed January 8, 2023, https://https://medlineplus.gov/fractures.html
73. "Tendinitis" MedlinePlus.gov. Accessed January 8, 2023, https://medlineplus.gov/ency/article/001229.htm
74. "Sprains and Strains" MedlinePlus.gov. Accessed January 8, 2023, https://medlineplus.gov/sprainsandstrains.html
75. Charak chikitsasthan 28/33 Vaidya Y. G. Joshi, Charak Samhita. Part 2: Vaidyamitra Publications, Pune:2016. pg.630.
76. Charak chikitsasthan 28/38 Vaidya Y. G. Joshi, Charak Samhita. Part 2: Vaidyamitra Publications, Pune:2016. pg.631.
77. Harit Trutiya 9, Prof. Anant Damodar Athavale, AyurvedVyadhivinishchay: Shridhar Damodar Prathisthan, Pune:1923. pg.267.
78. Asthang Hruday NidaanSthan 15/43 Acharya Vagbhat. Ashtang Hruday. Edited by Dr. G. K Garde. Pune: Rajesh Publication; pg.212.
79. Charak siddhisthan 2/12 Vaidya Y. G. Joshi, Charak Samhita. Part 2: Vaidyamitra Publications, Pune:2016. pg.799.
80. Sushrut Nidan 1/79 MaharsiSusruta. Susruta Samhita. Edited by Dr. Anant Ram Sharma. Varanasi: ChaukhambaSurbharati Prakashan;2004. pg.470.
81. Asthang Hruday NidaanSthan 15/52 Acharya Vagbhat. Ashtang Hruday. Edited by Dr. G. K Garde. Pune: Rajesh Publication; pg.212.
82. Sushrut Nidan SthanMaharsiSusruta. Susruta Samhita. Edited by Dr. Anant Ram Sharma. Varanasi: ChaukhambaSurbharati Prakashan;2004. pg.566.
83. Sushrut ChikitsaSthan 3 MaharsiSusruta. Susruta Samhita. Edited by Dr. Anant Ram Sharma. Varanasi: ChaukhambaSurbharati Prakashan;2004. pg.189.
84. Charak ChikitsaSthan 11/13 Vaidya Y. G. Joshi, Charak Samhita. Part 2: Vaidyamitra Publications, Pune:2016. pg.256.
85. Sushrut Nidan 1/74 MaharsiSusruta. Susruta Samhita. Edited by Dr. Anant Ram Sharma. Varanasi: ChaukhambaSurbharati Prakashan;2004. pg.468.
86. Nighant Ratnakar. Edited by Krisnasastri R. Navre, Pandurang Jawaji of Nirnaya-Sagar Press, Bombay:1934. pg.109.
87. Ashtang Hruday Sutrasthan 6/12 Acharya Vagbhat. Ashtang Hruday. Edited by Dr. G. K. Garde. Pune: Rajesh Publication; pg.24.
88. Ashtang Hruday Sutrasthan 6/14 Acharya Vagbhat. Ashtang Hruday. Edited by Dr. G. K Garde. Pune: Rajesh Publication; pg.24.
89. Ksemasarma, Ksemakutuhalam: Indian Institute of Ayurveda & Integrative Medicine, Bangalore: 2009. The Eleventh utasava. pg 337/26.

90. Ashtang Hruday Sutrasthan 6/110 Acharya Vagbhat. Ashtang Hruday. Edited by Dr. G. K Garde. Pune: Rajesh Publication; pg.31.
91. Charak Sutra Sthan 27/127 Vaidya Y. G. Joshi, Charak Samhita. Part 1: Vaidyamitra Publications, Pune:2017. pg.359.
92. Nighant Ratnakar. Edited by Krisnasastri R. Navre, Pandurang Jawaji of Nirnaya-Sagar Press, Bombay:1934. pg.9.
93. Nighant Ratnakar. Edited by Krisnasastri R. Navre, Pandurang Jawaji of Nirnaya-Sagar Press, Bombay:1934. pg.51.
94. Nighant Ratnakar. Edited by Krisnasastri R. Navre, Pandurang Jawaji of Nirnaya-Sagar Press, Bombay:1934. pg.191.
95. Ashtang Hruday Sutrasthan 5/58 Acharya Vagbhat. Ashtang Hruday. Edited by Dr. G. K Garde. Pune: Rajesh Publication; pg.21.
96. Sushrut ChikitsaSthan 3/3 MaharsiSusruta. Susruta Samhita. Edited by Dr. Anant Ram Sharma. Varanasi: ChaukhambaSurbharati Prakashan;2004. pg.189.
97. Sushrut ChikitsaSthan 3/5 MaharsiSusruta. Susruta Samhita. Edited by Dr. Anant Ram Sharma. Varanasi: ChaukhambaSurbharati Prakashan;2004. pg.189.
98. Sushrut ChikitsaSthan 3/4 MaharsiSusruta. Susruta Samhita. Edited by Dr. Anant Ram Sharma. Varanasi: ChaukhambaSurbharati Prakashan;2004. pg.189.
99. Sharangdhara Samhita, Madhyam kandh, Adhamalla, Deepika Teeka: Krishnadas Academy: Varanasi.1986 shloka169 pg.168.
100. Sharangdhara Samhita, Madhyam kandh Adhamalla, Deepika Teeka: Krishnadas Academy: Varanasi.1986 shloka 163 pg.167.
101. Ksemasarma, Ksemakutuhalam: Indian Institute of Ayurveda & Integrative Medicine, Bangalore: 2009. The Tenth Utsav pg.286.
102. Sheweita SA, Khoshhal KI. Calcium metabolism and oxidative stress in bone fractures: role of antioxidants. Curr Drug Metab. 2007 Jun;8(5):519–25. doi 10.2174/138920007780866852. PMID: 17584023.
103. Hughes MS, Kazmier P, Burd TA, Anglen J, Stoker AM, Kuroki K, Carson WL, Cook JL. Enhanced fracture and soft-tissue healing by means of anabolic dietary supplementation. J Bone Joint Surg Am. 2006 Nov;88(11):2386–94. doi: 10.2106/JBJS.F.00507. PMID: 17079395.
104. Charak Sutra Sthan 23/25 Vaidya Y. G. Joshi, Charak Samhita. Part 1: Vaidyamitra Publications, Pune:2017. pg.286.
105. Ashtang Hruday Sarvang Sundar Teeka 11/31 Acharya Vagbhat. Ashtang Hruday. Edited by Dr. G. K Garde. Pune: Rajesh Publication; pg.53.

Therapeutic Nutrition in Ayurveda for Otorhinolaryngology

Pravin M. Bhat

1. INTRODUCTION

Shalakyatantra is one of the eight branches of *Ayurveda* that deals specifically with diseases of the eye, ear, nose, and throat. The term can be roughly translated to mean "the science of treatment by instruments/ *Shalaka*." The discipline of *Shalakyatantra* has a long history that dates back to ancient India, and it has been widely practiced throughout the Indian subcontinent for thousands of years. Today, it remains an important field of study within *Ayurveda*, with practitioners using a range of specialized techniques and instruments to diagnose and treat a variety of eye, ear, nose, and throat disorders.

Otorhinolaryngology (ORL) is the branch that deals with diseases of the ear, nose, and throat. The branch has tremendously developed since the ancient era with newer techniques of diagnosis, medicines, and surgical procedures including lasers. In ORL a wide spectrum of diagnostic skills and treatment are available to mankind.[1] However, in spite of this development, otorhinolaryngology is lacking in the focus on nutrition and the literature about the same is silent.

Hearing loss,[2] allergic rhinitis,[3] and larynx cancer are the main streamline diseases in otolaryngology which have a significant global burden in terms of economy and quality of life.[4]

Therapeutic nutrition is a vital component of the management of patients with ENT (ear, nose, and throat) diseases. The role of nutrition in the healing and recovery process of these conditions cannot be overstated. ENT disorders can be caused by a variety of factors, including infection, inflammation, and trauma. However, regardless of the underlying cause, proper nutrition can help to optimize the body's natural healing mechanisms and support the recovery process.

One of the key principles of therapeutic nutrition in ENT diseases is the provision of adequate energy and protein intake.[5] These nutrients are essential for the repair and growth of tissues and are particularly important for patients who have suffered from trauma or infection. In cases of severe infection or inflammation, the body's energy demands may increase, and the patient may require a higher calorie intake to support the healing process. Similarly, patients with chronic ENT conditions, such as chronic obstructive pulmonary disease (COPD) or chronic rhinosinusitis, may require a higher protein intake to support the repair of damaged tissues.[6]

Another important aspect of therapeutic nutrition in ENT diseases is the provision of micronutrients. These include vitamins and minerals, which are essential for the normal functioning of the body's

DOI: 10.1201/9781003345541-15

metabolic processes. For example, vitamin A is essential for the normal functioning of the immune system and is important in the prevention and treatment of infections.[7] Similarly, vitamin C plays a vital role in the production of collagen, which is essential for the repair of damaged tissues. Zinc is another important micronutrient that plays a role in the healing process, particularly in the repair of wounds.[8]

In addition to energy, protein, and micronutrients, therapeutic nutrition in ENT diseases also involves the provision of appropriate fluids. Adequate hydration is crucial for the normal functioning of the body's metabolic processes, and it is particularly important for patients who have suffered from infection or inflammation.

Therapeutic nutrition in ENT diseases also involves the provision of appropriate diets. Patients with ENT conditions may have specific dietary requirements, such as a low-salt diet for patients with hypertension or a low-fiber diet for patients with inflammatory bowel disease. Additionally, some patients may require a special diet, such as a low-residue diet, to promote the healing of damaged tissues.[9–11]

2. NUTRITION IN OTORHINOLARYNGOLOGY FROM CLASSICAL TEXTS

सामान्यं कर्णरोगेषु घृतपानं रसाशनम् /
अव्यायामोऽशिरःस्नानम् ब्रह्मचर्यमकत्थनम्//

***Sushrut Samhita Uttaratantra* 21/3**

In diseases of the ear, *Acharya Sushruta* described the general line of treatment and he emphasized the role of *ghee* and *Mamsarasa* as a prime treatment in diseases of the ear. The *Ghrita/ghee* is the best *Vatashamak* dietary item having abundant medicinal properties. The ear is a seat of *Vata Dosha.*[12] The diseases of the ear are predominantly of *Vata Dosha. Ghrita* can be consumed orally after a meal to alleviate *Vata* and give strength to the *Vyana* and *Udana Vayu. Udana Vayu* plays an important role in the normal functioning of the *Indriya* (sensory organ).

3. NUTRITION IN OTORHINOLARYNGOLOGY FROM CONVENTIONAL SCIENCE

Age-related hearing loss (ARHL) is a significant and pervasive health issue that affects people all over the world, making those who have it disabled and socially isolated. A disabling degree of hearing loss (HL) affects around 466 million individuals worldwide, with men being more affected than women.[13] The environmental factors with eating habits and genetic factors influence the appearance of the disease. The association between food intake and hearing loss is an area of interest as reported publications highlighted the perspectives as prolonged nutritional deficiencies, the influence of dietary elements in the prevention of hearing loss, and the effect of antioxidants in the prevention of hearing loss. Restricting the consumption of a diet rich in saturated fats and cholesterol, ultimately increasing the consumption of fruits, vegetables, an antioxidant-rich diet, and polyunsaturated fatty acid (omega-3), will lead to protection from hearing loss, especially in old age.[14] The quality of life is affected by hearing loss and nutritional factors are neglected, which causes a significant global burden. Reported publications highlighted the potential of nutritional therapy in the protection of progressive hearing loss.[15] Studies have shown that recurrent otitis media leads to hearing loss in children who lack important dietary supplements and have poor nutritional status.[16]

3.1 Meniere's Disease

Endolymphatic hydrops, which is Meniere's disease, is an internal ear disorder that features tinnitus, sensorineural hearing loss (SNHL), and vertigo. The prevention of attacks, symptomatic relief during attacks, and prevention of aftereffects that may be permanent are the strategies available for the treatment of Meniere's disease. Dietary modification is one of the important aspects considered in the treatment methods for Meniere's disease. A first line of treatment includes a gluten-free diet, a reduction in daily caffeine and alcohol intake, a restricted sodium intake, and different preparations of specially processed grains (*Kritanna Kalpana*), which are all examples of dietary modification in Meniere's disease.[17]

3.2 Dysphagia

About 8% population of the world is affected by dysphagia and it's a major health concern across the globe.[18] Dysphagia majorly affects the nutritional status and leads to malnutrition in longstanding conditions. A longer hospital stay, a higher risk of complications, and a higher death rate are experienced by dysphagia patients who are malnourished and lack access to the appropriate therapy.[19] Dysphagia and malnutrition are closely associated.[20] It has been found that 13.6% of people at risk for malnutrition also have dysphagia and that 39.2% of patients with dysphagia are at risk for malnutrition.[21] A scoping review recommended using the minimum nutritional assessment tool to develop beneficial effective nutritional interventions.[22]

3.3 Allergic Rhinitis

The estimated prevalence of allergic rhinitis (AR) ranges between 10% and 40% and causes a substantial economic burden.[23]

Rhinitis, also known as nasal inflammation or nasal congestion, is a common condition characterized by symptoms such as nasal congestion, runny nose, and sneezing. It can be caused by a variety of factors, including allergens, viral infections, and irritants. However, research has also shown that certain dietary factors can trigger or worsen rhinitis symptoms.[24]

Histamine, a substance present in numerous food items, has been associated with rhinitis as one of the dietary factors. Histamine is a natural substance produced by the body that plays a role in the immune response. However, some people have a condition known as histamine intolerance, which occurs when the body is unable to break down histamine properly. It can lead to a buildup of histamine in the body, resulting in symptoms such as nasal congestion, sneezing, and itching. Foods that are high in histamine include fermented foods, such as cheese, wine, and yogurt, as well as citrus fruits and vegetables, such as tomatoes, strawberries, and spinach.[25] In *Ayurveda,* similar conditions of histamine intolerance may be correlated with *Dustha Pratishyaya* (chronic rhinitis/sinusitis) and different *Mukharaoga*-like autoimmune conditions of the oral cavity.

Another dietary factor linked to rhinitis is sulfites, a group of food preservatives commonly used in processed foods and drinks. Sulfites can cause allergic reactions in some people, including rhinitis symptoms such as nasal congestion, sneezing, and itching. Foods that contain sulfites include dried fruits, wine, and processed meats.[26]

A study investigated the relationship between food allergy and allergic rhinitis in 100 patients aged between 10 and 60 years with allergic rhinitis. The study found that 63% of patients were sensitized to common food allergens, with the highest response rate found in patients aged 21–40 years, and females were more prone to mediate allergic rhinitis induced by food allergies compared to males. Rice, citrus fruits, black grams, and bananas were identified as major allergens for inducing allergic rhinitis symptoms.[27]

Another dietary factor that has been linked to rhinitis is cow's milk. Cow's milk is one of the most common food allergens among children and can cause symptoms such as nasal congestion, runny nose, and sneezing.[28] Milk and milk products particularly cause vitiation of *Kapha,* which is responsible for these symptoms.

In a nutshell, dietary factors, such as histamine, sulfites, sugar, and cow's milk, can trigger or worsen rhinitis symptoms. Therefore, it is important for individuals with rhinitis to be mindful of their diet and to avoid foods that may exacerbate their symptoms.

In rhinitis, therapeutic nutrition can play a role in both preventing and managing symptoms. One nonpharmacological intervention that may be helpful for individuals with rhinitis is the *Nidan Parivarjana* (avoidance of triggering diet). It involves removing certain foods from the diet, such as those high in histamine or sulfites, to observe symptomatic improvements, if any.

Another dietary intervention in rhinitis is the consumption of anti-inflammatory foods, including foods with high antioxidants, such as fruits and vegetables, as well as omega-3 fatty acids found in fish and nuts. Probiotics have been found to be beneficial in reducing the symptoms of rhinitis. Probiotics are live microorganisms that can provide health benefits when consumed in adequate amounts. Studies have shown that probiotics can help to reduce inflammation in the nasal passages and improve immune function, which can lead to a reduction in rhinitis symptoms.[29]

It's important to note that people with rhinitis should avoid foods and drinks that may worsen the symptoms such as alcohol and spicy and hot foods.

4. GENERAL THERAPEUTIC NUTRITION IN ORL

In the context of ENT (ear, nose, and throat) diseases, *Ayurveda* emphasizes the need for a balanced diet that is tailored to an individual's unique body constitution or "*Dosha*."

According to *Ayurveda*, ENT diseases are often caused by an imbalance of the *Dosha*, particularly the *Vata* and *Kapha Dosha*. *Vata* governs the movement of air and fluids in the body and is associated with dryness and cracking in the ears, nose, and throat. *Kapha* governs the mucous and fluid secretions in the body and is associated with congestion and excess mucous production in the ears, nose, and throat. To balance the *Dosha* and prevent ENT diseases, the dietary recommendation is as follows.

Warm and cooked: Warm and cooked food is easier to digest and nourishes the body better.

Low in fat and oil: Excessive consumption of fat and oil can aggravate *Kapha* and increase the risk of congestion and excess mucous production.

Rich in Vitamin C: Vitamin C–rich foods like *Amalaki*, guava, and kiwi can be consumed to boost immunity and prevent infections.

Herbs and spices like ginger, turmeric, and black pepper are also recommended to help prevent ENT diseases.

4.1 Dysphagia

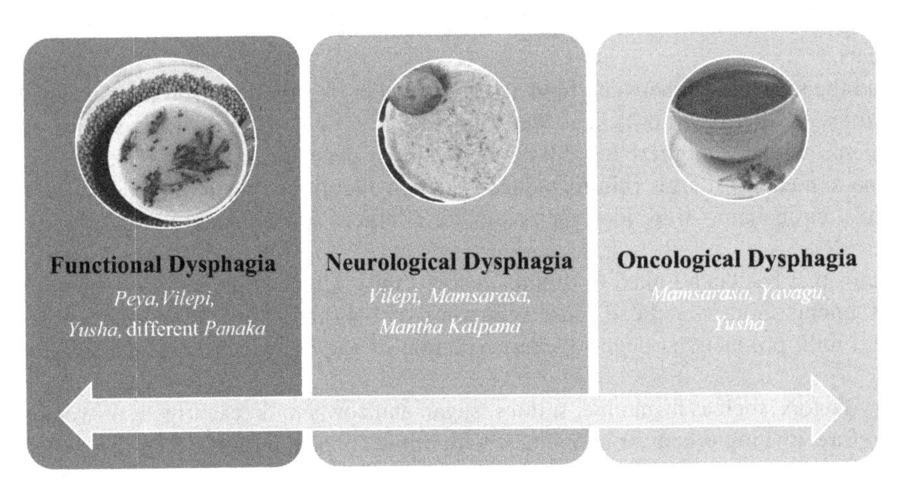

FIGURE 12.1 Dysphagia and diet recommendations as per types.*

* Image source https://main.ayush.gov.in/traditional-food-recipes-from-ayush-systems/

4.2 Hearing Loss and Other Ear Conditions

The therapeutic diet for ear diseases can include a variety of foods and herbs that support the balance of the body's three *Dosha – Vata, Pitta*, and *Kapha* – and improve overall health. The following are the therapeutic dietary items for ear disease management:

4.2.1 Ghee

Ghee, a type of clarified butter, is a mainstream treatment for systemic use in ear diseases (*Sushrut Samhita Uttaratantra 21/3*). It is considered to be an effective lubricant and moisturizer for the ear and may help to soothe and calm inflamed tissue.[30]

Sensorineural hearing loss (SNHL) is a common form of hearing impairment caused by damage to the inner ear (cochlea) or the auditory nerve. Despite the widespread occurrence of SNHL, the exact underlying mechanisms are still not well understood.

Ghee contains high levels of unsaturated fatty acids, including linoleic acid, which have been shown to have protective effects against oxidative stress, inflammation, and apoptosis.

4.2.2 Garlic (Allium sativum)

Garlic is a rich source of antioxidants and has anti-inflammatory properties, making it a beneficial ingredient for ear disease management.[31]

Garlic (*A. sativum*) has been studied for its potential role in reducing the risk of sensorineural hearing loss (SNHL). SNHL is the most common type of hearing loss and occurs due to damage to the inner ear (cochlea) or the nerve pathways that transmit sound signals to the brain. Studies have shown that garlic has antioxidant and anti-inflammatory properties, which can help to protect the inner ear and reduce oxidative stress, which is a contributing factor to SNHL.[32]

Garlic has been found to have a positive effect on blood flow to the inner ear.[33]

4.2.3 Ginger (Zingiber officinale)

Ginger has been used for centuries to treat ear-related problems. It has anti-inflammatory properties and is often recommended for reducing pain and swelling in the ear. Ginger can be consumed in the form of tea or added to meals for maximum benefit.[34]

Ginger has been identified as a potential therapeutic agent for the treatment of SNHL. Sensorineural hearing loss is characterized by damage to the inner ear (cochlea) or the auditory nerve and is a leading cause of hearing impairment in the world. Recent studies have shown that ginger contains anti-inflammatory, antioxidant, and neuroprotective properties[35] that may help protect against oxidative stress and inflammation in the inner ear, which are key factors in the development of sensorineural hearing loss. Ginger has also been shown to have a positive effect on the expression of antioxidant enzymes and the preservation of auditory function.

4.2.4 Til (Sesame Seeds)

Sesame seeds are believed to be beneficial for ear disease management due to their high content of calcium, magnesium, and healthy fatty acids. Sesame oil topically can be used as a lubricant and moisturizer for the ear and may help to reduce pain and inflammation.[36]

The role of sesame seeds in sensory-neural hearing loss has not been extensively studied and there is limited evidence available in the scientific literature. However, sesame seeds are a rich source of magnesium, which is known to play a role in maintaining normal hearing function. Magnesium is involved in the regulation of ion channels in the inner ear and is essential for proper auditory nerve function. Deficiencies in magnesium have been associated with an increased risk of age-related hearing loss and tinnitus (ringing in the ears).[37,38]

A study by Kim et al. (2020) investigated the effect of sesame seed supplementation on hearing function in rats with experimentally induced hearing loss. The results showed that supplementation with sesame seeds improved hearing thresholds and reduced oxidative stress in the cochlea, suggesting that sesame seeds may have a protective effect on hearing function.[39]

4.2.5 Turmeric (Curcuma longa)

Turmeric is a spice commonly used in traditional medicine for its various medicinal properties, including anti-inflammatory and antioxidant effects.[40]

In recent years, its role in treating sensorineural hearing loss has garnered attention. Several studies have demonstrated the protective effects of turmeric on the inner ear and auditory nerve. One study found that turmeric extract prevented oxidative stress–induced damage to cochlear hair cells in a mouse model.[41] Another study showed that turmeric extract reduced oxidative stress and inflammation in the cochlea of rats exposed to noise-induced hearing loss.[42]

Turmeric has also been shown to have neuroprotective effects on the auditory nerve. A study found that turmeric extract reduced oxidative stress and prevented nerve degeneration in a mouse model of cochlear nerve injury.[43]

4.2.6 Mamsarasa (Medicated Meat Soup)

Mamsarasa, or medicated meat soup in *Ayurveda*, is a traditional preparation that involves boiling meat from *Jangala* animal sources with herbal decoctions, juices, and spices to form a soup. The resulting preparation provides nourishment to the body and improves overall health.

Mamsarasa is considered to have properties such as improving digestion, strengthening the immune system, and providing energy. The specific herbs and spices used in the preparation are believed to have specific therapeutic effects, such as anti-inflammatory, antioxidant, and immunomodulatory properties.

However, there is limited scientific research available to support these statements.

Mamsarasa has been traditionally used as a therapeutic intervention for hearing loss. The concept of using meat-based remedies is based on the principles of *Samanya-Vishesh*,[44] which are believed to act as anabolic stimuli that can promote tissue regeneration and growth.

Hearing loss is described as a result of a *Vata* imbalance, leading to a decline in *Ojas* (vital fluid) and the accumulation of *Ama* (toxins) in the ear.[45] *Mamsarasa*, being a source of protein, provides the necessary nutrients for tissue repair and regeneration in the ear. It is helpful in removing the accumulation of *Ama* and promoting the proper functioning of *Vata*.[46]

However, the efficacy of *Mamsarasa* in the treatment of hearing loss has not been thoroughly studied using large-scale, well-designed clinical trials. Further research is required to establish its effectiveness in the treatment of hearing loss.

4.3 Allergic Rhinitis

AR is more or less correlated with *Pratishyay* with *Vata* Predominant features. *Pratishyay* has *Samavastha* (*Dosha* associated with *Aam*) and *Pakvavastha* (ripened stage of *Dosha*).[47] However, in *Samavastha*, *Deepan-Pachan* (stimulating digestive fire or appetizing action) diet is recommended, while in *Pakvavastha Ghritapana* is recommended.[48]

In *Samavastha* of *Pratishyay,* the following symptoms manifest as

- Heaviness in the head
- Loss of taste sensation
- Watery discharge from the nose
- Sneezing and many a time associated with fever.

The Dietary Recommendation in *Samavastha*

***Guda-Ardrak* (Jaggery-Ginger) medicated milk:** It is useful in *Samavastha* as jaggery alleviates the *Aam Dosha* in blood and purifies the blood. *Guda Ardrak Yog* works by reducing inflammation and improving nasal airflow, thus providing relief from these symptoms.

Sugarcane Juice With Ginger

Sugarcane has antioxidant properties[49] and ginger has anti-inflammatory properties that may alleviate the histamine-induced inflammation in the nasal mucosa and thus relieves symptoms. Ginger (wet form) is *Amapachak* and alleviates *Samavastha*.

Guda Shunthi Yog

Guda is a byproduct of sugarcane and has *Ushna* potency.[50] When it is combined with *Shunthi* (a dry form of ginger, *Z. officinale* R.) the combination becomes *Amapachak* and *Agnideepaka* (appetizer), which relives the symptomatology of allergic rhinitis.

Ajamamsa *(Goat Meat) Paste with* Trikatu Churna *(a Coarse Powder of* Shunthi, Marich, *and* Pippali*)*

Allergic rhinitis associated with complications can be prevented by the use of *Langhan* (therapeutic fasting) and then *Deepan-Pachana Ahar*. The goat meat soup prepared with *Trikatu*, *Dadima*, and rock salt is recommended for *Agnideepan* (appetizer).

Marichadi Yog: A combination of *Maricha* and *Guda* with curd is preferred in all seasons for rhinitis and associated symptoms.[51]

Shunthi Siddha Jala (medicated water with dry ginger powder): *Saptamushtik Yusha, Balmulak Yusha* (unripened radish soup), medicated boiled milk *Shunthi, Sattu, Erandmul, Jangal Mamsarasa,* and warm, light, and digestible food should be consumed. It is also recommended to drink medicated cow's milk with *Shunthi* (*Shunthisiddha Dugdha*).

घृतं क्षीरं यवाः शालिर्गोधूमा जांगला रसः ।

शीताम्लास्तिक्तशाकानि यूषा मूद्गाधिभिर्हिता : ।।

***Charak Chikitsasthana* 26/148**

Acharya Charaka described the following food items for the rhinitis[52]:

Ghrita (ghee), cow's milk, *Yava* (barley), *Raktashali* (rice), *Godhuma* (wheat), *Jangala Mamsarasa* (soup of animal meat from regions with dry forests and less rainfall), *Yusha* prepared from *Tikta shaka* (vegetables having bitter taste), and *Mudga* (preparations of *moonga* like *Mudga Yusha*).

The *Laghu* (light), *Amla* (sour), *Lavana* (salty), *Snigdha* (unctuous), and *Ushna* (warm) type of food is recommended for rhinitis patients.

4.4 Dietary Recommendations for Diseases of the Oral Cavity

Use of *Trina Dhanya* (group of grains produced by grass-like plants), *Yava* (barley), *Mudga* (green gram), *Kulattha* (horse gram), *Jangala Mamsarasa* (medicated meat soup), *Methika* (fenugreek), *Karvella* (bitter gourd), *Patola* (snake gourd), *Tambula sevana* (chewing of betel leaf along with nutmeg, clove, areca nut, etc.), *Karpurasiddha Jala* (camphor medicated water), lukewarm water, lentil soup mixed with oil and *ghee* and tampered with *Jeeraka*, and vegetables having bitter taste should be used on priority.[53]

5. NUTRITION IN OTORHINOLARYNGOLOGY AS PER *DWADASHA AAHAR VARGA*

TABLE 12.1 *Dwadasha Ahara Varga* (12 Classes of Food) and *Siddhanna* (Food Preparations) in the Context of Ophthalmic Diseases and Their Preparations in Brief According to the Classical Text

S. NO	*DIETARY ITEM*	*ENGLISH NAME*	*KRITANNA KALPANA*[54]	*PROPERTIES*[55]	*INDICATIONS*
	***Shali Dhanya* (should be used roasted)**				
1.	*Rakta Shali*	Brown rice	*Peya* (thin gruel), *Yavagu* (thick gruel), *Vilepi* (thick gruel), *Manda* (gruel water), *Odana* (rice), *Krushara* (*Khichadi*), *Shalisaktu*. It should be used after roasting, before making food items	*Santarpana* (rejuvenates and nourishes body tissue), increases *Kapha-Pitta*	Postsurgical diet, Rhinitis, Sinusitis, DNS, Hearing loss, Epistaxis
	***Shooka Dhanya* (corn with bristles) (should be used roasted)**				
2.	*Yava*	Barley	*Yava Saktu*	Alleviates *Kapha-Pitta*	*Peenasa* (chronic rhinitis)
3.	*Yavanaal* (*Jwari*)	Sorghum	Flatbreads, *Yavagu*	It alleviates all three *Dosha* and is light in digestion	*Peenasa* (chronic rhinitis), *Aruchi* (loss of taste sensation), postsurgical diet in ENT procedures
4.	*Mahayaavanaal*	Corn	Corn flakes made the traditional way, *Makki Rotika, Yusha* prepared from *Makki*	*Santarpak*	
	***Shimbi Dhanya* (should be used roasted)**				
5.	*Mudga*	Green gram	*Mudga Yusha* (soup of green gram), *Saptamushtik Yusha, Krushara,*	*Santarpaka*	Epistaxis, hearing loss, Aphthous ulcer
6.	*Masha*	Black grams	*Alikamaccha* (recipe prepared with black gram pasted betel leaves with steamed and fried process, *Masha Vetika, Masha Soup*	*Brihan*	Facial palsies

TABLE 12.1 *(Continued)* *Dwadasha Ahara Varga* (12 Classes of Food) and *Siddhanna* (Food Preparations) in the Context of Ophthalmic Diseases and Their Preparations in Brief According to the Classical Text

S. NO	*DIETARY ITEM*	*ENGLISH NAME*	*KRITANNA KALPANA*[54]	*PROPERTIES*[55]	*INDICATIONS*
7.	*Chanaka*	Horse gram	*Chanaka Rotika*, *Chanaka Yava Saktu*, *Vedhanika* (funnel cake), *Chanaka Yusha*	*Apatarpan*, alleviates *Kapha-Pitta*	Used as a coadminister with medicines in running nose cases having watery discharge from nose
8.	*Adhaki*	Red gram	*Adhaki Yusha*	Alleviates all three *Dosha*, improves digestive fire	Rhinitis, Sinusitis
	Truna Dhanya **(should be used roasted)**				
9.	*Ragi*	Finger millet	*Ragi Yusha*, *Yavagu*, flatbreads made of *Ragi*	Alleviates *Rakta* and *Pitta*, *Balya*	Epistaxis, fractures of nasal bone
	Sattu				
10.	*Yava Sattu*	Roasted barley flour	*Yava Sattu*	*Agnideepan*, light in digestion, alleviates *Kapha-Pitta*	Rhinitis, Sinusitis, Otorrhea, Epistaxis
11.	*Chanak Yava Sattu*	Roasted horse gram + barley flour	*Chanak Yava Sattu* with *ghee* and sugar	*Agnideepak*	Sinusitis
	Soup				
12.	*Kulattha Yusha*	Soup made from *Dolichos biflorus* L.	*Kulattha Yusha*	Alleviates *Kapha*	Rhinitis, Sinusitis
	Shaka Varga **(group of vegetables)**				
	Patra Shaka **(leafy vegetables)**				
13.	*Baalmulak*	Raw daikon radish	Soup made from daikon radish, *Saptamushtik Yusha*	Alleviates *Vata* and *Kapha*	Rhinitis
14.	*Rason*	Garlic	Used in tempering or *Chaunk* while making food especially vegetables, *Rason Siddha Kshirpaka* (medicated garlic milk)	Alleviates *Vata*	Beneficial in *Vata* predominant ear diseases, SNHL, Tinnitus
15.	*Punarnava*	*Boerhavia diffusa*	Food preparation		
16.	*Jeevanti Shaka*	*Leptadenia reticulate*	Steamed leaf boiled with buttermilk, tempered with asafoetida, oil, and buttermilk	Alleviates all three *Dosha*, *Rasayana*,	All degenerative conditions of ear

(*Continued*)

TABLE 12.1 *(Continued)* *Dwadasha Ahara Varga* (12 Classes of Food) and *Siddhanna* (Food Preparations) in the Context of Ophthalmic Diseases and Their Preparations in Brief According to the Classical Text

S. NO	*DIETARY ITEM*	*ENGLISH NAME*	*KRITANNA KALPANA*[54]	*PROPERTIES*[55]	*INDICATIONS*
17.	***Phala Shaka* (fruit vegetables)**				
18.	*Shigru*	Drumstick	*Shigru Yusha*	Alleviates *Kapha*	Diseases of oral cavity, Rhinitis, Sinusitis
19.	*Kharjur*	Dates	*Kharjuradi Mantha*	Alleviates *Vata*, *Pitta* and *Rakta*	Epistaxis
20.	*Draksha*	Black grapes	*Manuka* infusion	Alleviates *Pitta*	Epistaxis
21.	*Kushmanda*	*Benincasa hispida*	*Petha* (winter melon candy)	Alleviates *Vata* and *Pitta*	Headaches, epistaxis, SNHL
22.	*Udumbar Phala*	Cluster fig, *Ficus racemosa*	Medicated water with fruits of cluster fig	Alleviates *Pitta* and thereby purifies blood	Epistaxis
23.	*Badarbeeja*	Indian jujube seed, *Ziziphus mauritiana*	Powdered form	Alleviates *Vata* and *Kapha*	Rhinitis
	***Pushpa Shaka* (leafy vegetables)**				
24.	*Shatavha*	*Anethum graveolens*, Dill seed	*Yusha*	Alleviates *Kapha*	Rhinitis, sinusitis
	***Lavan Varga* (group of salts)**				
25.	*Saindhav*	Rock salt	Used in all food preparations for taste	Alleviates all three *Dosha*	Loss of taste sensation, diseases of the oral cavity, Rhinitis
	***Dugdha Varga* (group of milk) and its product**				
26.	*Godugdha*	Cow's milk	Cow's milk + *ghee*	*Rasayana*, alleviates *Vata Pitta*	*Rasayana*, SNHL, Epistaxis, degenerative conditions in ENT
27.	*Navaneet*	Butter	Can be used for tempering during the preparation of vegetables	Alleviates *Vata*	Degenerative conditions in ENT like SNHL, palsies
	***Ghrita Varga* (group of ghee)**				
28.	*Goghrita*	Cow's *ghee*	Used in food preparations, can be used for tempering, Cow's milk with ghee	*Rasayana*	Sensory neural Hearing Loss, Allergic Rhinitis
29.	*Mahish Ghrita*	Buffalo's ghee	Used in food preparations, can be used for tempering or chaunk		Hearing Loss

TABLE 12.1 *(Continued)* *Dwadasha Ahara Varga* (12 Classes of Food) and *Siddhanna* (Food Preparations) in the Context of Ophthalmic Diseases and Their Preparations in Brief According to the Classical Text

S. NO	DIETARY ITEM	ENGLISH NAME	KRITANNA KALPANA[54]	PROPERTIES[55]	INDICATIONS
30.	*Purana Ghrita*/ aged *ghee*	Old preserved ghee for 10, 20, or 100 years	Used for internal use	*Vranashodhak* (wound healing property), alleviates *Vata*	Rhinitis, Epistaxis, SNHL
	Taila Varga **(group of oils)**				
31.	*Rason Taila*	Oil medicated from garlic	External application	Alleviates *Vata*	Diseases of ear, hearing loss
	Mamsa Varga **(group of meat)**				
32.	*Aja Mamsa*	Meat of goat	Soup, different food preparations, *Rasodana* (rice with meat soup), *Trikatusiddha mamsa*	*Rasayana*, *Prasadak* (improves vision), *Balya* (strengthens the tissue)	Rhinitis, SNHL, Tinnitus
	Madhu Varga **(group of honey)**				
33.	*Madhu*	Honey	Used as single drug, can be mixed or spread over *Yava Rotica*, wheat breads, consumed with hot water	Alleviates *Kapha*	Rhinitis, Aphthous Ulcer, traumatic mouth ulcer

6. CLINICAL SIGNIFICANCE OF THERAPEUTIC NUTRITION IN ORL

Ayurveda, the ancient Indian system of medicine, has always recognized the importance of therapeutic nutrition in the management of various diseases. The practice of therapeutic nutrition in *Ayurveda* involves the use of specific foods and herbs to support the body's natural healing process and restore balance. Therapeutic nutrition is particularly significant in the treatment of chronic conditions such as SNHL, sinusitis, allergic rhinitis, and chronic ear infections and dysphagia, which require long-term management.

Ayurvedic clinical practice in ENT emphasizes the use of a diet that is tailored to the individual's specific *Dosha* (constitution) to promote healing and prevent further progression of the disease. For example, in *Kapha*-dominant individuals who are more prone to respiratory congestion and allergies/sinusitis, a diet that is light and warm is recommended. This diet may include spices such as ginger, black pepper, and cinnamon, as well as foods that are low in fat and sugar. Different *Yusha* tampered with asafetida, rock salt, *Hingvashtaka Churna*,[56] *Lavanabhaskar Churna*,[57] *Musta*, *Shunthi*, or other spices herbs can be used. The concept of tampering with food articles with ghee/oil and *Agnideepak* medicinal powders is useful as per *Dosha-Dushya,* which gives dual benefits of maintaining a balance of body humor and alleviation of disease process.

In *Pitta*-dominant individuals, who are more prone to inflammatory conditions such as allergies and ear infections, a diet that is *Sheet Veerya* (cool in potency) and anti-inflammatory is recommended. This may include foods such as *Kushmanda*, sugarcane, coconut water, *Kharjura Mantha*, and coriander, as well as herbs such as turmeric and *Shunthi*.

In *Vata*-predominant diseases, the main aim should be to alleviate *Vata Dosha* with *Snehapana* and *Mamsarasa* prepared with abundant ghee, tampered with *Jeeraka*, is useful. *Rason* is a very useful medicinal spice in *Vata*-predominant ear diseases. *Rason Kshirpaka* (medicated milk with garlic), used in SNHL and tinnitus, give good results.

Therapeutic nutrition is also used in *Ayurveda* to support the body's natural detoxification process, which can help to reduce inflammation and improve overall health. This is particularly relevant in the treatment of chronic ENT conditions, which are often associated with high levels of inflammatory changes in the body. Therapeutic fasting along with the use of *Saptamushtik Yusha* is used in clinical practice to alleviate *Ama* and thus relieve inflammatory changes in the body.[58] This approach is useful in allergic rhinitis, sinusitis associated with fever, acute suppurative otitis media, and tonsillitis.

7. SCOPE FOR FURTHER RESEARCH

The field of therapeutic nutrition in Ayurveda has gained significant attention in recent years for its potential to manage and prevent diseases. However, there is a limited understanding of the role of Ayurvedic nutrition therapy in the management of ear, nose, and throat (ENT) diseases.

The emerging scope of therapeutic nutrition in Ayurveda for ENT can lead to involve various research activities, including:

- Investigating the nutritional properties and bioactive compounds present in traditional Ayurvedic preparations used for therapeutic purposes in ENT.
- Future research in this area could focus on exploring the potential mechanisms underlying the effects of Ayurvedic nutrition therapy with various biochemical markers, inflammatory mediators, and immune function in ENT diseases.
- Studies could investigate the cultural and socioeconomic factors that impact the uptake and adherence to Ayurvedic nutrition therapy for ENT diseases.

REFERENCES

1 Weir N. Otorhinolaryngology. Postgrad Med J. 2000 Feb;76(892):65–9. doi: 10.1136/pmj.76.892.65. PMID: 10644381; PMCID: PMC1741503.

2 GBD 2019 Hearing Loss Collaborators. Hearing loss prevalence and years lived with disability, 1990-2019: findings from the Global Burden of Disease Study 2019. Lancet. 2021 Mar 13;397(10278):996–1009. doi: 10.1016/S0140-6736(21)00516-X. PMID: 33714390; PMCID: PMC7960691.

3 Bousquet J, Anto JM, Bachert C, Baiardini I, Bosnic-Anticevich S, Walter Canonica G, Melén E, Palomares O, Scadding GK, Togias A, Toppila-Salmi S. Allergic rhinitis. Nat Rev Dis Primers. 2020 Dec 3;6(1):95. doi: 10.1038/s41572-020-00227-0. PMID: 33273461.

4 Saunders JE, Rankin Z, Noonan KY. Otolaryngology and the global burden of disease. Otolaryngol Clin North Am. 2018 Jun;51(3):515–534. doi: 10.1016/j.otc.2018.01.016. PMID: 29773124.

5 Talwar B, Donnelly R, Skelly R, Donaldson M. Nutritional management in head and neck cancer: United Kingdom National Multidisciplinary Guidelines. J Laryngol Otol. 2016 May;130(S2):S32–S40. doi: 10.1017/S0022215116000402. PMID: 27841109; PMCID: PMC4873913.
6 Collins PF, Yang IA, Chang YC, Vaughan A. Nutritional support in chronic obstructive pulmonary disease (COPD): an evidence update. J Thorac Dis. 2019 Oct;11(Suppl 17):S2230–S2237. doi: 10.21037/jtd.2019.10.41. PMID: 31737350; PMCID: PMC6831917.
7 Huang Z, Liu Y, Qi G, Brand D, Zheng SG. Role of vitamin A in the immune system. J Clin Med. 2018 Sep 6;7(9):258. doi: 10.3390/jcm7090258. PMID: 30200565; PMCID: PMC6162863.
8 Lin PH, Sermersheim M, Li H, Lee PHU, Steinberg SM, Ma J. Zinc in wound healing modulation. Nutrients. 2017 Dec 24;10(1):16. doi: 10.3390/nu10010016. PMID: 29295546; PMCID: PMC5793244.
9 Schleimer RP, Kern RC. Nutrition and sinusitis. Otolaryngol Clin North Am. 2005;38(2):463–470.
10 Kalliomäki M, Salminen S, Poussa T, Arvilommi H, Isolauri E. Probiotics and prevention of atopic disease: 4-year follow-up of a randomised placebo-controlled trial. Lancet. 2003 May 31;361(9372):1869–71. doi: 10.1016/S0140-6736(03)13490-3. PMID: 12788576.
11 Science M, Johnstone J, Roth DE, Guyatt G, Loeb M. Zinc for the treatment of the common cold: a systematic review and meta-analysis of randomized controlled trials. CMAJ. 2012 Jul 10;184(10):E551–61. doi: 10.1503/cmaj.111990. Epub 2012 May 7. PMID: 22566526; PMCID: PMC3394849.
12 Gupta A. Ashtanga Hridaya of Vaghbhata; *Part-I, Vidyotini Hindi Commentary Sutrasthana.* Reprint ed. Ch. 12, verse 1–4. Varanasi: Chaukhamba Sanskrit Sansthan; 2005, p. 90.
13 Marlenga B, Berg RL, Linneman JG, Wood DJ, Kirkhorn SR, Pickett W. Determinants of early-stage hearing loss among a cohort of young workers with 16-year follow-up. Occup Environ Med. 2012 Jul;69(7):479–84. doi: 10.1136/oemed-2011-100464. Epub 2012 Mar 23. PMID: 22447644.
14 Rodrigo L, Campos-Asensio C, Rodríguez MÁ, Crespo I, Olmedillas H. Role of nutrition in the development and prevention of age-related hearing loss: A scoping review. J Formos Med Assoc. 2021 Jan;120(1 Pt 1):107–120. doi: 10.1016/j.jfma.2020.05.011. Epub 2020 May 28. PMID: 32473863.
15 Puga AM, Pajares MA, Varela-Moreiras G, Partearroyo T. Interplay between Nutrition and Hearing Loss: State of Art. Nutrients. 2018 Dec 24;11(1):35. doi: 10.3390/nu11010035. PMID: 30586880; PMCID: PMC6356655.
16 Jung SY, Kim SH, Yeo SG. Association of Nutritional Factors with Hearing Loss. Nutrients. 2019 Feb 1;11(2):307. doi: 10.3390/nu11020307. PMID: 30717210; PMCID: PMC6412883.
17 Oğuz E, Cebeci A, Geçici CR. The relationship between nutrition and Ménière's disease. Auris Nasus Larynx. 2021 Oct;48(5):803–808. doi: 10.1016/j.anl.2021.03.006. Epub 2021 Mar 26. PMID: 33773852.
18 Cichero JA, Lam P, Steele CM, Hanson B, Chen J, Dantas RO, Duivestein J, Kayashita J, Lecko C, Murray J, Pillay M, Riquelme L, Stanschus S. Development of International Terminology and Definitions for Texture-Modified Foods and Thickened Fluids Used in Dysphagia Management: The IDDSI Framework. Dysphagia. 2017 Apr;32(2):293–314. doi: 10.1007/s00455-016-9758-y. Epub 2016 Dec 2. PMID: 27913916; PMCID: PMC5380696.
19 Thomas MN, Kufeldt J, Kisser U, Hornung HM, Hoffmann J, Andraschko M, Werner J, Rittler P. Effects of malnutrition on complication rates, length of hospital stay, and revenue in elective surgical patients in the G-DRG-system. Nutrition. 2016 Feb;32(2):249–54. doi: 10.1016/j.nut.2015.08.021. Epub 2015 Sep 25. PMID: 26688128.
20 Gallegos C, Brito-de la Fuente E, Clavé P, Costa A, Assegehegn G. Nutritional Aspects of Dysphagia Management. Adv Food Nutr Res. 2017;81:271–318. doi: 10.1016/bs.afnr.2016.11.008. Epub 2016 Dec 23. PMID: 28317607.
21 Blanař V, Hödl M, Lohrmann C, Amir Y, Eglseer D. Dysphagia and factors associated with malnutrition risk: A 5-year multicentre study. J Adv Nurs. 2019 Dec;75(12):3566–3576. doi: 10.1111/jan.14188. Epub 2019 Sep 10. PMID: 31452231.
22 Ueshima J, Momosaki R, Shimizu A, Motokawa K, Sonoi M, Shirai Y, Uno C, Kokura Y, Shimizu M, Nishiyama A, Moriyama D, Yamamoto K, Sakai K. Nutritional Assessment in Adult Patients with Dysphagia: A Scoping Review. Nutrients. 2021 Feb 27;13(3):778. doi: 10.3390/nu13030778. PMID: 33673581; PMCID: PMC7997289.
23 Bernstein DI, Schwartz G, Bernstein JA. Allergic Rhinitis: Mechanisms and Treatment. Immunol Allergy Clin North Am. 2016 May;36(2):261–78. doi: 10.1016/j.iac.2015.12.004. Epub 2016 Mar 10. PMID: 27083101.
24 Wheatley LM, Togias A. Clinical practice. Allergic rhinitis. N Engl J Med. 2015 Jan 29;372(5):456–63. doi: 10.1056/NEJMcp1412282. PMID: 25629743; PMCID: PMC4324099.

25 Maintz L, Novak N. Histamine and histamine intolerance. Am J Clin Nutr. 2007 May;85(5):1185–96. doi: 10.1093/ajcn/85.5.1185. PMID: 17490952] [Shulpekova YO, Nechaev VM, Popova IR, Deeva TA, Kopylov AT, Malsagova KA, Kaysheva AL, Ivashkin VT. Food Intolerance: The Role of Histamine. Nutrients. 2021 Sep 15;13(9):3207. doi: 10.3390/nu13093207. PMID: 34579083; PMCID: PMC8469513

26 Vally H, Misso NL, Madan V. Clinical effects of sulphite additives. Clin Exp Allergy. 2009 Nov;39(11):1643–51. doi: 10.1111/j.1365-2222.2009.03362.x. Epub 2009 Sep 22. PMID: 19775253

27 Al-Rabia MW. Food-induced immunoglobulin E-mediated allergic rhinitis. J Microsc Ultrastruct. 2016 Apr-Jun;4(2):69–75. doi: 10.1016/j.jmau.2015.11.004. Epub 2015 Dec 14. PMID: 30023212; PMCID: PMC6014210

28 Caffarelli C, Baldi F, Bendandi B, Calzone L, Marani M, Pasquinelli P; EWGPAG. Cow's milk protein allergy in children: a practical guide. Ital J Pediatr. 2010 Jan 15;36:5. doi: 10.1186/1824-7288-36-5. PMID: 20205781; PMCID: PMC2823764

29 Zajac AE, Adams AS, Turner JH. A systematic review and meta-analysis of probiotics for the treatment of allergic rhinitis. Int Forum Allergy Rhinol. 2015 Jun;5(6):524–32. doi: 10.1002/alr.21492. Epub 2015 Apr 20. PMID: 25899251; PMCID: PMC4725706

30 Sindhuja S, Prakruthi M, Manasa R, Naik R S, Shivananjappa M. Health benefits of ghee (clarified butter) – A review from ayurvedic perspective. *IP J Nutr Metab Health Sci* 2020;3(3):64–72

31 Bayan L, Koulivand PH, Gorji A. Garlic: a review of potential therapeutic effects. Avicenna J Phytomed. 2014 Jan;4(1):1–14. PMID: 25050296; PMCID: PMC4103721

32 Wu X, Li X, Song Y, Li H, Bai X, Liu W, Han Y, Xu L, Li J, Zhang D, Wang H, Fan Z. Allicin protects auditory hair cells and spiral ganglion neurons from cisplatin – Induced apoptosis. Neuropharmacology. 2017 Apr;116:429–440. doi: 10.1016/j.neuropharm.2017.01.001. Epub 2017 Jan 3. PMID: 28062185

33 Kim SY, Lee JH, Kim HE, Kim JH (2018). "The effect of garlic supplementation on blood flow and hearing function in rats with noise-induced hearing loss". Journal of Otolaryngology. 47(3):121–127

34 Mashhadi NS, Ghiasvand R, Askari G, Hariri M, Darvishi L, Mofid MR. Anti-oxidative and anti-inflammatory effects of ginger in health and physical activity: review of current evidence. Int J Prev Med. 2013 Apr;4(Suppl 1):S36–42. PMID: 23717767; PMCID: PMC3665023

35 Arcusa R, Villaño D, Marhuenda J, Cano M, Cerdà B, Zafrilla P. Potential Role of Ginger (*Zingiber officinale* Roscoe) in the Prevention of Neurodegenerative Diseases. Front Nutr. 2022 Mar 18;9:809621. doi: 10.3389/fnut.2022.809621. PMID: 35369082; PMCID: PMC8971783

36 Narasimhulu CA, Selvarajan K, Litvinov D, Parthasarathy S. Anti-atherosclerotic and anti-inflammatory actions of sesame oil. J Med Food. 2015 Jan;18(1):11–20. doi: 10.1089/jmf.2014.0138. PMID: 25562618; PMCID: PMC4281857

37 Cevette MJ, Vormann J, Franz K. Magnesium and hearing. J Am Acad Audiol. 2003 May-Jun;14(4):202-12. PMID: 12940704.

38 Dawes P, Cruickshanks KJ, Marsden A, Moore DR, Munro KJ. Relationship Between Diet, Tinnitus, and Hearing Difficulties. Ear Hear. 2020 Mar/Apr;41(2):289–299. doi: 10.1097/AUD.0000000000000765. PMID: 31356390; PMCID: PMC7664714

39 Kim YH, Kim EY, Rodriguez I, Nam YH, Jeong SY, Hong BN, Choung SY, Kang TH. *Sesamum indicum* L. Oil and Sesamin Induce Auditory-Protective Effects Through Changes in Hearing Loss-Related Gene Expression. J Med Food. 2020 May;23(5):491–498. doi: 10.1089/jmf.2019.4542. Epub 2020 Mar 18. PMID: 32186941

40 Hewlings SJ, Kalman DS. Curcumin: A Review of Its Effects on Human Health. Foods. 2017 Oct 22;6(10):92. doi: 10.3390/foods6100092. PMID: 29065496; PMCID: PMC5664031

41 Soyalıç H, Gevrek F, Karaman S. Curcumin protects against acoustic trauma in the rat cochlea. Int J Pediatr Otorhinolaryngol. 2017 Aug;99:100–106. doi: 10.1016/j.ijporl.2017.05.029. Epub 2017 Jun 5. PMID: 28688549

42 İskender Emekli Z, Şentürk F, Bahadir O. Protective Effects of Curcumin and N-Acetyl Cysteine Against Noise-Induced Sensorineural Hearing Loss: An Experimental Study. Indian J Otolaryngol Head Neck Surg. 2022 Aug;74(Suppl 1):467–471. doi: 10.1007/s12070-020-02269-y. Epub 2020 Nov 11. PMID: 36032833; PMCID: PMC9411444

43 Haryuna TS, Riawan W, Nasution A, Ma'at S, Harahap J, Adriztina I. Curcumin Reduces the Noise-Exposed Cochlear Fibroblasts Apoptosis. Int Arch Otorhinolaryngol. 2016 Oct;20(4):370–376. doi: 10.1055/s-0036-1579742. Epub 2016 Mar 4. PMID: 27746842; PMCID: PMC5063744

44 Acharya JT. *Charaka Samhita* by *Agnivesha* with *Ayurved Dipika commentary* of *Chakrapanidatta. Sutrasthan Dirghanjivitiya Adhyay.* 1/28, 44. 3rd ed. Bombay: Nirnaysagar Press; 1941. p. 7,9
45 Srikantha Murthy KR. Sushruta Samhita, Vol. 1. Varanasi, India: Chaukhamba Sanskrit Pratishthan; 2002
46 Caldocott T. Ayurveda: The Divine Science of Life. Mosby Elsevier; 2006
47 Kunte AM, Paradkar H. *Ashtang Hridaya* by *Vagbhata* with *Ayurved Dipika commentries* of *Sarvangasundar* by *Arundatta* and *Ayurveda Rasayana* by *Hemadri. Uttaratantra Nasarogavidnyaniya Adhyay.* 19/13, 6th ed. Bombay: Nirnaysagar Press; 1939. p. 842
48 Acharya JT. *Sushrut Samhita* of *Sushruta* with *Nibandha Sangraha commentary* of *Shree Dalhanacharya. Uttaratantra Pratishyaya Pratishedha Adhyay.* 24/18–19, Revised 2nd ed. Bombay: Nirnaysaar Press; 1941. p. 594
49 Saska M, Chou CC. Antioxidant properties of sugarcane extracts. In Proceedings of First Biannual World Conference on Recent Developments in Sugar Technologies, Delray Beach, FL USA 2002 May 16.
50 BN AK, Sujatha K, Renjal PU. Pharmaceutical and Therapeutical Utility of Ikshu Varga Dravya: A Review. Journal of Ayurveda and Integrated Medical Sciences. 2016 Jun 30;1(01):73–7.
51 Yoga Ratnakara *Nasarogadhikar*, Yogaratnakara – Uttarardha edited with the commentary *Ratnagarbha* in by Kaviraj Vidyadhar Vidyalankar. Edition reprint. Motilal Banarasidas, Lahore; 1931. P 1237
52 Acharya JT. *Charaka Samhita* by *Agnivesha* with *Ayurved Dipika commentary* of *Chakrapanidatta. Chikitsasthan Trimarmiyachikitsitam Adhyay.* 26/148, 3rd ed. Bombay: Nirnaysagar Press; 1941. p. 607
53 Yoga Ratnakara *Mukharogadhikar*, Yogaratnakara – Uttarardha edited with the commentary *Ratnagarbha* in by Kaviraj Vidyadhar Vidyalankar. Edition reprint. Motilal Banarasidas, Lahore; 1931. P 1162
54 Pandey G. Kshemakutuhalam. Varanasi: Chaukhamba Krishnadas Academy; 2014.ISBN: 978–81-218-0350-2
55 Balkrishna A. Bhojankutuhal. Haridwar, Uttarakhanda: Divya Prakshan, Patanjali Yogapeeth; 2013. ISBN: 81-89235-90-7
56 Kunte AM, Paradkar H. *Ashtang Hridaya* by *Vagbhata* with *Ayurved Dipika commentries* of *Sarvangasundar* by *Arundatta* and *Ayurveda Rasayana* by *Hemadri. Chikitsasthana Gulmachikitsa Adhyay.* 14/35, 6th ed. Bombay: Nirnaysagar Press; 1939. p. 687
57 Tripathi B, *Sharangdhar Samhita, Dipika* Hindi Vyakhya. Varanasi: Chaukhamba Surbharati Prakashan; 2004. *Madhyama Khanda* 6/138–144
58 Momin, J., & E, G. V. (2020). Efficacy of *Langhana* with *Saptamushtika Yusha* in the Treatment of *Aama* in *Aamavata* with special reference to Rheumatoid Arthritis. *International Journal of Ayurvedic Medicine, 11*(1), 108–112. https://doi.org/10.47552/ijam.v11i1.1354

Therapeutic Nutrition in Ayurveda for Urology

Narayan Gangadhar Shahane and
Jui Narayan Shahane

INTRODUCTION

In Ayurveda the treatment has fourfold approaches: *Ahara Chikitsa* (therapeutic nutrition), *Vihaara Chikitsa* (lifestyle changes), *Shamana Chikitsa* (conservative/medicinal management), and *Shodhana Chikitsa* (*Panchakarma* treatments)[1] (*Charak Sutrasthana* 16/20, p. 97).

Ahara Chikitsa – Therapeutic nutrition – being the first and important part, plays a key role in Ayurvedic treatment.

It is obvious and frequently experienced that medicines alone are not able to restore health if they are not supplemented by therapeutic nutrition. Therapeutic nutrition is described in almost every *Adhyay* (chapter) of the *Samhita* – ancient texts[1] (*Charak Chikitsa sthana* 3/188, p. 415)

Therapeutic nutrition in urology is underestimated. Awareness regarding this concept in the masses is a necessity of the time. Proper guidelines are essential in this regard. A detailed description is found regarding the therapeutic nutrition in relation to different urological diseases in *Sushruta Samhita*[2] (*Sushruta Chikitsa Sthana* 7/35, p. 437). Description guides the patients suffering from urological diseases. It is likely that all urinary disorders have a close relationship with the food consumed.

The urinary system is supposed to be one of the most important excretory systems. It eliminates the wastes in urine. A small number of solids are also present in the dissolved form in the urine. Formation and elimination of the urine contributes a major portion of the daily output of fluids to maintain the fluid and electrolyte balance of the body.[3]

Recurrent urinary tract infection (UTI), urinary calculus, urethral stricture, interstitial cystitis (painful bladder syndrome), benign enlargement of the prostate, overactive bladder, hypotonic bladder, and neurogenic bladder are significant diseases of the urinary system with remarkable global impact on individual health.

Maharshi Sushruta, the father of surgery, elaborates anatomy and physiology of the urinary system in detail in *Sushruta Samhita*[2] (*Sushruta Nidana Sthana* 3/18–24, pp. 279–280)

Stools and urine are metabolic wastes of food. The quality of food or nutrition directly affects the nature of urine.

विन्मूत्रमाहारमला[2]:

Sushruta Sutrasthana 46/528, p. 253

किट्टमन्नस्यविन्मूत्रम[1]

Charak Chikitsa Sthana 15/18, p. 515

DOI: 10.1201/9781003345541-16

NUTRITION IN UROLOGY FROM CLASSICAL TEXTS

Ahara Varga (classes of food):

The food is classified into classes – *Ahara Varga*.

In urology, water has significance in terms of therapeutic nutrition. According to *Maharshi Sushruta* hot water has more therapeutic nutritional values than cold water.

Ushnodaka – Hot Water

कफमेदो अनिलामघ्नमदीपनम बस्तिशोधनम ।। 39 ।।
श्वासकास ज्वर हरम पथ्यमुष्णोदकमसदा ।। 40 ।। [2]

Sushruta Sutrasthana 45/39–40, p. 200

Hot water balances deranged *Kapha* and *Vata dosha* (two governing forces of the three), and it reduces *Meda* (fat) and *Ama* (an endotoxin generated due to weak biometabolic energy). It is *Deepanam* (strengthens the biometabolic energy) and *Bastishodhanam* (cleanses the urinary bladder). It is *Shwaasakaasajwaraharam* (very beneficial for asthma, cough and fever). Drinking hot water has significant health benefits.

Consumption of food without proper appetite results in *Ama* – an endotoxin generated due to weak *Agni* (biometabolic energy)[4] (*Ashtangahrudaya Sutrasthana* 13/25, p. 216). Urine is the metabolic waste of food. *Ama* may alter the composition of the urine and may result in the worsening of urinary symptoms.

TABLE 13.1 Classification of Food According to their Therapeutic Nutritional Properties

THERAPEUTIC NUTRITIVE PROPERTIES	FOOD (AHARA DRAVYA) S. NO.	AHARA *DRAVYA* (FOOD)	REFERENCE
Bastishodhanam/ Bastivishodhanam – Cleanses urinary bladder	1	*Ushnodaka* – Hot Water[2]	*Sushruta Sutrasthana* 45/39–40, p. 200
	2	*Narikelodaka* – Tender coconut water (*Cocos nucifera*)[2,4]	*Sushruta Sutrasthana* 45/43–44, p. 200 & *Ashtangahruday Sutrasthana* 5/19, p. 66
	3	*Godugdha* – Cow milk[4]	*Ashtangahruday Sutrasthana* 5/22–23, p. 68
	4	*Bala Kushmaanda* – Tender ash gourd (*Benincasa hispida*)[2]	*Sushruta Sutrasthana* 46/213, p. 230
	5	*Gokshoora* – Land-Caltrops/Puncture Vine (*Tribulus terrestris*)[6]	*Bhavprakash Nighantu, Guduchyadi Varga* 45–46, p. 292
	6	*Kushmanda* – Ash gourd (*Benincasa hispida*)[1,6,7]	*Charak Sutrasthana* 27/113, p. 159 & *Raaj Nighantu, Moolakadi varga*, 161, p. 218 & *Bhavarakash Nighantu, shaka varga*, 55, p. 679

(*Continued*)

TABLE 13.1 *(Continued)* Classification of Food According to Their Therapeutic Nutritional Properties

THERAPEUTIC NUTRITIVE PROPERTIES	FOOD (AHARA DRAVYA) S. NO.	AHARA *DRAVYA* (FOOD)	REFERENCE
Mootrakruchhranaashaka/ Mootrakruchchhrapaham/ Mootrakruchhrahara – Helps to improve dysuria	7	*Godugdha* – Cow milk[4]	*Ashtangahruday Sutrasthana* 5/ 22–23, p. 68
	8	*Dadhi* – Curd[2]	*Sushruta Sutrasthana* 45/65, p. 202
	9	*Takra* – Buttermilk[1,2]	*Sushruta Sutrasthana* 45/ 84, p. 203 & *Charak Sutrasthana* 27/229, p. 166
	10	*Masoora* – Red lentil (*Lens culinaris*)[7]	*Raaj Nighantu, Shalyadi varga*, 95, p. 547
	11	*Pindaalu* – Taro root (*Colocasia esculenta*)[7]	*Raaj Nighantu, Moolakadi varga*, 70, p. 201
	12	*Elaa* – Cardamom (*Elettaria cardamomum*)[6]	*Bhavaprakasha Nighantu, Karpooradi varga*, 63, p. 222
	13	*Gokshoora* – Land-Caltrops / Puncture Vine (*Tribulus terrestris*)[6]	*Bhavaprakash Nighantu, Guduchyadi Varga* 45–46, p. 292
	14	*Kushmanda* – Ash gourd (*Benincasa hispida*)[1,6,7]	*Charak Sutrasthana* 27/113, p. 159 & *Raaj Nighantu, Moolakadi Varga*, 161, p. 218 & *Bhavaprakash Nighantu, shaka Varga*, 55, p. 679
	15	*Karkati* – Cucumber (*Cucumis sativus*)[7]	*Raaj Nighantu, Moolakadi Varga*, 204, p 227
Mootragrahahara/ Mootraavarodhahara – Helps to relieve obstruction of urine	16	*Takra* – Buttermilk[1,2]	*Sushruta Sutrasthana* 45/ 84, p. 203 & *Charak Sutrasthana* 27/229, p. 166
	17	*Karkati* – Cucumber (*Cucumis sativus*)[7]	*Raaj Nighantu, Moolakadi Varga*, 204, p. 227
	18	*Mandajaata* – Improperly curdled curd[2]	*Sushruta Sutrasthana* 45/67, p. 202
	19	*Narikela Tailam* – Coconut oil (*Cocus nucifera*)[2]	*Sushruta Sutrasthana* 45/120, p. 206
	20	*Dagdhamrutjaat Shali* – Puffed rice (*Oryza sativa*)[6]	*Bhavaprakaash Nighantu, Dhanya Varga*, 8, p. 636

TABLE 13.1 ***(Continued)*** Classification of Food According to Their Therapeutic Nutritional Properties

THERAPEUTIC NUTRITIVE PROPERTIES	FOOD (AHARA DRAVYA) S. NO.	AHARA *DRAVYA* (FOOD)	REFERENCE
Mootraaghaatahara – Helps in oliguria – reduced urine production	21	*Ushahpaana* – Drinking warm water in the morning (empty stomach)[8]	*Bhavaprakash – Purva Khanda* 5/314, p. 104
	22	*Kushmanda* – Ash gourd (*Benincasa hispida*)[1,6,7]	*Charak Sutrasthana* 27/113, p. 159 & *Raaj Nighantu, Moolakadi Varga*, 161, p. 218 & *Bhavaprakash Nighantu, Shaka Varga*, 55, p. 679
Mootrala/Mootrakara – Diuretic; increases the quantity of urine	23	*Shali* – Rice (*Oryza sativa*)[6]	*Bhavaprakaash Nighantu, Dhanya Varga*, 7, p. 635
	24	*Ikshvaku* – Sugarcane (*Saccharum officinarum*)[2]	*Sushruta Sutrasthana* 45/148, p. 208
	25	*Ardraka* – Ginger (*Zingiber officinale*)[5]	*Bhojana Kutuhalam, Shak-prakaran* p. 105
	26	*Nishpav* – Flat beans (*Dolichos lablab*)[4]	*Ashtangahrudaya Sutrasthana* 6/20, p. 88
	27	*Karkati* – Cucumber (*Cucumis sativus*)[7]	*Raaj Nighantu, Moolakadi Varga*, 204, p. 227
	28	*Til tail* – Sesame oil (*Sesamum indicum*)[2]	*Sushruta Sutrasthana* 45/112, p. 205
	29	*Til* – Sesame seeds (*Sesamum indicum*)[4]	*Ashtangahrudaya Sutrasthana* 6/23, p. 88
Adhobhaagadoshaharam – Helps in the ailments of lower parts of the body like the bladder	30	*Eranda Tailam* – Castor oil (*Ricinus communis*)[2]	*Sushruta Sutrasthana* 45/114, p. 205
Mootrashodhana/ Mootravishodhana/ Mootrashuddhikrut – Cleanses urine	31	*Guda* – Jaggary[2]	*Sushruta Sutrasthana* 45/160, p. 209
	32	*Vaastukam* – Wild spinach/Lamb's quarter (*Chenopodium murale/ Beta vulgaris/ Chenopodium album*)[7]	*Raaj Nighantu, Moolakadi Varga*, 123, p. 210
Mootraamyaaghnam/ Mootradoshaapaham/ Mootradoshahara/ Toonipratitoonijit – Helps to relieve the diseases related to urine	33	*Draksha* – Grapes (*Vitis vinifera*)[7]	*Raaj Nighantu, Amradi Varga*, 101, p. 360
	34	*Haridmuga* – Moong beans or green gram (*Vigna radiata*)[7]	*Raaj Nighantu, Shalyadi Varga*, 77, p. 543
	35	*Kusunbhashaka* – Safflower (*Carthamus tinctorius*)[7]	*Raaj Nighantu, Moolakadi Varga*, 143, p. 214
	36	*Kulattha* – Horse gram legume (*Dolichos biflorus*)[4]	*Ashtangahrudaya Sutrasthana* 6/33, p. 92
	37	*Ervaaruka* – Musk melon (*Cucumis melo*)[7]	*Raaj Nighantu, Moolakadi Varga*, 208, p. 228

(*Continued*)

TABLE 13.1 ***(Continued)*** Classification of Food According to Their Therapeutic Nutritional Properties

THERAPEUTIC NUTRITIVE PROPERTIES	FOOD (AHARA DRAVYA) S. NO.	AHARA *DRAVYA* (FOOD)	REFERENCE
Ashmarihara/Ashmarichhedan – help for breaking and excretion of urinary stones	38	*Amashchanah* – Green chickpeas (*Cicer arietinum*)[7]	*Raaj Nighantu, Shalyadi Varga*, 86, p. 545
	39	*Kulattha* – Horse gram legume (*Dolichos biflorus*)[4]	*Ashtangahrudaya Sutrasthana* 6/33, p. 92
	40	*Gokshoora* – Land-Caltrops/Puncture Vine (*Tribulus terrestris*)[6]	*Bhavaprakash Nighantu, Guduchyadi Varga* 45–46, p. 292
	41	*Kushmanda* – Ash gourd (*Benincasa hispida*)[1,6,7]	*Charak Sutrasthana* 27/113, p. 159 & *Raaj Nighantu, Moolakadi Varga*, 161, p. 218 & *Bhavaprakash Nighantu, Shaka Varga*, 55, p. 679
	42	*Karkati* – Cucumber (*Cucumis sativus*)[7]	*Raaj Nighantu, Moolakadi Varga*, 204, p. 227
Mehaprashamanam – Helps to improve polyuria	43	*Madhu – Honey*[2]	*Sushruta Sutrasthana* 45/132, p. 207

NUTRITION IN UROLOGY FROM CONVENTIONAL SCIENCE

Fluid intake is one of the most important factors in reducing the risk of kidney stones, being directly associated with the incidence of nephrolithiasis. Along with fluid, a diet high in fruits and vegetables, low in animal proteins, balanced in low-fat dairy products, and with a reduced salt content is the best way to decrease the risk of kidney stone disease.[9]

A low intake of animal protein and a high intake of fruits and vegetables, lycopene, and zinc is a protective factor against benign prostatic hyperplasia.[10]

The use of cranberries has shown remarkable benefits in UTIs. Cranberry consumption may prevent bacterial adherence to uroepithelial cells and reduces UTI-related symptoms. Cranberry consumption could also decrease UTI-related symptoms by reducing inflammatory immunologic response to bacterial invasion.[11]

Foods rich in provitamin A carotenoids; vitamins C, D, and E; and certain spices (garlic, ginger, oregano, rosemary, and thyme) are found beneficial in controlling the inflammation of the urinary system, especially in patients suffering from interstitial cystitis.[12]

Citrus foods, tomatoes, tomato products, and alcohol are found to aggravate the inflammation in interstitial cystitis or bladder pain syndrome.[13]

CLASSIFICATION OF DISEASES IN UROLOGY ACCORDING TO *AVASTHA* (DISEASE CONDITIONS) AND GENERAL THERAPEUTIC NUTRITION IN UROLOGY

Recurrent Urinary Tract Infection

Increased daily water intake of more than 1.5 L helps in preventing recurrent urinary tract infections.[14] Drinking 250 mL hot water every half an hour is observed to be an effective nutritive remedy. It should be supplemented with one tender coconut water every couple of hours during the daytime only. Tender coconut water has numerous medicinal properties such as antibacterial, antifungal, antiviral, antiparasitic, antioxidant, hypoglycemic, and immunostimulant properties.[15] Patients suffering from compromised renal function and congestive cardiac disease should restrict their water or fluid intake as per the physician's advice.

The infection creates *Ama* – an endotoxin in the body. Hot water helps with the digestion of this endotoxin from the metabolism. It is observed that hot water causes mild perspiration and thus exhibits an antipyretic effect.

Ashta-guna Manda: This helps to cleanse the urinary bladder[15] (*Kaiyadev Nighantu, Krutanna varga shlok, 59*)

Ashmaree (Urinary Calculus)

Calculi are produced due to ions present in the urine. When the urine becomes highly concentrated due to these ions, then these ions clump together to form gravel and stone. As far as the urine is clear like water and smoothness of the whole urinary tract from the kidneys to the urethral outlet is maintained, no urinary stones can be formed. It is important to keep urine as clear as possible to prevent urinary calculus formation.[14]

In this regard, water, especially hot water, being *Bastishodhan*, may prove one of the best remedies.

The following food preparations are observed to help the patients with urinary calculus.

Shali-manda[16]**:** (*Kaiyadev Nighantu, Krutannavarg*, 52) proves beneficial in urinary calculus. *Shali* is stated to be diuretic in *Bhavprakash*. The preparation of *Shali-manda* is discussed later.

***Kulath Yusha*:** It is observed to be beneficial in *Ashma Sharkara* (crystalluria) and renal colic[16] (Kaiydeva Nighantu, Krutannavarga, 77).

It is observed that urinary stones are generally expelled in the morning.

Urethral Stricture

In patients suffering from urethral stricture disease, the urethral lumen is narrowed. Significant lower urinary tract symptoms (LUTS) are observed including burning, dysuria, reduced flow of urine, frequency, incomplete emptying of the bladder, and straining prolonged urination. Even mild inflammation inside the urethra further narrows the lumen making the passage of urine more difficult. It is observed that inflammation is caused mainly due to acidic urine and UTI.

Drinking hot water every half an hour during the daytime helps significantly to reduce urethral inflammation and relieves both obstructive and irritative lower urinary tract symptoms. It helps to lower the acidity and concentration of urine and thus may help to prevent recurrent episodes of urethral inflammation.

Tender coconut water is observed to be beneficial in urethral stricture disease with its specific properties such as antibacterial, antifungal, antiviral, antioxidant, hypoglycemic, and immunostimulant properties.[15]

Interstitial Cystitis/Painful Bladder Syndrome

Interstitial cystitis is a condition characterized by bladder pain and pressure with the constant urge to pass urine. Discomfort increases as the bladder is getting filled and is mildly relieved after passing urine. The patient passes urine in small amounts with very high frequency.

Tender coconut water is antioxidant, hypoglycemic, and immunostimulant.[15] Tender coconut water every couple of hours during the day helps to keep urine alkaline and helps the bladder to hold more quantity of urine. Sesamin, fat-soluble lignan derived from sesame seeds, has immunomodulatory and anti-inflammatory effects.[17] Eating five grams of sesame seeds by proper chewing four to five times a day is observed to be helpful to reduce the frequency. Having five grams of sesame seeds at bedtime reduces the nocturnal frequency significantly. Tender coconut water and sesame seeds together help to settle down the irritation of the bladder and help in reducing the pain up to a certain extent.

Benign Enlargement of Prostate (Prostate Enlargement)

Many patients with enlarged prostate suffer significant nocturia.[16] Drinking less fluid after sunset and hot water with dinner and after dinner is observed to be helpful to reduce nocturia. Chewing five grams of sesame seeds thrice a day is observed to reduce prostatic inflammation and thus may help to reduce frequency, nocturia, and episodes of near retention in patients with prostate enlargement.

Drinking hot water is observed to prevent sudden episodes of near retention or retention in patients with an enlarged prostate. *Ama* – an endotoxin generated in the system due to improperly digested food – causes congestion in the body. Whenever there is inflammation in the prostate, it results in either an episode of near-retention or retention.[16] Hot water helps to eliminate *Ama* and thus helps to prevent these episodes.

Overactive Bladder, Overactive Bladder at Menarche, and Overactive Bladder During Pregnancy

This condition is characterized by the frequency of the sudden urge to pass urine, which is difficult to control, and sometimes urge incontinence. Consumption of 10 mL cow ghee with hot cow milk before breakfast and before dinner is observed to be helpful to reduce frequency and urge. Sesame seeds have immunomodulatory and anti-inflammatory effects.[17] Chewing 5 g of sesame seeds four to five times a day is also observed to be helpful in this condition. In the first trimester of gestation, sesame seeds should not be consumed.

Hypotonic Bladder and Neurogenic Bladder

Consumption of 5 mL of cow ghee with hot water before food and 10 mL of cow ghee with hot water after food, thrice a day, is observed to be helpful up to a certain extent in hypotonic and neurogenic bladder.

CLINICAL SIGNIFICANCE OF THERAPEUTIC NUTRITION IN UROLOGY

Water and fluid intake: The need for water and fluid (milk, fresh fruit juices, and tender coconut water) in a person's body can never be decided by a doctor, but in fact, it is decided by the *Prukruti* of the individual. The need for water or fluid can never be decided in terms of liters. It is observed that the color of urine decides the need for water or fluid intake. Keeping the color of urine similar to water or straw yellow (a little yellowish) always helps to maintain the health of the urinary system. Yellow or dark urine is concentrated or more acidic and may cause inflammation in the urinary tract and thus may worsen urinary diseases in one or the other way. If the color of urine is near to water or straw yellow, it is less concentrated or less acidic and helps to improve recurrent UTI, urethral stricture, interstitial cystitis, and urinary calculus.

Considering this fact, drinking 4 to 5 L of water or fluid is unnecessary and may burden the kidneys and may affect digestion as well. So, for a person living in a cold country or air-conditioned atmosphere round the clock, around 2 L of water or fluid may help. At the same time, a person living in a hot country or working in heated surroundings requires larger quantities of water or fluid. The reason behind this is that the sweating and insensible loss of water is minimal in a cold atmosphere, and it is more in a hot atmosphere.

Patients suffering from compromised kidney function and congestive heart failure should restrict water and fluid intake as per the physician's advice, as it may cause fluid overload and may turn into an emergency.

Ushapaana (Drink Hot Water on an Empty Stomach)[8]

अर्श:शोथग्रहण्यो ज्वरजठरजराकुष्ठमेदोविकारा
मूत्राघातास्रपित्तश्रवणगलशिर: श्रौणिशूलाक्षिरोगा: ।।
ये चान्ये वातपित्तक्षतजकफकृता व्याधाय: सन्ति जन्तो:
तांस्तानभ्यासयोगदपहरति पय: पीतमन्ते निशाया: ।।

Bhavaprakash Poorva Khanda, 314, p 104

Water drunken regularly at the end of the night, that is, early morning, is beneficial in diseases like *arsha* (piles), *shotha* (edema), *grahani* (disorders of digestion), *jwara* (fever), *jathara* (disorders of stomach), *jara* (geriatric disorders), *kushtha* (skin disorders), *medovikara* (obesity), *mootraghata* (diseases of urinary tract), *asrapitta* (bleeding disorders), diseases of the ear, throat, head, eyes and flank pain. Along with this, it helps in diseases caused by deranged *vata, pitta, kapha, and kshata.*

Drinking 800 mL to 1000 mL of hot water empty stomach early morning helps in almost all urological conditions like urinary infections, interstitial cystitis, urethral stricture, and urinary calculus. In the morning, the activated metabolism aids in absorbing the water from the gut rapidly. It hydrates the body and maintains water and electrolyte balance. An extra quantity is flushed out of the body by the urinary system. Generally, after almost 60 to 90 minutes of drinking hot water, a person voids urine once or twice.

Ushapaana is observed to be very beneficial in patients suffering from a urethral stricture (reduces urethral inflammation) and interstitial cystitis (reduces inflammation and helps the urinary bladder to hold near the normal amount of urine – 400 to 500 mL and thus observed to reduce the frequency), helps to maintain the acidity of urine, and expels out the debris (tissue remnants, dead tissues, and cells and turbidity; lipids, calcium, and phosphates) of the urinary system and urinary stones of small size as well. *Ushapaana* is observed to facilitate the passage of urinary stones. It also helps to minimize the infection from the urinary tract.

Patients with calculus in the ureter and ureteric colic and even mild hydroureter should not practice *Ushapaana,* as it may increase the hydrometer and hydronephrosis.

Patients suffering from compromised kidney function and congestive heart failure should restrict water and fluid intake as per the physician's advice as it may cause fluid overload and may turn into an emergency.

***Ushnodaka* (hot water):** Hot water is *Bastishodhanam,* i.e., it cleanses the urinary bladder. It is observed clinically that drinking hot water clears the turbidity of the urine effectively.

Ama – It is an endotoxin generated in the body due to improper digestion and metabolism. It reversely affects the composition of urine. Hot water helps to clear *Ama* helps to improves the composition of urine[2].

***Anupana* (codrink or after drink):** A liquid consumed either with or after the medicine is *Anupana.* It is observed that *Anupana* increases the efficacy of the medicine.

Among different *Anupana* in urological conditions, hot water, tender coconut water, and sugarcane juice are observed to be very beneficial. The efficacy of Ayurvedic herbs increases remarkably if administered with tender coconut water.

Aushadha Sevana Kala (Dosage Timing)

Apana vayu governs the physiology of micturition. It holds urine in the bladder till the bladder is properly filled (400 to 500 mL) and once the bladder is full the same energy expels the urine out.

Dosage timing for the urinary system is *Apana kala,* i.e., *Jeernaanna Kala,* that is, when the consumed food is properly digested. Once the previously consumed food is digested, that person feels hungry. The exact dosage timing of medicine for the urinary system is before consuming the food.

In urinary conditions dosage timing called *Muhurmuhu* is also significant. *Muhurmuhu* means administering a medication again and again within a short interval of time, maybe half an hour, hourly, or every couple of hours.

It is observed that tender coconut water and sesame seeds are more effective in reducing the frequency of urine if consumed before food and four to five times a day.

Hot and spicy foods: Avoiding foods like black pepper, tamarind, tomato sauce, sour curd, deep-fried foods, frozen foods, stale foods, or acidic foods and consuming nonvegetarian food with dinner specifically increases the acidity of the blood and urine and may result into the aggravation of the urinary symptoms.

***Viruddha Ahara* (improper combination of foods):** Consuming milk with salty or sour food or nonvegetarian food causes the production of *Ama* in the body. While consuming fatty food, the best drink is hot water, as it helps with the emulsification and proper digestion of the fat. Consuming cold or fizzy drinks with fried or fatty food like cheesy pizza, burger, and pasta leads to improper digestion of food and *Ama* formation. Thus, it causes increased acidity in the urine and aggravates urinary symptoms.

Dadhi (curd/yogurt) is *mootrakruchchhrahara,* which helps to relieve dysuria[2] (Sushruta *Sutrasthana* 45/65). In practice, it is observed that consumption of curd without proper appetite or after evening or with dinner causes *Ama* and significant generalized congestion in the body that results in the altered composition of urine, may cause turbid urine, and may aggravate urinary symptoms.

AHAREYA KALPANA (DIETARY RECIPES) IN OPHTHALMOLOGY

Green gram possesses potential health benefits such as antioxidant, anticancerous, and anti-inflammatory activities.[18]

Hareedmudga Yusha (soup of Moong or green beans) and *Masoora Yusha* (soup of red lentils): Consumption of only Moong or *Masoora* soup for around 7 days helps significantly in almost all urinary disorders[19] (Kaiyadev Nighantu, Krutanna varga, 75). Both *Yusha* are easily digested and metabolized and help to clear *Ama* – the endotoxin from the body tissues – and at the same time curb the production of new *Ama.*

Hareedmudga Yusha Preparation

Cook 50 g of *moong* or *masoor* beans by adding one pinch of rock salt and turmeric. Once cooked properly process it as follows. Take one teaspoon of cow ghee in a pan, heat it properly, and add two pinches of cumin seeds and two pinches of asafetida. Add a quarter spoon paste of ginger, garlic, and chili. Fry it properly and add cooked moong or masoor beans to make it like soup. Boil it properly for 5 minutes and consume it hot.

Shali-manda[18]: Red rice and plain rice can be used to prepare *Shali-manda.* Take 50 g rice, and add 700 mL of water and 2 pinches of rock salt. Let the rice cook on medium flame. When the rice is cooked completely, the milky solution devoid of the rice grains is called *Shali-manda.*

Asht-guna Manda[19]: Take 15 g Moong and 30 g rice. Roast the ingredients for 3 to 4 minutes and stir intermittently such that the ingredients should not get burned. Add 700 mL of water and let the ingredients cook. Once properly cooked, separate the liquid *Mand* from the solid part. Season *Ashta-guna Manda* with 2 to 3 g coriander seeds, a pinch of dry ginger powder, pepper powder, and piper longum powder. Add a pinch of rock salt.

Kulatha Yusha – Horse Gram Legume Soup Preparation

Cook 50 g horse gram legume beans in 1800 mL of water by adding a pinch of rock salt and turmeric. After the *Yusha* is prepared, 5 mL of cow ghee can be added[18] (Kaiyadev Nighantu, Krutanna varga, 62).

Dhanyaka-jeeraka Hima (a drink prepared of coriander and black cumin seeds): Coriander seeds have antioxidant, anti-inflammatory, and antidiabetic properties.[20] Black cumin seeds have analgesic and anti-inflammatory effects.[21] *Dhanyaka-jeeraka hima* helps as a good systemic alkalizer, reduces the acidity of urine, reduces inflammation, and helps to relieve lower urinary tract symptoms like dysuria, burning, frequency, and urgency. So, it is observed to be beneficial in patients suffering from recurrent urinary tract infections, urethral strictures, interstitial cystitis, and overactive bladder.

Ideally, 200 mL of *Dhanyaka-jeeraka hima* should be consumed every half an hour. Patients suffering from compromised kidney function and congestive heart failure should restrict water and fluid intake as per the physician's advice as it may cause fluid overload and may turn into an emergency.

Dhanyaka-jeeraka Hima Preparation

Soak 50 g coriander seeds and 50 g cumin seeds in 5 L of drinking water at bedtime. Soak it for 8 hours. Crush the soaked seeds with clean hands in the water itself. Strain the water with a clean cloth. Keep the water in a clean container and drink it all day.

Patients suffering from compromised kidney function and congestive heart failure should restrict water and fluid intake as per the physician's advice as it may cause fluid overload and may turn into an emergency.

SCOPE FOR FURTHER RESEARCH AND NUTRAVIGILANCE

- It is observed that the consumption of hot water has considerable benefits in almost all urological conditions, but further clinical trials are necessary in this regard.
- *Dhanyaka-jeeraka Hima* is observed to be advantageous and there is significant scope for further studies, especially in conditions like recurrent UTI, interstitial cystitis, overactive bladder, and urethral stricture.
- Combining Ayurvedic and conventional nutrition to develop a novel range of beneficial therapeutic nutrition in urology.
- Use of sesame seeds can be studied in menarche and pregnancy-related overactive bladder.
- Developing new blends of foods based on Ayurvedic nutrition.
- Evolving conventional nutrition used in different regions of the world.
- Food like ash gourd, musk melon, and green chickpeas are stated to be beneficial in Ayurveda, but further research is required.

REFERENCES

1. Acharya YT. Charakasamhita [4th Edition]. Varanasi: Chaukhambha Sanskrit Sansthan; 1994. p. 187, 97, 415, 515, 159, 166. [Sanskrit].
2. Acharya YT. Sushrutasamhita [7th Edition]. Varanasi: Chaukhambha Orientalia; 2002. p. 437, 279, 280, 253, 200, 230, 202, 206, 208, 205, 209, 207. [Sanskrit].
3. Bijlani RL, Manjunath S. Understanding Medical Physiology [4th Edition]. New Delhi: Jaypee Brothers Medical Publishers P Ltd; 2011. Chapter 8.2, p. 431. [English].
4. Kunte AM, Navre KR. Ashtangahrdaya [6th Edition]. Varanasi: Chaukhamba Surbharati Prakashan; 1997. p. 216, 66, 68, 88, 92. [Sanskrit].
5. Raghunath Suri, Bhojankutuhalam [1st Edition]. Haridwar: Divya Prakashan, Patanjali Yogpeeth, 2013. p. 105. [Sanskrit & Hindi].
6. Chunekar KC, Pandey GS. Bhavaprakasha Nighantu [1st Edition]. Varanasi: Chaukhambha Bharti Academy; 1998. p. 292, 679, 222, 636, 104, 635. [Sanskrit & Hindi].
7. Tripathi I. Raj Nighantu [5th Edition]. Varanasi: Chowkhamba Krishnadas Academy; 2010. p. 218, 95, 201, 547, 227, 210, 360, 543, 214, 228, 545. [Sanskrit & Hindi].
8. Vaishya S. Bhavaprakash [1st Edition]. Mumbai: Khemraj Shrikrishnadas Prakashan; [Sanskrit & Hindi]. p. 104.
9. Ferraro PM, Bargagli M, Trinchieri A, Gambaro G. Risk of Kidney Stones: Influence of Dietary Factors, Dietary Patterns, and Vegetarian-Vegan Diets. Nutrients. 2020 Mar 15;12(3):779. PMID: 32183500; PMCID: PMC7146511. doi: 10.3390/nu12030779.
10. Cicero AFG, Allkanjari O, Busetto GM, Cai T, Larganà G, Magri V, Perletti G, Robustelli Della Cuna FS, Russo GI, Stamatiou K, Trinchieri A, Vitalone A. Nutraceutical treatment and prevention of benign prostatic hyperplasia and prostate cancer. Arch Ital Urol Androl. 2019 Oct 2;91(3). PMID: 31577095. doi: 10.4081/aiua.2019.3.139

11. Mantzorou M, Giaginis C. Cranberry Consumption Against Urinary Tract Infections: Clinical State-of-the-Art and Future Perspectives. Curr Pharm Biotechnol. 2018;19(13):1049–1063. doi: 10.2174/1389201020666181206104129. PMID: 30520372.
12. Gordon B, Blanton C, Ramsey R, Jeffery A, Richey L, Hulse R. Anti-Inflammatory Diet for Women with Interstitial Cystitis/Bladder Pain Syndrome: The AID-IC Pilot Study. Methods Protoc. 2022 May 18;5(3):40. doi: 10.3390/mps5030040. PMID: 35645348; PMCID: PMC9149882.
13. Shorter B, Ackerman M, Varvara M, Moldwin RM. Statistical validation of the Shorter-Moldwin Food Sensitivity Questionnaire for patients with interstitial cystitis/bladder pain syndrome. J Urol. 2014;191:1793–1801.
14. Hooton TM, Vecchio M, Iroz A, Tack I, Dornic Q, Seksek I, Lotan Y. Effect of Increased Daily Water Intake in Premenopausal Women With Recurrent Urinary Tract Infections: A Randomized Clinical Trial. JAMA Intern Med. 2018 Nov 1;178(11):1509–1515. doi: 10.1001/jamainternmed.2018.4204. PMID: 30285042; PMCID: PMC6584323.
15. DebMandal M, Mandal S. Coconut (*Cocos nucifera* L.: Arecaceae): in health promotion and disease prevention. Asian Pac J Trop Med. 2011 Mar;4(3):241–7. doi: 10.1016/S1995-7645(11)60078-3. Epub 2011 Apr 12. PMID: 21771462.
16. Bharadwaj JR, Deb P. Boyd's textbook of Pathology, [10th Edition], New Delhi, Wolters Kluwer (India) Pvt. Ltd. 2013, [English], p 1104, 1127.
17. Majdalawieh AF, Yousef SM, Abu-Yousef IA, Nasrallah GK. Immunomodulatory and anti-inflammatory effects of sesamin: mechanisms of action and future directions. Crit Rev Food Sci Nutr. 2022;62(18):5081–5112. doi: 10.1080/10408398.2021.1881438. Epub 2021 Feb 5. PMID: 33544009.
18. Mekkara Nikarthil Sudhakaran S, Bukkan DS. A review on nutritional composition, antinutritional components and health benefits of green gram (*Vigna radiata* (L.) Wilczek). J Food Biochem. 2021 Jun;45(6):e13743. doi: 10.1111/jfbc.13743. Epub 2021 May 1. PMID: 33934386.
19. Sharma P, Sharma G. Kaiyadev Nighantu [Reprint edition]. Varanasi, Chaukhamba Orientalia, 2009, [Sanskrit & Hindi], Krutannavarg, Sutra, p 409, 413, 411.
20. Mechchate H, Es-Safi I, Amaghnouje A, Boukhira S, Alotaibi A, Al-Zharani M, Nasr F, Noman O, Conte R, Amal EHEY, Bekkari H, Bousta D. Antioxidant, Anti-Inflammatory and Antidiabetic Proprieties of LC-MS/MS Identified Polyphenols from Coriander Seeds. Molecules. 2021 Jan 18;26(2):487. doi: 10.3390/molecules26020487. PMID: 33477662; PMCID: PMC7831938.
21. Amin B, Hosseinzadeh H. Black Cumin (Nigella sativa) and Its Active Constituent, Thymoquinone: An Overview on the Analgesic and Anti-inflammatory Effects. Planta Med. 2016 Jan;82(1–2):8–16. doi: 10.1055/s-0035-1557838. Epub 2015 Sep 14. PMID: 26366755.

Therapeutic Nutrition in Ayurveda for Neurological Disorders

Naresh Kore, Rahul Jadhav and Shinsha Puthiyottil

INTRODUCTION

Therapeutic nutrition in neurology encompasses a broad range of approaches because neurological diseases like Beriberi are caused by nutritional deficits themselves, while neurological conditions like Parkinson's make it physically difficult for patients to obtain enough nutrition. Depending on the extent of their etiology, neurological diseases can affect a patient's physical capacity, mental function, and metabolisms such as mental confusion, weariness, and early satiety.[1]

The *Vatadosha* (one of the regulatory functional factors of the body) is viewed as a key moving force in the human body[2] (*Charaka Samhita, Sutrasthana* 12/8). *Vata Dosha* regulates respiratory, gastrointestinal, and digestive homeostasis, all of which are functions of the nervous system. *Vata Dosha* also governs higher neuronal processes such as mental health and behavior. Dysfunction in *Vata Dosha* and the nervous system causes similar diseases. The majority of common neurological illnesses fall under one of two different *Vatavyadhi* pathologies, namely *Margavarodhajanya* (obstruction of pathways) or *Dhatukshayajanya* (tissue degeneration)[2] (*Charaka Samhita, Chikitsasthana*, 28/59–61).

When it comes to disease treatment in Ayurveda, therapeutic nutrition is just as essential as medicine.[3] Dietetic intervention in the management of *Vatavyadhi* aims to have therapeutic effects in addition to meeting nutritional needs.

THERAPEUTIC NUTRITION IN NEUROLOGY FROM CLASSICS

Most of the classical references advise routine use of *Godhuma* (wheat), *Masha* (black gram), *Kulatha* (horse gram), *Rakta Shali* (red rice), *Patola* (pointed gourd), *Shigru* (drum stick), *Vartaka* (brinjal),

DOI: 10.1201/9781003345541-17

Dadima (pomegranate), *Draskha* (grapes), *Parushaka* (*Grewia asiatica*), *Badara* (Indian jujube), *Ghruta* (clarified butter), *Dugdha Kilata* (condensed milk), *Jangalamamsa* (the meat of animals living in dry and arid regions), *Matsya* (fish), and *Lashuna* (garlic) for patients suffering from *Vata* (neurological) disorders.[4]

Therapeutic nutrition is described in the form of individual properties of food articles and in the form of specific *Pathya Kalpana* (dietetic formulations). Specific dietetic formulations for diseases such as *Pakshaghata* (paralysis) and *Kampa Vata* (Parkinson's disease) are very few and scattered but are mentioned for diseases such as *Shiroroga* (diseases related to the head).

In *Vataja Shiro Roga* (migraine, tension-type headache, trigeminal autonomic cephalalgias) the person should consume green gram, horse gram, or black gram pulses preparations in the night with *Ghruta* (clarified butter) or with milk. Milk is also consumed with *Ghruta* (clarified butter) or with sesame oil[4] (*Sushruta Samhita, Uttaratantra*, 26/3–4). *Ghruta* (clarified butter) ingestion after meals has been recommended by several classical texts for people who suffer from migraines[2] (*Charaka Samhita, Siddhi Sthana*, 9/81). It is advised to consume *Vataka* (fritter) prepared from black gram flour with butter for 7 days for those who have *Ardita* (facial paralysis)[5] (*Yogaratnakara, Purvardha*, 25/127).

MODERN THERAPEUTIC NUTRITION IN NEUROLOGICAL DISORDERS

Lesions in various areas of the central nervous system (CNS) can cause a variety of dysfunctions with varying nutritional implications.[6] Cognitive, metabolic, and physical abilities can all be impacted by neurological diseases. The body is unable to achieve its nutritional and metabolic needs when the CNS is damaged in any way. Weakness of the tongue and facial and masticator muscles can lead to prolonged feeding time and cough or choking while eating. Chewing and dysphagia could also arise. Emotional stress, metabolic stress, and trauma can compound both eating and nutritional problems, as these have an effect on nutritional requirements.[7] As a result, the dietician's involvement in neurology is multifaceted and covers everything from addressing specific dietary needs to providing feeding advice and strategies.

Common Neurological Disorders:

- Neurological disorders arising due to imbalanced nutritional intake – Common examples are the neurological manifestations of beriberi, pellagra, pernicious anemia, and Wernicke-Korsakoff syndrome due to nutrient deficits. Alcoholism and malabsorption could also be other causative factors.[8] Most of these disorders are reversible and nutritional intervention in them is focused on the individual nutrient.
- Neurological disorders of nonnutritional etiology – Some of the common disorders are paralysis, Parkinson's disease, migraine, and trigeminal neuralgia.[9]

Nutritional Management in Neurological Conditions

The effects of a healthy diet and a relative richness of nutrients on brain function are undeniable.[10] The main objectives of nutritional therapy in neurology are to complement disease management; avoid further impairment; restore or maximize the patient's physical, mental, and social capacities; and improve their overall quality of life.

The following nutritional intervention strategies can be adopted to achieve the objectives.[10]

- Evaluation of a person's dietary, physical, and other parameters and periodic assessment to enhance patient outcomes in terms of food consumption (both quantity and quality), weight changes, and clinical evaluation.
- Assessment of eating and feeding to determine the requirement for specialized foods, as well as textural and other food or feeding changes.
- Individualized suggestions for nutritional intervention. For example, since the neurotransmitters serotonin and dopamine depend on iron, dietary deficits such as anemia may need to be identified.[11]
- Restriction of certain nutrients should be weighed against the risk of their deficiency. For example, now people with Parkinson's disease are no longer encouraged to restrict their protein consumption, due to the risk of weight loss.[12]
- Nutrition counseling of the patient, family, and caregiver.

Specific Dietary Concerns

- Constipation is one of the most frequent nonmotor symptoms in neurological disorders such as Parkinson's and paralysis. It could be the result of an overall diet or a medication side effect.[13] It is advised to increase dietary fiber intake and drink enough water. For patients who require a soft/puree diet, fiber can be given in the form of oat porridge; pureed/mashed fruits like bananas, prunes, and dates; and thickened lentil-type soups.[14]
- Weight loss is linked to a lower quality of life in terms of health.[15] It is brought on by an insufficient intake, which may be brought on by a deteriorating capacity for food preparation, escalating tremors that make self-feeding difficult or swallowing difficult. Motor symptoms in the later stages may result in higher energy needs.[16]
- Old age and use of medications like diuretics, antihypertensive, and antidepressants may cause dry mouth.[17] Meals that are moist and served with sauce could be beneficial. The salivary flow rate is influenced by the type of taste stimuli. In general, a sour taste, elicited by citric acid or sour food like lemon and grapefruit, induces the highest flow rate.[18]

CLASSIFICATION OF NEUROLOGICAL DISORDERS

Vatavyadhi is a group of diseases that occur due to the involvement of *Vata*, one among three forces (*Tridosha* – three regulatory functional factors of the body)[19] (*Ashtanga Hrudaya, Sutrasthana* 1/6–7). The majority of neurological conditions are classified as *Vatavyadhis* in Ayurveda.[20] Neurological disorders are also explained in a broader context such as clinical features of various *Sannipata Jvara* (one of the types of fever), which are similar to that of infectious neurological diseases like meningitis and encephalitis. Syncope and coma can be correlated with the diseases like *Mada* (intoxication), *Murcha* (syncope), and *Sanyasa* (coma).[21] Similarly cerebrovascular accidents, facial palsy, and various neuropathies are part of the spectrum of *Vatavyadhis*.[22]

Dhatukshayajanya (Due to Diminution of Tissues) and *Avaranajanya* (Due to Occlusion of *Vata* Channels) *Vatavyadhi*

Generally, the aggravation of *Vata* can occur in two ways; one is due to *Dhatukshaya* (a condition where diminution of tissues occurs), and the other is due to *Avaranajanya* (some factors occlude the channels of

Vata to cause the obstruction in its movement and hence results in the development of disease)[2] (*Charaka Samhita, Chikitsasthana*, 28/15–19). *Dhatukshaya* means depletions of all *Sapta-Dhatus* (seven tissues), which would lead to *Vata* vitiation. The significance of this classification is that the treatment strategy, as well as the dietary recommendations, will vary depending on the underlying etiology[2] (*Charaka Samhita, Vimanasthana* 7/14). For example, in cases of tissue depletion, dietetic intervention would be focused on nourishment therapy[2] (*Charaka Samhita, Chikitsasthana* 28/104–106), and in cases of occlusion pathology, dietetic intervention would focus on pacifying the occluding *Dosha* followed by nourishment[2] (*Charaka Samhita, Chikitsasthana*, 28/183).

Concept of *Dhatukshayajanya Vatavydhi*

Dhatukshaya (tissue depletion) is one of the *Vatavaydhi* diseases. In *Dhatukshaya,* the status of the *Srotas* (channels that serve as a conduit for biological materials, nutrients, and waste products for their transportation) is atrophied or they are diminished in their functional capacity.[22] It will result in *Vata* acquiring these channels as its abode and results in further degeneration and functional loss[2] (*Charaka Samhita, Chikitsasthana* 28/18).

The etiological factors[2] (*Charaka Samhita, Chikitsasthana* 28/15–19):

- Excessive consumption of food items with *Ruksha* (an attribute that causes dryness in the body and tissues), *Laghu* (an attribute that results in the lightness of the body) properties
- Aging
- Chronic degenerative diseases and treatment that causes excessive impairment of the patient and his health.

Degenerative pathology in neurological diseases is evident, for example, neurological diseases like Alzheimer's and Parkinson's that occur due to aging.[23] Common neurological disorders associated with aging are stroke, epilepsy, dementia, Parkinson's disease, and demyelinating neuropathies.[24,25] Chronic hypertension and vascular diseases are raising the risk for hemorrhagic stroke in the elderly. Neurological diseases as a result of malnutrition are another important but underrecognized crisis, especially in the elderly.[26] *Dhatukshaya* (tissue depletion) can also occur due to excessive bleeding caused by trauma; however, the prognosis is poor in such cases[4] (*Sushruta Samhita, Nidanasthana* 1/63).

Avaranajanya Vatavyadhi (*Vata* Ailments Due to Occlusive Pathology)

This is the pathology of *Vatavyadhi,* where the vitiation of *Vata* occurs not because of its own vitiating factors but due to other *Doshas* or *Dhatus* (tissues) or *Malas* (metabolic waste). Obstruction of the normal movement of *Vata* leads to *Avaranajanya* (occlusive etiology) *Vatavyadhi*. Thus, *Vata* imbalance predominantly manifests in functional or qualitative aspects.[27]

Kevala Vatika and *Samsarga Doshaja Vatavyadhi*

Despite the presence of other *Doshas* in this type of pathology, *Vata* remains the primary and unaffected component. Hence based on the associated *Doshas* the *Vatavyadhi* can be grouped into two:

1. *Kevala Vatika* (only *Vata* is involved)
2. *Anya Dosha Samsargaja.*

DISEASES IDENTIFIED FOR THERAPEUTIC NUTRITION

Pakshaghata (Stroke)

Stroke is the second major cause of death and holds third place among the causes of death and lifetime disability (combined) worldwide.[28] According to World Health Organization (WHO) stroke is defined as "rapidly developed clinical signs of focal (or global) disturbance of cerebral function, lasting more than 24 hours or leading to death, with no apparent cause other than of vascular origin".[29]

On the basis of the clinical presentation, stroke can be correlated to the disease known as *Pakshaghata*, which falls in the spectrum of *Vatavyadhi* in Ayurveda. The term *Pakshaghata* means "paralysis of one half of the body" and refers to the loss or impairment of *Karmendriya* (motor system), *Dyanendriyas* (sensory system), and *Manas* (mind).[30]

Kampa Vata (Parkinson's Disease)

Sushruta has described *Kampa* as a feature of *Snayugata Vata* (*Vata Dosha* in muscle)[4] (*Sushruta Samhita, Nidansthana*, 1/27) and *Vepathu*, a synonym of *Kampa*, which means tremors/shakes are included in *Nanatmaja Vata Vikaras* (diseases occur by the involvement of *Vata* alone)[2] (*Charaka Samhita, Sutrasthana*, 20/11). This can be correlated with diseases such as Parkinson's disease, Parkinsonism, and progressive supranuclear palsy, where the tremor is a characteristic feature.[31] Based on the prevalence, Parkinson's disease (PD) is one of the most common diseases which can be correlated with *Kampa Vata*.[32]

Parkinson's disease is one of the most common neurodegenerative diseases, characterized by tremors, bradykinesia/akinesia, and postural instability.[33] Loss of dopaminergic neurons in the substantia nigra pars compact and accumulation of misfolded α-synuclein, which is found in intracytoplasmic inclusions known as Lewy bodies, are pathological hallmarks of Parkinson's disease.[34] From the sixth to the ninth decades of life, the incidence rises 5- to 10-fold.[35]

Shirashula (Headache)

Headache is known as *Shirashula* in Ayurveda. *Shirashula* itself is a disease and constitutes a symptom of diseases as well. According to Ayurved classics, *Shiroroga* is classified into five types, according to the involvement of *Dosha*[2] (*Charaka Samhita, Sutrasthana* 17/15) in pathology, and four types, namely, *Sankhaka, Anantavata, Ardhavabhedaka*, and *Suryavarta*, according to the characteristic of pain[2] (*Charaka Samhita, Siddhisthana* 9/70–86).

THERAPEUTIC NUTRITION

Food items are chosen based on their *Guna* (attributes). Before planning a therapeutic diet, food qualities, processing and combinations, food quantity and meal times, the patient's digesting ability, ancestral food habits, and the patient's personal food preferences are all taken into account.

Shuka Dhanya (Cereals)

Cereals have *Seeta* (cold) properties and slightly vitiate *Vata*. Some cereals, like a red variety of rice, on the other hand, contain a balance of qualities that allow them to be utilized therapeutically in *Vatavyadhi*[2]

(*Charaka Samhita, Sutrasthana*, 27/10–11). Cereals also make up a significant portion of a meal and are the primary source of energy; thus selecting the right cereal is critical.

Shali (rice) is one of the most important staple foods.[36] Rice is consumed in various forms as comfort food; however, the fact that it is high in simple carbohydrates may raise concerns for patients with comorbid conditions such as obesity and diabetes.[37] Red rice variants (*Oryza longistaminata* or *Oryza punctate*) are regarded to be the best of all rice varieties. Red rice balances all three *Doshas*, including *Vata*[2] (*Charaka Samhita, Sutrasthana*, 27/10). Red rice can be utilized in both *Margavarodhajanya* (obstructive pathology) and *Dhatukshayajanya* (degenerative) neurological illnesses in the form of *Peya* (soup), *Vilepi*, and *Krusara* (gruel). *Sashtika Shali* is another rice variety that is harvested in 60 days and has abilities to pacify three *Doshas*, including *Vata*. It also brings *Sthirata* (steadiness) in the body, which is beneficial in *Kampa* (tremors)[2] (*Charaka Samhita, Sutrasthana*, 27/13).

Godhuma (wheat) has properties that help to connect and mend bodily tissues. It also has characteristics that nourish the body, bring strength, and promote stability. All of these characteristics make wheat a good option among staple cereals for *Dhatukshayajanya* neurological diseases[2] (*Charaka Samhita, Sutrasthana* 27/21).

Comorbid conditions like obesity and diabetes are associated with neurological disorders, and stroke is one of them. In these cases, it's important to introduce variations in the consumption of cereals that are not energy dense, have excellent fiber content, have good mineral and protein content, and are low in calories. Examples include millets *Yavanala* (sorghum), *Nartaka* (finger millet), *Kangu* (foxtail millet), *Syamaka* (barnyard millet), *China* (proso millet), and *Gavedhuka* (adlay millet). These are called *Kshudra Dhanya* (small grains) and they have properties like *Ruksha* (dryness) and *Guru* (heavy); thus, they are ideal to counteract the effects of overnutrition.

Shami Dhanya (Pulses, Legumes, and Oil Seeds)

Shami Dhanya in general have properties like *Ruksha* (dryness) and *Sheeta* (coldness)[2] (*Charaka Samhita, Sutrasthana*, 27/23). These make them unsuitable for consumption in most of the *Vata* (neurological) disorders. Pulses are a good source of protein in the diet.[38]

Mudga (green gram) is said to be highly preferred of all pulses and is a *Pathya* (therapeutically useful dietary component) for ailments[19] (*Ashtanga Hrudaya, Sutrasthana* 6/18–19). Green gram also has a culinary advantage in that it is widely available and appetizing, so it can meet the needs of pulses/legumes in the diet.[39] Green gram is routinely advised in the form of *Yusha* (thick soup) to neurological patients who require a special feeding assistant.

Masha (black gram) has unique properties amongst the *Sami Dhanya*. Its *Ushna* (warm), *Snigdha* (unctuous), and *Guru* (heavy) properties are ideal for therapeutic use in *Vata* (neurological) disorders[2] (*Charaka Samhita, Sutrasthana* 27/24). It can be used more efficiently in *Dhatukshayajanya* (degenerative) *Vata* (neurological) disorders. Most of the *Aahariya Kalpana* (therapeutic food recipes) used in practice for neurological disorders have *Masha* as their key ingredients such as *Masha Vataka* (fritter) and *Masha Yusha* (soup).

Kulattha (horse gram) has properties such as *Ushna* (warm), which are favorable to pacify *Vata*[19] (*Ashtanga Hrudaya, Sutrasthana* 6/19–20). Horse gram is a good source of protein and dietary fiber. This makes it a good dietary choice for constipation which is a common complaint in neurological conditions such as paralysis and Parkinson's disease.[40] Due to its *Ushna* properties, horse gram is more useful in *Kapha Samsrushta* (*Vata* with *Kapha*) *Vatavyadhi*.[41]

Tila (Sesame) has *Ushna* (warm) and *Snigdha* (unctuous) properties[2] (*Charaka Samhita, Sutrasthana* 27/30), both of which have great therapeutic importance in neurological conditions. Sesame is more commonly used in the form of oil. Sesame oil is an integral part of the therapeutic management of neurological disorders in Ayurveda.

Mamsa (Meat)

Mamsa (animal meat) plays a crucial role as a therapeutic diet in all types of neurological conditions, whether it be *Dhatukshayajanya* (degenerative) or *Margavarodhajanya* (obstructive). Meat has properties like *Guru* (heavy), *Ushna* (hot), and *Snigdha* (unctuous); all of these properties make it therapeutically useful in *Vata* (neurological) disorders[2] (*Charaka Samhita, Sutrasthana 27/57*).

Animal meat can be very difficult to digest and overly fatty.[42] These issues can be resolved by consuming meat in the form of *Mamsa rasa* (meat soup) or by the usage of *Jangala Mamsa* (the meat of animals living in dry and arid regions), which has very little fat and *Laghu* (light for digestion)[2] (*Charaka Samhita, Sutrasthana* 27/59). *Jangala Mamsa* includes the meat of *Aja* (goat), *Charanayudha* (rooster), *Tittiri* (gray francolin), and *Lava* (common quail). Animal meat can be used in the form of *Mamsa rasa* (meat soup), *Bahurasa* (cooked with plenty of water), *Bhrushta* (roasted), and *Swinna* (steamed).

Phalavarga (Fruits)

Fruits contain polyphenols and flavonoid compounds, and these are functionally serving as antioxidants as well as neuroprotective.[43] Thus, generally, the dietary inclusion of fruits will be beneficial in neurodegenerative conditions and inflammatory conditions, where reduction of inflammation and resultant oxidative stress prevents progressive tissue damage.[44]

Mrudvika (dry black grapes) are considered the highly preferred fruits[2] (*Charaka Samhita, Sutrasthana*, 25/38). *Mrudvika* has *Madhura* (sweet) and *Snigdha* (unctuous) properties. This fruit will nourish the tissues and hence can be used in *Dhatukshayajanya Vatavyadhi* (degenerative neurological disorders) due to the deterioration of tissues[2] (*Charaka Samhita, Sutrasthana*, 27/126). The intake of *Mrudvika* is beneficial in diseases like *Raktapitta* (bleeding disorders) and thus can be advised as a food of choice in hemorrhagic stroke[2] (*Charaka Samhita, Sutrasthana*, 27/125). Scientific research has proved that grape peel and seed extract is effective in repairing neuronal damage in ischemic stroke.[45] Grapes are good laxatives and can be a good choice for patients having constipation. Dried *grapes* are more beneficial and are superior in quality to fresh or ripe grapes.

Kharjura (dates) is similar to *Mrudvika* in its properties like *Madhura* (sweet) and *Snigdha* (unctuous) and provides nourishment to the body tissues. *Kharjura* can be given to neurodegenerative diseases because of its *Brumhana* (nourishing the tissues) and *Vata-Pitta* alleviating properties[2] (*Charaka Samhita, Sutrasthana*, 27/127). Clinical and in vivo studies have suggested that date palm components, like hydroxycinnamates, gallic acid, ferulic acid, monohydroxybenzoic acids, flavones and α-synuclein, and anthocyanin, can cause inhibition of proinflammatory cytokines, amyloid beta peptides, and α-synuclein and increase brain ATP concentrations, and thus prevent the development of neurodegenerative diseases.[46] *Parushaka* (*G. asiatica*) and *Amrataka* (hog plum) have similar properties to *Kharjura* such as *Madhura* (sweet), *Snigdha* (unctuous), and *Brumhana* (provide nourishment to the body tissues). These can be given in conditions with *Vata-Pitta* aggravation, tissue degeneration, and constipation and to enhance body strength.

Amra (mango) has nourishing properties and enhances body strength[2] (*Charaka Samhita, Sutrasthana*, 27/139). The role of mango fruit extract in enhancing cognitive functions by means of increasing the cholinergic functions and reduction of oxidative stress damage of neurons has been proved in animal models.[47]

Dadima (pomegranate) has many varieties, of which those having a dominant sweet taste will be preferred in neurological conditions with cognitive decline[48] (*Bhavaprakasa, Purvakhanda*, 6/85). Studies have shown that pomegranate juice extract has a significant neuroprotective effect and antioxidant activity in CNS, which may help in the prevention and treatment of diseases like PD.[49]

Vrikshamla (kokum - *Garcinia cambogia*), *Amlika* (Tamarind – *Tamarindus indica* Linn.), and *Amlavetasa* (*Rheum emodi* Wall.) are sour taste–dominant fruits. They have properties to nourish tissues, enlighten the mind, increase body strength, improve the functions of sense organs, and alleviate *Vata Dosha*[2] (*Charaka Samhita, Sutrasthana*, 27/151), which are beneficial in stroke and Parkinson's disease. Kokum peel extract has been found to be beneficial in controlling obesity and hence its use is advisable in neurological diseases associated with obesity.[50] *Amlavetasa* (*Rheum emodi* Wall.) has laxative properties too[2] (*Charaka Samhita, Sutrasthana,* 27/151).

Amalaki (gooseberry) is a potent rejuvenator drug in Ayurveda that is used for prevention as well as therapy. This fruit is useful in bleeding disorders[48] (*Bhavaprakasa, Purvakhanda,* 1/37–39). Because of these properties, *Amalaki* is indicated in hemorrhagic stroke, degenerative and senile neurological disorders, and neurological cases associated with diabetes mellitus or other metabolic diseases.[51] Gooseberry has a significant neuroprotective effect which helps in the treatment and prevention of neurodegenerative diseases like Parkinson's disease.[52,53]

Panasa (ripe jackfruit) has properties like *Snigdha* (unctuous), *Seeta* (cold in potency), *Madhura* (sweet in taste), pacifying *Vata* and *Pitta* in nourishing the tissues, enhancing the body strength, and increasing muscle mass. It is indicated in the treatment of bleeding disorders and traumatic wounds. In *Dhatukshayaja Vatavyadhi* (neurological diseases developed due to diminution of tissue and strength) jackfruit can be included in the diet[48] (*Bhavaprakasa, Purvakhanda*, 6/22–23). Jackfruit contains functional biological compounds which have antidiabetic, antihypertensive, antioxidant, antihyperlipidemic, and anti-inflammatory properties.[54] Thus, it will be beneficial in the neurological diseases associated with diabetes mellitus, hypertension, and dyslipidemia such as stroke.

Shakavarga (Vegetable)

Shakavarga (groups of vegetables) includes different types of vegetables used for food. It is classified into six categories in Ayurveda, which are *Phala Shaka* (vegetable fruits), *Patra Shaka* (leafy vegetables), *Nala Shaka* (stalks), *Kanda Shaka* (tuberous vegetables), *Pushpa Shaka* (flowers), and *Samsvedaja* (saprophytic/parasitic, e.g., mushroom)[48] (*Bhavaprakasa, Purvakhanda*, 9/1). Consumption of unprocessed or raw vegetables is not advised. Most of the vegetables described in Ayurveda are *Ruksha* (dry) and *Guru* (heavy to digest). Intake of these will produce flatulence, produce excess stool, cause a diminution of strength in the body, and adversely affect the strength and function of bones and eyes[48] (*Bhavaprakasa, Purvardha*, 9/3–4). Instead of consuming raw vegetables, it is advised to process them with oil/ghee to reduce their adverse effect on the body.

Vastuka (*Chenopodium ambrosioides*) is a leafy vegetable that alleviates *Tridoshas* and has laxative properties also[2] (*CharakaSamhita, Sutrasthana*, 27/88). It contains carotenoids, vitamins, flavonoids, anthocyanins, and other phenolic components and has antioxidant action.[55] It is prescribed for stroke patients with constipation.

Kala Shaka (*Murraya koenigii*) is a leafy vegetable, known as curry leaf; it is carminative and has a therapeutic effect on *Shopha* (inflammation)[2] (*Charaka Samhita, Sutrasthana*, 27/89). Curry leaves have anti-inflammatory as well as antioxidant properties.[56] It will enhance cognitive functions and neuronal cholinergic functions. It is advised to be added in foods of patients suffering from memory loss associated with diseases like stroke, Parkinson's disease, or senile/degenerative decline of cognitive and motor functions.[56]

Vidarikanda (*Pueraria tuberosa*) is a tuberous vegetable. It possesses a sweet taste and cold potency and thus has good tissue-nourishing properties. It is a rejuvenating as well as an aphrodisiac herb[2] (*Charaka Samhita, Sutrasthana*, 27/121). It is advised for patients with reduced body strength or muscle power in stroke and Parkinson's disease.

Harita Varga (Which Can Be Used in Salad Form)

In this category of food items are ginger, *Jambira* (*Citrus medica* Linn.), *Mulaka* (radish), *Palandu* (*Allium cepa* Linn.), and *Lasuna* (*Allium sativum* Linn.).

Ginger and *Jambira* (lemon) are appetizers, enhance digestion, and alleviate *Kapha* and *Vata Doshas*; ginger gives relief from constipation[2] (*Charaka Samhita, Sutrasthana*, 27/166); and lemon has anthelminthic action and it helps in the food to digestion[2] (*Charaka Samhita, Sutrasthana*, 27/167). For patients with stroke, Parkinson's disease, or motor neuron diseases, where reduced appetite and constipation are present, ginger and lemon juice can be administered with food.

Grunjanaka (carrot – *Daucus carota* L.) It is advised to include carrot in the diet as it helps to maintain the normal functioning of the nervous system too.[57] Lutein, a carotenoid present in carrots, promotes neuronal growth in children and improves cognition in adulthood.[58]

Mulaka (radish) is a common vegetable used in salads. Ayurveda emphasizes to use of tender radishes, which can alleviate vitiated *Doshas,* whereas mature radishes will provoke all the *Doshas*; it has a therapeutic effect in *Vatavyadhi* when they are processed with oil/clarified butter. Dried radish, in particular, is advised for diseases due to *Vata* and *Kapha*[2] (*CharakaSamhita, Sutrasthana*, 27/168). It is advised for stroke patients having underlying cardiac issues.[59]

Rasona (garlic – *Allium sativum* Linn.) is a common spice used in India and a potential drug used in Ayurvedic formulations also. Garlic is unctuous, hot in potency, and pungent in taste and it will alleviate *Kapha* and *Vata*[2] (*Charaka Samhita, Sutrasthana*, 27/176). The use of garlic is advised in neurological diseases arising from *Avaranajanya* (occlusive) pathology such as stroke and stroke associated with diseases due to over nourishment such as obesity. Clinical studies proved that garlic will reduce peripheral vascular resistance[60] and has significantly lowered high blood pressure in hypertension patients.[61] Evidence from research has shown that garlic extract has significant fibrinolytic activity and can prevent thrombus formation.[62] Hence, garlic is advised for stroke cases with underlying cardiovascular diseases, hypertension, and hypercholesteremia.

Vartaka, known as "eggplant" (*Solanum Melongena* Linn.), is *Katu* (pungent), *Tikta* (astringent), and *Madhura* (sweet) in taste and hot in potency and good for the heart. It alleviates *Kapha-Vata,* while it will not aggravate *Pitta* and stimulate digestive power also[19] (*AshtangaHrudaya, Sutrasthana*, 6/81). *Vartaka* is a widely consumed vegetable. It's rich in flavonoids and has antioxidant and anti-inflammatory properties.[63,64]

Kusmanda (*Benincasa hispida*), known as ash guard, has *Madhura* (sweet) with a slightly alkaline taste, and *Seeta* (cold potency) and *Snigdha* (unctuous) alleviate *Vata, Pitta,* and *Rakta Doshas.* Tender ones are easy to digest, whereas mature ones are difficult to digest. Sweets made up of *Kusmanda* or fruit juice of *Kusmanda* will be beneficial in *Suryavarta* (a type of headache). *Suryavarta* is a type of headache that is associated with severe pain, increases with exposure to sunlight or heat, and is aggravated by the use of hot and spicy food, exertion, and stress. Being sweet in taste and cold and unctuous in attributes, *Kusmanda* will nourish the tissues, and thus *Kusmanda* is recommended for *Brumhana* (nourishment) in weak patients.

Kosataki (*Luffa acutangula*) – ridge gourd – is a vegetable that can enhance digestive power and is laxative in nature, and it is advised in patients with associated features of *Ama* (inflammation)[19] (*AshtangaHrudaya, Sutrasthana* 6/82).

Munjtataka (*Orchis latifolia* Linn.) is commonly known as *Salabmishri.* Its tuber is consumed. It is considered a good aphrodisiac drug and nervine tonic.[65] *Munjataka* is advised for lean/debilitated patients who require proper nourishment therapy[19] (*Ashtanga Hrudaya, Sutrasthana* 6/83).

THERAPEUTIC NUTRITION IN CLINICAL PRACTICE

Dietary interventions are planned on the basis of factors such as the stage of the disease, associated symptoms, etiological factors, age, and tolerance level of the patient.

Stroke

Amavastha (Initial inflammatory Stage)

Ischemic damage and inflammation are the main events that account for the pathologic progression of stroke. The reactive oxygen species and reactive nitrogen species generated due to the oxidative stress developed by the above pathologies produce further inflammation and cell injury. *Jvara* (fever) is also a condition that is marked by inflammation and tissue damage; hence dietary principles adopted in *Jvara Chikitsa* (treatment of fever) can be adopted here. The first line of treatment for fever is *Langhana* (fasting therapy)[2] (*Charaka Samhita, Chikitsasthana*, 3/139). Before the nutritional intervention, *Langhana* is a crucial step to improve the patient's impaired metabolic function. Fasting causes the "reorganization" of antioxidative defense lines and even increases plasma protective systems (total antioxidant capacity of plasma, plasma ceruloplasmin activity).[66] In this situation *Langhana* (fasting therapy) can be introduced as complete fasting or controlled fasting with a light diet like *Manda* (liquate gruel).[67] Fasting is followed by *Peyadi Krama* (gradually increasing the food, starting from easily digestible liquid to semisolid to solid). A protocol can be given for a period of 6 days or more till achieving the target (signs of reduction in inflammation)[2] (*CharakaSamhita, ChikitsaSthana*, 3/149–150).

Rigorous fasting is contraindicated in conditions such as persons with a history of addiction to alcohol if a stroke event happens during the summer season and in the case of *Urdhvaga Rakta-pitta* (a bleeding disorder where intracranial bleeding occurs). For example, in the initial stages of hemorrhagic stroke, internal bleeding into the brain tissues can be considered a *Urdhvaga Raktapitta* condition. In such cases, fasting is advised with an intake of *Tarpana*. *Tarpana* is satiating food article that provides nourishment to extremely weak individuals without putting strain on their *Jatharagni* (digestive capacity). *Tarpana* is prepared with powder of fried paddy mixed with honey, sugar, and fruit juice of fruits like *Draksha* (grapes), *Dadima* (pomegranate), *Parushaka* (*G. asiatica*), *Priyala* (*Bucchania lanzan*), and *Kharjura* (dates)[2] (*Charaka Samhita, Chikitsasthana*, 3/155–156).

For drinking purposes, hot water is advised, but in case of patients who have suffered a hemorrhagic stroke, patients with hypertension, and alcoholic patients, hot water is contraindicated, and here, cold water along with bitter herbs like *Nimba* (*Azadirachta indica*), *Patola* (*Trichosanthes dioica*), *Kantakari* (*Solanum xanthocarpum*), *Guduchi* (*Tinospora cordifolia*), and *Vasa* (*Adhatoda vasica*) is recommended[2] (*CharakaSamhita, ChikitsaSthana* 3/143).

Noninflammatory Stage

A good appetite is an indicator of adequate *Langhana*[19] (*Ashtanga Hrudaya, Chikitsasthana*, 1/3). At this stage a dietary regimen consisting of *Yusha* (soup prepared with pulses and vegetables) and *Mamsa rasa* (meat soup of animals living in dry and arid regions) should be given. In case of poor digestive power, the addition of sour items such as *Vrikshamla* (kokum), *Amlika* (tamarind), and *Amlavetasa* (*Garcinia pedunculata*) in the *Yusha* will be beneficial.

Recovery Stage

When there are features of aggravated *Vata* and no signs of associated *Ama* or other *Doshas*, nourishment therapy can be given to the patient; for that cow milk, clarified butter, sesame oil, chicken, or goat meat soup can be included in the diet. The food should always be sweet, sour, and salty in taste and unctuous in nature such as meat soup with oil/ghee, *Vyosha* (black pepper, dry ginger, and long pepper), and sour curd[2] (*Charaka Samhita, Chikitsasthana* 28/104–105). *Navanita* (butter) mixed with *Masha* (black gram flour) in the morning was found to be beneficial in hemorrhagic stroke patients aged 40–50 years, while cow ghee is used for patients who are above 50 years old. *Masha* is heavy, hot in potency, and unctuous and sweet in taste, and thus it will alleviate *Vata* and increase strength. *Navanita* is beneficial in *Pitta*-associated *Vata* and *Ghruta* (ghee) is good for *Vata* associated with *Pitta*. Humans have a physiologically *Pitta*-dominant nature in the fourth decade, followed by *Vata*-dominant nature as they step out in older age. The status of *Agni* in the elderly (older age) will be weak as compared to that in the 40s.[68] Being more *Pitta Samaka* (pacifying *Pitta Dosha*) and *Guru* (heavy to digest) in properties, *Navanita* is preferred in the fourth decade of life, and in later decades, ghee is used as it alleviates *Vata* and *Pitta* and ignites the digestive fire.

In chronic stroke cases or already lean/weak or elderly patients, *Vata Dosha* will be dominant with weak power of digestion and the patient has diminished body strength too. The diet advised in *Varsha Ritucarya* (the regimen for the rainy season), which is *Agni Dipana* (enhancing digestive and metabolic fire), easily digestible, *Vata* pacifying, and *Balya* (enhancing body strength), is to be used in this scenario. *Odana* (cooked rice) with old grains, *Yusha* (soup of pulses), *Mamsa Rasa* (meat soup), and *Mastu* (supernatant liquid separated from curd) are preferred in general during the rainy season. *Karkidaka Kanji* (porridge added with herbs), which is a special traditional recipe that has been used in southern India for hundreds of years during the month of *Karkidaka* (the month of the south-west monsoon, where peak rainfall will occur), has been found beneficial in nutritional and therapeutic aspects. This herbal porridge is prepared by processing porridge of *Sali* (rice)/*Sashtika* rice with herbs like *Dasamula* (roots of ten shrubs or herbs),[69] *Dasapuspa* (ten flowers),[70] *Jiraka* (cumin seeds), *Methika* (fenugreek seeds), *Candrasura* (common cress seeds), jaggery, and coconut milk. It will pacify aggravated *Vata Dosha*, increase digestive power, and enhance disease resistance and body strength. Porridge and *Yusha,* because of their liquid consistency, were recommended for stroke patients with difficulty in mastication and deglutition.

Specific conditions – examples:

A. In case of constipation/obstipation: *Peya* (rice soup) prepared with *Amalaka* (*Emblica officinalis*) and *Yava* (barley) added with ghee is given[2] (*Charaka Samhita, Chikitsasthana* 3/184).
B. Sleeplessness, excessive thirst: *Peya* prepared with *Nagara* (dry ginger) and *Amalaka* (gooseberry) fried in ghee and with sugar[2] (*Charaka Samhita, Chikitsasthana* 3/187).
C. In the case of dysphagia, diet must be highly individualized, depending on the patient's chewing and swallowing ability. Foods' texture and viscosity may be altered into that of thin nectar-like fluids by thorough cooking and the addition of clarified butter. Fluid should be thin enough to be sipped through a straw or a cup, but thick enough to fall off a tipped spoon slowly,[71] for example, thin *Yusha* (gruel) of *Mudga or Masha* with *Ghruta* and *Mamsa Rasa* (meat soup) with *Peya* (rice soup).

Parkinson's Disease

Though Parkinson's disease (PD) is a neurodegenerative disease, the contributive pathology for neurodegeneration is inflammation and resultant oxidative stress.[72] In such conditions *Rukshana* (dehydrating therapies) is advised instead of *Langhana* alone.[73] Beyond the localized inflammation in the central nervous system, inflammation in the gastrointestinal tract like inflammatory bowel disease also poses a great risk for the development of PD.[74] *Rukshana* and *Langhana* can be done by use of food articles such as barley, fried wheat grain powder, green gram, and green leafy vegetables. These have *Ruksha* (dry), *Ushna* (hot), and *Laghu* (light) properties.

After the first phase of *Rukshana* and *Langhana* nourishment diet should be given as explained in the recovery stage of stroke. Food made with *Masha* (black gram), such as *Masha Yusha* (black gram soup), *Mashadi Modaka* (fritters), *Idli*, and *Dosa*, has been found to be beneficial for Parkinson's disease patients with muscle rigidity. *Masha*, because of its *Snigdha*, *Guru*, and *Ushna* properties, will provide *Saithilya* (looseness) to the body, hence reducing rigidity in PD. *Masha* is recommended for spastic hemiplegic conditions also. One recipe, *Mashadi Modaka* (*Laddu*), made up of black gram, *Kapikacchu* (*Mucuna pruriens*), *Ghruta,* and *Marica* (black pepper), given in the morning, is a traditional recipe recommended for PD in various parts of India. *Kapikacchu* (*M. pruriens*) is a natural source of L-DOPA.[75] *Marica* (black pepper), because of its *Ushna*, *Teekshna* (sharp), and *Pramathi* (kind of eliminative therapeutic action which breaks the amalgamation of impaired *Dosha* and removes them out of the body) action, clearing the body channels, enables targeted delivery of the nutrients, because "piperine", the active ingredient in the *Marica,* enhances the absorption and bioavailability of thenutrients.[76]

Observational studies have shown that the consumption of fruits, vegetables, and marine fish is good for delaying the progression of PD.[77]

Headache

In Ayurveda, the pathology of *Shirashula* is due to the association of *Rakta* (blood) along with other *Doshas;* hence all *Pitta-Rakta* aggravating factors like foods having pungent, salty, sour, and hot properties and fatty food items (for example, spicy meals, horse gram, curd, pickles, fermented food) has to be restricted, especially in migraine headache associated with gastrointestinal symptoms. If the headache is aggravated/triggered by using excessive *Katurasa* (pungent), hot food items, exertion, and exposure to sunlight, then involved *Doshas* are dominantly *Pitta* and *Vata;* thus the diet should include food and drinks that have *Madhura* (sweet in taste), *Sheeta* (cold in potency), and *Snigdha* (unctuous) properties. *Payasa* (sweet gruel) prepared with rice, milk, sugar, and *Ghrita* (clarified butter) is given in this condition. *Jangala Mamsa Rasa* (meat soup) with boiled rice and *Ghruta is* also advised. Confectionaries made with wheat, sugar, ghee, and honey such as *Ghevar* (a common Indian sweet)[78] (*Chakradatta, Shiroroga Chikitsa*, 60/51) are also advised. In *Suryavarta* (a type of headache), *Ksheera Sarpi* (ghee extracted from milk)[78] (*Chakradatta, Shiroroga Chikitsa*, 60/52) is indicated for drinking purposes as it alleviates aggravated *Vata* and *Rakta* (blood), which is the causative factors of *Suryavarta*[2] (*CharakaSamhita, Siddhi Sthana* 9/79). Recipes with *Madhura* (sweet in taste) and *Snigdha* (unctuous) properties like *Kundalini* (*jalebi*) along with milk as well as *Petha* (a type of sweet made up of ash guard, milk, and sugar) in the morning time on an empty stomach are found to be beneficial for prevention of episodes of headache. In Northern India, the recipes have been in practice for more than 50 years. For *Ardhavabhedaka* type of headache, which has predominant features of *Vata, Morawala* (*Amlamurabba* – made from gooseberries and soaked in sugar syrup) taken on an empty stomach in the morning has shown promising results in treating *Pitta-Vata* dominant headaches and migraine. *Amlamurrabba,* by its sweet taste and cold potency, pacifies *Pitta* and, due to its unctuousness, alleviates *Vata* too.

Diet is considered a disease-modifying factor in migraine[79]and better outcomes in terms of frequency and severity of episodes are associated with modification of diet. An understanding of precipitating diet and avoidance of those factors is necessary. High-fat, spicy, fast food, and fasting are the most common aggravating factors.

NUTRAVIGILANCE

Barley, sorghum, and most of the millets are staple foods that should only be consumed cautiously because they are known to exacerbate *Vata* disorders[2] (*Charaka Samhita, Sutrasthana*, 27/16–19). Similarly, legumes like peas and chickpeas should also be avoided[2] (*Charaka Samhita, Sutrasthana*, 27/28–29). Excess use of kokum may cause an increase in anxiety or reduced sleep.[79,80] Use of *Rasona* (garlic) is contraindicated in bleeding disorders such as hemorrhagic stroke[19] (*Ashtanga Hrudaya Chikitsasthana* 22/70). Dietary products such as ice cream, tinned foods and drinks, red meat, dietary soda, and sugar-sweetened food items are associated with faster progression and severity of Parkinson's disease. Hence these are to be avoided in subjects of Parkinson's disease.[81]

Scope for Further Research

Nutraceuticals are widely used in neurological disorders. In cases where a patient isn't eating enough, nutritional supplements might help maintain their nutritional status. However, its use ought to be justified and ought to advance the therapeutic goal. Research may also be directed at the role that food plays as an epigenetic regulator of gene expression; a particular research area of interest is elaborating on how nutrients alter the regulation of genes whose expression is associated with neurological illnesses like Parkinson's disease. Furthermore, sensory and nutritional profiling of *Pathya Kalpana* mentioned in Ayurveda can be done to explore product development possibilities and establish quality control standards.

REFERENCES

1. Schwartz MW, Woods SC, Porte D Jr, Seeley RJ, Baskin DG. Central nervous system control of food intake. Nature. 2000 Apr 6;404(6778):661–71. doi: 10.1038/35007534. PMID: 10766253.
2. Agnivesa. CharakaSamhita. reprint edition. Varanasi: Chaukhamba Sanskrit Series office 2020; pp. 79, 619, 722, 617, 257,621, 624, 617,113, 100, 721-22, 153-155,157,130-131, 160-161,158, 162, 409-410, 414.
3. Tiwari P. editor, KashyapaSamhita of MariciKashyapa. Reprinted., Varanasi: ChaukhambaVishwabharati 2020;p.468.
4. Sushruta. Sushruta Samhita. Reprint edi. Varanasi: Chaukhambha Sanskrit Sansthan 2019; pp.656, 267, 261.
5. Asha Kumari, P V Tiwari. Editors. Yogaratnakara, reprint Edi.Varanasi: ChaukhambhaVisvabharati 2010; pp.589,621.
6. Dionyssiotis Y, Papachristos A, Petropoulou K, Papathanasiou J, Papagelopoulos P. Nutritional Alterations Associated with Neurological and Neurosurgical Diseases. Open Neurol J. 2016 Jul 26;10:32–41. doi: 10.2174/1874205X01610010032. PMID: 27563361; PMCID: PMC4962432.
7. Lopresti AL. The Effects of Psychological and Environmental Stress on Micronutrient Concentrations in the Body: A Review of the Evidence. AdvNutr. 2020 Jan 1;11(1):103–112. doi: 10.1093/advances/nmz082. PMID: 31504084; PMCID: PMC7442351.
8. Hammond N, Wang Y, Dimachkie MM, Barohn RJ. Nutritional neuropathies. NeurolClin. 2013 May;31(2):477–89. doi: 10.1016/j.ncl.2013.02.002. PMID: 23642720; PMCID: PMC4199287.
9. Thakur KT, Albanese E, Giannakopoulos P, et al. Neurological Disorders. In: Patel V, Chisholm D, Dua T, et al., editors. Mental, Neurological, and Substance Use Disorders: Disease Control Priorities, Third Edition (Volume 4). Washington (DC): The International Bank for Reconstruction and Development/The World Bank; 2016 Mar 14. Chapter 5. doi: 10.1596/978-1-4648-0426-7_ch5.
10. Mao XY, Yin XX, Guan QW, Xia QX, Yang N, Zhou HH, Liu ZQ, Jin WL. Dietary nutrition for neurological disease therapy: Current status and future directions. Pharmacol Ther. 2021 Oct;226:107861. doi: 10.1016/j.pharmthera.2021.107861. Epub 2021 Apr 23. PMID: 33901506.
11. Seth, Veenu. MFN 005 – Unit 17 Nutritional management of neurological disorders, Indira Gandhi National Open University, New Delhi, 2019, 430p.
12. National Institute for Health and Care Excellence (NICE) (2017). Parkinson's disease in adults. NG71. 22p. Available at: M www.nice.org.uk/guidance/ng71 (accessed on 12 December 2022).
13. Bassotti G, De Giorgio R, Stanghellini V, Tonini M, Barbara G, Salvioli B, Fiorella S, Corinaldesi R. Constipation: a common problem in patients with neurological abnormalities. Ital J GastroenterolHepatol. 1998 Oct;30(5):542–8. Erratum in: Ital J GasroenterolHepatol 1999 Apr;31(3).
14. J. Webster-Gandy, A. Madden, M. Holdsworth, Oxford Handbook of Nutrition and dietetics, 3rd Edition, United Kingdom, Oxford University Press, 2020, 921 P.

15. Aziz NA, van derMarck MA, Pijl H, Olde Rikkert MG, Bloem BR, Roos RA. Weight loss in neurodegenerative disorders. J Neurol. 2008 Dec;255(12):1872–80. doi: 10.1007/s00415-009-0062-8. Epub 2009 Jan 22. PMID: 19165531.
16. Amano Shinichi, Kegelmeyer Deborah, Hong S. Lee, Rethinking energy in parkinsonian motor symptoms: a potential role for neural metabolic deficits, Frontiers in Systems Neuroscience, VOLUME 8, YEAR 2015, www.frontiersin.org/articles/10.3389/fnsys.2014.00242. doi. 10.3389/fnsys.2014.00242. ISSN 1662-5137.
17. Leal SC, Bittar J, Portugal A, Falcão DP, Faber J, Zanotta P. Medication in elderly people: its influence on salivary pattern, signs and symptoms of dry mouth. Gerodontology. 2010 Jun;27(2):129–33. doi: 10.1111/j.1741-2358.2009.00293.x. Epub 2010 Mar 11. PMID: 20337727.
18. Spielman AI. Interaction of saliva and taste. J Dent Res. 1990 Mar;69(3):838–43. doi: 10.1177/00220345900690030101. PMID: 2182682.
19. Vaghbhata. Ashtanga Hrudaya. Reprint edition. Varanasi: ChaukhambhaSurbharatiPrakashan 2018; pp. 6, 87,102-103, 543, 734.
20. Mishra S, Trikamji B, Singh S, Singh P, Nair R. Historical perspective of Indian neurology. Ann Indian AcadNeurol 2013;16:467–77.
21. Ramteke RS, Patil PD, Thakar AB. Efficacy of Nasya (nasal medication) in coma: A case study. Anc Sci Life. 2016 Apr-Jun;35(4):232–5. doi: 10.4103/0257-7941.188188. PMID: 27621522; PMCID: PMC4995859.
22. Byadgi PS. Critical appraisal of DoshavahaSrotas. Ayu. 2012 Jul;33(3):337–42. doi: 10.4103/0974-8520.108819. PMID: 23723638; PMCID: PMC3665105.
23. Kowalska, M., Owecki, M., Prendecki, M., Wize, K., Nowakowska, J., Kozubski, W., & Dorszewska, M. L. a. (2017). Aging and Neurological Diseases. In J. Dorszewska, & W. Kozubski (Eds.), Senescence – Physiology or Pathology. IntechOpen.
24. Oksala NK, Oksala A, Pohjasvaara T, Vataja R, Kaste M, Karhunen PJ, Erkinjuntti T. Age related white matter changes predict stroke death in long-term follow-up. Journal of Neurology, Neurosurgery and Psychiatry. 2009;80:762–766.
25. Liu H, Yang Y, Xia Y, Zhu W, Leak RK, Wei Z, Wang J, Hu X. Aging of cerebral white matter. Ageing Res Rev. 2017 Mar;34:64–76. doi: 10.1016/j.arr.2016.11.006. Epub 2016 Nov 16. PMID: 27865980; PMCID: PMC5250573.
26. Prell T, Perner C. Disease Specific Aspects of Malnutrition in Neurogeriatric Patients. Frontiers in Aging Neuroscience. 2018;10:80. DOI: 10.3389/fnagi.2018.00080. PMID: 29628887; PMCID: PMC5876291.
27. Charaka Samhita chikitsasthana, available at www.Charakasamhitaonline.com/mediawiki-1.32.1/index.php?title=Vatavyadhi_Chikitsa#General_etiological_factors_and_basic_pathogenesis_of_vata_disorders (accessed on 26th December 2022).
28. Feigin VL, Brainin M, Norrving B, Martins S, Sacco RL, Hacke W, Fisher M, Pandian J, Lindsay P. World Stroke Organization (WSO): Global Stroke Fact Sheet 2022. Int J Stroke. 2022 Jan;17(1):18–29. doi: 10.1177/17474930211065917. Erratum in: Int J Stroke. 2022 Apr;17(4):478. PMID: 34986727.
29. Aho K, Harmsen P, Hatano S, Marquardsen J, Smirnov VE, Strasser T. Cerebrovascular disease in the community: results of a WHO collaborative study. Bull World Health Organ 1980; 58: 113–130.
30. Ediriweera ER, Perera MS. Clinical study on the efficacy of Chandra Kalka with MahadaluAnupanaya in the management of Pakshaghata (Hemiplegia). Ayu. 2011 Jan;32(1):25–9. doi: 10.4103/0974-8520.85720. PMID: 22131754; PMCID: PMC3215412.
31. Agarwal S, Gilbert R. Progressive Supranuclear Palsy. [Updated 2022 Apr 3]. In: StatPearls [Internet]. Treasure Island (FL): StatPearls Publishing; 2022 Jan.
32. Kampavata/ Vepathu (Parkinson's disease), available at www.nhp.gov.in/kampavata-vepathu-parkinsons-disease_mtl (accessed on 26 December 2022).
33. Lebouvier T, Chaumette T, Paillusson S, Duyckaerts C, Bruley des Varannes S, Neunlist M, Derkinderen P. The second brain and Parkinson's disease. Eur J Neurosci. 2009 Sep;30(5):735–41. doi: 10.1111/j.1460-9568.2009.06873.x. Epub 2009 Aug 27. PMID: 19712093.
34. Balestrino R, Schapira AHV. Parkinson disease. Eur J Neurol. 2020 Jan;27(1):27–42. doi: 10.1111/ene.14108. Epub 2019 Nov 27. PMID: 31631455.
35. Simon DK, Tanner CM, Brundin P. Parkinson Disease Epidemiology, Pathology, Genetics, and Pathophysiology. ClinGeriatr Med. 2020 Feb;36(1):1–12. doi: 10.1016/j.cger.2019.08.002. Epub 2019 Aug 24. PMID: 31733690; PMCID: PMC6905381.
36. Fukagawa NK, Ziska LH. Rice: Importance for Global Nutrition. J Nutr Sci Vitaminol (Tokyo). 2019;65(Supplement):S2–S3. doi: 10.3177/jnsv.65.S2. PMID: 31619630.

37. Hu EA, Pan A, Malik V, Sun Q. White rice consumption and risk of type 2 diabetes: meta-analysis and systematic review. BMJ. 2012 Mar 15;344:e1454. doi: 10.1136/bmj.e1454. PMID: 22422870; PMCID: PMC3307808.
38. Singh N. Pulses: an overview. J Food Sci Technol. 2017 Mar;54(4):853–857. doi: 10.1007/s13197-017-2537-4. Epub 2017 Feb 14. PMID: 28303036; PMCID: PMC5336460.
39. M. S, S, Bukkan, DS. A review on nutritional composition, antinutritional components and health benefits of green gram (*Vigna radiata* (L.) Wilczek). J Food Biochem. 2021; 45:e13743. https://doi.org/10.1111/jfbc.13743.
40. Johanson JF, Sonnenberg A, Koch TR, McCarty DJ. Association of constipation with neurologic diseases. Dig Dis Sci. 1992 Feb;37(2):179–86. doi: 10.1007/BF01308169. PMID: 1735333.
41. P V Sharma, G P Sharma. Editor. KaiyadevaNighantu (Pathya-ApathyaNibodhaka). 1st ed. Varanasi: ChaukhambaOrientalia 2013;p.316.
42. Valsta LM, Tapanainen H, Männistö S. Meat fats in nutrition. Meat Sci. 2005 Jul;70(3):525–30. doi: 10.1016/j.meatsci.2004.12.016. PMID: 22063750.
43. Haminiuk, C.W.I., Maciel, G.M., Plata-Oviedo, M.S.V. and Peralta, R.M. (2012), Phenolic compounds in fruits – an overview. International Journal of Food Science & Technology, 47: 2023–2044. https://doi.org/10.1111/j.1365-2621.2012.03067.x
44. Pandey KB, Rizvi SI. Plant polyphenols as dietary antioxidants in human health and disease. Oxid Med Cell Longev. 2009 Nov-Dec;2(5):270–8. doi: 10.4161/oxim.2.5.9498. PMID: 20716914; PMCID: PMC2835915.
45. Lie, Z. Y., Tandiono, M., Pricillia, L., Lukito, A. N., & Gasmara, C. P. (2016). Brain neuron regeneration in post-stroke rehabilitation treatment using grape peel and seed extract (*Vitis vinifera*) in inducing erk1 / 2 pathway. Malang Neurology Journal, 2(1), 19–23.
46. Essa MM, Akbar M, Khan MA. Beneficial effects of date palm fruits on neurodegenerative diseases. Neural Regen Res. 2016 Jul;11(7):1071–2. doi: 10.4103/1673-5374.187032. PMID: 27630684; PMCID: PMC4994443.
47. Wattanathorn J, Muchimapura S, Thukham-Mee W, Ingkaninan K, Wittaya-Areekul S. Mangifera indica fruit extract improves memory impairment, cholinergic dysfunction, and oxidative stress damage in animal model of mild cognitive impairment. Oxid Med Cell Longev. 2014;2014:132097. doi: 10.1155/2014/132097. Epub 2014 Jan 29. PMID: 24672632; PMCID: PMC3941952.
48. Amritpal Singh. Editor, Bhavamisra'sBhavaprakasha Nighantu. 1st edition, Delhi: ChaukhambaOrientalia2007; pp.195, 6, 177-178, 261, 271.
49. Braidy N, Selvaraju S, Essa MM, Vaishnav R, Al-Adawi S, Al-Asmi A, Al-Senawi H, AbdAlrahmanAlobaidy A, Lakhtakia R, Guillemin GJ. Neuroprotective effects of a variety of pomegranate juice extracts against MPTP- induced cytotoxicity and oxidative stress in human primary neurons. Oxid Med Cell Longev. 2013;2013:685909. doi: 10.1155/2013/685909. Epub 2013 Oct 3. PMID: 24223235; PMCID: PMC3816068.
50. Chuah LO, Ho WY, Beh BK, Yeap SK. Updates on Antiobesity Effect of Garcinia Origin (-)-HCA. Evid Based Complement Alternat Med. 2013;2013:751658. doi: 10.1155/2013/751658. Epub 2013 Aug 6. PMID:23990846; PMCID: PMC3748738.
51. Husain I, Zameer S, Madaan T, Minhaj A, Ahmad W, Iqubaal A, Ali A, Najmi AK. Exploring the multifaceted neuroprotective actions of *Emblica officinalis* (Amla): a review. Metab Brain Dis. 2019 Aug;34(4):957–965. doi: 10.1007/s11011-019-00400-9. Epub 2019 Mar 8. PMID: 30848470.
52. Teimouri E, Rainey-Smith SR, Bharadwaj P, Verdile G, Martins RN. Amla Therapy as a Potential Modulator of Alzheimer's Disease Risk Factors and Physiological Change. J Alzheimers Dis. 2020;74(3):713–733. doi: 10.3233/JAD-191033. PMID: 32083581.
53. Ranasinghe RASN, Maduwanthi SDT, Marapana RAUJ. Nutritional and Health Benefits of Jackfruit (*Artocarpus heterophyllus* Lam.): A Review. Int J Food Sci. 2019 Jan 6;2019:4327183. doi: 10.1155/2019/4327183. PMID:30723733; PMCID: PMC6339770. 10p.
54. Ajaib M, Hussain T, Farooq S, Ashiq M. Analysis of Antimicrobial and Antioxidant Activities of *Chenopodium ambrosioides*: An Ethnomedicinal Plant. Journal of Chemistry. 2016;2016:4827157. 1–11p.
55. Cabezas R, Fidel Avila M, Torrente D, Gonzalez J, Santos El-Bachá R, Guedes R, et al. Chapter 76 – Natural Antioxidants in Dementia: An Overview. In: Martin CR, PreedyVR, editors. Diet and Nutrition in Dementia and Cognitive Decline. San Diego: Academic Press; 2015. p. 827–36.
56. Mani V, Ahmad A, Ramasamy K, Lim SM, Abdul Majeed AB. Chapter 97 – Murrayakoenigii Leaves and Their Use in Dementia. In: Martin CR, Preedy VR, editors. Diet and Nutrition in Dementia and Cognitive Decline. San Diego: Academic Press; 2015. p. 1039–48.

57. Ahmad, T., Cawood, M., Iqbal, Q., Ariño, A., Batool, A., Tariq, R. M. S., Azam, M., &Akhtar, S. (2019). Phytochemicals in *Daucus carota* and Their Health Benefits-Review Article. *Foods (Basel, Switzerland)*, *8*(9),424. https://doi.org/10.3390/foods8090424.
58. Johnson E.J. Role of lutein and zeaxanthin in visual and cognitive function throughout the lifespan. *Nutr. Rev.* 2014;72:605–612. doi: 10.1111/nure.12133.
59. SooranadKunjanpillai. Editor, Raghunathasuri'sBhojanakutuhala. 1st Edition, Trivandrum: AnantashayanamVisvavidyalaya 1931;pp.59–60.
60. RashidA, KhanHH. The mechanism of hypotensive effect of garlic extract. *J PakMedAssoc.* 1985;35:357–362.
61. Ried K, Frank OR, Stocks NP. Aged garlic extract reduces blood pressure in hypertensives: a dose-response trial. *Eur J ClinNutr.* 2013a;67:64–70.
62. Bayan, L., Koulivand, P. H., &Gorji, A. (2014). Garlic: a review of potential therapeutic effects. *Avicenna journal of phytomedicine*, *4*(1), 1–14.
63. Satam, Namrata K, Parab, Lavu S, Bhoir, Suvarna I, HPTLC finger print analysis and antioxidant activity of flavonoid fraction of Solanum melongena Linn fruit (2013) International Journal of Pharmacy and Pharmaceutical Sciences, 5 (3), pp. 734–740, www.scopus.com/inward/record.uri?eid=2-s2.0-84879995097&partnerID=40&md5=fb3a7ae35ddeb62485d08a73b304d8ae
64. Das M., Barua N. Pharmacological activities of Solanum melongena Linn. (Brinjal plant) (2013), International Journal of Green Pharmacy, 7 (4), pp. 274–277, www.scopus.com/inward/record.uri?eid=2-s2.0-84890344989&doi=10.4103%2f0973-8258.122049&partnerID=40&md5=13acd532391f8388cf730b7023ac42ae
65. A., Aisha &Kotagasti, Tabassum & Ambar, S. (2019). Medicinal properties and uses of Salabmisri (*Orchis latifolia* Linn)– A literary review. The Journal of Phytopharmacology. 8. 18–20. 10.31254/phyto.2019.8105.
66. Gîlcă M, Soian I, Mohora M, Petec C, Muscurel C, Dinu V. The effect of fasting on the parameters of the anti-oxidant defense system in the blood of vegetarian human subjects. Rom J Intern Med. 2003;41(3):283–92. PMID: 15526512.
67. Rajani A, Vyas M K, Vyas HA. Comparative study of Upavasa and Upavasa with Pachana in the management of Agnisada. Ayu. 2010 Jul;31(3):351–4. doi: 10.4103/0974-8520.77166. PMID: 22131738; PMCID: PMC3221070.
68. Rémond D, Shahar DR, Gille D, Pinto P, Kachal J, Peyron MA, Dos Santos CN, Walther B, Bordoni A, Dupont D, Tomás-Cobos L, Vergères G. Understanding the gastrointestinal tract of the elderly to develop dietary solutions that prevent malnutrition. Oncotarget. 2015 Jun 10;6(16):13858–98. doi: 10.18632/oncotarget.4030. PMID: 26091351; PMCID: PMC4546438.
69. Aparna S, Ved DK, Lalitha S, Venkatasubramanian P. Botanical identity of plant sources of Daśamūla drugs through an analysis of published literature. Anc Sci Life. 2012 Jul;32(1):3–10. doi: 10.4103/0257-7941.113790. PMID: 23929986; PMCID: PMC3733204.
70. Vijayakumar N, Gangaprasad A. Preliminary phytochemical screening and antioxidant activity of *Emilia sonchifolia* (L.) DC., a member of 'Dashapushpa'. IJRAR. 2018;5(4):124–9.
71. Nikolaos K, Charilaos D, Meropi K, Evangelia M, Kalliopi A. Clinical Nutrition in Practice. 1st ed. United Kingdom. Blackwell Publishing Ltd. 2010.
72. Dias V, Junn E, Mouradian MM. The role of oxidative stress in Parkinson's disease. J Parkinsons Dis. 2013;3(4):461–91. doi: 10.3233/JPD-130230. PMID: 24252804; PMCID: PMC4135313.
73. Jindal N, Shamkuwar MK, Berry S. Importance of Rookshana Karma (dehydrating therapy) in the management of transverse myelitis. Ayu. 2012 Jul;33(3):402–5. doi: 10.4103/0974-8520.108852. PMID:23723649; PMCID: PMC3665107.
74. Li Y, Chen Y, Jiang L, Zhang J, Tong X, Chen D, Le W. Intestinal Inflammation and Parkinson's Disease. Aging Dis. 2021 Dec 1;12(8):2052–2068. doi: 10.14336/AD.2021.0418. PMID: 34881085; PMCID: PMC8612622.
75. Katzenschlager R, Evans A, Manson A, Patsalos PN, Ratnaraj N, Watt H, Timmermann L, Van derGiessen R, Lees AJ. *Mucuna pruriens* in Parkinson's disease: a double blind clinical and pharmacological study. J Neurol Neurosurg Psychiatry. 2004 Dec;75(12):1672–7. doi: 10.1136/jnnp.2003.028761. PMID: 15548480; PMCID: PMC1738871.
76. Anshuly Tiwari, Kakasaheb R. Mahadik, Satish Y. Gabhe. Piperine: A comprehensive review of methods of isolation, purification, and biological properties. Medicine in Drug Discovery. 2020(7);100027.
77. Laurie K. Mischley, Richard C. Lau, Rachel D. Bennett, "Role of Diet and Nutritional Supplements in Parkinson's Disease Progression", Oxidative Medicine and Cellular Longevity, vol. 2017, Article ID 6405278, 9 pages, 2017.

78. P V Sharma. Editor and translator, Chakradatta (Sanskrit text with English translation). Reprint edi..Delhi: Chaukhamba Orientalia 2007; Siroroga Chikitsa, Chapter-60, Verse 39-40. pp.520–521.
79. Ibrahim, M.K., Aboelsaad, M., Tony, F. et al. *Garcinia cambogia* extract alters anxiety, sociability, and dopamine turnover in male Swiss albino mice. SN Appl. Sci. 4, 23 (2022). 1–7p. https://doi.org/10.1007/s42452-021-04902-z
80. Kim D, Kim J, Kim S, Yoon M, Um M, Kim D, Kwon S, Cho S. Arousal-Inducing Effect of *Garcinia cambogia* Peel Extract in Pentobarbital-Induced Sleep Test and Electroencephalographic Analysis. Nutrients. 2021 Aug19;13(8):2845. doi: 10.3390/nu13082845. PMID: 34445005; PMCID: PMC8399249.
81. Vittorio Emanuele Bianchi, Laura Rizzi& Fahad Somaa (2022) The role of nutrition on Parkinson's disease: a systematic review, Nutritional Neuroscience, DOI: 10.1080/1028415X.2022.2073107.

SECTION C

Therapeutic Nutrition in Ayurveda for Gynecology and Obstetrics

Deepali Rajput

1. INTRODUCTION

Nutrition has been high on the international public health agenda after the United Nation's proclamation of the 'Decade of Action on Nutrition (2016–2025)'. As malnutrition due to poor-quality diet is becoming a global health problem, strong attention is required toward food and nutrition. Many families are unable to afford sufficient, diverse, and nutritious food such as fresh vegetables, fruits, legumes, nuts, and animal-source food, while energy-dense processed food and drinks that are high in fat, sugar, and/or salt are often cheaper and readily available.[1]

Kashyapa described '*Ahara*', i.e. food, as '*Mahabhaishaja*', i.e. supreme medicine, as health cannot be maintained with medicines only; it requires '*Pathya Ahara*', i.e. wholesome diet along with medicine.[2]

Nutrition influences not only the health of women but also the health of their progeny; hence adequate nutrition is especially critical for them.[3] Furthermore, women have distinct nutritional requirements throughout their life as they go through many hormonal and physical changes during puberty, pregnancy, breastfeeding, and menopause.[4]

During adolescence, growth and development are transformative and have profound consequences on an individual's health in later life as well as the health of any potential children. The current generation of adolescents is going through a time of unprecedented change in the food environment, where problems of micronutrient deficiency and food insecurity persist and obesity burgeoning.[5] It is widely recognized that optimum nutrition in early life is the foundation for long-term health. A healthy maternal dietary habit along with adequate maternal body composition, metabolism, and placental nutrient supply reduces the risk of maternal and fetal complications and long-term effects on the offspring.[6] During pregnancy, a poor diet lacking in key nutrients like iodine, iron, folate, calcium, and zinc can cause anemia, preeclampsia, hemorrhage, and death in mothers. They can also lead to stillbirths, low birth weight, wasting, and developmental delay in children. Overweight women who gain excess weight during pregnancy cause obesity, alteration in glucose metabolism, and increased cardiovascular risks in their children in later life, thus establishing intergenerational amplification of the obesity epidemic.[4]

Evidence suggests that the propensity to develop Noncommunicable diseases and obesity may be influenced during fetal life and infancy. The first 1000 days are crucial for the prevention of adulthood

DOI: 10.1201/9781003345541-19

diseases starting from conception. Hence special attention should be given to the nutrition of women during breastfeeding.[7] The same goes for the perimenopausal woman; diet around the menopausal transition has a vital effect on physical activity and health of the woman. As menopause is a period of hormonal transition, the reduction in estrogen levels leads to increased cardiovascular episodes, metabolic syndrome, and increased bone loss, leading to osteoporosis, reduction in muscle mass, increased fat deposition, and other psychosomatic discomforts. Hence there is an increased need for micronutrients and antioxidants to counteract the free radical injuries associated with aging.[8]

In *Ayurvedic* texts like *Brihatrayee* and *Laghutrayee*, dietary regimes for pregnancy, lactation, and postabortion are described in detail. There is a description of the dietary regime for the menstrual phase also. Though undernutrition is mainly an issue of low-income countries, malnutrition due to poor-quality food is becoming a global health issue.[6] Hence, creation of intersectoral and multidisciplinary collaboration to promote health throughout life through nutrition is necessary.

2. THERAPEUTIC NUTRITION FROM CLASSICAL REFERENCES

2A. Therapeutic Nutrition in Obstetrics from Classical References

2A.1 *Pregnancy*

Kashyapa has explained that the diet consumed by a pregnant woman leads to the nourishment of her own body, nourishment of the fetus, and nourishment of breasts.[9] It should be according to *Desha* (place), *Kala* (season), and *Agni* (Digestive fire).[10]

Monthly Diet Regime for a Pregnant Woman

TABLE 15.1 Dietary Recommendation during Pregnancy from Classical References

MONTH	*CHARAKA*[11]	*SUSHRUTA*[12]	*VAGBHATA*[13]	*HARITA*[14]
1st	*Asanskarita Ksheera* (milk) frequently in the desired quantity. Congenial diet twice a day.	Liquid Diet which has *Madhura* and *Sheeta* properties.	*Ksheera* (milk) is timely and in a specific quantity as per *Agni* (digestive fire).	*Navneeta* (Butter) and *Madhu* (Honey) with certain herbs. *Ksheera* (milk) processed with *Madhura Dravya.*
2nd	*Ksheera* (milk) is processed with Madhura herbs.	A liquid diet which has *Madhura* and *Sheeta* properties.	*Ksheera* (milk) processed with *Madhura* herbs.	*Ksheera* (milk) processed with *Kakoli.*
3rd	*Ksheera* (milk) with ghrita (*ghee*) and *Madhu* (Honey).	Liquid diet which has *Madhura* and *Sheeta* properties. *Shashtik* rice cooked with *Ksheera* (milk).	*Ksheera* (milk) with *Ghrita* (*ghee*) and *Madhu* (Honey).	*Krushera* (Olia prepared with Rice and pulses).

TABLE 15.1 ***(Continued)*** Dietary Recommendation during Pregnancy from Classical References

MONTH	*CHARAKA*[11]	*SUSHRUTA*[12]	*VAGBHATA*[13]	*HARITA*[14]
4th	Butter extracted from *Ksheera* (milk) ~12 g. *Ksheera* (milk) with butter.	*Shashik* rice with curd, Food processed with *Ksheera* (milk) and butter. Meat. Desired and Congenial food.	*Ksheera* (milk) with ~12 g butter.	*Sanskrit Odana* (processed Rice).
5th	*Ksheera* (milk) with *Ghrita*. *Ghrita* prepared from butter which is extracted from *Ksheera* (milk).	Cooked *Shashtik* rice with *Ksheera* (milk). Congenial food with meat soup *Ksheera* (milk) and *Ghee*.	*Ksheera* (milk) with *Ghrita*. *Ghrita* is prepared from butter which is extracted from *Ksheera* (milk).	*Payasam* [Sweetened rice gruel prepared with *Ksheera* (milk)].
6th	*Ghrita* prepared from *Ksheera* (milk) which is medicated with *Madhura* herbs.	Rice gruel or *Grita* medicated with *Tribulus terrestris*.	*Ghrita* prepared from *Ksheera* (milk) which is medicated with *Madhura* herbs.	*Madhura Dadhi* (Sweetened curd).
7th	*Ghrita* prepared from *Ksheera* (milk) which is medicated with Madhura herbs.	Medicated *Ghrita* with *Prithakparnyadi* Group of Herbs.	*Ghrita* prepared from *Ksheera* (milk) which is medicated with *Madhur* herbs.	*Ghritakhanda* (Sweet dish prepared with *ghee*).
8th	Gruel prepared with *Ksheera* (milk) with *Ghrita*.	*Snigdha* gruel Meat soup till delivery.	Rice Gruel prepared with *Ksheera* (milk).	*Ghritapurak* (Sweet dish prepared with ghee).
9th		*Snigdha* gruel Meat soup till delivery.	Rice cooked in meat soup with *Ghrita*.	Different varieties of cereals.

2A.2 Postpartum Period

TABLE 15.2 Dietary Recommendation for Postpartum Period from Classical References

CHARAKA15	*SUSHRUTA*[16]	*KASHYAPA*[17]
Considering the *Agni* (digestive fire) and strength of the patient. *Ghrita* with *Panchkola Churna* (powdered *Piper longum* fruits and roots, *Plumbago zeylanica*, *Piper chaba* and *Zingiber officinale*). Rice gruel with medicated *Ghrita* (ghee): 5–7 days. *Brihana* (Nourishing food).	Just after delivery, decoction of Bhadradaru (*Cedrus deodara*) followed by *Panchkola Churna* with Jaggery and lukewarm water for 2 days. Rice Gruel prepared with *Ksheera* (milk) or Rice Gruel with medicated *Ghrita* for 3 days. Rice with meat soup with *Yava* (*Hordeum vulgare*), *Kola* (*Ziziphus jujuba*, *Kulattha* (*Dolichos biflorus*) as per *Agni* (digestive fire).	*Manda pana*: 3–5 days considering the *Agni* (digestive fire). *Ghrita* (*ghee*). *Satmya* (Congenial food) which is *Laghu* (light to digest). Rice gruel without salt and without *Ghrita* (*ghee*) or any unctuous substance in small quantity seasoned with *Pippali* (*Piper longum*) and *Shunthi* (*Zingiber officinale*) *Kulattha* (*Dolichos biflorus*) soup with *Ghrita* and rock salt. Meat soup. Vegetables like *Kushmanda* (*Benincasa hispida*), Ervaruka (*Cucumis utilissimus*), *Balmulak* (*Raphanus sativus*) cooked in *Ghrita*. Lukewarm water for drinking.

2A.3 Abortions

a. Threatened Abortion
Ksheera with *Madhura* and *Sheeta Dravya*, such as Water Chestnuts (*Trapa bispinosa*), Cold water/tap for drinking, *Ksheera* (milk) medicated with '*Garbhostapaka Dravya*' such as *Shatawari (Asparagus racemosus)*, *Aja Ksheera* (milk), and Quail (*Coturnix coturnix*) meat.[18]

b. Complete Abortion
After the complete abortion, *Kulattha* (*Dolichos biflorus*) and *Haridra* (*Curcuma longa*) *Kwatha* (Decoction),[19] fat-free, salt-free rice gruel should be given. This gruel should be consumed according to the gestational age, i.e. if the pregnancy duration was 2 months, then it should be given for 2 days.[20]

2A.4 In Diseases in Obstetrics

TABLE 15.3 Dietary Recommendations for Diseases in Obstetrics from Classical References

DISEASES	*RECOMMENDED DIET*	*PROHIBITED DIET*
Emesis *Gravidarum:*	- *Yavasattu* (Flour of *Hordeum vulgare)* with *Shunthi* powder (*Zingiber officinale*) and *Bilva kwatha*[21] (*Aegle marmelos*). - *Tandulodaka* (Rice water) with *Sharkara* and *Dhanyak kalka* (crushed *Coriandrum sativum* leaves).[21] - *Lajambu* (Liquid prepared with parched paddy *Oryza sativa*) with *Bilwamajja (Aegle marmelos)*.[21] - *Matulunga rasa* (Citrus lemon juice), with *Laja* (parched paddy), *Kolmajja* (kernel of jujube *Ziziphus jujuba*), *Dadimasara* (*Punica granatum* seeds), *Madhu* (Honey) and *Sharkara*.[22] - *Ajamansa* (Goat Meat) cooked with *Amla Dadima* (*Punica granatum*) and without any salt.[22] - *Tandulodaka* (Rice water) with *Madhu* (Honey), *Sharkara, twaka* (*Cinnamomum zeylanicum*), *Ela* (*Elettaria cardamomum*), *Patra (Cinnamomum tamala)*, *Nagkeshara* (*Mesua ferrea*), and *Lajachurna* (parched paddy – *Oryza sativa*) [22] - *Laja Peya* (Drink prepared with parched paddy – *Oryza sativa*) with *Madhu* (Honey) and *sharkara*.[22] - *Mudga Yusha* (*Phaseolus aureus* soup) cooked with *Anardana* (*Punica granatum* seeds), *Lavana* (rock salt), and *Ghee*.[22] *Draksha* (*Vitis vinifera*), *Kapittha* (*Feronia elephantum*), *Narikel* (*Cocos nucifera*), *Dadima* (*Punica granatum*), *Amlaki* (*Emblica officinalis*)	
a. IUGR	- Meat soup with *Ghrita,* Meat of chicken, pork, iguana, goat, sheep, Eggs, Fish eggs, Meat grilled with skewer[23] - *Masha Yusha*[24] - *Ksheera* (milk) with *kashmariphala (Gmelina arborea)* and *sharker.*[25] - *Mastyanda* (fish eggs)[26]	*Upwas* (Fasting), *Pramitashana* (Eating in small quantity), *Kashaya* (Astringent) and *Ruksha* (dry food), Irregular food habits, stale food, *Vishtambi* (causing constipation) food, *Nitya Tiktarasa sevan* (Regular intake of bitter food articles)

2B. Therapeutic Nutrition in Gynecology from Classical References

2B.1 Menstrual Phase

During menstruation, women should consume *Havisya,* i.e. *Shali* Rice along with *ghee* and *Ksheera* (milk) or *Yava* (*Hordeum vulgare*) cooked in *Ksheera* (milk) in small quantities for 3 days. Women should use clay utensils or utensils made from leaves and should avoid food articles having *Amla* (sour), *Lavana* (salty), and *Ushna* (hot) properties.[27]

Vagbhata advised taking food made with *Yava* (*Hordeum vulgare*) which has *Koshthashodhana* and *Karshana* properties.[28]

2B.2 In Gynecological Disorders

a. **Artavakshaya (Oligomenorrhea, PCOS, Hypomenorrhea, Amenorrhea)**
   - *Tila* (*Sesamum indicum*), *Krishna tila* with jaggery, *Masha (Phaseolus mungo), Kulattha* (Dolichos biflorus)[26]
   - *Lashuna*[29]
   - *Ksheera* (milk), Buttermilk, Curd, Fish, Meat, meat soup[26]
   - *Amla* (Sour), *katu* (Pungent), and Tikshna, *Ushna* (hot) food articles[30]

b. **Raktapradar (Menorrhagia, DUB)**

Indicated Food[31]:

- *Sastika Shali* (Rice Variety which grows in 60 days), *Kodrava* (*Paspalum scrobiculatum*), *Yava* (*Hordeum vulgare*), *Mudga* (*Phaseolus aureus*), *Masura* (*Lens culinaris*), *Canaka* (*Cicer arietinum*), *Tuvari* (*Cajanus cajan*), *Makushta* (*Phaseolus aconitifolius*), *Lajasaktu* (Flour of parched paddy)
- Meat of Rabbit, Deer, pigeon, quail, Crab

Aja Ksheera (goatmilk), Goat *Ghee, GoKsheera* (cow milk), Cow *ghee*

- *Panasa* (*Artocarpus integrifolia*), *Priyala* (*Buchanania lanzan*), *Tanduliya* (*Amaranthus spinosus*), *Patola* (*Trichosanthes dioica*), *Kushmanda* (old *Benincasa hispida*), *Tadphal* (Ripe *Amomum subulatum* fruits and seeds), *Dadima* (*Punica granatum*), *Kharjura* (*Phoenix sylvestris*), *Amlaki* (*Emblica officinalis*), *Mishi* (*Anethum sowa*), *Narikela* (*Cocos nucifera*), Shringataka (*Trapa bispinosa*), Kapittha (*Feronia elephantum*), Shaluka (*Nelumbo nucifera* roots), Tumbi (*Lagenaria vulgaris*), *Draksha* (*Vitis vinifera*)
- Honey (Madhu), *Ikshu (Saccharum officinarum)*

Prohibited Food[32]:

- *Kulattha (Dolichos biflorus), Masha, Tila* (*Sesamum indicum*), *Lashuna* (*Allium sativum*), *Vartak* (*Solanum melongena*)
- Guda (jaggery), *Dadhi* (Curd), *Tambula* (*Piper betle*)
- *Katu* (Pungent), *Amla* (sour), and *Lavana* (salty) food articles

3. THERAPEUTIC NUTRITION FROM CONVENTIONAL SCIENCE

3A. Therapeutic Nutrition in Obstetrics

3A.1 Nutrition for Women during Pregnancy and Breastfeeding

Folic acid, Vitamin B12, Choline, Omega-3 fatty acids, Vitamin D, Calcium, and iron are some of the important micronutrients needed during childbearing age. During Pregnancy and Lactation, American College of Obstetricians and Gynecologists suggests taking around 340 kcal extra from second trimester, which is equivalent to around 1 glass of milk or 2–3 bread slices.[33] Pregnant women should have regular meals including food from all main groups, i.e. fruits and vegetables, carbohydrates, high-protein food, and dairy; 1900 mL (8 glasses) of liquids should be consumed as water, fresh juices without added sugar, and milk.[34]

Pregnant women should avoid alcohol, smoking, unpasteurized milk and milk products, raw or undercooked meat, liver, raw or partially cooked eggs, shellfish, and seafood. Pregnant women should also limit their caffeine intake up to 200 mg/dL and salt intake up to 6 g per day.[35] In breastfeeding period, women need around 330–400 extra kcal per day along with extra micronutrient supplementation. Smoking, alcohol, and caffeine should be avoided or restricted during breastfeeding.[36]

3A.2 Dietary Advice for Common Complaints during Pregnancy

a. Moring Sickness

Small and frequent meals with more complex carbohydrates such as dry toast, crackers, and breakfast cereals should be consumed. Food which increases nausea should be avoided. Consumption of fatty and sugary food should be reduced.

b. Constipation

Fiber-rich food such as whole grain breads, whole wheat pasta, etc., should be consumed.

c. Heart Burn

Small frequent meals should be taken.

Chocolates, fatty foods, alcohol, mint, and spicy and acidic food should be avoided especially before bedtime.

3B. Therapeutic Nutrition in Gynecology

3B.1 For Adolescent Girls

Adolescent girls are particularly vulnerable to malnutrition because this is the phase of growing faster than at any time after their first year of life. There is an increased need for protein, iron, and other micronutrients to support the growth spurt and increased iron demand due to menarche. Hence a healthy and balanced diet including all five food groups is recommended.[37]

3B.2 Women during the Perimenopausal Period

Diet must be rich in fruits and vegetables, whole grains, nuts, seeds and pulses, and sources of calcium and vitamin D such as dairy products. Unsaturated fats such as olive oil should be included.

Caffeine and alcohol trigger menopause symptoms such as hot flashes and night sweats; hence their intake should be limited. Lower intake of fatty meat, refined grains, sugar-sweetened food and beverages, etc., is recommended.[38]

After menopause, there are many changes in a woman's body. The requirement of Iron reduces as menstruation stops, but for some other nutrients it increases as absorption or metabolism also reduces

i) **Vitamin B12:** A diet abundant in fish, meat, and food fortified with B12 can supply adequate amount, but some may need supplements as absorption declines.
ii) **Calorie** requirement also reduces after menopause.

3B.3 Dietary Recommendation in Gynecological Disorders

There have been long debates regarding the role nutritional components and dietary habits play in modulating the risks of gynecological diseases such as leiomyoma, endometriosis, polycystic ovarian syndrome, and different gynecological malignancies.

To date, the contribution of diet and nutrition to gynecological Disorders remains a largely unexplored avenue that merits substantial future investigation. As most evidence is derived from epidemiological reports, experimental studies that consider potential confounding factors and accurately describe the independent effect of individual nutrients on the development and growth of gynecological diseases[39] are needed.

a. PCOS

Consumption of low calorie diet with limited intake of simple sugar, refined carbohydrates, low glycemic index food, and restricted intake of saturated and trans fatty acids helps in weight loss and further improvement in symptoms.[40]

b. Endometriosis

Alcohol consumption and diet high in trans fat have been shown to have negative impacts on occurrence of endometriosis. The results of the studies listed with regard to fruits, vegetables, dairy products, unsaturated fats, red meat, fibers, soy products, and coffee are not clear.[41]

4. CLASSIFICATION OF DISEASES IN OBSTETRICS AND GYNECOLOGY

4A. Classification of Diseases in Obstetrics

a. Garbhavyapada

Garbha = fetus, Vyapada = diseases, i.e. the diseases of the fetus such as abortion, Intrauterine growth restriction, oligohydramnios polyhydramnios.

b. Garbhopdrava

Garbha= fetus, *Upadrava* = complication

Another disease which develops/manifests after the main disease/event, i.e. diseases occurring in the pregnant woman due to the fetus such as Gestational Diabetes, Pregnancy-Induced Hypertension, and Hyperemesis Gravidarum which subsides on its own when the cause is removed. The main cause of these diseases is said to be the fetus and as the cause cannot be removed, it has to be treated symptomatically.

But if there is a threat to the mother's life, Sushruta advises to remove the root cause, i.e. fetus at the earliest possible.[42]

c. Garbhini Vyadhi

Garbhini = pregnant woman, *Vyadhi* = disease

The preexisting diseases of the pregnant woman or diseases occurring during pregnancy which are not related to the fetus come under this, for example, typhoid fever. Here the symptoms and pathology of the disease are the same as for the nonpregnant individual. However, the management approach differs, as the treatment not only should cure the patient but also shouldn't harm the fetus. Hence, *Charaka* described pregnant woman as a pot filled with oil up to the brim since the slightest disturbance/oscillation can harm the fetus.[43]

4B. Classification of Diseases in Gynecology

The Gynecological diseases can be classified on the basis of *Samprapti*[44] i.e. pathophysiology as follows:

Atipravriti – Excessive production/activity of the *Strotasa* as seen in menorrhagia, Dysfunctional uterine Bleeding, and Metrorrhagia.

Sanga – Mechanical or functional obstruction in the *Strotasa*, Polycystic ovarian syndrome, Amenorrhea, Oligomenorrhea.

Siragranthi – Formation of glandular structure in the *Strotasa*-like cysts.

Vimargaganana – Change in natural pathways of content of the *Strotasa* as seen in endometriosis.[45]

Not following advised dietary and lifestyle regime is the main cause of diseases like Premenstrual Syndrome and Perimenopausal Syndrome.

5. GENERAL THERAPEUTIC NUTRITION IN OBSTETRICS AND GYNECOLOGY

- Dietary recommendations for various stages of women's life along with a few gynecological diseases is found in *Ayurvedic* literature.
- Along with the Dietary recommendations for every month during pregnancy, postpartum, and postabortion, dietary modification for certain elements like Nausea and vomiting in pregnancy, threatened abortion, and IUGR are explained. Further modifications according *to Desha* (area of residence), *Kala* (season), period of gestation, and *Dushprajata* (patient who had abnormal labor) are also suggested.
- But certain conditions like PIH, GDM, moderate to severe anemia, and IUGR cannot be treated with dietary modifications alone. Here, diet acts as an adjacent therapy.
- Gynecological diseases like PCOS, Endometriosis, Adenomyosis, DUB, and Gynecological malignancies have complex *Samprapti* (pathology); hence they cannot be cured with dietary modification alone, but it can help to arrest further progression or recurrence of disease. There is a difference in *Samprapti* (pathology) of two phenotypes of PCOS, i.e. Obese and lean PCOS; hence the recommended diet also differs.
- Diseases like Premenstrual Syndrome and Perimenopausal syndrome occur due to not following the advised diet and lifestyle regimes. Here improvement can be seen with dietary and lifestyle modifications.

6. THERAPEUTIC NUTRITION AS PER *DWADASHA AHARA VARGA*

TABLE 15.4 Therapeutic Nutrition as per *Dwadasha Ahara Varga*

DISEASE		*HIGHLY INDICATED*	*INDICATED*	*CONTRA INDICATED*
***Raktapradar* (DUB, menorrhagia, metrorrhagia)**	*Shamidhyanya*	-	Green gram, Lentils, Chickpea, Moth Bean	Horse gram, Black gram
	ShakVarga	-	*Tanduliyaka*	-
	Ksheervarga	Goat milk	Cow milk	Curd, Buttermilk
	Ikshuvarga	-	*Sita*	Jaggery
Oligo/Hypomenorrhea/ Amenorrhea	*Shamidhyanya*	-	Horse gram	-
	Ksheervarga	-	Curd, Buttermilk	-
	Ikshuvarga	-	Jaggery	-
GDM	*Shukvarga*	Foxtail millet, Kodo Millet, Bamboo Seeds, Red Rice, barley	Wheat, Pearl Millet, Sorghum Millet	
	Shamidhyanya		Green gram, Cow pea, Chickpea, Pigeon Pea,	Flat Beans
	ShakVarga	Amaranthus, Bitter gourd, Moringa leaves, Ivy Gourd		Ash Gourd
	Phalavarga	Wood Apple, Indian Gooseberry		
	Ksheervarga			Milk, Curd
	Ikshuvarga			Jaggery, Honey, Sugarcane Juice

7. SIGNIFICANCE OF THERAPEUTIC NUTRITION IN CLINICAL PRACTICE

7A. Dietary Recommendations in Obstetrics Based on Clinical Experience

7A.1 Diet for Pregnant Woman

Kashyapa stated that the food the pregnant female consumes becomes congenial to the fetus.[46] We live in an era where the graph of the problems related to food allergies is on a constant rise. Therefore, it has become of utmost importance that a pregnant female must consume a diversified diet. This diet must be a well-balanced amalgamation of all six *Rasa* (tastes) along with a variety of cereals and congenial food.

a. In first trimester and for the patients who have recently undergone embryo transfer.
Apart from three fixed meals, pregnant woman is advised to have a flexible diet according to *Agni* (digestive fire).

First Meal
Puffed Rice/Fox nut parched paddy *chivada* (a dry snack prepared by frying it with a little amount of oil, cumin seed, turmeric, peanuts, curry leaves and green chili, and roasted Split chickpeas)

Breakfast/Snacks

Halwa (Indian sweet dish) – prepared with whole wheat flour/Green gram flour/Ragi Flour/Amaranthus seeds flour

Whole wheat vermicelli

Daliya (porridge prepared with coarse wheat flour)

Cheela (thin and big pancake-like dish) Sprouted green gram (*Phaseolus aureus*)/Finger Millet (*Eleusine coracana*)

Lukewarm milk frequently as per desire, throughout the day

Lunch and Dinner

Dal (thick soup prepared after soaking the pulses for half an hour and cooked with turmeric, Asafoetida, cumin seeds, ginger and curry leaves in ghee) – Split Green Gram (*Phaseolus aureus*)/ lentils (*Lens culinaris*)/pigeon peas (*Cajanus cajan*)

Roti: Finger millet (*Eleusine coracana*) flour, Amaranth (*Amaranthus caudatus*) Flour, Rice Flour, Sorghum Flour/Whole wheat Chapati

Ash gourd (old *Benincasa hispida*), Raw/Green banana (*Musa × paradisiaca* Fruit), Ivy gourd (*Coccinia indica*), Raw radish, Bitter gourd (*Momordica charantia*), Amaranthus leaves, Bottle gourd, Ridge Gourd, Sponge Gourd, Snake Gourd, Cabbage, Cauliflower, Broccoli, Raw Jackfruit, Spinach, Red Amaranthus

Sprout Salad prepared with well-cooked sprouts of green gram/Black Chana (*Cicer arietinum*)/Moth beans and curry leaves, coriander leaves, tomato, and onion Rice: Roasted and then cooked

Meat soup of chicken, goat

Khichadi (rice and split green gram cooked together along with ghee and spices)

Chutney (Indian side dish): made by grinding peanut/coconut, garlic, red chili powder and salt

Buttermilk, homemade butter made by churning the cream, ghee, milk, Shrikhanda (Sweet dish prepared with curd, sugar)

Koshimbir (Traditional Indian Salad): Side dish prepared with partially cooked vegetables like sponge gourd/ bottle gourd/Cucumber/radish/carrot/beetroot mixed with curd and coarsely ground roasted peanuts.

Ground Oil/Olive oil/Sunflower Oil should be used while cooking

Fruits: Should be consumed in the time only and not with Meal or Milk

Mango, jackfruit, Grapes (*Vitis vinifera*), guava, figs (*Ficus carica*), pomegranate (*Punica granatum*), oranges, Sweet Lime, Blueberries, Mulberries, Strawberries, Apple, Pears, Kiwi

b. Diet for Pregnant women for second and third trimester
Along with the diet advised for first trimester, following additions are recommended.

First Meal

Overnight soaked 4–5 Almonds (*Prunus dulcis*), 2–3 Walnuts (*Juglans regia*), 2–3 figs (*Ficus carica*)/ Laddu – Whole wheat/Black gram flour.

Breakfast

Kheer: a semisolid sweet dish prepared with Rice/Wheat coarse powder, ghee, sugar

Halwa: Water Chest Nuts (*Trapa bispinosa*), Whole wheat Halwa

Boiled Egg whites

Paratha: Type of Chapati stuffed with vegetables like spinach, potato, fenugreek leaves, and beetroot

Thalipeeth (chapati-like dish made up of mixed flour of roasted rice, wheat, sorghum, chickpea, pearl millet, and black gram and spices like coriander seeds, cumin seeds, turmeric)

Idli, Dosa, Uttapam (traditional Indian fermented food dishes)

Poha (flattened rice recipe), *Upitta* (dish prepared with semolina)

Lunch and Dinner

Panchamrut[47] – Madhu (Honey) (1/2 tsf), Curd (1/2 tsf), candy Sugar (1/2 tsf), Ghee (1 tsf), and *Ksheera* (milk) (5–6 tsf) mixed together and kept in a silver bowl overnight, can add 2–3 strands of saffron. It should be consumed daily with lunch except along with nonvegetarian food.

Meat – can have chicken, goat, sheep meat, eggs

Homemade Pickles of Mango, lemon, Indian gooseberry pickle

Whole wheat pasta, pizza occasionally

Nowadays, incidence of elderly primigravida has increased due to rising education levels, high career goals, and effective means of birth control.[48] Basic metabolic rate decreases with age which decreases the caloric need. Hence, while designing the diet for elderly primigravida, emphasis should be given to fitness and initial nutritional assessment.[49]

c. Prohibited Food during pregnancy:
Pregnant women should avoid:

Food articles having hot potency such as horse gram, sesame seeds, carom seeds, fenugreek seeds, and garden cress seeds (*Lepidium sativum*)

Reheating of cooked food, Fasting/dieting, irregular food habits, incompatible food items such as milkshakes and brownies

Stale food, undercooked beans and lentils, raw or undercooked eggs, meat and seafood

Liver and liver products, game meat such as goose, partridge, or pheasant

Chinese food items which contain MSG (Monosodium Gluconate)

Safflower oil, flax seed oil, hydrogenated refined oils, palm oil

Alcohol, carbonated drinks, diet soda, packaged fruit juices

Frozen vegetables, canned fruits, dried fruits, frozen meat

Ready-to-cook food items like instant noodles, soups

Ultraprocessed food such as breakfast cereals, biscuits, pastries, cakes, chips, ice creams

Intake of Food items like energy bars, caffeine, chocolates, panipuri, Chaat, misal (types of Indian snacks), and Garlic should be restricted.

7A.2 Dietary Recommendation for Common Diseases during Pregnancy Based on Clinical Experience

a. Nausea and Vomiting in Pregnancy:

Along with advised diet for first trimester, following additions are recommended.

Gulkand (jam-like sweet prepared with rose petals and sugar)

Whole wheat cookies, whole wheat dry toast in morning

Chewing Ginger candy – a preparation of ginger and sugar, *Amla* candy

Puffed Rice, Puffed Sorghum, Foxnut *Chivda*

Laja Manda: Laja (parched paddy soaked in water), then the water should be consumed with sugar and *Madhu* (Honey)

Panaka (a drink prepared by mixing certain fruits, water and candy sugar) – Plum, Indian *bael* fruit (*Aegle marmelos*)

Rice water with coriander (*Coriandrum sativum*) leaves pulp with sugar.

b. Constipation in pregnancy:

Overnight soaked black Raisins (*Vitis vinifera*) 5–6 in mornings

A glass of lukewarm cow's milk with 1 tsp cow's ghee at bedtime

6-8 glasses of liquids should be consumed approximately.

c. IUGR:

Eggs, Fish eggs, Meat soup, Chicken, Pork, Fish, Seafood

Soup prepared with black gram

Buffalo milk, ghee

Milk with *Kashariphala (Gmelina arborea* fruits*)*, *Sariva (Hemidesmus indicus),* and candy sugar[50]

d. Gestational Diabetes Mellitus

Recommended Diet

Cheela/Roti prepared with Wheat (*Triticum aestivum*), Nivara (A variety of wild rice - *Hygroryza aristata*), Foxtail Millet (*Setaria italica*), Barley (*Hordeum vulgare*), Vainava (*Bambuso arundinacea* seeds), Kodo Millet (*Paspalum scrobiculatum*), Shali (*Oryza sativa*), Raktshali (Red rice)

Cheela/Dal prepared with Green gram, Pigeon pea, Chickpea, cow pea along with spices such as cinnamon and cumin seeds.

Thalipeeth

Buttermilk

Meat of Chicken, Turkey, Rabbit, Goat, Deer, Quail, Partridge

Vegetables like Bitter gourd, lady finger, Red Amaranthus, Spinach, Ivy gourd, Drumsticks, raw banana, Cabbage, Cauliflower, Broccoli, Capsicum, Ridge gourd, sponge gourd, snake gourd, Elephant Yam, Colocasia roots

Fruits: Apple, Pear, Orange, Strawberries, pomegranate, Plum, Indian Gooseberry, Guava, Wood Apple, Mulberries, Indian Blackberry (*Eugenia jambolana*)

Dry Fruits: Almond, Walnut, Figs

Prohibited Diet

Pregnant Females having diabetes during pregnancy should avoid consuming freshly harvested cereals and pulses.

Milk, Curd should be consumed in moderation.

Sweet preparations like *Laddu, Kheer, Halwa* should be avoided.

Coconut water, Ash gourd, Potato, carrots, raisins, Jaggery, sugar, sugarcane, Honey, Ripped Mango, banana, Watermelon, grapes, Sapota should be avoided.

Intake of Fish, Crabs, Shrimp, prawns, lobster, pork, beef, squid, oyster, and clam should be restricted.

Popcorn, Chocolates, Rice crackers, Potato chips, packaged food, pastries, cakes, carbonated drinks should be avoided.

e. Pregnancy Induced Hypertension (PIH):
Pregnant women with PIH should follow the diet recommended for pregnant women. But consumption of Black gram, fermented food like *Idli, Dosa, Uttapam*, mayonnaise, butter, and mustard oil should be avoided, and salt should be consumed in restricted amounts.

7A.3 Diet in Postnatal Period

Just after delivery, there is *Sarvadhatu shithilata* (lethargy in all tissues), *Krishta* (weakness), *Pravahana Vedana* (Pain due to bearing down), *Kledarakta struti* (vaginal bleeding), and *Shunya shariratwa* (emptiness in the soul). Hence, following a proper diet regime helps the body to return to a prepregnant state comfortably.[51]

During the postpartum period, there is vitiation of *Vata* just after delivery and *Agnimandya* (reduction in digestive fire). Therefore, initially, unctuous, emmenagogue, and carminative food items should be included in the diet. As the strength of the digestive fire returns *Brihana* (nourishing) *Ahara* such as laddus should be introduced gradually. During this period, *Vata* vitiating or heavy-to-digest food should be avoided as this unwholesome food consumed in this time may lead to various diseases later.

a. After full-term delivery/After breaking NBM (Nil by mouth) in post-LSCS patients:
Ghee with powdered Black pepper (*Piper nigrum*) and dry ginger (*Zingiber officinale*) with jaggery should be given as a first meal.

For First Week

Whole wheat *Daliya* for breakfast

A glass of Milk with 1 tsf of Shatavari Kalpa (Sugary granules prepared with *Asparagus racemosus*) 2–3 times a day

Roti prepared with pearl millet (*Pennisetum glaucum*), Finger millet (*Eleusine coracana*), or old whole wheat chapati with ghee.

Vegetables should be consumed in curry form with lots of liquid content. Ivy gourd (*Coccinia indica*), bitter gourd (*Momordica charantia*), bottle gourd, pumpkin, cucumber, raw radish, Fenugreek leaves (*Trigonella foenum-graecum*), dill (*Anethum graveolens*) leaves, Brinjal (*Solanum melongena*), moringa (*Moringa oleifera*) leaves and fruits, spinach (*Spinacia oleracea*) cooked with garlic and cumin seeds, turmeric and asafoetida

Curry prepared with dry coconut (*Cocos nucifera*) and poppy seeds (*Papaver somniferum*)

Mukhwas is a postmeal supplement mixture of roasted fennel seeds (*Foeniculum vulgare*), flax seeds (*Linum usitatissimum*), carom seeds (*Trachyspermum ammi*), Sesame seeds (*Sesamum indicum*), dry coconut (*Cocos nucifera*), Dill seeds (*Anethum sowa*).

Lukewarm water should be used for drinking. In case of obstructed or prolonged labor or after LSCS, water at room temperature can be used.[52]

After the First Week

In addition to the first week's diet, fenugreek seed *Laddu* should be consumed in the early morning. Dietary preparations which contain legumes such as Dal and *Khichadi* can be added.

After 12 Days

Fruits like papaya (*Carica papaya*), mango (*Mangifera indica*), pineapple and poultry, goat meat, meat soup with ghee can be added.

b. Prohibited Food during the postpartum period:

- For the first week, food articles that are heavy to digest such as meat, eggs, dry fruits, *Paneer* (cottage cheese), and cheese should be avoided.
- Use of pungent food items such as Green/Red chilies is prohibited for first week.
- Bakery products and other food items made with refined flour should be avoided.
- Indian chat food, Burger, Pizza, Sausages, and Hamburgers should be avoided.
- Chickpea, flat beans, Green peas (*Pisum sativum*), potato, curd, fermented food, and other root vegetables like carrot and beetroot should be avoided.

These diet restrictions should be followed up to 1.5 months after delivery.[53]

c. To improve Lactation:
Among the diet recommended for postpartum period, following food items should be used abundantly:

- 1 tsf *Shatavari Kalp* with 1 cup milk, 2–3 times a day
- Garden cress seeds kheer, water chestnut *Halwa*
- Laddu prepared with garden cress seeds (*Lepidium sativum*), coconut (*Cocos nucifera*), seeds, jaggery, and ghee
- Sugar Cane (*Saccharum officinarum*)
- Sesame seed chutney, flax seed chutney

7B. Dietary Recommendations Based on Clinical Experience in Gynecology

7B.1 Diet during Puberty

During puberty, due to the complex major development in all *Saptdhatus*, the body evolves, resulting in rapid somatic growth, brain development, and maturation of reproductive organs. This period also marks the end of *Kapha* dominance and the commencement of *pitta* dominance in the body.[54] Hence the diet designed for an early phase in a pubertal girl, i.e. around thelarche should be *Kapha* balancing and *Rasaprasadaka* (forming good quality *Rasadhatu)*, which leads to the formation of healthy *Updhatu Artava* and consecutive *Dhatu*. Ghee, milk, meat soups, apricots, dates, and raisins should be consumed abundantly according to the strength of Agni (digestive fire).

In the later stage of puberty (Menarche), food articles with hot potency as well as pitta balancing should be consumed. The use of carminative spices such as cumin seeds, fennel seeds, asafoetida, and carom seeds is recommended.

Adolescent girls must be encouraged to avoid dietary control and low-calorie intake as it leads to further menstrual abnormalities. Also, post menarche, it is highly suggested to stick to the diet recommended during the menstrual phase as it minimizes the incidences of dysmenorrhea, PMS, and many other menstrual complaints.

Early Morning

About 5–6 overnight soaked Raisins (*Vitis vinifera*)/Apricot (*Prunus armeniaca*)/3–4 soaked Dates (*Phoenix sylvestris*).

Breakfast/Snacks

Halwa – Whole wheat flour, Water Chestnuts (*Trapa bispinosa*) flour

Daliya (porridge) – Whole wheat coarse flour/vermicelli

Cheela – Finger millet (*Eleusine coracana*), chickpea flour, Sorghum Cheela (*Sorghum bicolor*)

Fermented food like *Idli/Dosa* – once a week

Laddu – prepared with split black gram flour/Whole wheat flour

Rajgira Chikki – a snack bar prepared with jaggery and Peanuts and Amaranthus (*Amaranthus cruentus*) seeds

Foxnut/ puffed rice chivada

Cottage cheese/potato/beetroot/spinach paratha

Lunch/Dinner

Whole wheat *Chapati/pearl millet or sorghum flour Roti* with ghee

Meat and meat soups of poultry, goat, crab, seafood

Chutney – dry coconut, flax seed chutney, peanut sesame seed chutney

Homemade pickle – green mango, Amla, lemon/turmeric

Khichadi/Dal rice with ghee

All vegetables which are recommended during pregnancy can be consumed.

7B.2 Diet during Menstrual Phase

During the menstrual phase, there is *Vata* dominance and Agnimandya (reduction in digestive fire) is also observed. Hence, the female should consume light-to-digest food in small portions. On the other hand, junk food and bakery items should be avoided as they may lead to *Ama* (production of digestive toxins), *Rasadushti*, and vitiation of *Vata*. This can eventually lead to various menstrual abnormalities such as PMS dysmenorrhea and anovulation. The proliferative phase is of *Kapha* dominance, while the luteal phase is of pitta dominance. Hence, *Kapha* balancing diet is advisable in the proliferative phase, while Pitta balancing diet is preferred in the luteal phase.

Recommended Diet

Rice cooked in cow milk

Roasted green gram Cheela soup/prepared in ghee along with cumin seeds

Pearl Millet Roti

Barley (*Hordeum vulgare*) cooked with milk

Khichadi/Dal-rice

Prohibited Diet

Bakery Products

Fermented food

Paneer, Khoya (Solidified milk) sweets, Fruits, fruit juices

Excessive use of spices, green chilies

Indian chat food items like *Golgappe, Shevpuri*

Fermented food items like *Idli, Dosa, Uttapam*

7B.3 Diet for Perimenopausal Period

Menopause, a transitory phase in women's lives, has attained the form of the disease due to enhanced stress, poor lifestyle, and unhealthy dietary habits. According to Sushruta, it is a period of pitta dominance along with Dhatukshaya.[54]

Rasayana is a rejuvenating treatment described to achieve health with mental competence and to arrest the process of aging.[55] *Ajastrik*[56] type *of Rasayana,* i.e. regular intake of *Ksheera* (milk), ghee, and *Madhu* (Honey), should be practiced by a perimenopausal woman along with certain herbs like Indian gooseberries, Dill seeds, and Garden cress seeds.

Amla (*Emblica officinalis*) possesses antioxidant, immunomodulatory, and hypocholesterolemic properties. It is described as one of the best rejuvenating drugs.[57] Dill seeds (*Anethum sowa*) are described as one of the best drugs for gynecological disorders.[58] It is a good source of calcium, manganese, dietary fibers, and arginine, which prevents bone loss in the perimenopausal period. It contains phyto-estrogen which helps to relieve the dryness of the vagina and hot flashes. In addition, it also helps to relieve insomnia.[59] Garden cress seeds are an abundant source of calcium and phosphorus and have anti-inflammatory properties.[60]

After a meal, *Tambulsevana* is recommended as it contains *Piper betle* leaves with slaked lime, catechu, sesame seeds, dry coconut, dill seeds, fennel seeds, and *Gulkanda*. All of these ingredients help in digestion and also extend the support to balance *Vata, Pitta*, and *Kapha dosha*. Besides, it is a good source of calcium and essential fatty acids.

Morning

Laddu – Garden cress seed laddu

Shatawari Shatapushpa laddu[61]

Soaked nut and dry fruits – almond/Walnuts (*Juglans regia*), Dates (*Phoenix sylvestris*), Raisins (*Vitis vinifera*), apricots, cashews

For breakfast, lunch, and dinner, the dietary recommendation for Puberty can be followed with certain personal modifications.

7B.4 Diet for Various Diseases

a. Oligomenorrhea/Hypomenorrhea

Indicated Diet: Fasting once 1 week

Roti/Cheela: Pearl millet (*Pennisetum glaucum*), Finger millet (*Eleusine coracana*), Koda millet (*Paspalum scrobiculatum*), Old whole wheat, Barley (*Hordeum vulgare*), old hand pounded or single polished rice

Green Gram, lentils, pigeon peas

Vegetables: Moringo (*Moringa oleifera* fruits and leaves), bitter gourd (*Momordica charantia*), pointed gourd (*Tricosanthes dioica*), ridge gourd (*Luffa acutangula*), Fenugreek (*Trigonella foenum-graecum*) leaves, dill (*Anethum graveolens*) leaves, Brinjal (*Solanum melongena*), spiny gourd (*Momordica dioica*), yam (*Amorphophallus campanulatus*), *Colocasia* roots (*Colocasia esculenta*)

Fish, poultry

Ginger, Garlic

Buttermilk

Amaranths seeds, Garden cress seeds, black Sesame seeds (*Sesamum indicum*) with jaggery

Sesame seed Chutney, flax seed chutney

Prohibited Food

Meat of Pork, Duck

Refined flour food items such as cakes, pastries, bakery products

Milk, milk products such as cottage cheese and sweets

Fermented food

b. PCOS

In obese PCOS patients, there is *Sanga* (discontinuation of the flow of the content of *strotasa* i.e. unovulatary cycles, oligomenorrhea/amenorrhea) due to *Strotorodha* (obstruction). Here the diet recommended for oligomenorrhea/hypomenorrhea is advised.

While in lean PCOS patients, along with intake of 4–5 Soaked raisins, apricots, and dates in morning.

Breakfast, lunch, and dinner which is advised for pregnant woman in first trimester is recommended in lean PCOS.

c. Diet for Heavy Menstrual Bleeding

During bleeding phase, use of Goat milk, Ghee prepared with Goat milk, Water infused with sandalwood (*Santalum album*), *Ushira* (*Vetiveria zizanioides Linn*), *Shunthi* (dry *Zingiber officinale*), Meat soup of goat, Dates with Honey.

Raisins (*Vitis vinifera*), Amla (*Emblica officinalis*), pomegranate (*Punica granatum*), coconut, Wood Apple (*Feronia elephantum*), Sugarcane (*Saccharum officinarum*), sugar palm tree fruits.

Kodo millet, old rice, Barley, Green Gram, lentils, pigeon peas, chickpea, and parched paddy flour are recommended.

Prohibited diet: Horse gram, black gram, Sesame seeds (*Sesamum indicum*), jaggery, big mustard seeds, curd, *Piper betle*, garlic, fish and Sour, salty, spicy food items

8. BRIEF DESCRIPTION OF AHARIYA KALPANA

***1) Laja manda* for Vomiting in Pregnancy**

Laja Manda is prepared by adding *laja* to water in a ratio of 1:4 and cooking on low flame until *Laja* is fully cooked. Then the super dilute liquid made is called *Laja Manda.*

2) *Kashmaryadi Ksheerpaka* for IUGR:

- *Kashmarya phala* (fruit of *Gmelina arborea*) – 4 g
- *Sariwa* (*Hemidesmus indicus* roots) – 4 g
- *Sita* (Candy sugar) – 2 g
- Cow's *Ksheera* (milk) – 80 mL (coarse powder of each)
- Water – 320 mL

Boiled together on slow flame till complete water evaporates. Consume it freshly prepared always.

3) Garden cress seeds *Kheer (semisolid sweet dish)* – To improve lactation, during pubertal and perimenopausal period:

- 2–3 tsf garden cress seeds
- 2 cups milk
- Grated jaggery as per taste
- Almond, pistachio slivers

Method

Wash and soak garden cress seeds in water for 2–3 hours.

Heat the milk in a pan. When it starts boiling add the soaked garden cress seeds and cook it for 8–10 minutes. Add grated jaggery and cook for 2–3 minutes more, then add Almond and pistachio and turn off the stove.

4) **Garden Cress *Laddu*** – For postpartum, postmenopausal, and pubertal period:

- 2-3 cups garden cress seeds (soaked in milk for 2–3 hours)
- 2 cups milk
- 2 cups grated coconut
- 1 cup grated jaggery
- 3–4 tsf *Ghee*

Method

Roast soaked Garden seeds and coconut in a pan until coconut turns golden. Then add jaggery and cook it on low flame for 25–30 minutes more and turn off the stove. Let the mixture cool down.

In another pan, heat the *ghee,* add the above mixture, and cook it till it dries completely.

Hand roll the sphere-shaped *laddus* when the mixture is still warm.

5) ***Cheela***

- 1 cup Green gram/chickpea/ragi/sorghum flour
- Chili garlic paste as per taste
- ¼ tsf cumin seeds
- ¼ tsf Turmeric powder
- Finely chopped coriander leaves
- Salt as per taste
- Finely chopped vegetables like fenugreek/spinach/tomato/onion (optional)

Method

Take flour in bowl and add *chili* garlic paste, cumin seeds, finely chopped vegetables, turmeric powder, and salt. Add water and mix it well until smooth and flowing consistency is reached.

Heat a flat pan, grease bit of oil on it, pour the mixture, and spread into a thin circle.

Let it cook for a few minutes on either side.

6) ***Methi* Laddu – Fenugreek Seeds Laddu**

- 100 g Fenugreek Seeds
- 300 g grated jaggery
- 20 g Poppy Seeds, watermelon seeds, sunflower seeds each
- 10 g *Bhilawa Beej Magaj* (kernel of *Semecarpus anacardium*)
- 10 g Garden cress Seeds

100 g Dried Dates coarse powder:

- 100 g grated Dry coconut
- 100 g *Gond* (Edible Gum of *Acacia arabica*)
- 100 g Almond
- 100 g Cashew nut
- Ghee as per requirement

Method

Dry roast fenugreek seeds on low flame until a certain aroma is produced, then powder them.

Add garden cress seeds, kernel of marking nuts and other nuts, and seeds to Fenugreek powder.

Fry the *Gond* in *Ghee* until it pops out. Allow it some time to cool down, and then crush it with hands. Add the *Gond* and the grated Jaggery to the mixture.

Add melted Ghee. Hand-roll it in a sphere-shaped *Laddu.*

7) **Black gram flour/wheat flour *Laddu*:**

The ingredients and method are the same as those of *Methi Laddu*. Instead of fenugreek seed flour, flour of roasted skinless split black gram or flour of roasted whole wheat is used.

While making *Laddu* for pregnant women, Garden cress seeds, kernel of marking nut, and poppy seeds should be avoided.

9. SCOPE FOR FURTHER STUDIES

Nowadays, plant-based milk substitute consumption has spread rapidly around the globe due to lactose intolerance and cow milk allergy.[62] These milk substitutes are *Guru* (heavy on digestive fire) and *Sanskara Virudhha* (opposite to prescribed mode of consumption). According to *Kashyapa*, food consumed by pregnant woman become congenial to the fetus.[46] Hence experimental studies can be carried out to substantiate whether following *Charaka's* dietary regime[11] can reduce the incidence of cow milk allergy and lactose intolerance.

- Pregnant woman is advised to have diversified diet based on all six *Rasa* (tastes). Consumption of any particular *Rasa* (taste) food for a longer duration gives rise to various diseases in fetus in later life.[63] There is a need for observational studies to verify this concept of fetal origin of adult diseases.
- Breakfast cereals, though projected as healthy food, are ultraprocessed foods. They undergo intense processing and contain cosmetic additives and food dyes which causes a higher risk of noncommunicable diseases. Increase in the dietary share of ultraprocessed foods results in deterioration of the nutritional quality of overall diet.[64] Hence, there is an urgent need for public research combining epidemiological and experimental approaches to better understand the impact of food processing on women's health, fertility, pregnancy, and lactation.

REFERENCES

1. United Nations Decade of Action on Nutrition [Internet][Cited on 2022 Aug 30] Available on www.un.org/nutirtion/node/140
2. *Vṛddhajīvaka, ShriSatpalBhishakAacharya and Vṛddhajīvaka (*1953*) "Yushanirdeshiya," in Kās'yapa Samhitā = Vṛddhajivakīya Tantra*. Banaras: Chowkhamba, p.249. [Hindi-Sanskrit].

3. Elder, L. and Ransom, E. (2003) *Nutrition of Women and Adolescent Girls: Why it matters, PRB*. Population Reference Bureau. Available at: www.prb.org/resources/nutrition-of-women-and-adolescent-girls-why-it-matters/ (Accessed: September 8, 2022).
4. *Maternal nutrition* UNICEF. UNICEF for every Child. Available at: www.unicef.org/nutrition/maternal#:~:text=During%20pregnancy%2C%20poor%20diets%20lacking,and%20developmental%20delays%20for%20children. (Accessed: August 2022).
5. Norris, Shane A, et al. "Nutrition in Adolescent Growth and Development." The Lancet, vol. 399, no. 10320, Mar. 2022, pp. 172–184. https://doi.org/10.1016/s0140-6736(21)01590-7.
6. Cetin, I. and Laoreti, A. (2015) "The importance of maternal nutrition for health," *JPNIM*, 4(2). Available at: https://doi.org/10.7363/040220.
7. WHO Europe (2016) "Introduction to Maternal nutrition, prevention of obesity and noncommunicable diseases (NCDs): recent evidence," in Good maternal nutrition: The best start in life. Copenhagen: WHO Regional Office for Europe, pp. 2–11.
8. Fuke.R. and Rubberstampwala, F. (2017) in Practical approach to menopause management. Delhi: Jaypee Brothers Medical Publisher (P) Ltd., p. 159.
9. Vṛddhajīvaka, ShriSatpal BhishakAacharya and Vṛddhajīvaka (1953) "Lehadhyay," in Kās'yapa Samhitā = Vṛddhajivakīya Tantra. Banaras: Chowkhamba, p. 2. [Hindi-Sanskrit].
10. Vṛddhajīvaka, ShriSatpalBhishakAacharya and Vṛddhajīvaka (1953) "Lehadhyay," in Kās'yapa Samhitā = Vṛddhajivakīya Tantra. Banaras: Chowkhamba, p. 4[Hindi-Sanskrit].
11. Agnivesha (Re-print-2005) "Jatisutriyam Shariram," in T. Jadhavji (ed.) Charaksamhita. Banaras: Chaukhamba Surbharati Prakashan, p. 346 [Sanskrit].
12. Susruta (1980) "Garbhinivyakarana Sharir," in T. Jadhavji (ed.) Susrutasamhita of Susruta: With the NIBANDHASANGRAHA commentary of Sri Dalhanacharya and the nyayachandrika panjika of sri gayadasacharya on Nidanasthana. Varanasi: Chaukhambha Surbharati Prakashan, p.387 [Sanskrit].
13. Vāgbhaṭa and Gupta, K.A. (1964) "Garbhopacharaniyam Sharir" in Aṣṭāmgasamgraha. Varanasi: Chaukhamba Krishandas Academy, pp. 282–283 [Hindi-Sanskrit].
14. Pāṇḍeya Jayamīnī (2010) "Garbhopacharvidhi," in Hārīta saṃhitā: Saṃskṛta Mūla va nirmalā hindī ṭīkā. Vārāṇasī: Caukhambhā Viśvabhāratī, pp. 467–468 [Hindi-Sanskrit].
15. Agnivesha (Re-print-2005) "Jatisutriyam Shariram," in T. Jadhavji (ed.) Charaksamhita. Banaras: Chaukhamba Surbharati Prakashan, p. 349 [Sanskrit].
16. Susruta (1980) "Garbhinivyakarana Sharir," in T. Jadhavji (ed.) Susrutasamhita of Susruta: With the NIBANDHASANGRAHA commentary of Sri Dalhanacharya and the nyayachandrika panjika of sri gayadasacharya on Nidanasthana. Varanasi: Chaukhambha Surbharati Prakashan, p. 389 [Sanskrit].
17. Vṛddhajīvaka, ShriSatpalBhishakAacharya and Vṛddhajīvaka (1953) "Sutikopkramaniya," in Kās'yapa Samhitā = Vṛddhajivakīya Tantra. Banaras: Chowkhamba, p.306 [Hindi-Sanskrit].
18. Agnivesha (Re-print-2005) "Jatisutriyam Shariram," in T. Jadhavji (ed.) Charaksamhita. Banaras: Chaukhamba Surbharati Prakashan, p.345 [Sanskrit].
19. Sastri Laksmipati and Miśra Brahma Sankara (1955) "StrirogChikitsa," in Yogaratnakarah. Banaras: Chowkhamba Sanskrit Sansthan, p .417 [Hindi-Sanskrit].
20. Susruta (1980) "Garbhinivyakarana Sharir," in T. Jadhavji (ed.) Susrutasamhita of Susruta: With the NIBANDHASANGRAHA commentary of Sri Dalhanacharya and the nyayachandrika panjika of sri gayadasacharya on Nidanasthana. Varanasi: Chaukhambha Surbharati Prakashan, p.393 [Sanskrit].
21. Sastri Laksmipati and Miśra Brahma Sankara (1955) "StrirogChikitsa," in Yogaratnakarah. Banaras: Chowkhamba Sanskrit Sansthan, p .421 [Hindi-Sanskrit].
22. Vṛddhajīvaka, ShriSatpalBhishakAacharya and Vṛddhajīvaka (1953) "Antarvartamichikitsa," in Kās'yapa Samhitā = Vṛddhajivakīya Tantra. Banaras: Chowkhamba, p. 300[Hindi-Sanskrit].
23. Bhavmishra (1998) "Yonirogadhikar," in Bhavprakash. 2. Banaras: Chaukhambha Sanskrit Series Office, p. 813 [Hindi-Sanskrit].
24. Agnivesha (Re-print-2005) "Jatisutriyam Shariram," in T. Jadhavji (ed.) Charaksamhita. Banaras: Chaukhamba Surbharati Prakashan, pp. 344–345[Sanskrit].
25. Agnivesha (Re-print-2005) "Vatvyadhichikista," in T. Jadhavji (ed.) Charaksamhita. Banaras: Chaukhamba Surbharati Prakashan, p.621 [Sanskrit].
26. Susruta (1980) "Doshdhatumalkshayvruddhividdyananiya," in T. Jadhavji (ed.) Susrutasamhita of Susruta: With the NIBANDHASANGRAHA commentary of Sri Dalhanacharya and the nyayachandrika

panjika of sri gayadasacharya on Nidanasthana. Varanasi: Chaukhambha Surbharati Prakashan, p.70 [Sanskrit].

27. Susruta (1980) “Shukrashonitashuddhisharirопkram,” in T. Jadhavji (ed.) Susrutasamhita of Susruta: With the NIBANDHASANGRAHA commentary of Sri Dalhanacharya and the nyayachandrika panjika of sri gayadasacharya on Nidanasthana. Varanasi: Chaukhambha Surbharati Prakashan, pp. 346–347 [Sanskrit].
28. Vāgbhaṭa and Gupta, K.A. (1964) “Putrakamiyam Shariram” in Aṣṭāṃgasaṃgraha. Varanasi: Chaukhamba Krishandas Academy, p. 266 [Hindi-Sanskrit].
29. Vṛddhajīvaka, ShriSatpalBhishakAacharya and Vṛddhajīvaka (1953) “Lashunakalpadhyaya,” in Kās'yapa Samhitā = Vṛddhajivakīya Tantra. Banaras: Chowkhamba, p. 175[Hindi-Sanskrit].
30. Sastri Laksmipati and Miśra Brahma Sankara (1955) “Yonivyapadchikitsa” in Yogaratnakarah. Banaras: Chowkhamba Sanskrit Sansthan, p.406 [Hindi-Sanskrit].
31. Kavirāja Viśvanātha et al. (2018) “Part A Patha,” in Pathyāpathya vinirṇayah a decisive anthology on therapeutic dietetics of śrī viśvanātha kavirāja: A descriptive directory on diets, directions and drugs for wholistic health management. Varanasi: Chaukhamba Surbharati Prakashan, pp. 46–50 [Sanskrit-English].
32. Kavirāja Viśvanātha et al. (2018) “Part A Patha,” in Pathyāpathya vinirṇayah a decisive anthology on therapeutic dietetics of śrī viśvanātha kavirāja: A descriptive directory on diets, directions and drugs for wholistic health management. Varanasi: Chaukhamba Surbharati Prakashan, pp. 51–57 [Sanskrit-English].
33. *Nutrition during pregnancy ACOG.* The American College of Obstetricians and Gynecology. Available at: www.acog.org/womens-health/faqs/nutrition-during-pregnancy (Accessed: August 2022).
34. BDA. “Pregnancy and Diet.” Pregnancy and Diet | British Dietetic Association (BDA), www.bda.uk.com/resource/pregnancy-diet.html.
35. *Foods to avoid in Pregnancy NHS choices.* NHS. Available at: www.nhs.uk/pregnancy/keeping-well/foods-to-avoid/ (Accessed: August 2022).
36. *Maternal diet* (2022) *Centers for Disease Control and Prevention.* Centers for Disease Control and Prevention. Available at: www.cdc.gov/breastfeeding/breastfeeding-special-circumstances/diet-and-micronutrients/maternal-diet.html (Accessed: September 2022).
37. “Healthy Eating for Adolescents.” Nidirect, 27 Sept. 2022, www.nidirect.gov.uk/articles/healthy-eating-adolescents.
38. Schenker, Sarah. “Food Fact Sheet -Menopause.” Edited by Angie Jefferson, The Association of UK Dietitians, May 2019, www.bda.uk.com/resource/menopause-diet.html.
39. Afreen, S., Aiashquar, A. and Borahay, M. (2021) “Diet and Nutrition in Gynaecological disorders: a focus on clinical studies,” *Nutrients* [Preprint]. Available at: https://doi.org/10.3390/nu13061747.
40. Faghfoori Z; *et al.* (2017) *Nutritional management in women with polycystic ovary syndrome: A review study, Diabetes & metabolic syndrome.* U.S. National Library of Medicine. Available at: https://pubmed.ncbi.nlm.nih.gov/28416368/ (Accessed: September 2022).
41. Helbig, M. et al. (2021) Does nutrition affect endometriosis?, Geburtshilfe und Frauenheilkunde. Georg Thieme Verlag KG. Available at: www.ncbi.nlm.nih.gov/pmc/articles/PMC7870287/ (Accessed: September 2022).
42. Susruta (1980) “mudhagarbhachikitsadhyaya,” in T. Jadhavji (ed.) Susrutasamhita of Susruta: With the NIBANDHASANGRAHA commentary of Sri Dalhanacharya and the nyayachandrika panjika of srigayadasacharya on Nidanasthana. Varanasi: Chaukhambha Surbharati Prakashan, p. 462 [Sanskrit].
43. Agnivesha (Re-print-2005) “Jatisutriyaadhyaya,” in T. Jadhavji (ed.) Charaksamhita. Banaras: Chaukhamba Surbharati Prakashan, p. 344 [Sanskrit].
44. Agnivesha (Re-print-2005) “Strotovimana,” in T. Jadhavji (ed.) Charaksamhita. Banaras: Chaukhamba Surbharati Prakashan, p. 252 [Sanskrit].
45. Shivakumari, Shivakumari, et al. “A Study on SROTODUSHTI with Special Reference to Assessment of SROTOSHTI Lakshana in ARTAVAVAHA SROTAS through Clinical, Biochemical and Radiological Findings'- Survey Study.” *International Ayurvedic Medical Journal*, vol. 9, no. 10, 16 Oct. 2021, pp. 2339–2346. https://doi.org/10.46607/iamj0809102021.
46. Vṛddhajīvaka, ShriSatpalBhishakAacharya and Vṛddhajīvaka (1953) “Lehadhyaya,” in Kās'yapa Samhitā = Vṛddhajivakīya Tantra. Banaras: Chowkhamba, p. 4. [Hindi-Sanskrit].
47. Sharma, Neha, and Sujata Kadam. *PANCHAMRUT: NECTAR FOR A PREGNANT WOMAN*, vol. 7, no. 9, Sept. 2019, pp. 1612–1615. www.iamj.in/posts/images/upload/1611_1615.pdf.
48. Pradhan, Kumudini, et al. “Pregnancy Outcome in Elderly Primigravida.” *International Journal of Reproduction, Contraception, Obstetrics and Gynecology*, vol. 8, no. 12, 2019, pp. 4684–4689. https://doi.org/10.18203/2320-1770.ijrcog20195172.

49. Weigley, E S. "Nutrition and the older primigravida." *Journal of the American Dietetic Association* vol. 82,5 (1983): 529–30.
50. Rajput, D. and Dewaikar, S. "To Study the effect of *Ksheerbasti* in *Garbhakshaya* with special reference to IUGR." Swami Ramanand Teerth Marathwada University, Nanded, 2007.
51. Vāgbhaṭa and Gupta, K.A. (1964) "Garbhopacharniya Shariram" in Aṣṭāṃgasaṃgraha. Varanasi: Chaukhamba Krishandas Academy, p. 288 [Hindi-Sanskrit].
52. Vṛddhajīvaka, ShriSatpalBhishakAacharya and Vṛddhajīvaka (1953) "Sutikopkramaniya," in Kās'yapa Samhitā = Vṛddhajivakīya Tantra. Banaras: Chowkhamba, p.307[Hindi-Sanskrit].
53. Kavirāja Viśvanātha et al. (2018) "Part A Patha," in Pathyāpathya vinirṇayah a decisive anthology on therapeutic dietetics of śrī viśvanātha kavirāja: A descriptive directory on diets, directions and drugs for wholistic health management. Varanasi: Chaukhamba Surbharati Prakashan, pp. 190–191 [Sanskrit-English].
54. Susruta (1980) "Aaturopakramaniya," in T. Jadhavji (ed.) Susrutasamhita of Susruta: With the NIBANDHASANGRAHA commentary of Sri Dalhanacharya and the nyayachandrika panjika of sri gayadasacharya on Nidanasthana. Varanasi: Chaukhambha Surbharati Prakashan, p. 155 [Sanskrit].
55. Agnivesha (Re-print-2005) "Rasayanaadhyayana," in T. Jadhavji (ed.) Charaksamhita. Banaras: Chaukhamba Surbharati Prakashan, p.376 [Sanskrit].
56. Susruta (1980) "Sarvopghatshamniyarasayana," in T. Jadhavji (ed.) Susrutasamhita of Susruta: With the NIBANDHASANGRAHA commentary of Sri Dalhanacharya and the nyayachandrika panjika of sri gayadasacharya on Nidanasthana. Varanasi: Chaukhambha Surbharati Prakashan, p.498 [Sanskrit].
57. Agnivesha (Re-print-2005) "Yajjapurushiya," in T. Jadhavji (ed.) Charaksamhita. Banaras: Chaukhamba Surbharati Prakashan, pp. 131–132 [Sanskrit].
58. Vṛddhajīvaka, ShriSatpalBhishakAacharya and Vṛddhajīvaka (1953) "Shatavarishatapushpakalpadhyaya," in Kās'yapa Samhitā = Vṛddhajivakīya Tantra. Banaras: Chowkhamba, p.186[Hindi-Sanskrit].
59. Pradhan, Snehalata. "Utility of Shatapushpa (Indian Dill) in Kashyapa Samhita and Its Critical Analysis-A Review." *Utility of Shatapushpa (Indian Dill) in Kashyapa Samhita and Its Critical Analysis- A Review*, vol. 4, no. 3, 10 May 2016, pp. 44–51.
60. Dixit Jr Iii, Vinti, et al. "Lepidium Sativum: Bone Healer in Traditional Medicine, an Experimental Validation Study in Rats." *Journal of Family Medicine and Primary Care*, U.S. National Library of Medicine, 28 Feb. 2020, www.ncbi.nlm.nih.gov/pmc/articles/PMC7113932/.
61. Deore, Nilakshi. and Rajput, D. "An Open Labeled Single Armed Clinical Study Of *Shatapushpa Shatavari Modak* in perimenopausal syndrome." Maharashtra University Of Health Sciences, Nashik, 2020.
62. Aydar, Elif, et al. "Plant Based Milk Substitutes: Bioactive Compounds, Conventional and Novel Processes, Bioavailability Studies and Health Effects." Journal of Functional Foods, vol. 70, July 2020, www.sciencedirect.com/science/article/pii/s175646420301997. Accessed Oct. 2022.
63. Agnivesha (Re-print-2005) "Jatisutriyam Shariram," in T. Jadhavji (ed.) Charaksamhita. Banaras: Chaukhamba Surbharati Prakashan, p. 344 [Sanskrit].
64. Monteiro, Carlos A, et al. "Ultra-Processed Foods: What They Are and How to Identify Them." Public Health Nutrition, vol. 22, no. 5, 2019, pp. 936–941. https://doi.org/10.1017/s1368980018003762.

SECTION D

Therapeutic Nutrition in Ayurveda for Chronic Kidney Disease

Swarupa Bhujbal and Ganesh Malawade

INTRODUCTION

Basti is seat of *Udak* which balances functions of Water/fluid mechanics in body, maintaining homeostasis of urine expulsion.[1] It is *Maha Marma* (Vital point) exhibiting *Marma Anupalayan*[2] (Vital force of body functioning). The impact/insult to *Basti* functioning leads to *Mahagad* (Morbid disorders). However, *Swasthya* (Health) of *Basti* plays a pivotal role in controlling disorders of kidney in the early stage. The kidney as an organ is not explained in *Basti* and *Mutra* formation directly but *Pakvashyastha Nadi*[3] (Channels in large intestine) absorbs and circulates throughout the body and influences *Mutravahi-Dhamani* (Renal artery) which further divides into *Sukhma* (Minute), *Sahashatra* (Millions) of *Mukhani* (Tubal openings)[4] resembling nephrons and their functions. Thus, understanding the formation of the kidney, its functions, and diseases is important to define chronic kidney disease (CKD) staging and its dietary treatment to conserve the health of each nephron from injuries/insults and barotrauma from nutrients.

A kidney is the basis of *Medovahasrotas*[5] (Fat tissue system) and its formative aliments are *Rakta* (Blood) and *Meda* (Fat) *Dhatu Prasad* portion (Nourished portion). However, the pathway of *Ahara* (Diet) divides into *Sara* (Nourished) and *Asara* (waste portion), which is the main contributor to maintaining functions like homeostasis, hemodynamics, water and fluid, and acid-base balance of the kidney. *Asara* (Waste portion) contributes to formation of major excreta urine and feces with minor metabolic waste such as creatinine and urea.[6] Therefore, *Apathya ahar* (unwholesome food), which disturbs formative components (*Rakta* and *Meda*), is responsible for the dysfunction of *Vrukka* (Kidney). This leads to pathologies such as *Mrudu* (Gentle), *Darun* (Morbid), *Kshipra* (Accelerated), *Ashukari* (Acute), and *Chirakari* (Chronic) and can be correlated as CKD staging. *Ayurveda* specialty elaborated CKD as *Anukta Vikara (Unexplained Disease)*[7] which comprises correlations with conditions such as *Grahani* and *Ajirna* (metabolic distress due to improper diet habits), *Mutrachrucchra* (Dysuria), and *Mutraghat* (Oliguria and anuria).[8]

Consideration of *Ahara* (diet) is the main contributor to CKD and its staging. Its epidemiology and prevalence are ambiguous as asymptomatic patients remain unidentified in society.[9] The prevalence of

DOI: 10.1201/9781003345541-21

CKD affects 10% of the population worldwide, ranked 16th in the mortality index (2016).[10] It is expected to be upgraded to 5th rank in the coming decade.

CKD is defined as kidney damage or GFR < 60 mL/min/1.73 m for more than 3 months or irrespective of the cause.[11] The albuminuria overt to proteinuria persistently has to be considered chiefly for renal insufficiency. Thus, renal damage can be explored minutely with the algorithm of dietary causes/dynamics of obesity, and HTN and DM can be precursors of CKD causation too. There is a need for definitive dietary control of hemodynamic autoregulation of salt and water [Renin angiotensin system (RAAS)][12,13] and precise control of glucose metabolism to avoid the progression of asymptomatic CKD stages. However, poor nutritional outcome of CKD staging leads to ESRD.

The acid-ithomeostasis is the key physiologic function of the kidney, such as the reabsorption of filtered carbonates and replacement of bicarbonates which are consumed by pathologic acids.[14] However, it is an organ of filtration, modulation, and homeostasis of ions. Its functions revolve around hemodynamic flow, ionic exchange, and nutrients. The suboptimal/maladaptation of nutrients exhibits multiple biochemical alterations (Metabolic acidosis), gut dysbiosis (Urea and Ammonia), and hormonal dysregulation (PTH), which could promote varied CKD staging altering GFR.[15]

The challenges to minimize CKD progression lie in the algorithm of nutrients.

The standard-of-care recommendations are medicinal therapy with dialysis along with dietary supplements. The management of potassium, calcium, and phosphate from dietary sources with poor assimilation leads to inverse electrolytes/solutes/ions proportion in blood, disturbing energy homeostasis, i.e. creatine phosphate from muscles.[16] It needs to be addressed with the required nutrient and its assimilation. Therapeutic nutrition in *Ayurveda* (TNA) focuses on basic fundamental principles of renal injury and its repair mainly.

There is a lack of nutrigenetics' individualized approach to conserving renal mass/managing GFR in spite of advancement in structural, physiological, and pathological kidney disease aspects. *Ayurveda* fulfills the pitfalls of progressive trends in CKD. TNA provides specificity of characteristics and diverse attributes of *Ahara* (Food) to define renal dietary designing as per the condition of *Dosha* and *Dhatu* vitiation involved in pathogenesis. Renal diet designing comprises *Aharia Hetu* (Dietary causes) and *Aharia Vidhi* (Dietary strategy) with eight directives of ingestion of food AVV (*Ashta Vidhi Visesh Ayatananai*) to arrest renal injuries. It includes *Santarpana* (Nourishing) or *Aptarpana* (Reducing) principle by selecting food items from *Dwadasha Varga* (12 food groups). The TNA multifold dietary strategies can conserve renal mass to check CKD progression.

Vrukka (Kidney)

Vrukka (kidney) is formed from *Rakta* (Blood) and *Meda Prasad* (Nourished portion) of *Dhatu*.[17] The *Medovahasrotas* has a system of kidney, ureters, bladder, and *Vapavahana* (Omentum).[18] *Apan Vayu* is the potential dynamics of the expulsion of metabolic waste, faces, and urine.[19] However, *Vrukka* functional unit depends upon the homeostasis of *Rakta* and *Meda Dhatu* along with *Apan Vayu*.

Vrukka Vikara (Diseases of the Kidney)

Anukta category of *Vikaras*/diseases is not explained by their name in *Samhitas* but a similar condition is mentioned in classics. Considering this concept, CKD staging can encompass various diseases like *Pandu* (Anemia), *Shotha* (Edema), *Ajirna* (Metabolic dysfunction), *Grahani*, *Mutrakrucchtra* (Dysuria), and *Mutraghata* (Oliguria). *Kledavahana* can be a part of solutes/osmolarity management reflected as a component of urine formation.[20]

TABLE 16.1 Ayurveda Perception of CKD Staging – Dietary Etiologies

DISEASES	*DIETARY ETIOLOGIES*	*TREATING PRINCIPLE*
Pandu[21] (Anemia)	*Kshara* (Alkali), *Amla* (Sour), *Lavan* (Salt), *Atyushna Bhojana* (Excessive hot diet)	*Virechna* (Purgation)
Shoth/shwayathu (edema)	*Amla* (Sour), articles of food and drinks having *tikshna* (Sharp) attribute, *Ushna* (hot) food and drinks and *Guru* (heavy food); curd, uncooked food	*Amaja-langhana* (Fasting Therapy), *Pachana* (Appetizer) and *Shodhana* (Detoxification therapies)
Ajeerna/Grahanee (Metabolic dysfunction of gut)	*Abhojan* (Excessive fasting) *Ajirna* (indigestion) *Ati Bhojana* (overeating) *Vishamashan* (irregular eating) *Asatmya* (intake of unwholesome) *Guru* (heavy), *Sheeta* (cold) *Ruksha* (excessively unctuous) *Dushta Bhojana* (polluted food)	*Dipan* (Digestive[22])
Prameh-General (Diabetes mellitus)	Curds Arid and aquatic meat soup Excess of milk preparations, newly harvested grains Excess preparation of jaggery	*Vaman* (Emesis) *Virechana* (Purgation)
Mutrakrichra/Mutraghat (Dysuria)	Excessive *Teekshnaushadh* (Strong Medication), *Rooksh Madya* (Dry property Alcohol), *Anoop Mamsa* (Living near water bodies and eating fish)	As Per *Doshik* condition

Details of dietary causes and their attributes.

Pandu (Anemia)[23]

The dietary causes (Table 16.1) attenuate the functions of *Pitta* and *Vayu,* disturbing the functions of *Hruday* (Heart). The symptoms are *Aruchi* (Anorexia), *Chhardi* (Vomiting), *Saad* (Exhaustion), and *Klama* (Fatigue).

Shotha (Oedema)[24]

The dietary etiologies (Table 16.1) can produce endogenous vitiation of *Vayu* which lodges in *Twak* (Skin), *Sira* (Capillaries), and *Pakvashya* (Particular area of body pelvic), disturbing *Pitta, Kapha,* and *Rakta.* This vitiation of *Doshas* exhibits edema with specific symptoms such as *Ushnasparsha* (Warm), *Siratanutva* (Fragility capillary/tubular), and *Vaivarnya* (Discoloration).

Ajirna (Metabolic dysfunction)[25]

The causation (Table 16.1) of *Ajirna's* has contributed to understand CKD symptoms. The *Ajirna/Atibhojna/Vishmashana* (Malpractices/Maladaptation), *Abhojana* (Inadequate), and *Asatamya* (Incompatible) food items were meant to form metabolic toxins. The contributing diseases and symptoms have been mentioned as *Prameha* (DM) and *Mutraroga* provoking CKD pathogenesis further.[26]

Grahani (Metabolic Dysfunction of Gut)[27]

The *Grahani* as a functional unit plays a pivotal role in the elaboration of CKD correlation from the *Ayurveda* perspective.[28] *Grahani* dealt with major task of conversion of consumed diet into *Saar* (Nourished) and *Kitta* (Metabolic waste), *Kitta* further followed by distribution and differentiation in two portions, i.e. *Sthula* (Faces) and *Drav* (Urine). The dietary etiologies (Table 16.1) of *Grahani* vitiates *Agni*. The common symptoms are *Chhardi* (Vomiting) and *Shotha* (Edema). CKD early stage is relevant to *Kaphaj Grahani*, reflecting symptoms of *Hrullas* (Nausea), *Avipak* (Delayed digestion), and *Klam* (Fatigue).[29]

Prameha (Diabetes Mellitus)

The dietary etiologies (Table 16.1) vitiate *Kapha/Pitta/Vata Dosha*, respectively, disturbing *Meda* and *Mansa* and leading to *Kleda dushti* in *Basti*. The types of *Prameha* (Diabetes mellitus) as per *Doshas* are based upon the urine appearance, color, and turbidity flow and its form.[30]

Mutrakrucchra (Dysuria)[31]

The difficulty in micturition along with referred pain exhibited with physical causes such as excessive exercise and exhaustive physical activity along with dietary etiologies, e.g. meat consumption, alcohol, and *Ati-Tikshna Aushadi* consumption[32] (Excess use of medication/polypharmacy regimen). The etiologies vitiate *Vata* dynamics disturbing functions of *Basti*.

Mutraghata (Oliguria and Anuria Conditions)[33]

The *Rooksha Deha* (Emaciation) and *Klanta* (Exhaustion) are prone conditions to vitiate *Vata* and *Pitta* at *Mutravaha Strotas*. The vitiated *Doshas* impact the kidneys/*Basti*, manifesting anuria as a symptom.[34]

CLASSIFICATION OF CKD AS PER DIETARY PATHOGENESIS

The diseases of the kidney can be classified as per condition: *Sama/Nirama*, *Ashukari/jirna* (Acute and Chronic), *Santarpana* (Nourishing), *Aptarpana* (Reducing), and *Anukta* (Unexplained Disease).

TABLE 16.2 CKD Staging – Dietary Pathologic Conditions[35]

SR.NO	*CONVENTIONAL*	*TYPE OF STAGE/CONDITION OF VITIATION*	*POSSIBLE CORRELATION FROM AYURVEDA*
1	*CKD 1* (Asymptomatic)	*Santarpana* (nourishing)	*Pandu* (Anemia)
2	CKD *2*	*Santarpana* (nourishing)	*Shoth* (edema), *Ajirna* (Metabolic dysfunction)
3	CKD 3a and 3b	*Santarpana/Apatrapna/jirna* (Nourishing/reducing/Chronic)	*Metabolic dysfunction (Ajirna)/ Grahani*

TABLE 16.2 ***(Continued)*** CKD Staging – Dietary Pathologic Conditions

SR.NO	*CONVENTIONAL*	*TYPE OF STAGE/CONDITION OF VITIATION*	*POSSIBLE CORRELATION FROM AYURVEDA*
4	CKD 4 and 5	*Jirna* (Chronic)	*Grahani* (Metabolic dysfunction of the gut) Dysuria *(Mutrakruhtra/Mutraghata)*

Categorization of kidney diseases as per Classics/*Samhita*.

Chronic Kidney Disease

CKD progression with declining GFR can be with various complications such as cardiovascular diseases, hyperlipidemia, severe anemia, metabolic acidosis, and malnourishment.[36] CKD staging as early stages 1–3a and b (nondialysis dependent) and end stages 4 and 5 (dialysis dependent) configured with age, weight, and estimated GFR as per guidelines of National Kidney Foundation (NKF) and Kidney Disease Outcomes Quality Initiative (KDOQI). Early stages of CKD can be monitored by medication and a kidney-friendly diet, but for end stages 4 and 5 (ESRD), dialysis is preferred, needing RRT further.[37]

The nutritional and metabolic derangements can be an outcome of exaggerated protein degradation than synthesis.[38] The catabolic stage demands high-level protein and high-energy enzymes food, producing metabolic end products such as urea and ammonia. The oxidative stress is mainly from the endogenous production of acidification or calcification and produces reactive oxygen species (ROS)–enhancing inflammatory conditions.[39]

The alteration in the metabolism of protein, water, salt, calcium, and phosphate has affected chiefly. However, vigilance of diet is needed to avoid lodging of metabolic waste such as ammonia and urea to arrest further progression into albuminuria and proteinuria.[40] Higher consumption of sugar/soda/cold beverages enhances progression of CKD.[41] High-sodium food can induce hyperfiltration, leading to renal damage. Malnourishment in CKD may be result of systemic inflammation, stress (Oxidative and metabolic), and imbalance between anabolism and catabolism of protein.[42]

CKD nondialysis-dependent patients are recommended a low-protein plant-based diet. Dialysis-dependent patients are advised dairy and lean meat functional food[43] to manage energy homeostasis enhancing function of creatine phosphate.

End-Stage Renal Disease

Subjective Global Assessment (SGA) is used as a nutritional assessment tool in various CKD stages. It proves an effective measure to assess nutrition in peritoneal and hemodialysis.[44] Nutritional requirements in ESRD focus on maintaining optimal nutritional status, minimizing uremic symptoms, and avoiding malnourishment outcome sarcopenia. The palatability of the patient's nutritional plan is important for further renal replacement therapy fitness also.[45]

CONCEPT CONFLUX

Charak postulation is based on minute observations of symptoms in relation to time/season, form, method of preparation, and environmental conditions of food along with the lifestyle of the individual. The basis of kidney disease is governed by *Rakta* and *Meda dhatu* considering systems *Annavaha Srotas* (Digestive tract) and *Medavahasrotas* too.[46]

CKD dietary strategies are divided mainly in two groups labelled nondialysis-dependent and dialysis-dependent. The nondialysis-dependent patients suffer from diseases such as *Shotha*, *Pandu*, *Ajirna*, and *Grahani*. It follows nutritional principles such as *Apatarpana* (Reducing/Depleting), *Dipana* (Digestives), *Sara* (Motility enhancer), and *Srushta Mutra Pravartana* (Easy elimination of urine) mainly followed by *Tarpana* (Replenish energy). This pacifies vitiation of *Tridosha* and pathologic *Kleda* in early stages. Dialysis-dependent patient involves oliguric emaciated condition due to *Mansa* and *Meda Kshaya* (Depletion of muscle and fat) recommended with *Tarpana/Santarpna* (Nourishing), *Rasayana* (Antioxidant/rejuvenation), *Vrushya* and *Viryavardhak* (Tonicity and strength), *Hridya* (Cardiotonic), and *Mutral* (Diuresis). Dietary designing for each stage is based upon selection of food items from liquid or solid groups, respectively (Tables 16.3–16.6).

Renal functional food recommendations are categorized into two main groups: *Drava Dravya Vidyaniya* (Liquids) and *Annaswarup* (Solid form). Water/liquid and salt are the main contributors to kidney diseases, so quantification of water and salt is important to arrest the progression of disease. The functions of food items are categorized for benefit of renal filtration/perfusion aided by digestives, motility, easy elimination, and diuresis for reducing oxidative and metabolic stress, clearing *Asara Dhatu* (Nonnourished portion).

TABLE 16.3 Liquid Food Groups (*Drava-dravyavidnyaniya Varga*)[47]

PROPERTIES	*DEEPAN (DIGESTION AND METABOLISM ENHANCING)*	*SARA (MOBILITY)*	*SRUSHTA VIT AND MUTRA (EASY ELIMINATION)*	*MUTRAL (DIURETIC)*
Water (Jala)	*Kupajala* (Well water), *Ushnodaka* (Hot water), *Narikelodak* (Coconut water) (*Cocos nucifera* L.)	*Godugdha* (Cow milk), *Aja dugdha* (Goat milk), *iksuvarga* (Sugarcane) (*Saccharum officinarum*) * **advisable to nondiabetes patients only**	-	*Ushnodak* (Hot water), *Narikelodak* (Coconut water)
***Ksheera* (Dairy)**	*Navaneet* (Butter), *Takra* (Buttermilk), *Ghruta* (Ghee), *Ushtra ksheera* (Camel milk) (*Camelus dromedarius*)	-	-	-
***Ikshu* (Group of Sugars)**	-	*Madhu* (Honey)	*Dhaut Gud* (Clean Jaggary)	*Panak* *(Beverage), *Iksu* (Sugarcane)

Categorization of water/liquids with their properties, quantity of liquid varies in each individual as per hunger and metabolic demands.

TABLE 16.4 Solid Food Groups (*Ahara Varga*)[48]

PROPERTIES	DEEPAN *(DIGESTION AND METABOLISM ENHANCING)*	SARA *(MOBILITY)*	SRUSHTA VIT AND MUTRA *(EASY ELIMINATION)*	MUTRAL *(DIURETIC)*
Shuka **(Grains)**		*Shali varga* (All Rice Varieties), *yavak* (Barley) (*Hordeum vulgare*), *Godhuma* (Wheat) *(Triticum aestivum)*	*Yavak* (Barly) (Hordeum vulgare)	*Shuk dhanya* (Group of cereals), *shali* (Rice) (*Oryza nivara*)
Shimbi **(Pulses/ Legume)**		Black gram (*Vigna mungo*)		*Rajmash* (Rajma) (*Phaseolus vulgaris* L.)
Krutanna **(Processed Food items)**	*Peya (*Thin gruel of rice*)*, *Viilepee* (thick gruel of rice), *Him* (Cold infusion)	*Laja* (Puffed rice) (*Alpinia galanga*)		*Gud panak* (Jaggery beverage)
Shaak **(Vegetables)**	*Changeri* (yellow sorrel) (oxalis corniculate), *Karvellka* (bitter gourd) (*Momordica charantia*), *Koshataki* (Ridge gourd) (*Luffa acutangula*), *Kushmand* (Ash gourd) (*Benincasa hispida*), *Varshabhu* (Desert horse) (*Trianthema decandra*), *Chirbilwankur* (Tender leaves of Ponagam Tree) (*Pongamia pinnata*),	*Kaakamachi* (Black nightshade) (*Solanum nigrum*), *Chirabilwa* (Indian elm tree) (*Holoptelea integrifolia*), *Kusumbha* (Safflower) (*Carthamus tinctorius*)	*shak varga* (All vegetables)	*Trapus* (Cucumber) (*Cucumis sativus*), *Vidarikanda* (Kudzu) (*Pueraria tuberosa*), *Kuashmand* (Ash gourd**)**
Phal **(Fruits)**	*Draksh* (Grapes*)* (Vitis vinifera), *Amra* (Mango*)* (*Mangifera indica*), *Bilwa* (Stone Apple) (*Aegel marmelos*), *Vrukshamla* (Goa butter tree) (*Garcinia indica*), *Amlika (*Tamarind*)* (*Tamarindus indica*)	*Badar* (Full moon*)* (*Ziziphus jujuba*), *Vaatam* (Almond) (*Prunus dulcis*), *Karamard* (Karonada) (*Carissa carandas*)	*Draksha (Grapes) (Vitis vinifera)*	*Tal phala* (Toddy palm*)* (*Borassus flabellifer*)
Haritak **(Fresh Tubers/Spices/ Vegetables)**	All Spices, *Lasuna* (Garlic) (*Allium sativum*), *Suran* (Elephant foot yam) (*Amorphophallus campanulatus*)	*Lasuna* (Garlic)		
Aharopayogi **(Adjuvant Food)**				*Romak lavan* (Sambhar Salt)

Classification of food items – with their characteristic features in conserving renal perfusion/mass.

Pathya-ahara (Wholesome) dealt with *Matra* (quantity), *Vidhi* (procedure of food preparation and consumption) and *Jalpana* (Water Consumption).[49] *Ahitakar Ahar* such as processed food and meat, *Virudh* (Sprouts) having tannins, oxalates, and antinutrients phytics which directly impact renal functions,[50] e.g. filtration and ultrafiltration of nutrients and metabolic waste. However, CKD stages 1–3 (early stage) can be mainly ingrained in *Ananavaha* and *Medvaha Srotas*. Thus, *Vata* dynamics leads to decline in filtration rate gradually.[51] Morbid complication of declined filtration rate due to vitiation of *Rakta* and *Med dhatu prasad* (Nourished) portion mitigates in muscle and bone tissue such as sarcopenia and bone mineralization.[52]

Grahani is a site of *Agni* exhibiting differentiation of *Sara* (Nourished) and *Asara* (Nonnourished) portion of diet/nutrients assay to form further healthy *Dhatus. Eventually, Grahani* homeostasis helps in maintaining normalcy of kidney functions and conservation of renal mass.[53] In turn, *Rakta* and *Meda* nourished portions maintain renal mass and their functions. Primarily, TNA is based on understanding the underlying diseases and dietary etiologies of CKD staging. However, TNA principle focuses on arresting CKD staging progression and deals with poor morbid outcome with its customized nutritional guidelines.

TABLE 16.5 Liquid Food Groups Based on Functional Attributes[54]

PROPERTIES	*TARPAN (NOURISHING)*	*VRUSHYA/ VIRYAVARDHAK (APHRODITIC/TONICITY)*	*HRIDYA (CARDIOTONIC)*	*RASAYAN (ANTIOXIDANT)*
Jala **(Water)**	*Jala* (Water)	*Narikelodak* (Coconut water)		
Ksheera **(Dairy)**	Urial (*avik dugdha*) (*Ovis vignei*)	*Dugdha varga* (All dairy varieties)		*Godugdha* (Cow milk)
Ikshu ***(Group of Sugars)**		*Sarkara* (Sugar), *Ekshu* (All sugarcane) common guna, *Madhu* (honey)	*Gud* (Jaggery)	

Essentials of nutritional functions required in CKD staging and its complication.
* Nondiabetes mellitus

TABLE 16.6 Solid Food Groups Based on Functional Attributes[55]

PROPERTIES	TARPAN (*NOURISHING*)	VRUSHYA (*APHRODITIC/ TONICITY*)	HRIDYA (*CARDIOTONIC*)	RASAYAN (*ANTIOXIDANT*)
Shuka **(Grains)**			*Shukta* (Vinegar), *Shirnvrintak* (Right brinjal)	
Krutanna **(Processed Food items)**	*Dhana* (Bhrushta dhanya)	*Shaali* (rice common guna), *Madhulika* (finger millet) (*Eleusine coracana*), *Godhum* (wheat) (*Triticum sativum*), *Rasala* *Raag*, *Shadav*, *Yava* (Barley)	*Panak* (Prepared from rice) (*Dhanyamala*), *Vilepi* (Thick gruel of rice)	

TABLE 16.6 ***(Continued)*** Solid Food Groups Based on Functional Attributes

PROPERTIES	TARPAN (*NOURISHING*)	VRUSHYA (*APHRODITIC/ TONICITY*)	HRIDYA (*CARDIOTONIC*)	RASAYAN (*ANTIOXIDANT*)
Jangal Mamsa **(Arid Meat)**			*Jangal Mamsa ras* (Gruel prepared with Arid meat) green gram, pomegranate	
Shaak **(Vegetables)**		*Jeevanti* (Cork swallow) (*Leptadenia reticulata*), *Kushmand* (Ash gourd),	*Patola* (Snake gourd) (*Trichosanthes cucumerina* Linn.), *Vartak* (Brinjal)	*Kakmachi* (Black nightshade), *Jivanti* (Cork swallow)
Phal **(Fruits)**		*DrakSha* (Grape), *Priyal* (Chirongi) (*Buchanania lanzan* Spreng)	*Lavli fala* (Gooseberry tree) (*Cicca acida* Merrill), *Draksa* (Grape), *Narikel* (coconut), *Badar* (Full moon) (*Ziziphus jujuba*), *Amra* (Mango) (*Mangifera indica* Linn.)	*Vidarikand* (Kadzu) (*Pueraria tuberosa*)
Haritak **(Spices)**			*Lahsun* (Garlic) (allium sativum)	*Lahsun* (Garlic)

Major contributory Ayurveda dietary functional attributes for management of GFR, CKD, and its complications.

CKD 1 to 3a chiefly define the following symptoms: fatigue, malaise, edema, nausea, and anorexia.[56] The nutritional principle and recommendation of *Pandu* and *Shotha Chikista* is applicable to treat nondialysis dependent stage considering above symptoms. The condition of nausea and malaise can be treated with *Laghu Santarpana* and *Virya Vardhak* food (Tables 16.5 and 16.6). The food groups used are digestives, motility, *Tarpana,* and *Hridya* primarily (Tables 16.5 and 16.6). The *Shuk Varga,* having property of *Sara* and *Rasayana*, are selected for preparing gruel of *Rajmash* (*Phaseolus vulgaris* L.), soup of green gram and lentil. Diabetic patients can have *Badar Panak* to alleviate *Agni*, while nondiabetes CKD patients can have *Mrudvika Panak* (Raisins) as a *Tarpak* (Replenishing energy). The use of ash gourd with ghee or milk manages patient's fatigue.[57]

CKD 3b, complete alteration of renal insufficiency, leads to vomiting, pruritus, oliguria, and dyspnoea.[58] It is a condition of *Vata* dominance reducing *Kapha Dosha* and leading to *Mutrkruchtra.*[59] The *Mutracrucchra* principle *Snehana* can help in reducing *Vata* dominance and use of *Pitta* pacifying *Dravyas.* End-stage 3b patients can follow recommendation with nutritional principles and food items from *Sadya Santarpana*, *Bruhana*, *Vrushaya*, *Hrudya,* and *Mutral* food groups (Tables 16.5 and 16.6). Nondialysis individuals are recommended *Tarpana* and *Mantha Kalpna,* mainly where restriction of water is not a prerequisite. Promoting diuresis (Table 16.7) can be incorporated into diet. The acidosis, which leads to many complications, can be buffered well with *Shaka* to maintain bicarbonate load of patient.[60] The uremic metabolic waste can be managed with *Dipnaiya* and *Pachaniya dravyas*. The gut dysbiosis reduced using *dipaniya* and *pachaniya* principle to overcome insufficiency of nutrient absorption. Thus, it helps in reducing further load of metabolic waste to kidney. The progression can be checked with antioxidant plant-based *Rasayana* recipes. The energy homeostasis can be maintained with use of *Mansa Varga* food items and recipes. Vigilance is needed while using *Mansa* and *Shak Varga* to alleviate effect of deficiency of vitamin E and ascorbic acid.

Dialysis-dependent patients are prone to dryness of skin and emaciation resembling symptoms of *Mutarkshaya*. Dialysis patients may gain weight, i.e. intra dialectic weight gain (IDWG) due to ultrafiltration. While during dialysis large volume depletion of many micronutrients is displayed with hypotension, headaches, and fatigue. Apathy toward meal is due to extreme dietary and water restrictions. Primarily, addressing nutritional recommendation for apathy to emaciation needs meal serving with *Pachana,* preceding *Diapan* and followed by *Bruhana* specifically. IDWG addressed by using *Mutral* food item. The ash gourd, *Mrudvika,* and *Kshiraamruta* can be used as *Bruhana.*

Malnourishment in CKD is a morbid complication. The preparations mentioned in (Tables 16.5 and 16.6) help in *Mansa Dhatu Upachya* (Enhancement of Muscle tissue). The outcome of uremic waste impact is depletion of muscle mass leading to malnourishment. The *Mansa* and *Gorasa Varga* preparations can be used (Table 16.7), e.g. *Rajika Mansa*, *Kshiramurta,* and *Khandkushmand.*

Malnourishment: Chart No. 1

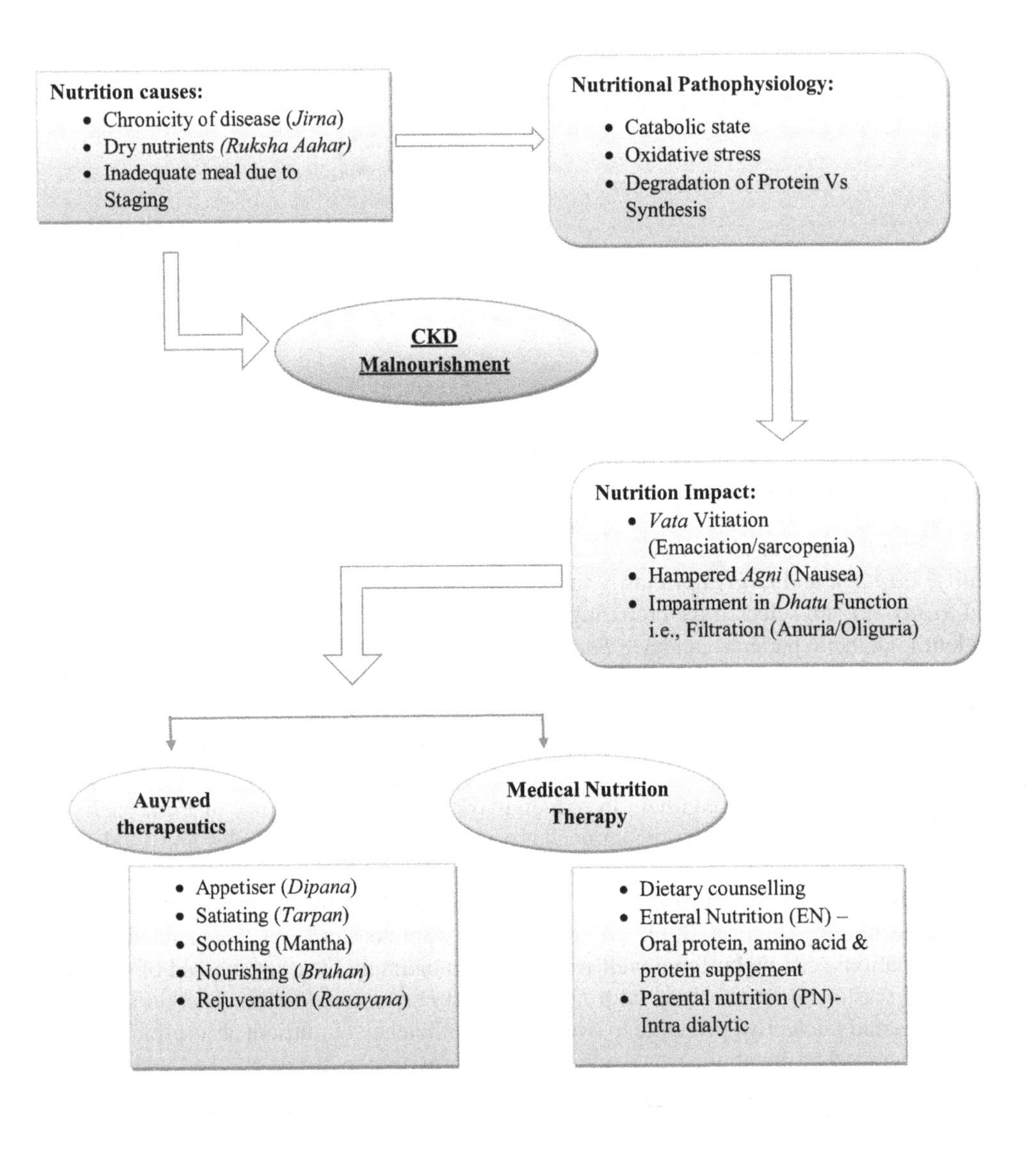

ESRD is a poor outcome of nutrition with array of nutritional insufficiencies and its complication. However, maintaining gut flora is key feature to avoid further deterioration. The nutritional principle to manage ESRD is *Laghu Santarpna*, *Mansa Varga, Hrudya,* and *Rasayana* (Tables 16.5 and 16.6).[61] The *Rasayana* (Immune nutrients) helps in reducing intradialytic fatigue, and periodic dialysis leads to maximum micronutrient loss through permeable membrane of dialyzing compartment and dialyzing agent too. Micronutrient insufficiency can be arrested with use of nutrient-rich *Panak Kalpna* effectively (Table 16.7). The *Ghrita Siddha* preparations are highly recommended as PUFAs (polyunsaturated fatty acids) mainly deal with slowing disease progression, decreasing matrix production and proliferation, and enhancing endothelial functions.[62]

AHARIYA KALPANA (RENAL FUNCTIONAL FOOD)

The *Ahariya kalpan*'s is a concept of understanding dietary etiologies, and food attributes which causes depletion of renal functions and renal mass in CKD. The renal dietary strategy goal is to maintain renal functions and mass. Eventually, framing nutritional strategies is based upon selection of food items from required functional food groups from *Dwadasha Ahara Varga* (12 food groups) and food matrices such as *Dipaniya*, *Mutral*, etc. The renal functional food provides essentials of renal function–conserving properties.

TABLE 16.7 CKD Stages Essential Renal Functional Food

CATEGORIES	*MAIN INGREDIENTS (NAME OF THE FOOD ITEMS)*	*PREPARATORY METHODS*	*BENEFITS/BIOACTIVE FEATURES*[63]
Shaka Varga (Fruit vegetable)	*Khanda - Kushmanda (Ash Gourd)*[64]	Pieces boiled in milk sauté in ghee add sugar and spices pepper and Cardamom	Protein, Carbohydrate, Fiber-rich, Fat, Vit C Riboflavin, Zinc.
Shaka Varga (Fruit vegetable)	*Karvellak* (Bitter Gourd[65])	Pieces of Bitter gourd sauté in oil with pinch of Asafoetida, paper powder, and rock salt[66]	Rich in vitamins A, E, C and B6, iron, Calcium, magnesium, Phosphorous, and zinc.
Phal shaak Varga (Fruit Vegetables)	*Vruntaak* (Brinjal/ Eggplant)	Ground pieces of brinjal, dry ginger, and coriander; fry it in melted butter sauté with cumin piper powder and asafoetida incense[67]	fiber, copper, manganese, vitamin B6, and thiamine.
Mamsa varga (Meat)	*Rajika Mamsa* (Mustard Meat)	Meat boiled in buttermilk, sauté in mustard seed and powder and pink salt, spices[68]	Protein, Amino Acids

(Continued)

TABLE 16.7 ***(Continued)*** CKD Stages Essential Renal Functional Food

CATEGORIES	*MAIN INGREDIENTS (NAME OF THE FOOD ITEMS)*	*PREPARATORY METHODS*	*BENEFITS/BIOACTIVE FEATURES*[63]
Jalavarga (Water)	*Dhanyak him* (Coriandrum *Cold Infusion*)[69]	Sock coriander seed in water overnight and smash and strain it in morning	Useful for thirst, relieves burning sensation, *pitta* disorder, indigestion, abdominal pain, clears *srotas* (bodily channels), in fever, indigestion, worm infestation
Shuk Varga (Grains)	*Yavasaktu* (Roasted Barley flour)[70]	Roasted Rice, Barley, Rye, Varae (Proso millet): grind it and use the Flour.	Calories, fat, protein, carbohydrate. Dietary fiber, saturated fat, cholesterol, iron, calcium.
Phala Varga (Fruits)	*Dadim Yush* (Pomegranate Gruel)[71]	Boiled green gram and pomegranate seeds, sauté in clarified butter	Calcium, Vitamin C, total ascorbic acid, Folate, Vitamin K (phylloquinone), Magnesium, Phosphorus.

Various renal functional food with their bioactive properties.

NUTRAVIGILANCE

Use of *Shak Varga* (Vegetables) without blanching and leaching leads to hyperkalemia. Thus, it is mandatory to follow the rule of a preparatory method for each meal/day.[72]

CONCLUSION

CKD is a gradually progressive condition that needs periodic TNA intervention as per dietary regimen recommended with reassessment of dietary intervention for nondialysis- and dialysis-dependent patients. The dietary design has to be prepared for each stage of nondialysis and dialysis patients, considering their prime diseases such as obesity, DM, and HTN to avoid progression. CKD end-stage (4 and 5) recommendation of *Rasayana* (immune nutrients) can help in reducing malnourishment and varied infections during dialysis. TNA can be used to overcome the challenges of nutritional supplements with a variety of natural plant–based/arid lean meat preparations with high biological value protein, essential amino acids, and vitamins too.

IMPLICATIONS

1. Exploration of *Ayurveda* perspective of Intradialytic renal functional food.
2. Evaluating tailored sequencing of dietary regimen in CKD staging to avoid progression.

3. Exploring renal functional food nutrients assay analysis which binds excess of potassium, phosphates and maintains bicarbonates.

ABBREVIATIONS

Advanced Glycation End Products (AGEs)
Reactive Oxygen Species (ROS)
Acute kidney injury (AKI)
Astaouahar vidi visheshayatananai (AVV)
Ayurvedic Nutrition Therapy (ANT)
Chronic kidney disease (CKD)
Blood urea Nitrogen (BUN)
Diabetes Mellitus (DM)
dietary energy intake (DEI)
dietary protein intake (DPI)
End-Stage Renal Disease (ESRD)
Enteral Nutrition (EN)
Glomerular Filtration Rate (GFR)
Hypertension (HTN)
interleukin (IL)
Intradialytic weight gain (IDWG)
Kidney Disease Outcomes Quality Initiative (KDOQI)
Life Expectancy (LE)
Low-Density Lipids (LDL)
Medical Renal Disease (MRD)
National Kidney Foundation (NKF)
Nuclear factor (NF)
Over-the-Counter Drug (OTC)
Parental nutrition (PN)
Peritoneal Dialysis (PD)
Polycystic Kidney Disease (PCKD)
Polyunsaturated Fatty Acid (PUFA)
Protein energy wasting (PEW)
Quality of Life (QoL)
Renin angiotensin aldosterone System (RAAS)
Serum Creatinine (Sr. Cr.)
Serum Electrolyte (Sr. Electrolyte)
tumor necrosis factor (TNF)
Ahara Vidhi (AV)

REFERENCES

1 Yadavaji Trikramji Acharya; Fifth edition 2001; Charaka samhita by agnivesha; Choukhamba Sanskrit Bhawan, Varanasi, siddhi sthan 9/4.
2 Yadavaji Trikramji Acharya; Fifth edition 2001; Charaka samhita by agnivesha; Choukhamba Sanskrit Bhawan, Varanasi, Charkara pani – Charak samhita chikista 26/3-4.
3 Yadavaji Trikramji Acharya; edition Reprint 2019; Sushrut samhita by Sushruta with Nibandhasangraha commentary by Dalhana; Choukhamba Sanskrit Sansthan, Varanasi, Dallhana sushruta samhita nidana sthan 3/18-24, p. 279.
4 Yadavaji Trikramji Acharya; edition Reprint 2019; Sushrut samhita by Sushruta with Nibandhasangraha commentary by Dalhana; Choukhamba Sanskrit Sansthan, Varanasi, Dallhana sushruta samhita nidana sthan 3/18-24, p.-279.
5 Yadavaji Trikramji Acharya; Fifth edition 2001; Charaka samhita by agnivesha; Choukhamba Sanskrit Bhawan, Varanasi, viman 5/8.
6 Yadavaji Trikramji Acharya; Fifth edition 2001; Charaka samhita by agnivesha; Choukhamba Sanskrit Bhawan, Varanasi, sutra sthan 28/3, 4, 5.
7 Pt. Shyamsundar Sharma, bhanumati commentary of Sushruta Samhita sutrasthana; published by; Registrar Agra University & Secretary, Swami Lakshmi Ram trust, sutrasthan 35/19, p. 237.
8 P.V. Sharma; 1 April 2005; Charaka samhia, text with English translation, Chaukhamba Orientalia, Varanasi, Chikitsa sthan 15/49.
9 Thomas, R., Kanso, A., & Sedor, J. R. (2008b). Chronic Kidney Disease and Its Complications. *Primary Care: Clinics in Office Practice*, *35*(2), 329–344. https://doi.org/10.1016/j.pop.2008.01.008

10 Elshahat, S., Cockwell, P., Maxwell, A. P., Griffin, M., O'Brien, T., & O'Neill, C. (2020b). The impact of chronic kidney disease on developed countries from a health economics perspective: A systematic scoping review. *PLOS ONE*, *15*(3), e0230512. https://doi.org/10.1371/journal.pone.0230512

11 Akchurin, O. M. (2019). Chronic Kidney Disease and Dietary Measures to Improve Outcomes. *Pediatric Clinics of North America*, *66*(1), 247–267. https://doi.org/10.1016/j.pcl.2018.09.007

12 Wesson, D. E., Buysse, J. M., & Bushinsky, D. A. (2020b). Mechanisms of Metabolic Acidosis–Induced Kidney Injury in Chronic Kidney Disease. *Journal of the American Society of Nephrology*, *31*(3), 469–482. https://doi.org/10.1681/asn.2019070677

13 Bidani, A. K., Polichnowski, A. J., Loutzenhiser, R., & Griffin, K. A. (2013). Renal microvascular dysfunction, hypertension and CKD progression. *Current Opinion in Nephrology and Hypertension*, *22*(1), 1–9. https://doi.org/10.1097/mnh.0b013e32835b36c1

14 Hamm, L. L., Nakhoul, N., & Hering-Smith, K. S. (2015). Acid-Base Homeostasis. *Clinical Journal of the American Society of Nephrology*, *10*(12), 2232–2242. https://doi.org/10.2215/cjn.07400715

15 Thomas, R., Kanso, A., & Sedor, J. R. (2008b). Chronic Kidney Disease and Its Complications. *Primary Care: Clinics in Office Practice*, *35*(2), 329–344. https://doi.org/10.1016/j.pop.2008.01.008

16 Post, A., Tsikas, D., & Bakker, S. J. (2019). Creatine is a Conditionally Essential Nutrient in Chronic Kidney Disease: A Hypothesis and Narrative Literature Review. *Nutrients*, *11*(5), 1044. https://doi.org/10.3390/nu11051044

17 P.V. Sharma, first edition, 2000; English translation of text and Dalhana commentary, along with critical notes of Susruta samhita; with; volume 1, 2 and 3; Choukhamba Visvabharati, Varanasi, Sharirasthan 4/31.

18 Chakrapanidatta; edition 2019; Ayurveda Deepika commentary of Charak Samhita, Choukhamba Surabharati Prakashan, Varanasi, Vimansthan 5/8.

19 Dr. Anna Kunthe; reprint 2002 Astanga Hridaya of Vagbhata with the commentaries: sarvangasundara of arundatta & ayurvedarasayana of hemadri; Choukhamba Surabharati, Varanasi; Sutrasthan 12/9.

20 Dr. Anna Kunthe; reprint 2002 Astanga Hridaya of Vagbhata with the commentaries: sarvangasundara of arundatta & ayurvedarasayana of hemadri; Choukhamba Surabharati, Varanasi; Sutrasthan 11/5.

21 Dr. Anna Kunthe; reprint 2002 Astanga Hridaya of Vagbhata with the commentaries: sarvangasundara of arundatta & ayurvedarasayana of hemadri; Choukhamba Surabharati, Varanasi.

22 Prof. K.R. Srikanth Murthy; 4th edition 2005; Astanga sangraha of vagbhata (text, English translation, notes, Appendices and index) Vol. I, II, III; Choukhamba Orientalia, Varanasi.

23 P.V. Sharma; 1 April 2005; Charaka samhia, text with English translation, Chaukhamba Orientalia, Varanasi Yadavaji Trikramji Acharya; Fifth edition 2001; Charaka samhita by agnivesha; Choukhamba Sanskrit Bhawan, Varanasi.

24 P.V. Sharma, first edition, 2000; English translation of text and Dalhana commentary, along with critical notes of Susruta samhita; with; volume 1, 2 and 3; Choukhamba Visvabharati, Varanasi, Uttartantra 40/182.

25 Chakrapanidatta; edition 2019; Ayurveda Deepika Commentary of Charak Samhita, Choukhamba Surabharati Prakashan, Varanasi, chikitsasthan 16/10.

26 Chakrapanidatta; edition 2019; Ayurveda Deepika Commentary of Charak Samhita, Choukhamba Surabharati Prakashan, Varanasi, Chikitsa 12/8.

27 Chakrapanidatta; edition 2019; Ayurveda Deepika Commentary of Charak Samhita, Choukhamba Surabharati Prakashan, Varanasi, Chikitsasthan 15/42-49.

28 Chakrapanidatta; edition 2019; Ayurveda Deepika Commentary of Charak Samhita, Choukhamba Surabharati Prakashan, Varanasi, Chikitsasthan 15/49.

29 Chakrapanidatta; edition 2019; Ayurveda Deepika Commentary of Charak Samhita, Choukhamba Surabharati Prakashan, Varanasi, Chikitsasthan 15/70.

30 Dr. Anna Kunthe; reprint 2002 Astanga Hridaya of Vagbhata with the commentaries: sarvangasundara of arundatta & ayurvedarasayana of hemadri; Choukhamba Surabharati, Varanasi nidan sthan 10/1-5.

31 Chakrapanidatta; edition 2019; Ayurveda Deepika Commentary of Charak Samhita, Choukhamba Surabharati Prakashan, Varanasi, Chikitsasthan 26/33.

32 P.V. Sharma, first edition, 2000; English translation of text and Dalhana commentary, along with critical notes of Susruta samhita; with; volume 1, 2 and 3; Choukhamba Visvabharati, Varanasi sutrasthan 39/10.

33 P.V. Sharma, first edition, 2000; English translation of text and Dalhana commentary, along with critical notes of Susruta samhita; with; volume 1, 2 and 3; Choukhamba Visvabharati, Varanasi Uttartantra 58/17.

34 P.V. Sharma, first edition, 2000; English translation of text and Dalhana commentary, along with critical notes of Susruta samhita; with; volume 1, 2 and 3; Choukhamba Visvabharati, Varanasi Uttartantra 58/17.

35 P.V. Sharma; 1 April 2005; Charaka samhia, text with English translation, Chaukhamba Orientalia, Varanasi, Sutrasthan 23/5-6-7.

36 Thomas, R., Kanso, A., & Sedor, J. R. (2008c). Chronic Kidney Disease and Its Complications. *Primary Care: Clinics in Office Practice*, *35*(2), 329–344. https://doi.org/10.1016/j.pop.2008.01.008

37 Ikizler, T. A., Burrowes, J. D., Byham-Gray, L. D., Campbell, K. L., Carrero, J. J., Chan, W., Fouque, D., Friedman, A. N., Ghaddar, S., Goldstein-Fuchs, D. J., Kaysen, G. A., Kopple, J. D., Teta, D., Yee-Moon Wang, A., & Cuppari, L. (2020). KDOQI Clinical Practice Guideline for Nutrition in CKD: 2020 Update. *American Journal of Kidney Diseases*, *76*(3), S1–S107. https://doi.org/10.1053/j.ajkd.2020.05.006

38 Rapa, S. F., Di Iorio, B. R., Campiglia, P., Heidland, A., & Marzocco, S. (2019b). Inflammation and Oxidative Stress in Chronic Kidney Disease—Potential Therapeutic Role of Minerals, Vitamins and Plant-Derived Metabolites. *International Journal of Molecular Sciences*, *21*(1), 263. https://doi.org/10.3390/ijms21010263

39 Tirichen, H., Yaigoub, H., Xu, W., Wu, C., Li, R., & Li, Y. (2021). Mitochondrial Reactive Oxygen Species and Their Contribution in Chronic Kidney Disease Progression Through Oxidative Stress. *Frontiers in Physiology*, *12*. https://doi.org/10.3389/fphys.2021.627837

40 Fouque, D., & Laville, M. (2009). Low protein diets for chronic kidney disease in nondiabetic adults. *Cochrane Database of Systematic Reviews*. https://doi.org/10.1002/14651858.cd001892.pub3

41 Matovinović, M. S. (2009). 1. Pathophysiology and Classification of Kidney Diseases. *EJIFCC*, *20*(1), 2–11.

42 Rapa, S. F., Di Iorio, B. R., Campiglia, P., Heidland, A., & Marzocco, S. (2019d). Inflammation and Oxidative Stress in Chronic Kidney Disease—Potential Therapeutic Role of Minerals, Vitamins and Plant-Derived Metabolites. *International Journal of Molecular Sciences*, *21*(1), 263. https://doi.org/10.3390/ijms21010263

43 Kalantar-Zadeh, K., Joshi, S., Schlueter, R., Cooke, J., Brown-Tortorici, A., Donnelly, M., Schulman, S., Lau, W. L., Rhee, C., Streja, E., Tantisattamo, E., Ferrey, A., Hanna, R., Chen, J., Malik, S., Nguyen, D., Crowley, S., & Kovesdy, C. (2020). Plant-Dominant Low-Protein Diet for Conservative Management of Chronic Kidney Disease. *Nutrients*, *12*(7), 1931. https://doi.org/10.3390/nu12071931

44 Thomas, R., Kanso, A., & Sedor, J. R. (2008d). Chronic Kidney Disease and Its Complications. *Primary Care: Clinics in Office Practice*, *35*(2), 329–344. https://doi.org/10.1016/j.pop.2008.01.008

45 Zha, Y., & Qian, Q. (2017). Protein Nutrition and Malnutrition in CKD and ESRD. *Nutrients*, *9*(3), 208. https://doi.org/10.3390/nu9030208

46 Yadavaji Trikramji Acharya; Fifth edition 2001; Charaka samhita by agnivesha; Choukhamba Sanskrit Bhawan, Varanasi, Viman sthan 5/8.

47 Prof. K.R. Srikanth Murthy; 4th edition 2005; Astanga sangraha of vagbhata (text, English translation, notes, Appendices and index) Vol. I, II, III; Choukhamba Orientalia, Varanasi; Sutrasthan 6.

48 Prof. K.R. Srikanth Murthy; 4th edition 2005; Astanga sangraha of vagbhata (text, English translation, notes, Appendices and index) Vol. I, II, III; Choukhamba Orientalia, Varanasi; Sutrasthan 7.

49 Prof.K.R. Srikanth Murthy; 4th edition 2005; Astanga sangraha of vagbhata (text, English translation, notes, Appendices and index) Vol. I, II, III; Choukhamba Orientalia, Varanasi, sutra sthan, 6./16.

50 Carbas, B., Machado, N., Oppolzer, D., Ferreira, L., Queiroz, M., Brites, C., Rosa, E. A., & Barros, A. I. (2020b). Nutrients, Antinutrients, Phenolic Composition, and Antioxidant Activity of Common Bean Cultivars and their Potential for Food Applications. *Antioxidants*, 9(2), 186. https://doi.org/10.3390/antiox9020186

51 Yadavaji Trikramji Acharya; Fifth edition 2001; Charaka samhita by agnivesha; Choukhamba Sanskrit Bhawan, Varanasi, chikitsa sthan 15/48,49.

52 Prof.K.R. Srikanth Murthy; 4th edition 2005; Astanga sangraha of vagbhata (text, English translation, notes, Appendices and index) Vol. I, II, III; Choukhamba Orientalia, Varanasi Sutrasthan 19/9.

53 Yadavaji Trikramji Acharya; Fifth edition 2001; Charaka samhita by agnivesha; Choukhamba Sanskrit Bhawan, Varanasi, chikitsa sthna, 15/56-57.

54 Dr. Anna Kunthe; reprint 2002 Astanga Hridaya of Vagbhata with the commentaries: sarvangasundara of arundatta & ayurvedarasayana of hemadri; Choukhamba Surabharati, Varanasi; Sutrasthan 6.

55 Prof. K.R. Srikanth Murthy; 4th edition 2005; Astanga sangraha of vagbhata (text, English translation, notes, Appendices and index) Vol. I, II, III; Choukhamba Orientalia, Varanasi; Sutrasthan 7.

56 Anderson, R. K. (2022). *Solution for chronic kidney failure: A straightforward manual to comprehending the causes, symptoms, diagnosis, prevention, and effective treatment of chronic renal failure.* Independently published.

57 Palamthodi, S., Kadam, D., & Lele, S. S. (2018). Physicochemical and functional properties of ash gourd/bottle gourd beverages blended with jamun. *Journal of Food Science and Technology*, *56*(1), 473–482. https://doi.org/10.1007/s13197-018-3509-z

58 *Krause's Food & Nutrition Therapy by Janice L Raymond MS RD CD (2007-10-01)*. (2023). Saunders.
59 Yadavaji Trikramji Acharya; Fifth edition 2001; Charaka samhita by agnivesha; Choukhamba Sanskrit Bhawan, Varanasi, chikitsa sthana, 26/36.
60 Rodrigues Neto Angéloco, L., Arces de Souza, G. C., Almeida Romão, E., & Garcia Chiarello, P. (2018). Alkaline Diet and Metabolic Acidosis: Practical Approaches to the Nutritional Management of Chronic Kidney Disease. *Journal of Renal Nutrition*, *28*(3), 215–220. https://doi.org/10.1053/j.jrn.2017.10.006
61i Dr. Anna Kunthe, (reprint 2002). Astanga Hridaya of Vagbhata with the commentaries: sarvangasundara of arundatta & ayurvedarasayana of hemadri; Choukhamba Surabharati, Varanasi; Sutrasthan 5 & 6;
61ii Prof. K.R. Srikanth Murthy, 4th edition 2005; Astanga sangraha of vagbhata (text, English translation, notes, Appendices and index) Vol. I, II, III; Choukhamba Orientalia, Varanasi; Sutrasthan 6 & 7;
61iii Yadavaji Trikramji Acharya; 5th edition 2001; Charaka samhita by agnivesha; Choukhamba Sanskrit Bhawan, Varanasi Chakrapanidatta; edition 2019; Ayurveda Deepika Commentary of Charak Samhita, Choukhamba Surabharati Prakashan, Varanasi, Sutrasthan 26 & 27.
62 Baggio, B., Musacchio, E., & Priante, G. (2005). Polyunsaturated fatty acids and renal fibrosis: pathophysiologic link and potential clinical implications. *Journal of Nephrology*, *18*(4), 362–367.
63 Jk, D. (2023). *Food Biochemistry (Pb 2020)*. CBS.
64 Dr Indradev Tripathi; 2078; Shrikshema Sharma's Kshemakutuhal with Manjula commentary, Chaukhamba Orientalia, Varanasi. 8/53 p. 119.
65 Dr Indradev Tripathi; 2078; Shrikshema Sharma's Kshemakutuhal with Manjula commentary, Chaukhamba Orientalia, Varanasi. 8/67 p. 122.
66 Dr Indradev Tripathi; 2078; Shrikshema Sharma's Kshemakutuhal with Manjula commentary, Chaukhamba Orientalia, Varanasi. 8/68 p. 123.
67 Dr Indradev Tripathi; 2078; Shrikshema Sharma's Kshemakutuhal with Manjula commentary, Chaukhamba Orientalia, Varanasi., 8/17 p. 112.
68 Dr Indradev Tripathi; 2078; Shrikshema Sharma's Kshemakutuhal with Manjula commentary, Chaukhamba Orientalia, Varanasi.
69 Sarangadharacharya; 7th edition 2008; Sarngadara samhita with the commentary Adhamalla's Dipika and kashirama's Gudhartha-Dipika; choukhamba orientalia, Varanasi, Madhyam khand, 2/7-8.
70 Shri Brahm Shankar Mishra; 12th edition 2016; Shastri vidyotini commentary of Bhavprakasha, 1st part, Choukhamba Sanskrit bhavan, Varanasi Krutanna Varga 167 p. 892.
71 *KaiyadevanighaNTu*. (n.d.). https://niimh.nic.in/ebooks/e-Nighantu/kaiyadevanighantu/?mod=read (Kaiyadeva Nighantu 5 krutanna varga 88).
72 Shri Brahm Shankar Mishra; 12th edition 2016; Shastri vidyotini commentary of Bhavprakasha, 1st part, Choukhamba Sanskrit bhavan, Varanasi, shaka varga.

Therapeutic Nutrition in Ayurveda for Asthma

Satwashil Desai

INTRODUCTION

According to *Ayurveda*, *Dosha*, *Dhatu*, and *Mala* are the three main components of the human body. *Prakrit Vayu* (*Vata Dosha*) regulates the process of respiration.[1] The diseases of *Pranavaha Srotas* are known as respiratory system diseases. *Shwasa*, *Rajayakshma*, and *Hikka* are diseases related to *Pranavaha Srotas*.[2] For the formation of those diseases, *Agni* (*Pachak Pitta*) plays an important role. *Agnimandya* is the main etiological factor for the development of diseases and *Agni* depends on *Ahara*, i.e., a diet that we consume as food.[3]

In *Ayurveda*, *Ahara* is said to be *Prana* (life). A healthy diet is necessary for a healthy mind and body[4]. Diseases of *Pranavaha Srotasa* like *Shwasa* and *Kasa* are *Amashaya Samudbhava Vyadhis*[5]. The vitiated *Kapha* is responsible for the pathogenesis of respiratory disorders as *Hikka* and *Shwasa*[2]. *Kapha Vata Prakopak Ahar* is responsible for *Pranavaha Srotodushti*. Foods containing *Nishpav* (flat beans – Phaseolus vulgaris), *Masha* (black gram – Vigna mungo), *Pinyak* (Sesamum indicum oil cakes), *Til* oil (sesame oil), *Pishtanna* (food prepared from flour), *Guru* (*Vishtambhi Ahar*, *Jalaj – Anupa Mamsa* (Seafood), milk and milk products, *Abhishyandi Ahara* are heavy to digest, slimy in nature and obstruct through *Srotas*. These vitiate *Kapha* and lead to developing *Vyadhis* like *Shwasa*[6].

For the nourishment of *Pranavaha Srotas Kapha-Vatahar Ahara* is necessary, which includes *Snigdha* (Oily); *Ushna* (hot); *Laghu* (light); *Drava* (liquid); *Anulomak*, *Deepak* (carminative), and *Pachak* (digestive) type of *Ahar*. *Ahardravyas* includes *Jangal* dried meat) *Mamsa Varg* – *Avishkir* (birds), *Pratud* (group of animals that use their beak for eating food), *Kravyad* (carnivores animals), *Shuka Dhanyas* (grains) like *Raktashali* (red variety of rice), *Yava*, *Shimbi Dhanya* (pulses) like *Mudga* (green gram), *Kulathha* (horse gram), *Shaka* (vegetables) like *Mulaka* (radish), *Methika* (fenugreek), *Phala* (fruits) like *Draksha* (grape), *Shobhanjan* (drumstick), etc. Dietary preparations like *Manda* (Rice water), *Yusha* (Soup), *Mamsarasa* (Meat Soup), and *Utkarika* (Wheat Bread/*Roti*) prepared from these *Dravyas* are good for the nourishment of the respiratory system.

Consuming foods like flat beans (*Nishpav* – *Phaseolus* vulgaris), black gram (*Masha* – *Vigna mungo*), sesame oil cakes (*Pinyak* – *Sesamum indicum*), sesame oil (*Til oil*), flour-based dishes (*Pishtanna*), hearty meals (*Guru*), aquatic and marshy meat (*Jalaj* – *Anupa Mamsa*), as well as dairy products and foods that are heavy to digest and have a slimy nature (*Abhishyandi Ahara*), can disrupt the balance of *Kapha dosha*, potentially leading to the onset of conditions such as respiratory problems like *Shwasa*[6].

For the nourishment of *Pranavaha Srotas Kapha-Vatahar Ahara* is necessary, which includes *Snigdha* (oily), *Ushna* (hot), *Laghu* (light), *Drava* (liquid), *Anulomak*, *Deepak* (carminative), and *Pachak* (digestive) type of *Ahar*. Various food items from different *Ahara vargas* are incorporated into the diet. The *Mamsa*

DOI: 10.1201/9781003345541-22

varga (meat category) includes items like arid meat *(Jangal)*, birds (*Vishkir*), animals that use their beaks for eating (*Pratud*), carnivorous animals (*Kravyad*), grains like the red variety of rice (*Raktashali*) and barley *(Yava)*, pulses like green gram (*Mudga*) and horse gram *(Kulathha)*, vegetables such as radish *(Mulaka)* and fenugreek (*Methika*), and fruits like grapes (*Draksha*) and drumstick (*Shobhanjan*). Dietary preparations like *Manda* (Rice water), *Yusha* (Soup), *Mamsarasa* (Meat Soup), and *Utkarika* (Wheat Bread/*Roti*) prepared from these Ahara *dravyas* are good for the nourishment of the respiratory system.

Medicines are used to cure or to break the pathogenesis of diseases, but the *Ahariya Dravya* not only helps to cure diseases but also strengthens the body elements essential for that system to work properly[7].

Generally, our food includes carbohydrates in the form of grains, rice, and bread; proteins in the form of meat, eggs, and cereals; fats in the form of milk and butter; glucose and fructose in the form of sugar and fruits; fibers in the form of vegetables; and salt as mineral and drinking water.[8]

The metabolism of carbohydrates produces more CO_2 than the amount of oxygen used, whereas the metabolism of fat and proteins produces less amount of CO_2[8]. Healthy lungs require a low-carbohydrate, high-protein, low-fat, and high-fiber diet and low-sodium diet. Plenty of fluids are recommended to keep the respiratory tract moist (mucus) for healthy Respiration.

Asthma is a common chronic inflammatory respiratory disorder with high morbidity and mortality all over the world, affecting over 330 million children and adults and the affected population is anticipated to increase to 4000 million by 2025[9]; hence to control the morbidity and mortality ratio therapeutic nutrition is necessary.

THERAPEUTIC NUTRITION FROM CLASSICAL REFERENCES

Classical books *Charak Samhita, Sushrut Samhita, Yogratnakar Ashtanga Hridaya, Bhojan Kutuhal,* and *Kshemkutuhal* advised routine use of dietary products prepared from *Shashtishali* (Rice), *Rakta Shali* (Red rice), *Utkarika* prepared from *Godhum* (Wheat), *Yusha* (soup), *Kulattha* (Horse Gram), and *Mudga* (Green Gram). They also advise the consumption of *Mamsarasa* (Meat soup) prepared from *Jangal Mamsa along* with *Harit Varga Dravyas* like dry ginger. Additionally, they recommend Soups prepared from vegetables like *Kasamard* (*Cassia sophera* Linn.), Drumsticks, and dry Radish as well as the inclusion of vegetables like *Patol* (Pointed gourd), Brinjal Garlic and fruits like *Bimbi* (*Coccinia indica*), *Jambeer* (Lemon) and Grapes Leafy vegetables like *Tanduliya* (*Amaranthus spinosus*), Goat milk and Butter, Honey, and warm water are also suggested for managing Ashtma.[10,11,12] In *Vegawastha* (Acute Asthmatic stage) dietary preparations like *Ushnajala* (hot water) and *Yusha* (Soup) are useful to clear the obstructed phlegm. During *Avegavastha* (Chronic Asthmatic conditions) dietary products like *Jangal Mamsrasa* (High proteins), Goat milk and Butter (Fats), leafy vegetables (High fibers), and Fruits (Fibers and low sugar) are useful.

THERAPEUTIC NUTRITION FROM CONVENTIONAL SCIENCE

Gas exchange is the primary function of the respiratory system. The cellular metabolic demands are fulfilled by the lungs by providing oxygen and removing carbon dioxide produced. Healthy lung structure, elasticity, and functions, respiratory muscle mass, strength, and endurance are maintained by good nutrition. Good nutrition is also needed for the lungs immune defense system.

A high level of hemoglobin is required for good oxygen-carrying capacity and proteins; iron is required to maintain a high hemoglobin level. The function of respiratory muscles is maintained by minerals, calcium, magnesium, phosphorus, and potassium. Proteins are required to avoid hypo albuminuria to avoid pulmonary edema in pulmonary diseases.

Ascorbic acid is required for the synthesis of cognitive tissues of the lungs, which are composed of collagen fibers; an adequate amount of fluid intake is required to maintain moisture in airway mucus. Mucus consists of water, glycoproteins, and electrolytes.[13]

Microbial dysbiosis enhances lung inflammation and asthma-related symptoms. The bacterial genera *Lactobacillus* and *Bifidobacterium* are among the most commonly used probiotic bacteria. They exhibit anti-inflammatory analgesic effect and regulate the gut-lung axis. The probiotic bacteria from the intestinal tract modulate the gut microbial composition and enhances microbial metabolite levels, specifically short-chain fatty acids (SCFAs), and modulate host immunity.[9]

The gut microbiota gets affected by unsaturated fats and high-fiber diet, which helps to reduce systemic inflammation. Frequent consumption of sweets and dairy products enhances the onset of asthmatic odds.[14]

Prebiotics contain dietary fibers. The fructose and galactose-oligosaccharides are derived from dietary fibers and are subsequently fermented by intestinal bacteria. The use of prebiotics in the treatment of asthma reduces exercise-induced bronchoconstriction. A prebiotic diet decreases allergic airway inflammation (AAI) and airway hyperresponsiveness (AHR). The soluble fibers from the prebiotic diet are prone to have acute anti-inflammatory effects in asthmatic airways.

Lower odds of asthma by lowering levels of interleukin 17 were associated with high consumption of grains.[15]

High consumption of grains was associated with lower odds of asthma, attributed to the reduction in levels of interleukin 17.[15]

TABLE 17.1 Showing Dietary Sources Their Properties and Action on Respiratory System

NAME OF DIETARY SOURCE	*PROPERTIES*	*ACTION ON RESPIRATORY SYSTEM*
Grains	High fiber (soluble) content. Prebiotics.	Anti-inflammatory action on airways.
Pulses	Protein supplements	Prevents albuminuria, prevents pulmonary edema.
Meat (Lean Meat)	Protein supplement, Vitamin B12 supplement.	Prevents albuminuria. Prevents pulmonary edema
Vegetables	Soluble dietary fibers, Prebiotics	Maintains gut microbiomes, anti-inflammatory.
Fruits	soluble dietary fibers, prebiotics	Maintains gut microbiomes, anti-inflammatory
Milk and Milk products	Low-density unsaturated fats, Vitamin D, Vitamin B12	Enhances SCFAs and microbial metabolites. Anti-inflammatory.
fermented milk products	Probiotics	Anti-inflammatory.

CLASSIFICATION OF ASTHMA

Stages of asthma:

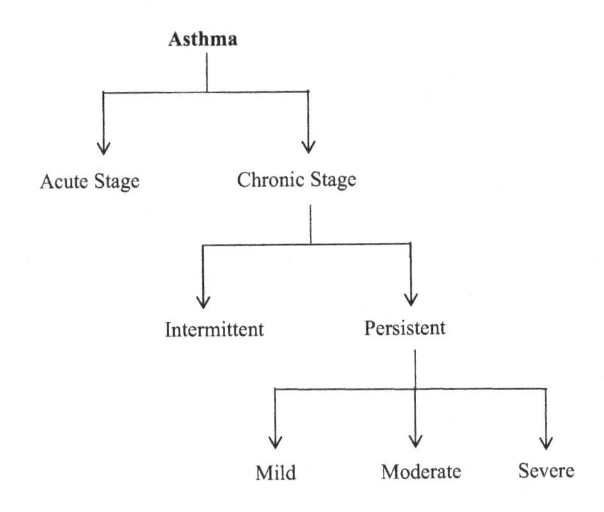

Asthma

Asthma is an obstructive lung disease that occurs due to airway obstruction; during an asthmatic attack three things can happen:

Bronchospasm – The muscle around the airways gets constricted, which leads to being narrowing of airways and causes difficulty in breathing.
Inflammation – The inflammation of the linings of airways leads to difficulty in breathing.
Mucus Production – During the asthmatic attack bronchus secretes excessive mucus that clogs the airways and leads to difficulty in breathing.

Classification of *Pranavaha Srotasa Vyadhi* (diseases of the respiratory tract) as per Etiological Factors[16]:

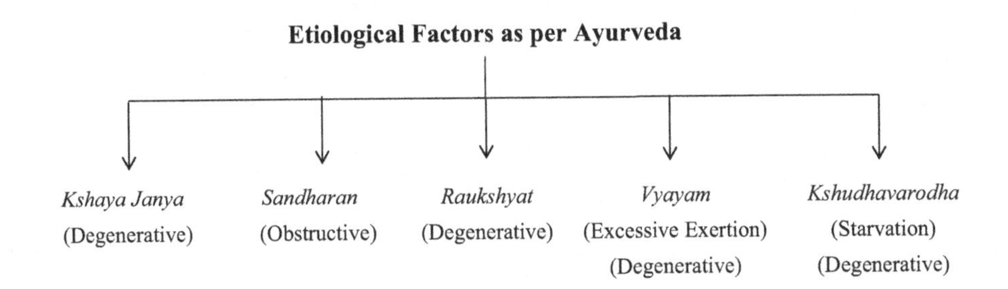

The etiological factors for Asthma are classified as:

1. *Dhatukshyajanya* (Degenerative Pathology)
2. *Margavrodhjanya* (Obstructive Pathology)
   - Due to a nonnutritious diet, excessive hard work, and excessive starvation degeneration of respiratory channels takes place, leading to the diseases *Shwasa*, *Kasa*, and *Kshaya*.
   - Due to improper diet, vitiation *Kapha Vata Dosha* takes place, and changes in properties of *Kapha Dosha* and *Vata Dosha* occur, which subsequently obstructs the channels of the respiratory system and leads to diseases like *Shwasa*, *Kasa*, and *Kshaya*.

Shwasa (Asthma):

Classification of *Shwasa*[17]:

a. *Vega Avastha* (Acute onset state)
b. *Avega Avastha* (Chronic state)

THERAPEUTIC NUTRITION IN ASTHMA:

The Asthmatic case gets represented in the following stages:

1. Acute stage (Inflammatory stage)
2. Chronic stage (Intermittent and Persistent stage)

In the acute stage bronchodilators, mucolytic agents and controllers[18] (corticosteroids and antihistamines) are to be used as emergency management in modern science. Ayurveda classics recommend the use of *Ushna* (Hot) and *Kapha-Vataghna* (balancing of vitiated *Kapha and Vata Dosha*) medicines and dietary approaches for the treatment of Asthma.[19]

Considering that the diet should be *Kaph-Vataghna* (balancing of vitiated *Kapha* and *Vata Dosha*), it is advisable to consume food with *Ushna* (Hot) properties in this condition. This includes items like *Ushna Jala* (Hot Water), *Jangal Mamsarasa, Kultha Yusha, Makustha Yusha, Sura, Madhu, Mulak Yusha,* which are *Ushna* (Hot), *Kapha-Vataghna* (balances vitiated *Kapha* & *Vata Dos*ha), as well as possess *Ruksha* (Dry) properties.

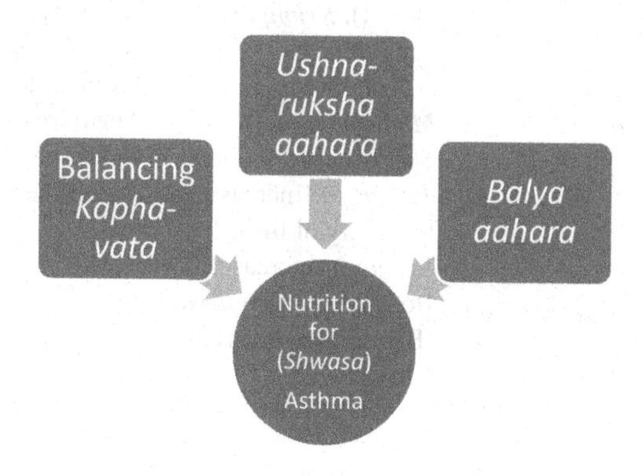

In Intermittent and chronic conditions the patient is managed by *Pathyakar* (Good for health diet), *Kapha-Vataghna* (balances vitiated *Kapha* and *Vata Dosha*), and *Balya* strengthening) *Ahar*. The use of *Jangal Mamsarasa, Shastishali, Yavarotika, Angarkarkati, Mudga Yusha*, Vegetables like *Patol, Vartak* and Fruits like *Draksha, Akhrot* are considered *Kapha Vataghna* (balances vitiated *Kapha* & *Vata Dosha*), and *Balya* (strengthening) elements.

NUTRITION IN ASTHMA AS PER *DWADASH AHAR VARGA*:

1. *Shuka Dhanya Varga:*

Cereals are included in *Shuka Dhanya Varga.*[20] The nutritive recipes of *Shuka Dhanya Varga* are used as a therapeutic diet in *Shwasa* (asthma).

Chart showing the classification according to its group of *Shuka Dhanya*, its properties, and its *Kritanna Kalpana*

TABLE 17.2

S. NO.	AHARA *DRAVYA*	LATIN NAME	PROPERTIES	*KRITANNA KALPANA*	ACTION ON THE RESPIRATORY SYSTEM
1.	*Shali* (Rice)[21]	*Oryza longistaminata*	*Tridoshaghna* (Balances all *Tridosha*), *Balya* (Strengthens body), *Madhur* (Sweet), *Kashaya* (Astringent), *Sheeta* (Cold in Potency).	*A) Bhakta* (rice) *B) Anna*(rice) *C) Aatidrava Bhakta*(rice with rice water) *D) Bhrushta Tandula Bhakta.*[22] (rice from roasted paddy)	Lowers the odds of Asthmatic onset
2.	*Yava* (Barley)[23]	*Hordeum vulgare*	*Ruksha* (Dry) *Sheeta*(Cold), *Sthira* (Steadiness) in body, *Guru* (Heavy to Digest), *Krimighna* (antihelminthic)	*A)Yavadi Vatya*[24]*(Gruel prepared from Barley) (B) Yavarotika*[25] (Bread prepared From barley flour)	Anti-inflammatory action on Airways. Lowers the odds of Asthmatic onset
3.	*Godhuma*[26] (Wheat)	*Triticum*	*Madura* (Sweet), *Sheeta* (Cold), *Jeevaniya* (Increases longevity of life), *Bruhana* (Increases Body Mass), *Sthairya* (Brings Steadiness in body)	*Angarkarkati*[27] (bread prepared from Wheat flour)	Anti-inflammatory action on Airways. Lowers the odds of Asthmatic onset

2. *Shamidhanya Varga* –

Pulses/legumes/oil seeds are included in *Shamidhanya Varga.*

Chart showing the classification according to its group of *Shami Dhanya*, its properties, and its ***Kritanna Kalpana***

TABLE 17.3

S. NO.	AHAR DRAVYA	*LATIN NAME*	*PROPERTIES*	KRITANNA KALPANA	*ACTION ON RESPIRATORY SYSTEM*
1.	*Kulattha*[28] (Horse Gram)	*Macrotyloma uniflorum*	*Ushna* (Hot), *Kapha-Vataghna, Pittakara*	*Kulattha Yusha*[29]	Protein supplement, Prevent Pulmonary edema
2.	*Makustha*[30] (Aconite Kidney Bean)	*Phaseolus aconitifolius*	*Ruksha* (Dry), *Sheeta* (Cold), *Laghu* (light in digest)	*Makustha Yusha*	
3.	*Chanak*[31] (chickpea) *Yusha*	*Cicer arietinum*	*Ruksha* (Dry), *Sheeta* (Cold), *Laghu* (light in digest)	*Chanak Yusha*	

3. *Mamsa Varga* (Meat)

TABLE 17.4 Chart Showing the Classification According to Its Group of Animals, Its Properties, and Therapeutic Nutritive Values

S.N.	MAMSA VARGA	MAMSA DRAVYA	*PROPERTIES*	KRITANNA KALPANA	*ACTION ON RESPIRATORY SYSTEM*
1.	*Vishkira Mamsa* (Birds)	*Tittar*[32] (Gray francolin)	*Guru* (Heavy in Digestion), *Ushna* (Hot), *Agnivardhak* (Carminative)	*Mamsa Rasa*	Good source of dietary animal protein, a Rich source of vitamin B12, and Prevents pulmonary edema.
		Kukkut Mamsa[33] (Rooster)	*Snigdha* (fatty) *Ushna*(Hot) *Balya*(Gives Strength), *Vatanashak* (Reduces Vitiated *Vata*)	*Mamsa Rasa*	
2.	*Bilesheya* (Animals that live in Burrows)	*Shasha Mamsa*[34] (Rabbit)	*Madhur* (Sweet) *Pitta-Kapha Nashak* (Reduces Vitiated *Kapha-Pitta*)	*Mamsa Rasa*	
3.	*Gramya Mamsa* (Domestic Animals)	*Chagal Mamsa*[35] (Goat)	*Pitta-Kaphaghna* (Reduces Vitiated *Pitta-Kapha*)		

4. *Shaaka Varga*[36] Leafy Vegetables:

It includes six types of vegetable categories *Phalashaaka* (Vegetable Fruit), *Patra Shaaka* (Leafy Vegetables), *Nala Shaaka* (Stalks), *Kanda Shaaka* (Tubers), *Pushpa Shaak* (Flowers) and *Samavedaja* (Parasitic Vegetables)

TABLE 17.5 Chart Showing the Classification According to Its Group of *Shaaka*, Its Properties, and Its *Kritanna Kalpana*

S.N	*AHARA DRAVYA*	*LATIN NAME*	*PROPERTIES*	*KRITANNA KALPANA*	*ACTION ON RESPIRATORY SYSTEM*
1.	*Karavellak*[37] (Bitter Gourd)	*Momordica charantia*	*Laghu* (Light in digestion), and *Sheet* (cold), Reduce vitiated *Kapha* and *Pitta krimighna* (Antihelminthic)	*Shaka*	Dietary fibers – Prebiotics, anti-inflammatory effects on Airways in Asthma
2.	*Katutumbi*[38] (Bottle Gourd)	*Lagenaria siceraria*	*Sheet* (cold), Bitter in taste, reduces vitiated *Vata* and *Pitta*	*Shaka*	Dietary fibers – Prebiotics, anti-inflammatory, effects on Airways in Asthma
3.	*Shobhanjana Phala*[39] (Drumstick)	*Moringa oleifera*	*Madhur* (Sweet), reduces vitiated *Kapha* and *Pitta*, good carminative.	*Yusha*	Dietary soluble fibers – Prebiotics, Relieves bronchospasm in acute Asthma.
4.	*Kasamarda*[40] (Senna)	*Occidentalis*	*Madhur* (Sweet), *Tikta* (Bitter), *Laghu* (Light in digestion), *Ruksha* (Dry), *Ushna* (Hot potency), alleviator of *Kapha* and *Vata*, Carminative	*Yusha*	Dietary soluble fibers – Prebiotics, Relieves bronchospasm in acute Asthma.
5.	*Methi shaaka*[41] (Fenugreek)	*Trigonella foenum-graecum*	*Ushna* (Hot in potency), *Katu* (Pungent), *Tikta* (Bitter), *Laghu* (Light to digest), and *Snigdha* (oily), balances *Kapha* and *Vata Dosha*	*Shaka*	Dietary fibers – Prebiotics, anti-inflammatory
6.	*Kakamachipatra Shaaka*[42] (Black nightshade)	*Solanum nigrum*	*Tikta* (Bitter), *Laghu* (light to digest), *Snigdha* (oily), and *Anushna* (not too hot in potency), balance all *Tridosha*	*Yusha*	Dietary soluble fibers – Prebiotics, Relieves bronchospasm in acute Asthma.

TABLE 17.5 ***(Continued)*** Chart Showing the Classification According to Its Group of Shaaka, Its Properties, and Its Kritanna Kalpana

S.N	*AHARA DRAVYA*	*LATIN NAME*	*PROPERTIES*	*KRITANNA KALPANA*	*ACTION ON RESPIRATORY SYSTEM*
7.	*Prapunnad Patra Shaaka*[43] (Ringworm plant/coffee pod)	*Cassia tora*	*Tikta* (Bitter), *Laghu* (light to digest), *Snigdha* (oily), *Anushna* (not too hot in potency)	*Shaka*	Dietary fibers – Prebiotics, anti-inflammatory, effects on Airways in Asthma
8.	*Kanchnar Pushpa Shaaka*[44] (Orchid)	*Tree Bauhinia Variegate*	*Kashaya* (Astringent), *Laghu* (Light to digest), *Ruksha* (Dry), *Sheeta* (Cold in potency), balances *Kapha* and *Pitta Dosha*	*Shaka*	Dietary fibers – Prebiotics, anti-inflammatory, effects on Airways in Asthma
9.	*Agasti Pushpa Shaaka*[45] (Vegetable hummingbird)	*Sesbania grandiflora*	*Sheeta* (Cold in potency), *Tikta* (Bitter) in taste, balances *Vata* and *Kapha Dosha*	*Shaka*	Dietary fibers – Prebiotics, anti-inflammatory, effects on Airways in Asthma
10.	*Surana kand Shaaka*[46] (Elephant yam)	*Amorphophallus campanulatus*	*Kashay* (Astringent), *Katu* (Pungent) in taste, *Ushna* (Hot in potency), reduces *Vata* and *Kapha Dosha*	*Shaka*	Dietary fibers – Prebiotics, anti-inflammatory, effects on Airways in Asthma
11.	*Aardrak Shaak*[47] (Ginger)	*Zingiber officinalis*	*Katu* (Pungent) in taste, *Ushna* (Hot in potency), reduces *Vata* and *Kapha Dosha*	*Ardrak Modak*	Mucolytic.
12.	*Karchur*[48] (White turmeric)	*Cucurma zedoaria*	*Katu* (Bitter), *Katu* (Pungent) in taste, *Ushna* (Hot in potency), reduces *Vata* and *Kapha Dosha*	*Shaaka*	Dietary fibers – Prebiotics, anti-inflammatory, effects on Airways in Asthma
13.	*Sunishanna*[49] (Dwarf waterclover)	*Marsilea minuta*	*Madhur* (Sweet), *Kashay* (Astringent) in taste, *Sheeta* (Cold in potency), *Laghu* (light to digest), *Ruksha* (Dry), *Deepan* (kindles digestive fire) and balances *Tridosha*	*Shaaka*	Dietary fibers – Prebiotics, anti-inflammatory, effects on Airways in Asthma
14.	*Baala Vruntak*[50] (Brinjal)	*Solanum melongena*	*Madhur* (Sweet), *Kashay* (Astringent) in taste, balances *Vata* and *Kapha*	*Yusha*	Dietary fibers – Prebiotics, anti-inflammatory, effects on Airways in Asthma

5. *Phalavarga*

TABLE 17.6 Chart Showing the Classification According to Its Group of *Phala*, Its Properties, and Its *Kritanna Kalpana*

S.N.	*AHAR DRAVYA*	*LATIN NAME*	*PROPERTIES*	*KRITANNA KALPANA*	*ACTION ON RESPIRATORY SYSTEM*
1.	*Draksha*[51] (Grapes)	*Vitis vinifera*	*Madhur* (Sweet) in taste, *Sheeta* (Cold in potency), *Guru* (Heavy to digest), *Snigdha* (oily), and balances *Vata Pitta Dosha*	-	Dietary fibers – Prebiotics. Anti-inflammatory action on Asthmatic airways.
2.	*Akshotaka*[52] (*Akarot*) (Walnut)	*Juglans regia*	*Madhur* (Sweet) in taste, *Ushna* (Hot in potency), *Guru* (Heavy to digest), *Snigdha* (Oily), balances *Kapha* and *Vata Dosha*	-	

6. *Harita Varga*[53]

It includes the food items which are from spices group having carminative properties.

TABLE 17.7 Chart Showing the Classification According to Its Group of Harita, Its Properties, and Its *Kritanna Kalpana*

S.N.	*AHAR DRAVYA*	*LATIN NAME*	*PROPERTIES*	*KRITANNA KALPANA*	*ACTION ON THE RESPIRATORY SYSTEM*
1.	*Lashoona*[54] (Garlic)	*Allium Sativum*	*Katu* (Pungent) in taste, *Tikshna* (Sharp), *Guru* (Heavy to digest), *Snigdha* (Oily), *Ushna* (Hot in potency), balances *Kapha* and *Vata Dosha*[84]	-	Anti-inflammatory effect on inflamed bronchus[55]. Removes the oxidative stress in Asthma[56].
2.	*Mulak*[57] (Radish)	*Raphanus sativus*	*Katu* (Pungent) in taste, *Laghu* (Light to digest), *Ruksha* (Dry), *Ushna* (Hot in potency), Digestive in nature, balances all *Tridosha*	*Mulak Yusha*	Dietary soluble fibers – Prebiotics, Relieves bronchospasm in acute Asthma.

7. *Jala Varga*

TABLE 17.8 Chart Showing the Classification According to Its Group of *Jala*, Its Properties, and Its *Kritanna Kalpana*

S.N.	AHAR DRAVYA	*PROPERTIES*	*KRITANNA KALPANA*	*ACTION ON RESPIRATORY SYSTEM*
1.	*Ushnodak*[58] (Hot Water)	*Deepan* (Digestive) in nature, removes vitiated *Kapha*, *Meda* (Mucus)	*Ushnodaka*	Mucolytic action

8. *Goras Varga*[59]

TABLE 17.9 Chart Showing the Classification According to Its Group of *Goras,* Its Properties, and Its *Kritanna Kalpana*

S.NO.	GORAS VARGA DRAVYA	*ENGLISH NAME*	*PROPERTIES*	KRITANNA KALPANA	*ACTION ON RESPIRATORY SYSTEM*
1.	*Go Dugdha*[60]	Cow Milk	*Madhur* (Sweet), *Sheeta* (Cold), *Guru* (Heavy to digest), *Snigdha* (Fatty), *Bruhana* (Increases Body mass), and *Balya* (strengthening Body), Balances *Vata-Pitta Dosha*.	*Siddha Ksheera*	Low unsaturated fat – enhances gut microbiota. Anti-inflammatory action on asthmatic airways.
2.	*Aja Dugdha*[61]	Goat Milk	*Katu* (Pungent), *Tikta* (Bitter), *Agnidipak* (Stimulates Hunger), *Laghu* (Light to Digest),	*Siddha Ksheera*	
3.	*Bhedi Dugdha*[62]	Sheep Milk	*Madhur* (Sweet), *Sheeta* (Cold), *Guru* (Heavy to digest), *Snigdha* (unctuous). Balances *Vata-Kapha Dosha*.	*Siddha Ksheera*	
4.	*Go Dadhi*[63]	Cow milk Curd	*Madhur* (Sweet), *Sheeta* (Cold), *Agnidipak* (Stimulates Hunger), *Snigdha* (unctuous), *Bruhana* (Increases Body mass), *Balya* (strengthens Body). Balances *Vata Dosha*.	*Rasala Kalpana*	Probiotics. Enhances microbial metabolite level, regulates Gut-Lung axis, and Anti-inflammatory action on asthmatic airways.

(*Continued*)

TABLE 17.9 ***(Continued)*** Chart Showing the Classification According to Its Group of *Goras,* Its Properties, and Its *Kritanna Kalpana*

S.NO.	GORAS VARGA DRAVYA	*ENGLISH NAME*	*PROPERTIES*	KRITANNA KALPANA	*ACTION ON RESPIRATORY SYSTEM*
5.	*Aja Dadhi*[64]	Goat Milk Curd	*Laghu* (Light to digest), *Deepan* (Carminative), *Tridoshaghna* (Balances all *Tridosha*), *Grahi*	-	
6.	*Aja Ghrita*[65]	Goat Milk Clarified Butter	*Deepan* (Carminative), *Balavardhak* (Strengthen Body), *Laghu* (Light to digest)	-	Enhances SCFAs and microbial metabolites. Anti-inflammatory.
7.	*Takra*[66]	Cow Butter Milk	*Kshudhavardhak* (Increase Appetite), *Balya* (Strengthen body), *Kapha-Vataghna* (Balances *Kapha* and *Vata Dosha*)	*Kadhi Kalpana*	Probiotics. Enhances microbial metabolite level, regulates Gut-Lung axis, Anti-inflammatory action on asthmatic airways.

9. *Ekshu Varga*

TABLE 17.10 Chart Showing the Classification According to Its Group of *Ekshu Varga*, Its Properties, and Its *Kritanna Kalpana*

S.N.	*AHARA DRAVYA*	*PROPERTIES*	*KRITANNA KALPANA*	*ACTION ON THE RESPIRATORY SYSTEM*
1.	*Madhu*[67]	*Madhur* (Sweet), *Laghu* (Lite), *Kaphavatghna* (Balances *Kapha* and *Vata Dosha*)	-	As an adjuvant therapy increases potency of other drugs used in Asthma.[68]

10. *Kritanna Varga*

In *Kritanna Varga* the therapeutic nutrition preparation recipes are mentioned; the following recipes are used for Therapeutic Nutrition in *Shwasa* (Asthma).

TABLE 17.11 Chart Showing the Classification of Various Kritanna Kalpana and Its Properties

S.N.	KRITANNA KALPANA	*PROPERTIES*	*ACTION ON THE RESPIRATORY SYSTEM*
1.	*Mamsa* Rasa (Meat Soup)[69]	*Ushna* (hot), *Kaphavataghna* (balances *Kapha* and Vata *dosha*), *Balya* (strengthens body elements)	Good source of dietary animal protein, a Rich source of vitamin B12, and Prevents pulmonary edema.
2.	*Kulattha Yusha* (Soup)[28]	*Ushna* (Hot Potency), *Kapha Vataghna* (Balances *Kapha* and *Vata Dosha*), *Bhedi* (Penetrative in Action), *Saraka* (Purgative in action)	Protein supplements, Prevents Pulmonary edema
3.	*Yavadi Vaatya Manda*[24]	*Brihan, Preenan, Vrishya.*	Protein supplements, Prevents Pulmonary edema
4.	*Ati Dravam Bhakta*[22] (Rice soup)	*Shwasakasaghna*, *Balya* (Strengthens Body)	Lowers the odds of Asthmatic onset
5.	*Bhrushta Tandula Bhakta*[22] (Roasted paddy)	*Ruchikarak* (tasty), *Laghu* (Lite for Digestion), *Kaphaghna* (controls vitiated *Kapha Dosha*), *Grahi* in Nature.	Lowers the odds of Asthmatic onset
6.	*Chanak Yusha*[31]	*Anushna* (mild Hot in potency), *Kashaya* (Astringent), *Laghu* (lite for Digestion), *Pitta* and *Kaphanashak* (Balances Vitiated *Pitta* and *Kapha Dosha*)	Protein supplements, Prevent Pulmonary edema
7.	*Vaidala Yusha*[70]	*Balya* (Strengthens Body), *Laghu* (Lite for Digestion), *Kantha Kaphapaham* (Removes Phlegm from Throat)	Protein supplements, Prevent Pulmonary edema
8.	*Makustha* (Aconite Kidney Bean) *Yusha*[30]	*Laghu* (lite for Digestion), *Santarpak* (Strengthens Body elements in a short time), *Hridya* (Healthy for Heart), *Pittashleshm Ahara* (balances Vitiated *Pitta* and *Kapha*).	Protein supplements, Prevent Pulmonary edema
9.	*Angaar Karkati*[27]	*Laghu* (lite for Digestion), *Deepan* (Stimulates Digestive Fire), *Balya* (Strengthens body), *Brihan* (increases body mass)	Anti-inflammatory action on Airways. Lowers the odds of Asthmatic onset
10.	*Yavarotika*[25]	*Madhur* (Sweet), *Laghu* (lite for Digestion), *Kaphaghna* (Balances Vitiated *Kapha*), *Balya* (Strengthens Body)	Protein supplements, Prevent Pulmonary edema
11.	*Vedhani*[71]	*Balya* (Strengthens body), *Brihan* (Increases body mass), *Vataghna* (Controls vitiated *Vata*)	Reduces the inflammation of asthmatic bronchus.
12.	*Kadhi*[72]	*Deepan* (Carminative), *Hridya* (Healthy for the heart), *Ruchikarak* (tasty), *Vata Kaphaghna* (Balances vitiated *Vata* and *Kapha Dosha*).	It is a good prebiotic that contains lactobacilli to enhance microbiomes in the gut. Anti-inflammatory action on Airways.

(*Continued*)

TABLE 17.11 *(Continued)* Chart Showing the Classification of Various Kritanna Kalpana and Its Properties

S.N.	KRITANNA KALPANA	*PROPERTIES*	*ACTION ON THE RESPIRATORY SYSTEM*
13.	*Shuska Mulak Yusha*[73]	*Ushna* (Hot in potency), *Kapha Vata*ghna (Balances vitiated *Kapha* and *Vata Dosha*).	Dietary soluble fibers – Relieves bronchospasm in acute Asthma.
14.	*Ardraka Modaka*[74]	*Ruchikarak* (Tasty), *Agnijanak* (carminative), removes Phlegm from Throat.	Mucolytic.
15.	*Vaidala Poorika*[75]	*Madhur* (Sweet), *Balya* (Strengthens Body), *Brihan* (Increases Body mass), Vataghni (controls vitiated *Vata*)	Good dietary protein supplement in Asthma
16.	*Kasamarda Patra Yusha*[76]	*Ushna* (Hot Potency), Kaphaghna (controls vitiated *Kapha*)	Dietary soluble fibers – Relieves bronchospasm in acute Asthma.
17.	*Shobhanjana Yusha*[77]	*Ushna* (Hot Potency), *Deepan* (carminative), *Bhedan, Kapha Vataghna* (Balances vitiated *Kapha* and *Vata Dosha*)	Dietary soluble fibers – Relieves bronchospasm in acute Asthma.

CLINICAL SIGNIFICANCE OF THERAPEUTIC NUTRITION IN ASTHMA

The recommendation of a nutritive diet for Asthma depends upon the type of Asthma, the patient's physical condition, and the availability of nutritive products.

Patients with acute asthmatic stage are initially treated with a medicinal line of treatment. In this stage it is advised to drink hot (*Ushna jala*) water (*Muhurmuhu*), i.e. sip by sip, which removes the obstructed phlegm from inflamed bronchus. Hot soup prepared from rooster meat gives strength to the body, and hence it strengthens the exhausted respiratory muscle. Soups of Horse Gram, soup of *Kakamachipatra* dry Radish, and *Makushtha* (Aconite Kidney Bean), soup of *Kasamarda*, and soup of drumstick contain soluble fibers that act as prebiotics; they decrease allergic AAI and also decrease Airway Hyper responsiveness (AHR) which leads to reduce inflammation and bronchoconstriction of Inflamed airways and so are beneficial as medicine and nutrition. Chewing of *Ardrak modak* (candy prepared from ginger) controls coughing and sneezing during an asthmatic onset by its mucolytic action.

In chronic asthmatic conditions, the motto of therapeutic nutrition is to avoid dietary triggers and reduce airway inflammation. This type of nutrition increases the time interval and severity of acute asthmatic onset. Oily, spicy diet restriction prevents GERD. Airway inflammation is restricted by advising the Mediterranean diet. If rice is major food, then eating *Bhrusht Tandul Bhakt* (rice prepared from roasted paddy) and *Atidravam Bhakt* (Rice soup) is advised. These grains lower the odds of asthma.

Vegetables and fruits indicated in *Shaaka Varga* are rich in fiber content and act as prebiotics and hence act as anti-inflammatories on asthmatic bronchus.

Curd, *Takra*, and *Kadhi* are advised as a part of a diet that acts as probiotics, enhances gut micro biomes, and regulates the gut-lung axis to reduce inflammation of asthmatic bronchus.

Pulses are a good source of dietary proteins; an adequate amount of proteins provides calories for the repair of exhausted respiratory muscles. Pulses indicated in *Shamidhanyavarga* must be provided in the form of *Yusha* (soup), *Vaidal Poorika*, and *Vedhini*.

Meat is a good source of dietary protein *Mamsarasa* (Meat soup) prepared from *Jangal Varg* should be included in the diet, as well as *Shullya mamsa* (Boneless meat pieces marinated well with curd, and spices and pierced in an iron rod, and roasted over burning charcoal heat). It is very light and easy to digest.

Milk and milk products having low unsaturated fat are very important to enhance metabolites and SCFAs to modulate the immunity of an Asthmatic person. Cow milk, Goat milk, and Cow ghee must be a part of the diet.

Ahariya Kalpana (Dietary Recipe) preparation specifically for Therapeutic Nutrition in Asthma:

Yavarotika[25]:

The Dough prepared from *Yava* flour is converted into slightly thick, round bread and then baked in an oven of charcoal; properly baked bread is known as *Yavarotika.* It gives strength to body elements and controls vitiated *Vata Dosha*; it is *Guru* (heavy to digest) and *Bruhan* (increases body mass) in the property. *Yavarotika* contains soluble dietary fibers and so acts as prebiotic, enhances gut micro biomes, and regulates gut-lung axis to prevent inflammation of the asthmatic airway. It is *Brihan* in nature and so gives strength to respiratory muscles.

Vedhani[71]:

The thick macerated paste of *Masha* (Bengal Gram) and *Mudga* (Green Gram) is rolled in Dough of *Godhuma* (Wheat) flour and balls of it are prepared, which are then fried into oil/*ghrita*; the well-fried balls are named *Vedhani.* It is the combination of Protein, fibers, and low unsaturated fat that fulfills the requirement of prebiotics, proteins, and unsaturated fat, which is useful to reduce the inflammation of asthmatic bronchus.

Kadhi[72]:

The *Takra* (Buttermilk) is boiled with bolus prepared with the powder of *Marich* (Black Pepper), *Shunthi* (Dry Ginger), *Chavya* (*Piper chaba*), *Pippali* (*Piper longum*), *Pippalii Mula*, *Dhanyak* (Coriander seeds), *Saindhav* (Rock salt), *Dadim* (Dried Pomegranate seeds), *Amalaki* (*Emblica officinalis*), and *Hingu* (Asafoetida), all in equal quantity; the well-cooked *Takra* is known as *Kadhi.* It is a good probiotic that contains lactobacilli, which enhance the microbiomes in the gut. The sour nature of *kadhi* is managed by other carminative drugs from *harit Varg*, so it doesn't increase *Kapha Dosha*.

Ardrak Modak[74]:

Ardrak (Ginger) is boiled in water and processed with *Chincha Patra* (Tamarind) leaves; the well-cooked *Ardrak* (Ginger) is then cooked in *Go Dugdha* (Cow Milk), then the cooked *Ardrak* (Ginger) is cut into small pieces and rolled in Rice flour, then Fry in *Go Ghrit* (Cow Clarified Butter) the well fried *Ardrak* pieces then deep into sugar syrup and with *Ela* (Cardamom) powder and *Ardrak Modak* is prepared.

Ardrak (Ginger) is boiled in decoction of *Chincha Patra* (Tamarind) leaves. The well-cooked *Ardrak* (Ginger) is then further cooked in Go *Dugdha* (Cow Milk). Afterward, the cooked Ardrak (Ginger) is cut into small pieces, rolled in Rice flour, and fried in Go *Ghrit* (Cow Clarified Butter). The well-fried *Ardrak* pieces are then dipped into a sugar syrup and scented with *Ela* (Cardamom) powder, resulting in the preparation of *Ardrak Modak*.

This tasty nutritional recipe is used to enhance appetite in chronic asthma; it removes the sticky phlegm from the mouth and pharynx and thus reduces coughing.

Vaidal Poorika[75]:

The well-cooked and drained *Chana dal* (Split Chickpea lentils) is mixed with Jaggery and is macerated well by adding Rock salt, *Jeera* powder (Cumin seed), and *Hingu* (Asafetida) powder; bolus of this is prepared, and the bolus is then wrapped in *Godhuma* (wheat flour) dough and fried in oil and *Vaidal Poorika* is prepared. This recipe is a combination of proteins, fibers, fructose, salt, and carminative dietary items used as good dietary protein supplements for Asthma.

SCOPE FOR FURTHER RESEARCH AND NUTRAVIGILANCE

The occurrence of asthma symptoms is linked to food allergies, specifically to constituents found in certain foods. Anaphylactic reactions are caused by shellfish and nuts, and these reactions can exacerbate wheezing in patients with asthma. Additionally, certain food preservatives, such as metabisulphite, have been found to increase the likelihood of developing asthma. Tartrazine, a food coloring agent, is also a known trigger for sudden onset of asthma attacks.[78] Asthma in bakers is triggered by fungal amylase in wheat flour (Occupational Hazards).[79] Food processors like potassium meta sulfate and sodium sulfide have been found to be triggering for Asthmatic Patients.

As per classical references of Ayurveda, specific dietary nuts such as *Nishpav* (Flat beans – *Dolichos lablab*), *Mash* (Black Gram – *Phaseolus mungo* Linn.), *Pinyak* (sesame cake), and *Til* (Sesame – *Sesamum indicum*) have been identified as causative factors for Asthma[6]. Additionally, Asthma can also be triggered by consuming food made from flour, seafood (*Jalaj*), *Anup Mamsa* (high-fat meat), and milk products.

REFERENCES

1. *Vagbhat. Sutrasthan Doshadividnyaniyam Adhyay.* In: Dr. Garde G K, editor. *Sartha Vagbhat*. Varanasi, Chaukhamba Surbharati Prakashan; 2011, Verse 01. P. 51.
2. *Maharshi Sushrut. Sutrasthan Annapanvidhi Adhyay.* In: Shastri K A, editor. *Sushruta Samhita.* Varanasi, Chaukhamba Sanskrit Sansthan; 2012. Verse 12. P. 96.
3. *Maharshi Agnivesh. Chakrapanidatta Chikitsa Sthan GrahaniDosha Chikitsa Adhyay.* In: Vd. Acharya Y T, editor. *Charak Samhita*. Varanasi, Chaukhamba Surbharati Prakashan; 2000. Verse 3–4. P. 512.
4. *Maharshi Agnivesh. Sutrasthan Annapanvidhi Adhyay* In: Dr. Dwivedi L, editor. *Charak Samhita*. Varanasi, Chowkhamba Krishnadas Academy; 2013. Verse 03. P. 512.
5. *Vagbhat. Nidan Sthan Shwasa – Hidma Nid anam Adhyay.* In: Dr. Garde G K, editor. *Sartha Vagbhat*. Varanasi, Chaukhamba Surbharati Prakashan; 2011, Verse 3–4. P. 174.
6. *Maharshi Agnivesh. Chakrapanidatta Chikitsa Sthan Hikka Shwasa Chikitsa Adhyay.* In: Vd. Acharya Y T, editor. *Charak Samhita*. Varanasi, Chaukhamba Surbharati Prakashan; 2000. Verse 14–15. P. 533
7. *Maharshi Agnivesh. Sutrasthan Annapanvidhi Adhyay* In: Dr. Dwivedi L, editor. *Charak Samhita*. Varanasi, Chowkhamba Krishnadas Academy; 2013. Verse 03-04. P. 513.
8. Association AL. Nutrition and COPD [Internet]. American Lung Association. [cited 2023 Jan 28]. Available from: www.lung.org/lung-health-diseases/lung-disease-lookup/copd/living-with-copd/nutrition

9. Liu A, Ma T, Xu N, Jin H, Zhao F, Kwok LY, Zhang H, Zhang S, Sun Z. Adjunctive Probiotics Alleviates Asthmatic Symptoms via Modulating the Gut Microbiome and Serum Metabolome. Microbiol Spectr. 2021 Oct 31;9(2):e0085921. doi: 10.1128/Spectrum.00859-21. Epub 2021 Oct 6. PMID: 34612663; PMCID: PMC8510161.
10. *Maharshi Agnivesh. Chakrapanidatta Chikitsa Sthan Hikka Shwasa Chikitsa Adhyay*. In: Vd. Acharya Y T, editor. *Charak Samhita*. Varanasi, Chaukhamba Surbharati Prakashan; 2000. Verse 89–93. P. 537.
11. *Maharshi Sushrut. Uttartantra Shwasapratishedha Adhyay*. In: Shastri K A, editor. *Sushruta Samhita*. Varanasi, Chaukhamba Sanskrit Sansthan; 2012. Verse 47. P. 485.
12. Vd. Shastri L *Shwasa Chikitsa Adhyay*. In: Shastri B, editor. *Yogaratnakar*. Varanasi, Chaukhamba Prakashan; 2018. Verse 2–3. P. 436.
13. Krause's Food and Nutrition Care Process, L. Kathleen Mahan, Janice L. Raymond, 14th Edition, 34 chapters. P. 688.
14. McAleer JP, Kolls JK. Contributions of the intestinal microbiome in lung immunity. Eur J Immunol. 2018 Jan;48(1):39 49. doi: 10.1002/eji.201646721. Epub 2017 Aug 31. PMID: 28776643; PMCID: PMC5762407.
15. Guan ZW, Yu EZ, Feng Q. Soluble Dietary Fiber, One of the Most Important Nutrients for the Gut Microbiota. Molecules. 2021 Nov 11;26(22):6802. doi: 10.3390/molecules26226802. PMID: 34833893; PMCID: PMC8624670.
16. Association AL. Nutrition and COPD [Internet]. American Lung Association. [cited 2023Jan28]. Available from: www.lung.org/lung-health-diseases/lung-disease-lookup/copd/living-with-copd/nutrition
17. *Maharshi Agnivesh. Chakrapanidatta Viman Sthan Strtovimaniya Adhyay*. In: Vd. Acharya Y T, editor. *Charak Samhita*. Varanasi, Chaukhamba Surbharati Prakashan; 2000. Verse 10. P. 251.
18. *Maharshi Agnivesh. Chakrapanidatta Chikitsa Sthan Hikka-Shwasa Chikitsa Adhyay*. In: Vd. Acharya Y T, editor. *Charak Samhita*. Varanasi, Chaukhamba Surbharati Prakashan; 2000. Verse 55–62. P. 535.
19. Harrison. Diseases of respiratory system. In: Fauci, Braunwald, Kapser, Hauser, Longo, Jameson, Loscalzo, editor. Harrison's Principles of Internal Medicine 17th Edition, volume 2. New York, McGraw-Hill Medical Publication; 2008. Chapter 248 Asthma, PP. 1602–1603.
20. *Maharshi Agnivesh. Chakrapanidatta Chikitsa Sthan Hikka-Shwasa Chikitsa Adhyay*. In: Vd. Acharya Y T, editor. *Charak Samhita*. Varanasi, Chaukhamba Surbharati Prakashan; 2000. Verse 147. P. 539.
21. *Maharshi Agnivesh. Sutrasthan Annapanvidhi Adhyay*. In: Dr. Dwivedi L, editor. *Charak Samhita*. Varanasi, Chowkhamba Krishnadas Academy; 2013. Verse 08–10. P. 516.
22. *Vagbhat. Sutrasthan Annaswaroop vidnyaniyam Adhyay*. In: Dr. Garde G K, editor. *Sartha Vagbhat*. Varanasi, Chaukhamba Surbharati Prakashan; 2011, Verse 03. P. 24.
23. Vd. Shastri L. *Siddhaannadipak Guna Adhyay*. In: Shastri B, editor. *Yogaratnakar*. Varanasi, Chaukhamba Prakashan; 2018. Verse 05. P. 41.
24. *Maharshi Agnivesh. Sutrasthan Annapanvidhi Adhyay* In: Dr. Dwivedi L, editor. *Charak Samhita*. Varanasi, Chowkhamba Krishnadas Academy; 2013. Verse 19–20. P. 519.
25. *Maharshi Agnivesh. Sutrasthan Annapanvidhi* In: Dr. Dwivedi L, editor. *Charak Samhita*. Varanasi, Chowkhamba Krishnadas Academy; 2013. Verse 265. P. 566
26. Raghunath Suri Virachita. *Bhakshyani Adhyay. Bhojana Kutuhal*. P. 35
27. *Maharshi Agnivesh. Sutrasthan Annapanvidhi* In: Dr. Dwivedi L, editor. *Charak Samhita*. Varanasi, Chowkhamba Krishnadas Academy; 2013. Verse 21–22. P. 519.
28. *Shri Kshemsharma. 10th Utsav* In: Dr. Tripathi I, editor. *Kshemkutuhalam Manjula hindi Vyakhya visushitam*. Varanasi, Chaukhamba orientaliya; 1978. Verse 80–81. P. 184.
29. *Maharshi Agnivesh. Sutrasthan Annapanvidhi* In: Dr. Dwivedi L, editor. *Charak Samhita*. Varanasi, Chowkhamba Krishnadas Academy; 2013. Verse 26. P. 520.
30. Vd. Shastri L. *Siddhaannadipak Guna Adhyay*. In: Shastri B, editor. *Yogaratnakar*. Varanasi, Chaukhamba Prakashan; 2018. Verse 07. P. 44.
31. Vd. Shastri L. *Siddhaannadipak Guna Adhyay*. In: Shastri B, editor. *Yogaratnakar*. Varanasi, Chaukhamba Prakashan; 2018. Verse 06. P. 43.
32. *Maharshi Sushrut. Sutrasthan Annapanvidhi Adhyay*. In: Shastri K A, editor. *Sushruta Samhita*. Varanasi, Chaukhamba Sanskrit Sansthan; 2012. Verse 63. P. 248.
33. Vd. Shastri L. *Shwasa Chikitsa Adhyay*. In: Shastri B, editor. *Yogaratnakar*. Varanasi, Chaukhamba Prakashan; 2018. Verse 1–2. P. 436.
34. Vd. Shastri L. Jangal *Mamsa Guna Adhyay*. In: Shastri B, editor. *Yogaratnakar*. Varanasi, Chaukhamba Prakashan; 2018. Verse 3. P. 36.

35. Raghunath Suri Virachita. *Mamsa Prakaran. Bhojana Kutuhal.* P. 171.
36. Bhavmishra. *Shaaka Varga.* In: Dwivedi V, *Bhaavprakash Nighantu.* Delhi, Motilal Banarasidas; 1998. Verse 01. P. 401.
37. Vd. Shastri L. *Dhanya phaladi Shaaka Prakaran.* In: Shastri B, editor. *Yogaratnakar.* Varanasi, Chaukhamba Prakashan; 2018. Verse 12. P. 26.
38. Vd. Shastri L. *Dhanya phaladi Shaaka Prakaran.* In: Shastri B, editor. *Yogaratnakar.* Varanasi, Chaukhamba Prakashan; 2018. Verse 18. P. 26.
39. Vd. Shastri L. *Dhanya phaladi Shaaka Prakaran.* In: Shastri B, editor. *Yogaratnakar.* Varanasi, Chaukhamba Prakashan; 2018. Verse 28. P. 27.
40. *Shri Kshemsharma. 8th Utsav.* In: Dr. Tripathi I, editor. *Kshemkutuhalam Manjula hindi Vyakhya visushitam.* Varanasi, Chaukhamba orientaliya; 1978. Verse 108–109. P. 131.
41. *Shri Kshemsharma. 8th Utsav.* In: Dr. Tripathi I, editor. *Kshemkutuhalam Manjula hindi Vyakhya visushitam.* Varanasi, Chaukhamba orientaliya; 1978. Verse 123. P. 136.
42. *Shri Kshemsharma. 8th Utsav.* In: Dr. Tripathi I, editor. *Kshemkutuhalam Manjula hindi Vyakhya visushitam.* Varanasi, Chaukhamba orientaliya; 1978. Verse 126. P. 136.
43. *Shri Kshemsharma. 8th Utsav.* In: Dr. Tripathi I, editor. *Kshemkutuhalam Manjula hindi Vyakhya visushitam.* Varanasi, Chaukhamba orientaliya; 1978. Verse 130. P. 137.
44. *Shri Kshemsharma. 8th Utsav.* In: Dr. Tripathi I, editor. *Kshemkutuhalam Manjula hindi Vyakhya visushitam.* Varanasi, Chaukhamba orientaliya; 1978. Verse 167. P. 146.
45. *Shri Kshemsharma. 8th Utsav.* In: Dr. Tripathi I, editor. *Kshemkutuhalam Manjula hindi Vyakhya visushitam.* Varanasi, Chaukhamba orientaliya; 1978. Verse 171. P. 147.
46. *Shri Kshemsharma. 8th Utsav.* In: Dr. Tripathi I, editor. *Kshemkutuhalam Manjula hindi Vyakhya visushitam.* Varanasi, Chaukhamba orientaliya; 1978. Verse 200. P. 154.
47. Raghunath Suri Virachita. *Shaaka Prakaran. Bhojana Kutuhal.* P. 65.
48. Raghunath Suri Virachita. *Shaaka Prakaran. Bhojana Kutuhal.* P. 66.
49. Raghunath Suri Virachita. *Shaaka Prakaran. Bhojana Kutuhal.* P. 75.
50. Raghunath Suri Virachita. *Harita Prakaran. Bhojana Kutuhal.* P. 109.
51. *Maharshi Agnivesh. Sutrasthan Annapanvidhi* In: Dr. Dwivedi L, editor. *Charak Samhita.* Varanasi, Chowkhamba Krishnadas Academy; 2013. Verse 125–126. P. 536.
52. Raghunath Suri Virachita. *Haritaki Prakaran. Bhojana Kutuhal.* P. 102.
53. *Maharshi Agnivesh. Sutrasthan Annapanvidhi* In: Dr. Dwivedi L, editor. *Charak Samhita.* Varanasi, Chowkhamba Krishnadas Academy; 2013. Verse 166. P. 542.
54. *Maharshi Sushrut. Sutrasthan Annapanvidhi Adhyay.* In: Shastri K A, editor. *Sushruta Samhita.* Varanasi, Chaukhamba Sanskrit Sansthan; 2012. Verse 244. P. 263.
55. Hsieh CC, Peng WH, Tseng HH, Liang SY, Chen LJ, Tsai JC. The Protective Role of Garlic on Allergen-Induced Airway Inflammation in Mice. Am J Chin Med. 2019;47(5):1099 1112. doi: 10.1142/S0192415X19500563. Epub 2019 Jul 31. PMID: 31366207.
56. Sánchez-Gloria JL, Rada KM, Juárez-Rojas JG, Sánchez-Lozada LG, Rubio-Gayosso I, Sánchez-Muñoz F, Osorio-Alonso H. Role of Sulfur Compounds in Garlic as Potential Therapeutic Option for Inflammation and Oxidative Stress in Asthma. Int J Mol Sci. 2022 Dec 9;23(24):15599. doi: 10.3390/ijms232415599. PMID: 36555240; PMCID: PMC9779154.
57. *Maharshi Agnivesh. Chakrapanidatta Chikitsa Sthan Hikka-Shwasa Chikitsa Adhyay.* In: Vd. Acharya Y T, editor. *Charak Samhita.* Varanasi, Chaukhamba Surbharati Prakashan; 2000. Verse 131. P. 419.
58. Vd. Shastri L. *Sheetoshnavari Guna.* In: Shastri B, editor. *Yogaratnakar.* Varanasi, Chaukhamba Prakashan; 2018. Verse 01. P. 93.
59. *Maharshi Agnivesh. Sutrasthan Annapanvidhi Adhyay* In: Dr. Dwivedi L, editor. *Charak Samhita.* Varanasi, Chowkhamba Krishnadas Academy; 2013. Verse 217. P. 553.
60. Vd. Shastri L. *Dugdha Guna.* In: Shastri B, editor. *Yogaratnakar.* Varanasi, Chaukhamba Prakashan; 2018. Verse 02. P. 96.
61. Vd. Shastri L. *Shwasa Chikitsa.* In: Shastri B, editor. *Yogaratnakar.* Varanasi, Chaukhamba Prakashan; 2018. Verse 1–2. P. 436.
62. *Maharshi Sushrut. Sutrasthan Dravyadravyavidhi Adhyay.* In: Shastri K A, editor. *Sushruta Samhita.* Varanasi, Chaukhamba Sanskrit Sansthan; 2012. Verse 54. P. 223.
63. *Maharshi Sushrut. Sutrasthan Dravyadravyavidhi Adhyay.* In: Shastri K A, editor. *Sushruta Samhita.* Varanasi, Chaukhamba Sanskrit Sansthan; 2012. Verse 49. P. 222.

64. *Maharshi Sushrut. Sutrasthan Dravyadravyavidhi Adhyay.* In: Shastri K A, editor. *Sushruta Samhita.* Varanasi, Chaukhamba Sanskrit Sansthan; 2012. Verse 68. P. 225.
65. *Maharshi Sushrut. Sutrasthan Dravyadravyavidhi Adhyay.* In: Shastri K A, editor. *Sushruta Samhita.* Varanasi, Chaukhamba Sanskrit Sansthan; 2012. Verse 98. P. 228.
66. Vd. Shastri L. *Takra Guna.* In: Shastri B, editor. *Yogaratnakar.* Varanasi, Chaukhamba Prakashan; 2018. Verse 01. P. 104.
67. *Maharshi Sushrut. Sutrasthan Dravyadravyavidhi Adhyay.* In: Shastri K A, editor. *Sushruta Samhita.* Varanasi, Chaukhamba Sanskrit Sansthan; 2012. Verse 132. P. 232.
68. Abbas AS, Ghozy S, Minh LHN, Hashan MR, Soliman AL, Van NT, Hirayama K, Huy NT. Honey in Bronchial Asthma: From Folk Tales to Scientific Facts. J Med Food. 2019 Jun;22(6):543 550. doi: 10.1089/jmf.2018.4303. Epub 2019 May 24. PMID: 31135254.
69. *Maharshi Sushrut. Sutrasthan Annapanvidhi Adhyay.* In: Shastri K A, editor. *Sushruta Samhita.* Varanasi, Chaukhamba Sanskrit Sansthan; 2012. Verse 363. P. 273.
70. Vd. Shastri L. *Siddhaannadi Paka Guna.* In: Shastri B, editor. *Yogaratnakar.* Varanasi, Chaukhamba Prakashan; 2018. Verse 01. P. 43
71. Raghunath Suri Virachita. *Bhakta Guna. Bhojana Kutuhal.* PP. 26–27.
72. Raghunath Suri Virachita. *Bhakta Guna. Bhojana Kutuhal.* P. 27.
73. Raghunath Suri Virachita. *Bhakshyani. Bhojana Kutuhal.* P. 36.
74. Raghunath Suri Virachita. *Vatak. Bhojana Kutuhal.* P. 41.
75. Vd. Shastri L. *Kasachikitsa Adhyay.* In: Shastri B, editor. *Yogaratnakar.* Varanasi, Chaukhamba Prakashan; 2018. Verse 01. P. 424.
76. *Shri Kshemsharma. 10th Utsav.* In: Dr. Tripathi I, editor. *Kshemkutuhalam Manjula hindi Vyakhya visushitam.* Varanasi, Chaukhamba orientaliya; 1978. Verse 48–50. P. 177.
77. Vd. Shastri L. *Siddhaannadipaka Guna Adhyay.* In: Shastri B, editor. *Yogaratnakar.* Varanasi, Chaukhamba Prakashan; 2018. Verse 06. P. 47.
78. Tinsley Randolph Harrison. Diseases of respiratory system. In: Fauci, Braunwald, Kapser, Hauser, Longo, Jameson, Loscalzo, editor. Harrisons Principles of Internal Medicine 17th Edition, volume2. New York, McGraw-Hill Medical Publication; 2008. Chapter 248 Asthma, P. 1597.
79. Tinsley Randolph Harrison. Diseases of respiratory system. In: Fauci, Braunwald, Kapser, Hauser, Longo, Jameson, Loscalzo, editor. Harrisons Principles of Internal Medicine 17th Edition, volume2. New York, McGraw-Hill Medical Publication; 2008. Chapter 248 Asthma, P. 1601.

Therapeutic Nutrition in Ayurveda for Epilepsy

Vikas Autade

INTRODUCTION

Epilepsy is a chronic noncommunicable disease of the brain that affects people of all ages and is characterized by recurrent seizures, which are brief episodes of involuntary movement that may involve a part of the body (partial) or the entire body (generalized) and are sometimes accompanied by loss of consciousness and control of bowel or bladder function.

The seizures or convulsions are the output of excessive electrical discharges in the brain cells, which may be generated from single or different parts of the brain. The severity of convulsions may vary from the short distraction of attention or muscle jerks to severe and prolonged convulsions, with loss of consciousness also. The frequency of seizures can also vary from a few episodes in the year to multiple episodes in a single day. The diagnosis of epilepsy depends on the establishment of two or more episodes of convulsions, separated by more than 24 hours duration. (Up to 10% of people worldwide have one seizure during their lifetime.) Epilepsy is one of the world's oldest recognized conditions, and its written records can be found in many old kinds of literature, dating back to 4000 BCE.[1]

Epilepsy – and Ayurveda Viewpoint

Ayurveda defines epilepsy as *Apasmara: Apa*, meaning negation or loss of; *Smara*, meaning recollection or consciousness. *Apasmara* is considered a severe disease that is chronic and difficult to treat. Several causative factors are mentioned. Treatments included correcting the etiological factors and dietary regimen and avoiding dangerous places that may result in injuries.

The onset of *Apasmara* includes falling down; shaking of the hands, legs, and body; rolling up of the eyes; grinding teeth; and foaming at the mouth. Four major types of epilepsy or *Apasmara* in *Ayurveda* are mentioned, based on the disturbance of *Doshas*. The *Doshas* govern the physiological and physiochemical activities of the body and are also responsible for pathological changes in the body. These *Doshas* also form the base for the classification and diagnosis of epilepsy and are important in the management.[2]

DOI: 10.1201/9781003345541-23

RATE OF DISEASE AND SOCIOECONOMIC IMPACTS

Epilepsy or seizures cause a significant proportion of the world's disease burden and affect around 50 million people worldwide. The overall estimated worldwide population living with active epilepsy means those who are continuing seizures or with the need for treatment at a given time between 4 and 10 per 1000 people.[3] Epilepsy is causing a significant social and economic burden to the epileptic person and his family, in terms of loss of life years due to premature mortality and inadequate health. Around 50 million people worldwide have epilepsy, making it one of the most common neurological diseases globally.[4]

APPROACH FROM THE THERAPEUTIC NUTRITION IN AYURVEDA

According to recent studies, there might be a relationship between specific diet and seizure occurrence. Casein is an important protein of milk, leading to some type of hypersensitivity turning into epileptogenic focus in the brain when they cross the blood-brain barrier. On the other hand, several studies represent full-fat milk or higher-fat dairy products as an effective anti-inflammatory factor that elevates seizure threshold and is helpful in controlling seizures.[5] Some of the food additives are also leading to increased susceptibility to seizure and some are not,[6,7] but very limited experimental evidence is there to explain the phenomena. The research about its efficacy and limitations is going on to prove its role in the management of epilepsy.

Ayurveda also enumerates many dietary aspects in the etiology and pathology as well as the management of diseases. This chapter is a genuine effort to find out the coded suggestions with references about dietary prescriptions that are coming across the *ayurvedic* texts and literature, in the context of therapeutic nutrition in the management of epilepsy.

Causes of Epilepsy

Epilepsy has no identifiable cause in many of the people with the condition. In the others, the causes may be searched with various pathological abnormalities like structural brain abnormalities, genetic influence, and infection of the brain and its coverings, metabolic, immune, and prenatal trauma and head injury.

Causes of Epilepsy From Ayurveda Perspective

Charaka Samhita mentions the following points in relation to *Apasmara*.[8] *Charaka* opines that persons with the following morbidities are more susceptible to epilepsy:

- Persons who have a mind afflicted by *Rajas* and *Tamas* which means persons aggravated by emotions such as passion, anger, fear, greed, attachment, excitement, grief, anxiety, and perturbation.
- Individuals with imbalanced and excessively aggravated conditions of *Tridosha*, the three bodily humor maintaining homeostasis.
- Individuals consuming food that is impure, untimely, decomposed, possessing antagonistic properties, or touched/cooked by unclean hands. The persons might be suffering from improper digestion, malnourishment, and infections due to these reasons.
- An individual follows improper methods (of diet and lifestyle, conduct) and neglects prescribed rules. Ayurveda defines many rules for social behavior, personal hygiene, and dietary practices in the *Dinacharya*[9] and *Sadvrutta*[10] might be leading to dosha vitiation, stressful life incidences, and in term predisposing to *Apasmara*.
- An individual observes improper techniques (of treatment) and resorts to unhealthy regimens and behavior. *Charaka* has prescribed seasonal purificatory practices for *Dosha Shodhan* to maintain the balance of *Tridosha* and term health.
- An individual suffers from excessive degeneration or debility, which may be due to faulty food habits, or incorrect nutrition, leading to multiple micro and macro nutritional deficiencies and predisposing to epilepsy.

Pathogenesis of *Apasmara*

Charaka Samhita explains that the *Sharir Doshas* are excessively aggravated due to habitual intake of improper/harmful and impure/contaminated food articles and the mind is affected by *Rajas* and *Tamas* due to stressful emotions such as passion, anger, fear, greed, attachment, excitement, grief, anxiety, perturbation. These aggravated *Doshas* are leading to obstruction in the pathways of consciousness resulting in the presentation of *Apasmara* as a disease. *Charaka* ensures that physical and psychological factors have an important role in the pathogenesis of epilepsy.[11]

THERAPEUTIC NUTRITION IN EPILEPSY – CLASSICAL REFERENCES

In the Ayurvedic classical texts like *Charaka Samhita*, *Susruta Samhita*, and *Yogaratnakar* much of the dietary advice are written to be advocated for the management of *Apasmara*. These dietary food articles are to be used as supportive to the treatment and also help to prevent and correct the *Dosha* vitiation in the person suffering from epilepsy.

Table 18.1 contains the available references of the classical food and dietary items described in connection with epilepsy in *Charaka Samhita*, *Susruta Samhita*, and *Yogaratnakar*. Among the cereals, *Shali* (rice), specifically the red variety of rice, 60-day variety of rice, and older rice, are generally healthier options and maintain the dosha balance. *Godhuma* (wheat) is also important as a part of daily food and has *Vata* pacifying and nourishing properties.

Among the pulses, only *Mudga* (green gram) enters the list of prescribed food items. Contrary to the other pulses *Mudga* maintains the dosha balance and can be consumed as *Mudga Yusha* (soup) or as a part of food recipes.

Meats are *Vata* pacifiers and nourishing but heavy to digest. Among all, only *Jangala Mamsa* is advised as it is easy to digest and preferably consumed like meat soup.

TABLE 18.1 Wholesome Diet and Habits to Be Followed in Epilepsy[12–17]

FACTORS ADVISED *(SANSKRIT)*	FACTORS ADVISED *(ENGLISH)*
Shuka Dhanya Varga	**Millets and Cereals**
Puraan Shali	Old rice
Rakta Shali	Red rice
Shastika Shali	60-Day-old rice
Godhuma	Wheat
Shami Dhanya Varga	**Pulses**
Mudga Yusha	Green gram soup
Mamsa Varga	**Meats**
Kurma Mamsa	Tortoise meat
Dhanwarasa	*Jangala Mamsa*
Shaka Varga	**Leafy Vegetables**
Shigru	Drumstick leaves
Brahmi	*Bacopa monnieri* leaves
Vacha	*Acorus calamus* leaves
Patola	Snake gourd leaves
Vaastuka	*Chenopodium murale* leaves
Phala Varga	**Fruits**
Draksha	Grapes
Falgu	Fig
Puraan Kushmanda	Old wax gourd
Dadima	Pomegranate
Kapittha	Elephant apple
Panasa	Jack fruit
Parushaka	Indian sherbet berry
Amalaka	Gooseberry
Narikela Jala	Coconut water
Gorasa Varga	**Milk and Dairy Products**
Goksheers	Fresh cow's milk
Puraan Ghrita	Aged ghee
Goghrita	Cow's ghee
Kritanna Varga	**Cooked Foods**
Shali Tandula Payasam[18]	Sweet rice pudding

Shigru and *Vacha* leaves are recommended as dietary items, though some of them have medicinal properties also.

The nourishing fruits and *Vata Dosha* pacifiers and not lead to indigestion are advised in neurological disorders, and in epilepsy fruits like grapes, jackfruit, pomegranate, wax gourd, and Indian gooseberry are mentioned.

Among the milk and dairy products cow's milk, aged as well as fresh cow's ghee, is good for neurocognitive disorders and can be consumed alone or may be utilized in different recipes. In the cooked

TABLE 18.2 Unwholesome Diet and Habits to Be Avoided in Epilepsy[19]

FACTORS TO AVOID *(SANSKRIT)*	FACTORS TO AVOID *(ENGLISH)*
Guru Aahar	Foods heavy to digest
Viruddha Aahar	Incompatible food
Teekshna, Ushna Bhojanam	Excessive bitter and spicy food items
Madya	Alcohol
Matsya	Fish
Aashadhaki Phalam	Black gram
Bimbiphala	*Coccinea grandis*
Kshut Trit Vega Dhaaran	Suppression of thirst and appetite urges
Anidra	Sleeplessness
Ativyaayam, Atiaayas	Excessive work and exertion
Chinta	Mental stress or anxiety

According to *Yogratnakar*, the food items and habits mentioned in the table should be avoided by people who suffer from epilepsy.

food recipes *Shali Tandula Payasam* (sweet rice pudding) is *Vata Dosha* pacifying and prescribed as useful for epileptic patients.

Food articles causing disturbances in digestion and metabolic activity are recommended to restrict when a person is suffering from *Dosha* vitiation. Similarly, excessive exertional work, dehydration, risky adventures, distress, starvation, and sleepless nights are known to trigger the convulsion and put the epileptic patient's life in danger.

EPILEPSY CLASSIFICATION AND DEFINITION OF TERMS

The classification of seizures and epilepsies by the International League Against Epilepsy (ILAE), 2017, is the most recent classification model which aimed to simplify terminologies so that patients and their caregivers can easily understand and identify seizures that have both focal and generalized onset and incorporate missing seizures. Epilepsy is classified into mainly two types – generalized epilepsy and partial epilepsy. This is the most recent classification, proposed by ILAE [International League Against Epilepsy] and is dependent on clinical presentation as well on EEG findings.[20]

Partial Seizures

A partial seizure is also known as a focal seizure since abnormal brain signals begin from a small area of the brain. Its clinical presentation differs as per the local area of the brain involved in an abnormal electrical discharge.

Generalized Seizures

A generalized seizure begins when both sides of the brain are involved in discharging abnormal electrical signals and leading to loss of consciousness.

Tonic, Clonic, and Tonic-Clonic (Formerly Called Grand-Mal) Seizures

Tonic-clonic seizures can evolve from any of the focal or generalized seizure types.

CLASSIFICATION OF *APASMARA* IN AYURVEDA

The classification of epilepsy or *Apasmara* in Ayurveda is based on the disturbance of *Doshas* (humors) that govern the physiological and physiochemical activities of the body.

The *Apasmara* is classified into four types by *Charaka* based on *Tridosha*.[21]

1. *Vataja Apasmara* – characterized by trembling, grinding of teeth, frothing from the mouth, and gasping, more frequent episodes of *Apasmara*, loss of consciousness for a shorter duration.
2. *Pittaja Apasmara* – characterized by frequent episodes of *Apasmara*, instant loss of consciousness and regaining it quickly, yellowish, reddish froth, body, face, and eyes.
3. *Kaphaja Apasmara* – characterized by whitish froth (from the mouth), body, face, and eyes, less frequent episodes of *Apasmara*, gradual loss and late regain of consciousness.
4. *Sannipataja Apasmara* – mixed presentation.

This *Dosha*-dependent classification is important in the management of epilepsy. The treatment modalities in Ayurveda are based on *Dosha*-dependent type of epilepsy, for the selection of medicine as well as therapeutic nutrition.

Apasmara or epilepsy is such a disorder in which the patient presents with episodes of seizure or convulsion and after getting recovered from the episodes leads a normal life. In *Ayurveda,* it is called *Vegavastha* (status epilepticus phase) and *Avegavastha* (nonepisodic phase). The principle of the management of epilepsy is to control the episodic phase without complication and try to prevent the maximum generation of further episodes. The major focus of the management is on the *Avegavstha* of epilepsy, during which medication or nutritional therapy is applied to prevent forthcoming convulsions.[22] During this nonepisodic phase itself, the physician has to collect all the relevant information from the patient and relatives for clearer diagnosis and *Dosha* involvement in the *Apasmara* pathology, as well as any specific dietary habits which are predominant in the person's lifestyle.

Treatment and Prevention

The important goal of epilepsy treatment is to control the convulsions and to prevent the forthcoming episodes of seizures to the maximum. Absolute control over seizures and improvement in a person's quality of life is the prime aim of epilepsy management.[23]

Predictions of prognosis and recurrence of seizures are dependent on any specific etiology or abnormal finding of electroencephalography (EEG). Epileptic convulsions can be controlled well with modern antiepileptic medication and the majority (70%) of the people suffering from epilepsy could lead a seizure-free life. Alterations or discontinuation of the doses of antiepileptic medication is only considered after 2 years of absolute control of epilepsy. Neurosurgery is an option for nonresponding patients only.

The prevention of epilepsy is very similar to lifestyle disorders and vascular disorders, such as the control or prevention of hypertension, diabetes mellitus, obesity, and many of the addictions like tobacco or alcohol. An estimated 25% of epilepsy cases are preventable.[24]

Nonpharmacological Treatment of Epilepsy

Antiepileptic medications are good enough to control the majority of epileptic seizures, but the patients are not satisfied with them due to some troublesome adverse effects, long-standing treatment, and some misconceptions or females willing for pregnancy are ready to undergo alternative treatment or nonpharmacological treatment options.[25]

Common nonpharmacological treatments are *Yoga,* nutrition from *Ayurveda*, biofeedback techniques, physiotherapy exercises, music therapy, acupuncture, transcranial magnetic stimulation, and traditional Chinese medicine.

Nonpharmacological treatments or alternative therapies, though, are not for a replacement to the antiepileptic medication but to be considered complementary to it, either to improve the efficacy or to reduce the untoward effects of treatment. Some of the treatments like Yoga and music therapy reduce distress and work on the psychological aspect of epilepsy and improve the quality of life.[25]

SIGNIFICANCE OF NUTRITION IN EPILEPSY

Many of the patients suffering from epilepsy and treated with antiepileptic drugs and still not getting optimum control over seizures benefit significantly just by avoiding the dietary items which are asked to be avoided in unwholesome preparations in Ayurveda classical references, which are tabulated in this chapter above (Table 18.2).

Along with the antiepileptic treatment adopting simple changes like adding cow's milk and ghee to the regular diet also benefits well to get control over the upcoming seizures. It's clinically observed that just adding cow's ghee 5–10 mL per day to the routine diet of the patient also decreased the seizure frequency and gave better control over disease. *Goghrita* (cow ghee) might be working as either a *Vata* pacifier or nutritively in the aspect of the epilepsy treatment.

According to *Charaka Samhita,* neurological as well as psychological disturbances are at the core of *Apasmara* pathogenesis, and other neurological disorders are described under the *Vata Vyadhi* chapter, where *Vata Dosha* leads to paralysis or neurological disorders. Convulsive disorders leading to loss of consciousness are explained as *Apasmara*, while many of the conditions like *Apatantraka* and *Apatanaka* having convulsions with minimal loss of consciousness are also described under *Vata Vyadhi* chapter, which is especially devoted to neurological diseases.[26] In these convulsive disorders (*Apatantrak* and *Apatanaka*) loss of orientation and consciousness is evidenced to be very limited and presents as seizures associated with other physical diseases.

While describing etiological factors of epilepsy, *Charaka Samhita* mentioned vitiated *Doshas* and habitual intake of improper/harmful and impure/contaminated food as one of the most important factors.[27] The treatment of *Apasmara* is dependent on the *Dosha* involvement in the selection of drugs and special therapeutic procedures. The prescription of therapeutic dietary preparations has to be synergistic with the *Dosha* involvement in *Apasmara*.[28]

Charaka Samhita elaborates on management protocol for the epilepsy, where it is explained the importance of purificatory methods to unblock the channels of the heart and the mind that may be clogged by the excess of *Doshas* or humor. The opening of the channels is practiced using various concoctions and *Dosha*-dependent purificatory methods like *Virechana, Vamana,* and *Basti.*[29] The drugs to be taken orally are cooked with oils and ghee and, in addition, external oil applications, massages, and baths are also part of the treatment. However, the treatment selected for one patient with epilepsy may not be applicable for another as it is. Treatment modalities that include strong elimination purgatives are used to alleviate the symptoms, depending upon specific requirements, and are mentioned as being useful for epilepsy patients.

Charaka considers the involvement of *Vata Dosha* to have a crucial role in the pathogenesis of epilepsy and attains importance in the management as well as in therapeutic nutrition. The general management principles and therapeutic nutritional advice like *Ghritas* and *Medhya Rasayana* recommended in other neurological disorders may be applied to epilepsy also.[30] A wide variety of medicated *Ghrita* has been recommended for internal use in the treatment of *Vata Vyadhi* and *Apasmara* also. The use of mixtures of *Ghrita* (ghee) and *Taila* (oil) cooked with drugs has also been mentioned. Oil cooked with various herbal and animal products has been recommended for the massage of the body of the patient. Disease-specific dietary restrictions and therapeutic nutrition are advised during the *Avegavastha* of *Apasmara* as a prophylactic tool along with the specific treatment to improve the outcome of medication as well as limit the adverse effects.

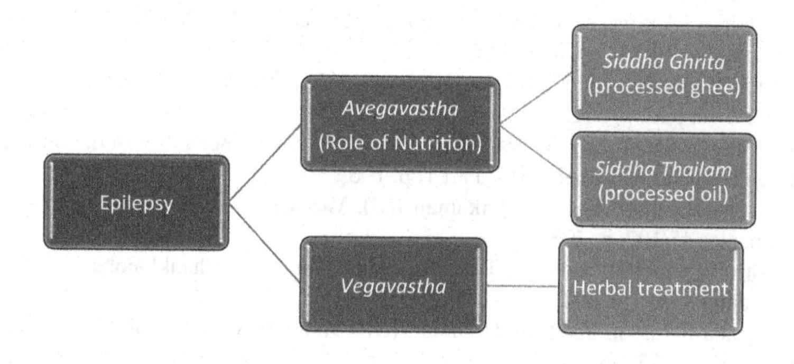

SCOPE FOR FURTHER RESEARCH

Medicated ghee preparations are very common in the treatment of epilepsy and other neurological disorders as well, which invites the attention of further research to explore its role in the treatment of epilepsy.

Adding cow's ghee or milk to the regular diet does make any changes in the seizure trigger mechanism to get control over epilepsy. According to recent studies, there might be a relationship between specific diet and seizure occurrence; e.g., casein is an important protein in milk that often causes hypersensitivity, whereas full-fat milk improves the seizure threshold, making space for further research.[31]

There is no specific research or literature about different fruits, such as grape, fig, wax gourd, pomegranate, elephant apple, jackfruit, Indian sherbet berry, gooseberry, or coconut water, having any antiepileptic activity or supportive role in increasing its efficiency and this needs to be researched further.

REFERENCES

1. World Health Organization. Epilepsy Fact Sheet [Internet]. World Health Organization; [cited 2023 Mar 10]. Available from: www.who.int/news-room/fact-sheets/detail/epilepsy

2. Trikamaji Y, editor. Agnivesha Charaka Samhita. Varanasi: Chaukhamba Surbharati Prakashan; 2019. Nidaansthan chapter 8/4, p. 226.
3. World Health Organization. Epilepsy Fact Sheet [Internet]. World Health Organization; [cited 2023 Mar 10]. Available from: www.who.int/news-room/fact-sheets/detail/epilepsy#:~:text=Globally%
4. World Health Organization. Epilepsy Fact Sheet [Internet]. World Health Organization; [cited 2023 Mar 10]. Available from: www.who.int/news-room/fact-sheets/detail/epilepsy#:
5. Inaloo S, Pirsalami F, Dastgheib M, Moezi L. The effects of dairy products on seizure tendency in mice. Heliyon. 2019 Mar 12;5(3):e01331. doi: 10.1016/j.heliyon.2019.e01331. PMID: 30911694; PMCID: PMC6416732.
6. Rigo FK. Propionic acid induces convulsions and protein carbonylation in rats. Neurosci Lett. 2006;408(2):151–154. doi: 10.1016/j.neulet.2006.08.004. PMID: 16973254.
7. Arauz-Contreras J, Feria-Velasco A. Monosodium-L-glutamate-induced convulsions--I. Differences in seizure pattern and duration of effect as a function of age in rats. Gen Pharmacol. 1984;15(5):391–395. doi: 10.1016/0306-3623(84)90088-7. PMID: 6386987.
8. Trikamaji Y, editor. Agnivesha Charaka Samhita. Varanasi: Chaukhamba Surbharati Prakashan; 2019. Sutrasthan chapter 5/5-8, pp. 37–38.
9. Trikamaji Y, editor. Agnivesha Charaka Samhita. Varanasi: Chaukhamba Surbharati Prakashan; 2019. Sutrasthan chapter 8/18,19, pp. 58–59.
10. Trikamaji Y, editor. Agnivesha Charaka Samhita. Varanasi: Chaukhamba Surbharati Prakashan; 2019. Chikitsasthan chapter 10/5,6, pp. 474–475.
11. Shastri S, editor. Yogratnakar Vidyotini commentary by Vd. Shrilakshmipati Shastri. Varanasi: Chaukhamba Prakashan; 2017. 1st part, p. 502.
12. Shastri SL, Shastr SB (eds). Yogratnakar Vidyotini commentary by Vd. Shrilakshmipati Shastri. Varanasi: Chaukhamba Prakashan; 2017. Part 1, p. 106.
13. Sushruta. Sushruta samhita. Yadavaji Trikamaji (ed). Varanasi: Chaukhamba Surbharati Prakashan; 2018. Sutrasthana chapter 45/107, p. 205.
14. Sushruta. Sushruta samhita. Yadavaji Trikamaji (ed). Varanasi: Chaukhamba Sanskrit Sansthan; 2018. Sutrasthana chapter 45/96, p. 204.
15. Agnivesha. Charaka Samhita. Yadavaji Trikamaji (ed). Varanasi: Chaukhamba Surbharati Prakashan; 2019. Sutrasthana chapter 27/231, p. 166.
16. Agnivesha. Charaka Samhita. Yadavaji Trikamaji (ed). Varanasi: Chaukhamba Surbharati Prakashan; 2019. Sutrasthana chapter 27/233, p. 166.
17. Sushruta. Sushruta samhita. Yadavaji Trikamaji (ed). Varanasi: Chaukhamba Surbharati Prakashan; 2018. Uttartantra chapter 61/38, p. 802.
18. Shastri SL, Shastr SB (eds). Yogratnakar Vidyotini commentary by Vd. Shrilakshmipati Shastri. Varanasi: Chaukhamba Prakashan; 2017. Part 1, p. 502.
19. Sarmast ST, Abdullahi AM, Jahan N. Current Classification of Seizures and Epilepsies: Scope, Limitations and Recommendations for Future Action. Cureus. 2020 Sep 20;12(9):e10549. PMID: 33101797; PMCID: PMC7575300. doi: 10.7759/cureus.10549.
20. Agnivesha. Charaka Samhita. Yadavaji Trikamaji (ed). Varanasi: Chaukhamba Surbharati Prakashan; 2019. Nidaansthan chapter 8/8, p. 226.
21. Joshi YG. Kayachikitsa. Pune: Pune Sahitya Vitarana; 1989. p. 737.
22. Ashrafi MR, Heidari M. General Principles of the Medical Management of Epilepsy in Children: A Literature Review. Research in Clinical Medicine. 2018;5(2):49–53.
23. World Health Organization. Epilepsy. Fact Sheet. Updated February 2019. Available from: www.who.int/news-room/fact-sheets/detail/epilepsy.
24. World Health Organization. Epilepsy [Internet]. Geneva (CH): World Health Organization; 2021 [cited 2023 Mar 10]. Available from: www.who.int/news-room/fact-sheets/detail/epilepsy
25. Saxena VS, Nadkarni VV. Nonpharmacological treatment of epilepsy. Ann Indian Acad Neurol. 2011 Jul;14(3):148–52. doi: 10.4103/0972-2327.85870. PMID: 22028523; PMCID: PMC3200033.
26. Agnivesha. Charaka Samhita. Edited by Yadavaji Trikamaji. Varanasi: Chaukhamba Surbharati Prakashan; 2019. Chikitsasthan chapter 28/50, p. 618.
27. Agnivesha. Charaka Samhita. Edited by Yadavaji Trikamaji. Varanasi: Chaukhamba Surbharati Prakashan; 2019. Nidaansthan chapter 8/4, p. 226.

28. Agnivesha. Charaka Samhita. Edited by Yadavaji Trikamaji. Varanasi: Chaukhamba Surbharati Prakashan; 2019. Chikitsasthan chapter 10/15, p. 475.
29. Agnivesha. Charaka Samhita. Edited by Yadavaji Trikamaji. Varanasi: Chaukhamba Surbharati Prakashan; 2019. Chikitsasthan chapter 10/14, p. 475.
30. Agnivesha. Charaka Samhita. Edited by Yadavaji Trikamaji. Varanasi: Chaukhamba Surbharati Prakashan; 2019. Chikitsasthan chapter 10/17,62, p. 475, 477.
31. Inaloo S, Pirsalami F, Dastgheib M, Moezi L. The effects of dairy products on seizure tendency in mice. Heliyon. 2019 Mar 12;5(3):e01331. doi: 10.1016/j.heliyon.2019.e01331. PMID: 30911694; PMCID: PMC6416732.

Therapeutic Nutrition in Ayurveda for Neurocognitive Conditions

Yogesh Shamrao Deole

INTRODUCTION

Cognitive functions are an integral part of behavior and adaptations in human beings. As per the new classification in *Diagnostic Statistical Manual–5* (DSM-5), a category named 'neurocognitive disorder' is introduced. It was formally known as 'dementia, delirium, amnestic, and other cognitive disorders.'[1] As per a study in France, the prevalence of these disorders is about 40/1,000 people after 60 years of age and gradually increases to 180/1,000 after 75 years of age, reaching almost 1 in 2 people after 90 years of age.[2] The incidence is increasing so rapidly that there is an urgent need to prevent the disorders.

In *Ayurveda*, cognition is regulated by the faculties like *Buddhi* (intellect), *Dhriti* (restraint), and *Smriti* (memory). Therefore, the principles of nutrition of these faculties described in ancient Indian classics are applied to therapeutic nutrition for neurocognitive conditions. This chapter describes the principles, their applications, and recipes highlighted in Indian tradition used to prevent and treat these conditions.

AYURVEDIC PERSPECTIVE OF NEUROCOGNITIVE CONDITIONS

As per the *Ayurveda* perspective, neurocognitive conditions can emerge from overnutrition as well as undernutrition. The disorders like confusion (*Buddhi Moha*) and drowsiness (*Tandra*) are caused due to overnutrition. Whereas delirium (*Pralapa*) is caused by undernutrition. [3] The conditions due to overnutrition are predominantly caused by *Ama*. These disorders are mainly due to the aggravation of *Kapha Dosha* and the vitiation of *Rasa Dhatu*, *Rakta Dhatu,* and *Majja Dhatu*.[4] Food, beverages, and medicines with hot potency and pungent postdigestive effects are preferred for the treatment of these disorders.

DOI: 10.1201/9781003345541-24

Undernutrition or malnourishment can lead to degeneration and depletion of body tissues. These disorders are due to the aggravation of *Vata*[5] and *Pitta Dosha* and *Majja Dhatu*. The food, beverage, and medicines with cold potency and sweet postdigestive effect are prescribed for treatment of the same.

The faculties like *Dhi*, *Dhriti*, and *Smriti* are responsible for all cognitive functions. *Prana Vayu*, *Udana Vayu*, *Sadhaka Pitta*, and *Tarpaka Kapha* are responsible for physiological functions related to cognition. [6]

COMMON NEUROCOGNITIVE CONDITIONS

The most common neurocognitive conditions are dementia and delirium.[7] *Ayurveda* focuses on the causes (*Hetu*), pathophysiology (*Samprapti*), clinical features (*Lakshana*), and prognosis (*Sadhya-Asadhyatva*) to know the disease and its management. Discussing all the causes and prognosis will be out of the context of this chapter. Breaking the pathogenesis is vital for treatment. Therefore, important factors related to pathogenesis and the role of therapeutic nutrition in it are discussed below:

1. Dementia (*Smriti bhrimsha*):

Dementia is a general term that includes an impaired ability to remember, think, or make decisions. Alzheimer's disease is the most common cause of dementia.[8] Dementia is mostly observed in older adults and increases with the progression of age. Ayurveda classics term '*Smritibrimsha*' (impairment of memory) indicates dementia.

Smriti (Memory)

Smriti is the capacity of remembering objective experiential knowledge. It is the real knowledge of subjects/experiences that happened in the past.[9]

Dietary Causes Affecting Memory:

- Eating excess sour, saline, pungent
- Excess alkaline food
- Dried vegetables, meat
- Sesame, sesame paste, and preparation of (rice) flour
- Germinated or fresh, awned or leguminous cereals
- Food incompatible, unsuitable to one's body constitution
- Decomposed, heavy to digest, putrefied, and stale food items
- Indulging in an irregular diet or eating while the previous food is undigested
- Wine and alcohol preparations[10]

These types of food items vitiate *Dosha* and cause blockages in channels of transportation and transformation (*Srotas*). This leads to various diseases including impairment of memory.

Types of Dementia and Its Pathophysiology:

As per the contemporary understanding, dementia is categorized into two main types:

1. Alzheimer's dementia
2. Non-Alzheimer's dementia

1. **Alzheimer's dementia**:
This is the most common cause of dementia, accounting for 60 to 80 percent of cases.[11] It is related to the loss of cholinergic neurons in the brain.

2. Non-Alzheimer's dementia:
It is further categorized as vascular dementia, Lewy body dementia, frontotemporal dementia, and mixed dementia. These types of dementia are caused by the loss of serotonergic neurons. Vascular dementia accounts for about 10 percent of dementia cases. These are due to strokes or impaired blood circulation in the brain. Diabetes, high blood pressure, and high cholesterol are risk factors. In Lewy body dementia, memory loss, movement, or balancing problems like stiffness or trembling are seen. In frontotemporal dementia, changes in personality and behavior are observed. Mixed dementia is presented with more than one type of dementia. Reversible causes such as the adverse effect of medication, increased pressure in the brain, vitamin deficiency, and thyroid hormone disorders are observed in some cases. The causes of dementia based on advanced diagnostic and clinical features are important to know. The objectives of therapeutic nutrition can be fixed based on these causes. For example, the utility of *Tinospora* will be more to treat vascular dementia based on its pharmacological activity profile.[12]

Confusion (*Sammoha*) or Delirium (*Pralapa*):

Acute confusion state or delirium is a critical condition showing severe deterioration of cognitive functions. It is a mental and behavioral state of reduced comprehension, coherence, and capacity to reason. Ayurvedic classics describe it as '*Mada*' and '*Pralapa*'.

Vata and *Pitta Dosha* dominance is observed in these disease conditions.

Ayurvedic Perspective of Neurocognitive Conditions:

For a better understanding of neurocognitive conditions, it can be classified into two types:

1. Neurocognitive conditions type 1 due to *Vata* and *Pitta Dosha* dominance [Fig.01]
2. Neurocognitive conditions type 2 due to *Pitta* and *Kapha Dosha* dominance [Fig.02]

The following Figures 01 and 02 show pathophysiological events in these two conditions:

The treatment and therapeutic nutrition differ in both types.

The principle of treatment of type 1 neurocognitive condition includes nourishment and replenishment (*Santarpana*), whereas type 2 will be treated with depleting therapies and restricted nourishment therapies (*Apatarpana*).

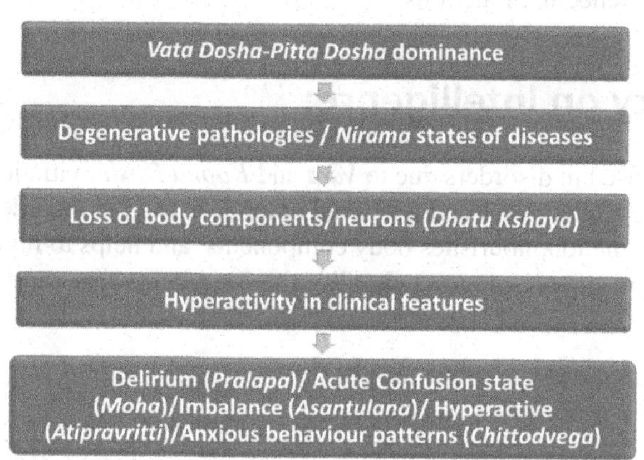

FIGURE 19.1 Pathogenesis of neurocognitive disorders due to undernutrition

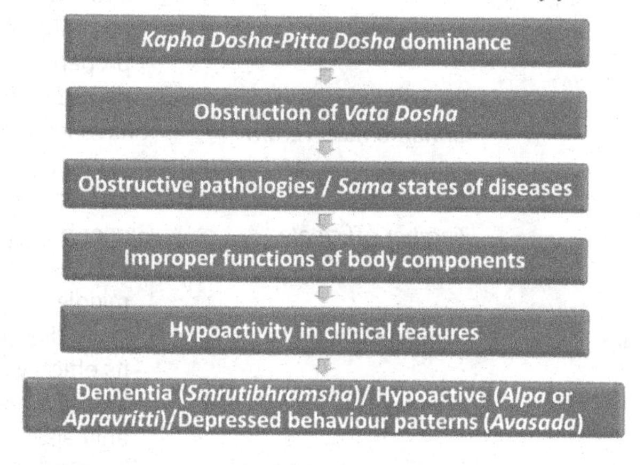

FIGURE 19.2 Pathogenesis of neurocognitive disorders due to overnutrition

THERAPEUTIC NUTRITION FOR NEUROCOGNITIVE DISORDERS

The nutrition profile of a food depends on its taste, potency, and properties. It is also important to know the interactions of body physiology with food to produce its nutritive effects.

Effects of Tastes on Intelligence:

The tastes of food have diverse effects on the mind and intellect. Sweet taste produces a blissful state of mind and all senses. A sour taste stimulates mental functions. Bitter taste pacifies the inflammatory pathologies and is used in the treatment of syncope.[13] Food having these tastes is used for nourishment of the mind and enhancing intellectual functions.

Effect of Potency on Intelligence:

Food with hot potency is used in disorders due to *Vata* and *Kapha Dosha* vitiation. It removes obstruction in pathways and stimulates functions. Food with cold potency is used in disorders due to *Pitta* and *Rakta* vitiation. It pacifies inflammation, nourishes body components, and helps to replenish tissue nutrients.

The following categorization is done to simplify the food items that can be used in neurocognitive conditions. (Table 01)

TABLE 01 Food items in therapeutic nutrition for neurocognitive conditions.

CATEGORY	*FOOD ITEM*	*BOTANICAL NAME*	*RATIONALE*
Spices	*Haridra*	*Curcuma longa* Linn.	Hot potency and dry properties help to reduce inflammatory swellings and blockages in microchannels. It has antioxidant and antimutagenic properties with efficacy in age-related cognitive decline.[14]
	Dalachini	*Cinnamomum verum*	It has shown an effect on neuronal loss and cognitive decline due to traumatic brain injuries.[15]
	Ginger	*Zingiber officinale*	Ginger has shown an effect on improving cognitive functions in middle-aged women.[16]
Pulses:	Mudga	*Vigna radiata*	Light to digest, mung bean has shown its efficacy in preventing medicine induce amnesia in experimental animals.[17]
Beverages: Milk	Milk Cow milk	Cow milk	Cow milk proteins have a positive impact on brain growth and cognitive development.[18]
	Buffalo milk		Buffalo milk has an impact on intelligence.[19] It showed efficacy to alleviate fatigue in young mice, rescue oxidative stress damage and promote learning and memory in aging mice.[20]
	Goat milk		Fermented goat milk consumption increased dopamine, oxytocin, serotonin, synaptophysin, and α-MSH and decreased MAO-A and MAO-B, suggesting a potential neuroprotective effect on brain functions. This could enhance brain molecular functions.[21]
Milk-products:	Ghee		Milk with ghee is best among *Rasayana Dravya*.[22]

Dietary Items

Butter:

Butter promotes activity for intellectual functions.[23]

Cow Ghee:

Ghee is indicated for persons who are indulged in intellectual activities and who wish to enhance memory, cognition, and intellectual capacity.[24] Ghee possesses antioxidant and adaptogenic activities. [25] Research suggests that ghee has potent antioxidant effects and can be used in heart diseases, diabetes, and stress too.[26]

Ghee carries the therapeutic properties of substances to all the body's tissues. It is an excellent vehicle for transporting food items to the deeper tissue layers of the body.[27] Proper digestion, absorption, and delivery to a target organ system are crucial in obtaining the maximum benefit from any therapeutic formulation; the lipophilic action of ghee facilitates transportation to a target organ and final delivery inside the cell since the cell membrane also contains lipids.[28] Vitamins A and E present in ghee are antioxidants that help in preventing oxidative injury to the body. Ghee is well acceptable to the internal environment of the body, even higher doses due to its palatability.

Ghee is an excellent medium for cooking to prevent and treat neurocognitive disorders due to the vitiation of *Vata* and *Pitta Dosha*. The food for patients with neurocognitive conditions is prepared in ghee, instead of oil. Generally, 20 grams of ghee per serving is advised to be consumed either in the curry or in dal or taken with rice and whole wheat bread in patients with neurocognitive conditions. The taste of the food prepared in ghee is delicious, which enhances the perception to take food in these conditions.

Honey:

Honey has the potency to improve cognitive functions [Kaiyadeva Nighantu 1/176]. Honey is reported to possess antioxidant and adaptogenic activities.[29, 30, 31] The antioxidant activity by having constituents like propolis (rich in flavonoids and polyphenolic compounds) and gluconic acid; ascorbic acid; hydroxymethyl furaldehyde; and activities of glucose oxidase, catalase, and peroxidase. Thus, the antioxidant capacity of honey appears to be a result of the combined activity of a wide range of compounds including phenolics, peptides, organic acids, enzymes, Maillard reaction products, and possibly other minor components. Honey may have potential use in reducing the effects of acute and chronic free radical–mediated diseases, such as atherosclerosis, diabetes, and cancer.[32] Antioxidant properties of honey act as an antidepressant during high emotional, physical, and intellectual stress.[33,34]

The following dietary vegetables, oils, and meats are referred to as having therapeutic action on cognitive functions. However, research is needed to study their efficacy in various nutritive forms.

Vegetables:

- *Vastuka* (*Chenopodium murale*) is effective to improve intellectual functions.[23] It is used to make curry.
- *Sthauneyaka* (*Taxus baccata*) increases intellectual functions. [23] Its dried powder is used in curry.
- *Chanchu Shaka* (*Corchorus fascicularis*) can improve the intelligence.[19] Its leaves are used in the diet.
- *Yava* (*Hordeum vulgare* L.) improves stability and cognitive functions.[23]

Oil:

- *Tila* (*Sesamum indicum*)[23] and its oil taila (sesame oil)[23]
- *Eranda taila* (castor oil)[23]
- *Yavatikta taila* (*Andrographis paniculata* [Burm. fil.]) [23]

Mamsa (Meats)

- Meat of common myna (*Sarika Mamsa*)[23]
- Meat of demoiselle crane (*Krakar Mamsa*)[23]
- Meat of partridge (*Tittiri Mamsa*)[23]
- Meat of snake (*Sarpa Mamsa*)[23]

RESEARCH EVIDENCE ON FOOD ITEMS AND HERBS

The *Rasayana* diet and herbs are used to prevent and treat these disorders. Mainly used *Rasayana* herbs are *Mandukaparni* (*Centella asiatica* Linn.), *Yashtimadhu* (*Glycyrrhiza glabra* Linn.), *Guduchi* (*Tinospora cordifolia* [Wild] Miers), and *Shankhapushpi* (*Convolvulus pluricaulis* Chois). The neuroprotective activities and nootropic effects of these herbs are mainly due to their free radical scavenging antioxidant effects.[35] The herbs and food items are used for therapeutic nutrition in neurocognitive conditions. Precise utilization of these food items and herbs in different pathologies produces significant benefits.

Garlic (*Allium sativum*)

Garlic is extensively used in Indian kitchens in preparing recipes. It has a pungent and hot pharmacotherapeutic effect. It is beneficial for improving cognitive functions and has the *Rasayana* effect. [23] It is used in disorders due to *Kapha* and *Vata* dominance.

It possesses antihypertensive, cholesterol-lowering, antiatherosclerotic, antithrombotic, hypoglycemic, and antioxidant effects.[36] These effects favor the use of garlic as a neuroprotective food to prevent neurocognitive conditions. Garlic is useful in vascular pathologies.

Fresh buds of garlic are chewed or mixed with food. It is used after boiling in milk to reduce its pungent effect. Ghee processed with garlic is beneficial in *Vata*-related abdominal and neurological disorders.[37] Garlic paste is added to various food items.

Shatavari (*Asparagus racemosus*)

Shatavari, wild asparagus, has a bitter, sweet taste with cold potency and a sweet postdigestive effect. It is beneficial in improving cognitive functions. *Shatavari* is known to enhance cognitive functions. [23] It is used in disorders due to *Vata* and *Pitta Dosha* dominance with vitiation of blood.[38]

It possesses antioxidant and adaptogenic activity.[36] *Shatavari* is specifically useful in women and degenerative conditions.

Fresh juice, granules, processed ghee, and processed milk of *Shatavari* are administered for the treatment of disorders due to *Vata* and *Pitta* aggravation.

Haridra (Curcuma longa)

Haridra (turmeric) is a popular ingredient in Indian food. It has a pungent postdigestive effect with hot potency. It is known for its anti-inflammatory properties; hepatoprotective activity; immunostimulant effect; and anticancer, antitumor, and proliferative effects. It is also potent hypoglycemic, antihyperlipidemic, and antioxidant. [36] Thus, it can be used in dementia related to a metabolic syndrome caused due to over nutrition and inflammation.

Amalaki (Emblica officinalis)

Amalaki (Indian gooseberry) fruit is commonly used for its rejuvenation and antiaging effect. It has a sweet postdigestive effect with cold potency. It is specifically indicated in bleeding disorders and obstinate urinary diseases including diabetes.[39]

Amalaki has antihypercholesterolemic, immunomodulatory, and anticarcinogenic activities. It is a potent antioxidant with significant adaptogenic, antistress, immunopotentiation, and memory-facilitating effects. It mitigates chronic stress-induced effects, normalizes SOD activity, and increases the activity of catalase and glutathione peroxidase. It shows potent free radical scavenging activity. [36] Hence it is used in many stress-related disorders.

Amalaki fruit is used in food as many preparations like candy, pickles, sweet jam, etc. *Amalaki* is a beneficial food for preventing age-related neurocognitive decline. It is useful in reversing the adverse effects of stress.

Yashtimadhu (Glycyrrhiza glabra)

Yashtimadhu (liquorice) is a commonly used herb with a sweet postdigestive effect and cold potency. It is listed among the four important intellect-promoting medicines (*Medhya Rasayana*).[40] It has a sweet postdigestive effect with cold potency. It is proven to have anti-inflammatory, immunomodulatory, and antimicrobial activity. It shows decreased lipid peroxidation, suggesting an antioxidant effect. [36]

Yastimadhu sticks and powder are used as natural sweeteners in food. It is also used in mouth fresheners. Ghee processed with *Yashtimadhu* is used in pharmaco-therapeutics.

Tulasi (Ocimum sanctum)

Holy basil or *Tulasi* has hot potency and a pungent postdigestive effect. It has proven anti-inflammatory and anxiolytic activity.[36] Fresh leaves are used in the preparation of ghee. It has potent antistress activity, too.

Kushmanda (Benincasa hispida)

Kushmanda (*Benincasa hispida*) has intellect-promoting activity.[41] Ghee processed with *Kushmanda* has shown significant antidepressant effects and improvement in mood and behavior.[42] *Kushmanda* is used in treating neurocognitive disorders due to the vitiation of *Vata* and *Pitta Dosha*. *Kushmanda* fruit is used in vegetable curry. Fresh juice is also used for dietary purposes.

CLINICAL EXPERIENCES AND PROTOCOLS

Whenever an individual with apparent signs or confirmed diagnosis of neurocognitive conditions comes to Ayurveda treatment, a thorough clinical examination is essential. The causes (*Hetu*), clinical features (*Linga*), and management (*Aushadha*) are three fundamental approaches for the patient. The diagnosis of the pathogenic event of obstruction or degeneration is confirmed based on the clinical history and examination of the individual. Nutrition is a vital aspect of management. Eradicating the causes like intoxicating food items (*Madakari*), alcohol (*Madya*), and food that causes sluggishness and heaviness (*Tamasika* food) is the first principle to prescribe. The food items with higher efficacy to break either pathogenesis are advised. In obstructive pathogenesis, food items with hot potency, pungent postdigestive effect, and purification properties are advised. In degenerative pathogenesis, food items with cold potency, sweet postdigestive effect, and pacification properties are advised. Therapeutic nutrition works well as an adjunct therapy with main therapies like nasal administration of *Ayurveda* medicines and other medicines.

Rasayana Diet for Prevention and Management of Neurocognitive Conditions Type 1:

The diagnosis of degenerative pathologies due to the vitiation of *Vata Dosha* and *Pitta Dosha* is made based on clinical examination and assessment of the patient. Therapeutic nutrition protocol along with the standard pharmacological management protocol provides additional health benefits. Cow milk mixed with cow ghee in the early morning, porridge prepared with wheat, porridge of rice, porridge of black gram, curry of *Kushmanda* prepared in ghee, and meat soup are recipes advised for these patients. This diet is practiced for 7 days to 3 months depending on the severity of the conditions and strength of the patient. This regimen is stopped or modified if the patient shows signs of vitiation of *Kapha Dosha,* excess weight gain, excess sleep, etc.

Rasayana Diet for Prevention and Management of Neurocognitive Conditions Type 2:

Diagnosis of obstructive pathologies in neurocognitive conditions type 2 is done after assessment of the patient. The dietary protocol advised for the treatment of *Kapha Dosha* and *Pitta Dosha* is useful in this condition. A soup of green gram added with spices like garlic, turmeric, black pepper, and rock salt is useful to remove the obstruction by *Kapha Dosha.* This dietary regimen is prescribed for a minimum of 7 days and a maximum of up to 1 month. Lightness in the body, enthusiasm, and higher energy levels are observed in patients after following this diet. If signs of weakness, vertigo, or muscle pain are observed, then the diet is stopped.

SCOPE FOR FURTHER RESEARCH AND NUTRAVIGILANCE

The pathophysiology of neurocognitive disorders shall be studied with special emphasis on their relationship with the *Dosha* disequilibrium. Nutrition with various food items, processing media, and spices in variable quantity and their effect on cognition needs to be elaborated further. Personified therapeutic nutrition protocol shall be developed to prevent and treat neurocognitive conditions. Nutravigilance is required to monitor the pharmacotherapeutic effects of nutritional supplements taken by individuals. A trend to take vitamins and mineral supplements and health supplements is observed in society. Ayurveda advises taking any health supplement or medicine as per the digestive and metabolic capacity (*Agni*) of the person. If the capacity to metabolize and absorb vitamin supplements is poor, then this may cause untoward effects leading to neurocognitive disorders of type 2. Therefore, nutravigilance is required to check and prescribe health supplements based on personified needs and metabolic capacity. This checks and can control the trend of taking unnecessary vitamin and mineral supplements under the alluring labels of nutraceuticals for health.

REFERENCES

1 Sachs-Ericsson N, Blazer DG. The new DSM-5 diagnosis of mild neurocognitive disorder and its relation to research in mild cognitive impairment. Aging Ment Health. 2015 Jan;19(1):2–12. doi: 10.1080/13607863.2014.920303. Epub 2014 Jun 10. PMID: 24914889.

2 Rochoy M, Chazard E, Bordet R. Épidémiologie des troubles neurocognitifs en France [Epidemiology of neurocognitive disorders in France]. Geriatr Psychol Neuropsychiatr Vieil. 2019 Mar 1;17(1):99–105. French. doi: 10.1684/pnv.2018.0778. PMID: 30907374.

3 Sabnis M., Deole Y. S.. Santarpaniya Adhyaya. Verse 07,28. In: Dwivedi R.B., Basisht G., eds. Charak Samhita New Edition. 1st ed. Jamnagar, Ind: CSRTSDC; 2020. www.carakasamhitaonline.com/index.php?title=Santarpaniya_Adhyaya&oldid=41153. Accessed December 10, 2022.

4 Anagha S., Deole Y.S.. Majja dhatu. In: Basisht G., eds. Charak Samhita New Edition. 1st ed. Jamnagar, Ind: CSRTSDC; 2020. www.carakasamhitaonline.com/index.php?title=Majja_dhatu&oldid=41301. Accessed December 10, 2022.

5 Bhojani M. K., Tanwar Ankur Kumar. Vata dosha. In: Deole Y.S., eds. Charak Samhita New Edition. 1st ed. Jamnagar, Ind: CSRTSDC; 2020. www.carakasamhitaonline.com/index.php?title=Vata_dosha&oldid=41957. Accessed December 10, 2022.

6 Vagbhata. Ashtanga Hridaya. Doshabhediya adhyaya. Edited by H.S. Paradakara. Chaukhamba Sanskrit Sansthan, Varanasi. pp.192–210

7 Rochoy M, Chazard E, Bordet R. Épidémiologie des troubles neurocognitifs en France [Epidemiology of neurocognitive disorders in France]. Geriatr Psychol Neuropsychiatr Vieil. 2019 Mar 1;17(1):99–105. French. doi: 10.1684/pnv.2018.0778. PMID: 30907374.

8 Thomas D. Bird, Bruce L. Miller. Dementia. In Harrison's Principles of Internal Medicine. McGraw Hill publications. 17th edi. 2008. Pp.2536

9 Deole Y.S. Smruti (memory). In: Basisht G., eds. Charak Samhita New Edition. 1st ed. Jamnagar, Ind: CSRTSDC; 2020. www.carakasamhitaonline.com/index.php?title=Smruti_(memory)&oldid=41313. Accessed December 10, 2022.

10 Singh R.H.,Sodhi J.S., Dixit U.. Rasayana Adhyaya. In: Dixit U., Basisht G., eds. Charak Samhita New Edition. 1st ed. Jamnagar, Ind: CSRTSDC; 2020. www.carakasamhitaonline.com/index.php?title=Rasayana_Adhyaya&oldid=41198. Accessed December 10, 2022.

11 Dementia available from www.cdc.gov/aging/dementia/index.html cited on December 10, 2022

12 Charak. Sutra Sthana, Cha.25 Yajjapurushiya Adhyaya verse 40. In: Jadavaji Trikamji Aacharya, Editor. Charak Samhita.1st ed. Varanasi: Krishnadas Academy;2000.
13 Dubey S.D., Singh A.N., Singh A., Deole Y. S. Atreyabhadrakapyiya Adhyaya. In: Sirdeshpande M.K., Basisht G., eds. Charak Samhita New Edition. 1st ed. Jamnagar, Ind: CSRTSDC; 2020. www.carakasamhitaonline.com/index.php?title=Atreyabhadrakapyiya_Adhyaya&oldid=41156. Accessed December 10, 2022.
14 Sarker MR, Franks SF. Efficacy of curcumin for age-associated cognitive decline: a narrative review of pre-clinical and clinical studies. Geroscience. 2018 Apr;40(2):73–95. doi: 10.1007/s11357-018-0017-z. Epub 2018 Apr 21. PMID: 29679204; PMCID: PMC5964053.
15 Qubty D, Rubovitch V, Benromano T, Ovadia M, Pick CG. Orally Administered Cinnamon Extract Attenuates Cognitive and Neuronal Deficits Following Traumatic Brain Injury. J Mol Neurosci. 2021 Jan;71(1):178–186. doi: 10.1007/s12031-020-01688-4. Epub 2020 Sep 8. PMID: 32901372.
16 Saenghong N, Wattanathorn J, Muchimapura S, Tongun T, Piyavhatkul N, Banchonglikitkul C, Kajsongkram T. Zingiber officinale Improves Cognitive Function of the Middle-Aged Healthy Women. Evid Based Complement Alternat Med. 2012;2012:383062. doi: 10.1155/2012/383062. Epub 2011 Dec 22. PMID: 22235230; PMCID: PMC3253463.
17 Kaura, Sushila, Milind Parle. Anti-Alzheimer potential of green moong bean. Int. J Pharm. Sci. Rev. Res.2015;37(2):178–182.
18 Mennella JA, Trabulsi JC, Papas MA. Effects of cow milk versus extensive protein hydrolysate formulas on infant cognitive development. Amino Acids. 2016 Mar;48(3):697–705. doi: 10.1007/s00726-015-2118-7. Epub 2015 Oct 26. PMID: 26497857; PMCID: PMC4754137.
19 Tripathy Indradev. Kshemakutuhala.Chaukhambha Orientalia. Varanasi: page 228, verse 101, 8/138 page 139
20 Xinglv Liao, Muhammad Jamil Ahmad, Chao Chen et al. Buffalo Milk Can Rescue Fatigue in Young Mice, Alleviate Oxidative Stress and Boost Learning and Memory in Aging Mice, 03 August 2020, PREPRINT (Version 1) available at Research Square [https://doi.org/10.21203/rs.3.rs-48866/v1]
21 Moreno-Fernández J, López-Aliaga I, García-Burgos M, Alférez MJM, Díaz-Castro J. Fermented Goat Milk Consumption Enhances Brain Molecular Functions during Iron Deficiency Anemia Recovery. Nutrients. 2019 Oct 7;11(10):2394. doi: 10.3390/nu11102394. PMID: 31591353; PMCID: PMC6835798.
22 Dwivedi B.K., Deole Y. S. Yajjah Purushiya Adhyaya. In: Dwivedi R.B., Basisht G., eds. Charak Samhita New Edition. 1st ed. Jamnagar, Ind: CSRTSDC; 2020. www.carakasamhitaonline.com/index.php?title=Yajjah_Purushiya_Adhyaya&oldid=41155. Accessed December 10, 2022.
23 Kaiyadeva Nighantu 4/253, 1/623, 1/1380, 3/38, 3/82, 4/302, 4/314, 4/326, 6/131, 6/136, 6/153, 6/175, 6/192, 1/1221, 1/1064 Available from https://niimh.nic.in/ebooks/e-Nighantu/kaiyadevanighantu/?mod=read accessed on 13/01/2023
24 Thakar A. B., Auti S.. Snehadhyaya. In: Mangalasseri P., Basisht G., eds. Charak Samhita New Edition. 1st ed. Jamnagar, Ind: CSRTSDC; 2020. www.carakasamhitaonline.com/index.php?title=Snehadhyaya&oldid=41143. Accessed December 10, 2022.
25 Savrikar S.S. 'Comparative Study of Physico-chemical characteristics of *'Accha Sneha'* and *'Siddha Sneha'* with reference to *'Guduchi Ghrita'*, the butter fat, medicated with *Tinospora cordifolia* (Willd.) Miers.' 2006, PhD thesis submitted to Swami Ramanand Teerth Marathwada University, Nanded, Maharashtra, India.
26 Sharma H, Zhang X, Dwivedi C. The effect of ghee (clarified butter) on serum lipid levels and microsomal lipid peroxidation. Ayu. 2010 Apr;31(2):134–40. doi: 10.4103/0974-8520.72361. PMID: 22131700; PMCID: PMC3215354.
27 Lad V. The Complete Book of Ayurvedic Home Remedies. New York: Harmony Books; 1998.
28 Sharma HM. Butter oil (ghee) – Myths and facts. Indian J Clin Pract 1990;1:31-2.
29 Marcucci MC. Propolis; chemical composition, biological properties and therapeutical utility, Apidologie, 1995, 26, 83–89.
30 Viuda Martos M, Ruiz-Navajas Y, Fern´Andez-L´Opez J And P´Erez-´Alvarez JA. Functional Properties of Honey, Propolis and Royal Jelly, Journal of food science, 2008, Vol. 73, 116–124.
31 Aonan A, AL-Mazrooa, Mansour I. Sulaiman, Effects of honey on stress-induced ulcers in rats, *J KAU Med Sci*, 1999, 7 (1), 115–122.
32 *Tahira Farooqui*; Honey: An Anti-Aging Remedy to Keep you Healthy in a Natural Way *on The Ohio State University* www.scienceboard.net/community/perspectives.228.html accessed on 03/03/2011
33 Jaganathan SK, Mandal M. Antiproliferative Effects of Honey and of Its Polyphenols: A Review. J Biomed Biotechnol 2009; 2009: 830616.

34 M.I. Khalil, S.A. Sulaiman and L. Boukraa, Antioxidant Properties of Honey and Its Role in Preventing Health Disorder; *The Open Nutraceuticals Journal,* 2010, *3,* 6–16
35 Kulkarni R, Girish KJ, Kumar A. Nootropic herbs (Medhya Rasayana) in Ayurveda: An update. Pharmacogn Rev. 2012 Jul;6(12):147–53. doi: 10.4103/0973-7847.99949. PMID: 23055641; PMCID: PMC3459457.
36 Sabnis M. Chemistry and Pharmacology of Ayurvedic Medicinal Plants. Chaukhambha Surabharati Prakashan, Varanasi. First Edition.2006. pages 87–99, 119, 178-79, 189-93, 201-203, 268
37 Vagbhata. Ashtanga Hridaya. Doshabhediya adhyaya. Edited by H.S. Paradakara. Chaukhamba Sanskrit Sansthan, Varanasi. Chikitsa Sthana 14/22–25
38 Bhavaprakasha. Shatavari. Guduchyadi varga. Verse 159 E-Nighantu Available from https://niimh.nic.in/ebooks/e-Nighantu/bhavaprakashanighantu/?mod=read&h=shatAvarI accessed on 12/12/2022
39 Bhavaprakasha. Amalaka. Haritakyadi varga. Verse 36 E-Nighantu Available from https://niimh.nic.in/ebooks/e-Nighantu/bhavaprakashanighantu/?mod=read&h=Amalaka accessed on 12/12/2022
40 Singh R.H.,Sodhi J.S., Dixit U.. Rasayana Adhyaya. In: Dixit U., Basisht G., eds. Charak Samhita New Edition. 1st ed. Jamnagar, Ind: CSRTSDC; 2020. www.carakasamhitaonline.com/index.php?title=Rasayana_Adhyaya&oldid=41198. Accessed December 10, 2022.
41 Bhavaprakasha. Kushmanda. Shaka varga. Verse 46 E-Nighantu Available from https://niimh.nic.in/ebooks/e-Nighantu/bhavaprakashanighantu/?mod=read accessed on 12/12/2022
42 Chandre R, Upadhyay BN, Murthy KH. Clinical evaluation of Kushmanda Ghrita in the management of depressive illness. Ayu. 2011 Apr;32(2):230–3. doi: 10.4103/0974-8520.92592. PMID: 22408308; PMCID: PMC3296346.

Appendix I

NUTRITIVE VALUE AND MEASURE GUIDE FOR INDIAN CUISINE

The nutritive and caloric values have been studied and documented by the National Institute of Nutrition, which can be referred to in link below, for the food preparations and recipes described in chapters of this book.

www.nin.res.in/downloads/DietaryGuidelinesforNINwebsite.pdf

COMPARISON BETWEEN TEASPOON AND TABLESPOON

	TEASPOON	*TABLESPOON*
Abbreviation	TSP	TBSP
Definition (Oxford)	A small spoon used typically for adding sugar.	A large spoon for serving food.
Size	Smaller	Larger
Quantity	5 milliliters	15 milliliters
Conversion	1 Tablespoon = 3 Teaspoons (US, UK, or Metric] 1 Tablespoon = 4 Teaspoons (Australia]	

Appendix II

TERMINOLOGIES FOR THERAPEUTIC NUTRITION IN AYURVEDA

TERM IN SANSKRIT	*APPROXIMATE TERMINOLOGY IN ENGLISH*
Ahara	Diet
ahara matra	Optimal digestible quantity of food
ama	Endotoxin
amashaya	Stomach
amapakvashaya	Gastrointestinal tract
amisha	Nonvegetarian diet
anaha	Distention of abdomen due to incomplete evacuation of stool, urine, and flatus
anoopa mamsarasa	Soup of animal meat obtained from animals of water source or from the region having excess rainfall
apa	Water
adhamanam	Distention of abdomen due to gas
aleham	Licking
asvasakara	Solace, assurance
atopa	Painful distension of abdomen accompanied by rumbling noise
ayurveda	The Science and Knowledge of life
ayurveda	Science of life
abhyavaharana shakti	Power of ingestion
adhikamatravat ahara	Intake of food more than required quantity
agni	Digestive/metabolic energy
agnimandya	Lack of digestive power
agnimardava	Lack of digestive power
agninasha	Loss of digestive power
agniparikshina	Loss of digestive power
agnipranashtam	Loss of digestive power
agnisada	Lack of digestive power
agnisadana	Impairment of digestive power
alpahara	Less amount of diet
alpabala	Loss of physical strength or weakness
amatravat ahara	Intake of inappropriate quantity of food
ambu	Water
amla	Sour
amla rasa	Sour taste
amla skandha	Group of sour substances
anulomanam	Physiological direction of various flows within the body like flatus and stool
anupana	Compliant

TERM IN SANSKRIT	APPROXIMATE TERMINOLOGY IN ENGLISH
anurasa	Taste of a substance which is perceived toward the end or with less intensity
anushna	Lukewarm
apaka	Indigestion
Apana Vayu	One of the five subtypes of *Vayu (Vata)*
apatarpana	Under-nutrition
apathya	Unwholesome
arochaka	Loss of taste
ashana	Eatables
audaka mamsa	Flesh of aquatic variety
audbhida jala	Artesian well – water
audbhida lavana	Salts derived from plants
aushadha	Medicine/drug
aushadhi yukta bhojana	Medicine mixed with food
aushnya	Lukewarm
avastha paka	Intermediate metabolic transformation
avastha viruddha	Regimens which are unwholesome to health status
bahubhuk	Intake of more and frequent quantity of food
bala	Physiological energy
balya	Strength, stamina
bhajana	Utensils
bhutagni	Agni of Panchamahabhuta
bileshaya	Animals living in burrows
brunhana	Body tissue strengthening treatment
brinhaniya	Anabolic/promoters of tissue growth
chaksushya	Beneficial for ocular health
dadhimastu	Supernatant liquid of curd/yoghurt
dadhivarga	Categories of curds
dadhikurchika	Solid portion of curdled milk
dhatu	Functional tissue
dhatvagni	Metabolic energy of Dhatu
dipana	1. Stimulating/promoting digestion; 2. To increase appetite
dipaniya /deepaniya	Stimulating metabolic energy
dosha	Bioregulators
drava	Liquid
dravya	Material/matter/substance used for therapeutics
dravya	Matter
dravya	Substance used for therapeutic purpose
dugdha	Milk, dughdha/Kshira/Payasa
dushya	Saptadhatu
dvikala bhojana	Twice a day meal
ekakala bhojana	Once a day meal
falam	Fruits
gamga jala	Type of water which is not contaminated with dust, soot, and toxic
gandha	Smell
gaurava	Feeling of heavyness
ghrita	Ghee/clarified unsalted butter
ghirta varga	Group of different ghees
gorasavarga	Group of cow milk products
harita varga	Group of freshly harvested tubers/spices/vegetables/herbs, consumed for gustatory stimulation (taste stimulation)
hetu	Etiology
hima	Cold infusion
hinamatra ahara	Suboptimum quantity diet

TERM IN SANSKRIT	*APPROXIMATE TERMINOLOGY IN ENGLISH*
indriya	Senses
ikshu varga	Group of sugar cane products
ikshuvikara	Sugarcane products
jatharagni	Metabolic energy for digestion
jala	Water
jala varga	Types of water bodies/sources
jalachara vihamga vasa	Aquatic bird's fat
jarana shakti	Individual digestive capacity
jatharam	Stomach to first part of duodenum
jangama	Substance of animal origin
jangala mamsarasa	Soup of animal meat from region with dry forests and less rainfall
jirnam ashniyata	Subsequent meals should be consumed after the previous food is digested completely; this makes it easier to digest the food
kalantara vipaki	Delayed biotransformation
kanji	Fermented gruel
kanjika	Fermented gruel
kanda	Rhizome
kanda varga	Group of rhizome/bulb/tuber
kapha	A synonym for Slesma
kashaya	Decoction of herbs
kashaya	Astringent
kasaya skandha	Group of Drugs having astringent taste
katu	Pungent
katu rasata	Pungent taste
katu skandha	Group of drugs having pungent taste
kilata	Coagulated milk
kleda	Moistened/moisture
krsara	A rice and pulses recipe
krita	Material is fried by using ghee or oil with addition of pungents, mustard, asafoetida, turmeric, etc.
kritanna varga	Group of processed foods
kshara	
kshara varga	Group of alkaline substances
kshaudra	Variety of honey
kshina	Weak, feeble
kshira varga	Group of milk obtained from different animals
kshirabham	Milky
kshirastaka	Class of eight milk-producing animals: cow, buffalo, sheep, goat, horse, elephant, human.
kshirapaka	Processed medicated milk
kudhanya varga	Group of inferior quality grain
kurchika	Semifermented curd
kvatha	Medicated decoction
laghava	Lightness
laja manda	A thin gruel prepared from parched rice
laja	Parched rice
laghu panchamula	Combination of five roots obtained from small plants: Shalparni *(Desmodium gangeticum),* Prishnparni *(Uraria picta),* Brahti *(Solanum indicum),* Kantkari *(Solanum surattense),* and Gokshura *(Tribulus terrestris).*
langhana	Therapeutic fasting
lavana rasa	Salty taste
lavana skandha	Class of different salts
lavana varga	Group of salts

TERM IN SANSKRIT	APPROXIMATE TERMINOLOGY IN ENGLISH
lavanam	Salt
makshika	Variety of honey
mamsa varga	Group of meats sourced from different animals
mamsarasa	Soup prepared with meat by adding 8 times water and reducing it to 1/4th
matra viruddha	Type of incompatibility associated with quantity
matra	Specific quantity
matravata ahara	Specific quantity of food
matravat ashniyata	Food intake as per specific quantity
madhu varga	Group of honey
madhura	Sweet
madhura skandha	Class of substances having sweet taste
Pancha mahabhuta	Five protoelements
mala	Metabolic waste
manda	Thin gruel of rice
mandagni	Diminished digestive energy
mantha	Drink made by churning
manthana	The process of churning
matravat ashan	Specific quantity of diet
medhya	Cognition
mithyayoga	Inaccurate perception by sense organ
modaka	A variety of Indian recipes like laddu
mridu	Physiological and pharmacological softness
mrdu virya	Mild potency
mutra	Urine
mutra varga	Group of urine
odana	Cooked rice preparation
pachaka pitta	One of the five subtypes of *Pitta*
pachana	Digestion and assimilation
pachana	Digestion and assimilation
paka	Vipaka- complete metabolic transformation
paka virudda	Incompatible food processing
panchamruta	Group of following five substances: Cow ghee, cow yoghurt, cow milk, Madhu (honey), and Sharkara (sugarcane candy)
pathya	Wholesome
paya	Milk
peya	Thick gruel of rice
phamta	Hot infusion Hima- cold infusion
phanita	Molasses
phala	Fruits
phala varga	Group of fruits
pitta	Pitta Dosha
pragbhakta	Just before meal
prana	Vitality
prakrti	Phenotype
prasaha	Animals and birds who take their food by snatching
prithuka	Flattened rice
pupalika	Kind of sweet cake fried in ghee or oil
pushpa	Flower
puspa varga	Group of different flowers
rakta	Blood tissue
rasa	Taste
rasayana	Rejuvenation

TERM IN SANSKRIT	*APPROXIMATE TERMINOLOGY IN ENGLISH*
rehcanam	Purgation
rutucharyaa	Seasonal regimen
ruchiḥ	Taste
ruksha	Physiological and pharmacological dryness
satmya	Easily assimilable
satmya viruddha	Habit unwholesome of diet and drugs
sabhakta	Administration of medicine mixed with food
sadaṅgapaniya	Medicated decoction with specific six drugs
saindhava lavana	Rock salt
sakthu	dietetic preparation
samagni	Optimum metabolism
samana vayu	One of the five subtypes of Vata (Vayu)
samashana	Admixture of desirable and undesirable food items
samskara	Pharmaceutical process for desirable transformation
samskara	Pharmaceutical process for desirable transformation
samskara virudda	Opposite to the standard process of preparation of food
santarpana	Repleting restorative nourishment
saptadhatu	Bodily tissue
saptamusti	Quantity which is equal to seven fistfuls
shaka	Vegetables
shaka varga	Group of vegetables
shamana	Alleviates/pacification
shami dhanya	Pulses
shimbi dhanya	Group of pulses
shita	Physiological and pharmacological coolness
suka dhanya	Group of grains with monocotyledons
suka dhanya	Cereals
shukta	Vinegar
snigdha	Physiological and pharmacological sliminess
snigdha- asniyat	Eating unctuous food
srotas	Biochannels
stambhanam	Restraining, impeding
suksmaḥ	minuteness
supya dhanya	Roasted and dehusked grains
svastha	Healthy
svinna	Process of fomentation
taila	Oil
taila varga	Group of oil
takra	Buttermilk
takra varga	Group of buttermilk
tandula vari	Water obtained by washing rice
tanmana bhunjit	Mindful eating
tarpana	Instant replenishment
tikshna	Physiological and pharmacological quickening of processes
tikshnagni	Excessive metabolic energy
tikta	Bitter
tikta skandha	Group of plants having bitter taste
triguna	The three primary attributes of universe
ushnodaka	Water is boiled and reduced to 1/8th or 1/4th or 1/2th of the quantity after boiling
usthrakshira	The milk of camel
utkarika	Poultice
vatya	Barley
vatya manda	Gruel prepared with roasted barley

TERM IN SANSKRIT	APPROXIMATE TERMINOLOGY IN ENGLISH
vahni /agni	Metabolic energy
vesavara	Meat which is devoid bone
vesvara	Soup prepared with boneless meat
vida lavana	Red granular salt
vidaha	Glycation
vilepi	Thick gruel of rice
vipaka	Complete metabolic transformation
vishamagni	Erratic state of agni
virya /veerya	The principal component important for the therapeutic action
virya	Energy
vishamashana	Inappropriate intake of food
vishkira	Gallinaceous
vishkira shakuni vasa	Gallinaceous bird fat
vrishyam	phrodisiacs
vyadhikshamatva	Immunity
yavagu	Thick gruel of cereals
Yusha	Soup of vegetables and/or pulses

Index

Note: Page locators in **bold** and *italics* represents tables and figures, respectively.

Printed in the United States
by Baker & Taylor Publisher Services

Printed in the United States
by Baker & Taylor Publisher Services